Handbook of Multiple Comparisons

Chapman & Hall/CRC
Handbooks of Modern Statistical Methods

Series Editor:
Garrett Fitzmaurice, Department of Biostatistics, Harvard School of Public Health, Boston, MA, U.S.A.

The objective of the series is to provide high-quality volumes covering the state-of-the-art in the theory and applications of statistical methodology. The books in the series are thoroughly edited and present comprehensive, coherent, and unified summaries of specific methodological topics from statistics. The chapters are written by the leading researchers in the field and present a good balance of theory and application through a synthesis of the key methodological developments and examples and case studies using real data.

Handbook of Methods for Designing, Monitoring, and Analyzing Dose-Finding Trials
John O'Quigley, Alexia Iasonos, and Björn Bornkamp

Handbook of Quantile Regression
Roger Koenker, Victor Chernozhukov, Xuming He, and Limin Peng

Handbook of Statistical Methods for Case-Control Studies
Ørnulf Borgan, Norman Breslow, Nilanjan Chatterjee, Mitchell H. Gail, Alastair Scott, and Chris J. Wild

Handbook of Environmental and Ecological Statistics
Alan E. Gelfand, Montserrat Fuentes, Jennifer A. Hoeting, and Richard L. Smith

Handbook of Approximate Bayesian Computation
Scott A. Sisson, Yanan Fan, and Mark Beaumont

Handbook of Graphical Models
Marloes Maathuis, Mathias Drton, Steffen Lauritzen, and Martin Wainwright

Handbook of Mixture Analysis
Sylvia Frühwirth-Schnatter, Gilles Celeux, and Christian P. Robert

Handbook of Infectious Disease Data Analysis
Leonhard Held, Niel Hens, Philip O'Neill, and Jacco Walllinga

Handbook of Meta-Analysis
Christopher H. Schmid, Theo Stijnen, and Ian White

Handbook of Forensic Statistics
David L. Banks, Karen Kafadar, David H. Kaye, and Maria Tackett

Handbook of Statistical Methods for Randomized Controlled Trials
KyungMann Kim, Frank Bretz, Ying Kuen K. Cheung, and Lisa Hampson

Handbook of Measurement Error Models
Grace Yi, Aurore Delaigle, and Paul Gustafson

Handbook of Multiple Comparisons
Xinping Cui, Thorsten Dickhaus, Ying Ding, and Jason C. Hsu

For more information about this series, please visit: https://www.crcpress.com/Chapman–HallCRC-Handbooks-of-Modern-Statistical-Methods/book-series/CHHANMODSTA

Xinping Cui, Thorsten Dickhaus, Ying Ding, Jason C. Hsu (Eds.)

Handbook of Multiple Comparisons

CRC Press is an imprint of the
Taylor & Francis Group, an **informa** business
A CHAPMAN & HALL BOOK

First edition published 2022
by CRC Press
6000 Broken Sound Parkway NW, Suite 300, Boca Raton, FL 33487-2742

and by CRC Press
2 Park Square, Milton Park, Abingdon, Oxon, OX14 4RN

Library of Congress Cataloging-in-Publication Data
Names: Cui, Xinping, editor.
Title: Handbook of multiple comparisons / [edited by] Xinping Cui, Thorsten
Dickhaus, Ying Ding, Jason Hsu.
Description: First edition. | Boca Raton : CRC Press, 2021. | Series:
Chapman & Hall/CRC handbooks of modern statistical methods | Includes bibliographical references and index. | Summary: "Written by experts that include originators of some key ideas, chapters in the Handbook of Multiple Testing cover multiple comparison problems big and small, with guidance toward error rate control and insights on how principles developed earlier can be applied to current and emerging problems. Some highlights of the coverages are as follows. Error rate control is useful for controlling the incorrect decision rate. Chapter 1 introduces Tukey's original multiple comparison error rates and point to how they have been applied and adapted to modern multiple comparison problems as discussed in the later chapters. Principles endure. While the closed testing principle is more familiar, Chapter 4 shows the partitioning principle can derive confidence sets for multiple tests, which may become important as the profession goes beyond making decisions based on p-values. Multiple comparisons of treatment efficacy often involve multiple doses and endpoints. Chapter 12 on multiple endpoints explains how different choices of endpoint types lead to different multiplicity adjustment strategies, while Chapter 11 on the MCP-Mod approach is particularly useful for dose-finding. To assess efficacy in clinical trials with multiple doses and multiple endpoints, the reader can see the traditional approach in Chapter 2, the Graphical approach in Chapter 5, and the multivariate approach in Chapter 3. Personalized/precision medicine based on targeted therapies, already a reality, naturally leads to analysis of efficacy in subgroups. Chapter 13 draws attention to subtle logical issues in inferences on subgroups and their mixtures, with a principled solution that resolves these issues. This chapter has implication toward meeting the ICHE9R1 Estimands requirement. Besides the mere multiple testing methodology itself, the handbook also covers related topics like the statistical task of model selection in Chapter 7 or the estimation of the proportion of true null hypotheses (or, in other words, the signal prevalence) in Chapter 8. It also contains decision-theoretic considerations regarding the admissibility of multiple tests in Chapter 6. The issue of selected inference is addressed in Chapter 9. Comparison of responses can involve millions of voxels in medical imaging or SNPs in genome-wide association studies (GWAS). Chapter 14 and Chapter 15 provide state of the art methods for large scale simultaneous inference in these settings"—Provided by publisher.
Identifiers: LCCN 2021020343 (print) | LCCN 2021020344 (ebook) | ISBN
9780367140670 (hardback) | ISBN 9781032111551 (paperback) | ISBN
9780429030888 (ebook)
Subjects: LCSH: Multiple comparisons (Statistics)
Classification: LCC QA278.4 .H36 2021 (print) | LCC QA278.4 (ebook) | DDC
519.5/35—dc23
LC record available at https://lccn.loc.gov/2021020343
LC ebook record available at https://lccn.loc.gov/2021020344

ISBN: 9780367140670 (hbk)
ISBN: 9781032111551 (pbk)
ISBN: 9780429030888 (ebk)

DOI: 10.1201/9780429030888

Visit the companion website/eResources: https://www.routledge.com/Handbook-of-Multiple-Comparisons/Cui-Dickhaus-Ding-Hsu/p/book/9780367140670

Typeset in Times
by codeMantra

To our families, friends,
and all the dedicated essential
Covid-19 workers and caregivers.

Contents

Preface

Statistics is a science, and multiple comparisons is a part of it. This handbook will treat the topics of multiple comparisons, simultaneous and selective inference from a variety of different perspectives. The handbook will be useful for (a) researchers, (b) students/lecturers, (c) practitioners. The need for such a systematic treatment of the field originates from the relevance of multiple comparisons in many applications (medicine, industry, and economics), and from the diversity of approaches and developments, which shall be described here in a coherent manner.

With chapters on fundamental multiple comparison concepts and principles, as well as approaches and applications in complex clinical trials and large scale inference, this handbook connects multiple comparisons from its hand-drawing roots in the 1950s to its current high-tech applications.

This handbook has two parts, with the first part laying the foundations of general methodology, while the second part deals with applications in medicine.

These parts are not disjoint. On the contrary, having them together in a handbook makes it convenient to read about the theoretical basis of various approaches, and to see examples of applications of the approaches.

We the co-editors are grateful to our colleagues who have contributed to this handbook.

Diverse in areas both geographical and technical, it is clear that they did not treat writing a chapter for this handbook as just adding to their publications.

Instead, they gave their *best* in trying to transfer their knowledge and unique insights to the readers, for the greater good!

To the readers, we hope you find this handbook useful, and perhaps even a bit enjoyable to read, as much as we have enjoyed putting it together.

We are grateful to John Kimmel of Chapman & Hall/CRC Press for working in collaboration and partnership with us on this handbook project, almost a labor of love for us.

Xinping Cui, Thorsten Dickhaus, Ying Ding, Jason C. Hsu
January 2021

Editors

Xinping Cui is Professor and Chair of the Department of Statistics at the University of California, Riverside, USA. Her interdisciplinary research focuses on multiple testing, statistical genomics, precision medicine, and system biology.

Thorsten Dickhaus is a full Professor of Mathematical Statistics at the University of Bremen, Germany. He is a (co-) author of approximately 50 journal articles and four books. For more than 15 years, his research focuses on simultaneous statistical inference and multiple testing.

Ying Ding is an Associate Professor in the Department of Biostatistics at the University of Pittsburgh. Her research focuses on survival analysis, large-scale genomics and proteomics analysis, multiple testing and precision medicine.

Jason C. Hsu is an Emeritus Professor in Statistics at the Ohio State University. His research areas of interests include multiple comparisons, logic-respecting estimands, and targeted therapies for personalized/precision medicine.

Contributors

Gilles Blanchard
Department of mathematics,
Université Paris-Saclay, CNRS, Inria,
Orsay, France

Taras Bodnar
Department of Mathematics
Stockholm University
Stockholm, Sweden

Malgorzata Bogdan
Department of Mathematics
University of Wroclaw
Wrocław, Poland
and
Department of Statistics
Lund University Lund, Sweden

Björn Bornkamp
Statistical Methodology, Global Drug Development
Novartis AG
Basel, Switzerland

Frank Bretz
Statistical Methodology, Global Drug Development
Novartis AG
Basel, Switzerland

Arthur Cohen
Department of Statistics
Rutgers University
New Brunswick, New Jersey

Xinping Cui
Department of Statistics
University of California, Riverside
Riverside, California

Thorsten Dickhaus
Institute for Statistics
University of Bremen
Bremen, Germany

Ying Ding
Department of Biostatistics
University of Pittsburgh
Pittsburgh, Pennsylvania

Helmut Finner
Institute of Biometrics and Epidemiology
German Diabetes Center
Düsseldorf, Germany

Florian Frommlet
Center for Medical Statistics, Informatics and Intelligent Systems (Institute of Medical Statistics)
Medical University Vienna
Vienna, Austria

Ekkehard Glimm
Novartis Pharma AG
Basel, Switzerland

Jelle J. Goeman
Department of Biomedical Data Sciences
Leiden University Medical Centre
Leiden, Netherlands

Jiangtao Gou
Department of Mathematics and Statistics
Villanova University
Villanova, Pennsylvania

Lisa V. Hampson
Novartis Pharma AG
Basel, Switzerland

Jason C. Hsu
Department of Statistics
The Ohio State University
Columbus, Ohio

Yi Liu
Data Sciences and Systems
Nektar Therapeutics
San Francisco, California

André Neumann
Institute for Statistics
University of Bremen
Bremen, Germany

Pierre Neuvial
Institut de Mathématiques de Toulouse
Université de Toulouse, CNRS
Toulouse, France

Etienne Roquain
Laboratoire de Probabilités Statistique et Modélisation, Sorbonne Université,
Université de Paris, CNRS

Jonathan D. Rosenblatt
Department of Industrial Engineering & Management
Ben-Gurion University of the Negev
Beersheba, Israel

Harold Sackrowitz
Department of Statistics
Rutgers University
New Brunswick, New Jersey

Aldo Solari
Department of Economics, Management and Statistics
University of Milano-Bicocca
Milan, Italy

Ajit C. Tamhane
Department of Industrial Engineering and Management Sciences
Northwestern University
Evanston, Illinois

Szu-Yu Tang
Worldwide Research and Development
Pfizer
Andover, Massachusetts

Hong Tian
Global Statistics
BeiGene
Beijing, China

Bushi Wang
Clinical Data Science
Boehringer Ingelheim Pharmaceuticals, Inc.
Ingelheim am Rhein, Germany

Xinjun Wang
Department of Biostatistics
University of Pittsburgh
Pittsburgh, Pennsylvania

Yue Wei
Department of Biostatistics
University of Pittsburgh
Pittsburgh, Pennsylvania

Dong Xi
Statistical Methodology, Global Drug Development,
Novartis Pharmaceuticals
Basel, Switzerland

Part I

General Methodology

1

An Overview of Multiple Comparisons

Xinping Cui
University of California Riverside

Thorsten Dickhaus
University of Bremen

Ying Ding
University of Pittsburgh

Jason C. Hsu
The Ohio State University

CONTENTS

In this chapter, we introduce some basic concepts and quantities, which will be referred to in several of the subsequent chapters. Furthermore, we provide an overview of the other chapters of this handbook and point the readers to online supplementary material.

1.1 Basics of Multiple Comparison Error Rates

In the beginning, there was Tukey (1953), which has been reprinted as Tukey (1994). With the inferences being confidence intervals, not tests of null hypotheses, he defined the following error rates:

$$\textbf{error rate per comparison} = \frac{\text{number of erroneous statements}}{\text{number of comparisons}}$$

$$\textbf{error rate per family} = \frac{\text{number of erroneous statements}}{\text{number of families}}$$

$$\textbf{error rate family-wise} = \frac{\text{number of erroneous families}}{\text{number of families}}$$

DOI: 10.1201/9780429030888-2

TABLE 1.1
Illustration of Definition of Multiple Comparison Error Rates in Tukey (1953) Each study results in a family of 100 confidence intervals (CIs).

	No. of CIs Not Covering True Value				
	Study 1	**Study 2**	**Study 3**	**Study 4**	
	0	5	7	2	Rate
Per comparison errors	0	5	7	2	$\frac{14}{400}$
Per family errors	0	5	7	2	$\frac{14}{4}$
Family-wise error	0	1	1	1	$\frac{3}{4}$

Table 1.1 illustrates what they are. If each study or experiment results in a family of confidence intervals, then error rates *per family* and *family-wise* both consider each study as a unit, with *per family* error rate being the average *number* of confidence intervals not covering their true values, averaged over the studies, while *family-wise* error rate is the *proportion* of studies with *at least one* confidence interval not covering its true value.

1.1.1 Family-wise Error Rate, and Danger of Weak Control

John W. Tukey was against testing equality null hypotheses (see Chapter 4 of this Handbook), so his definitions did not differentiate between true and false null hypotheses: a confidence interval either covers its true value or it does not.

When Tukey's error rates are applied to tests of hypotheses, we differentiate between true and false null hypotheses, and Type I error refers to incorrectly rejecting one or more true null hypotheses.

An unfortunate early practice in multiple comparisons was to control the so-called *experimentwise* Type I error rate. Paraphrasing a bit, it is defined by the American Psychological Association (APA) for example, as follows.

> In multiple comparisons, the probability of making at least one Type I error over an entire research study is called the experimentwise Type I error rate.

This definition is not well-defined because this probability may well depend on which null hypotheses are true and which are false. In practice, the experimentwise Type I error rate is typically computed under the scenario that *all* the null hypotheses are true. This is unfortunate because methods such as Newman-Keuls' multiple range test, for example, controls this probability only when all the null hypotheses are true but not otherwise.

For any statistical error rate to be relevant, its control needs to translate into a meaningful control of a rate of an incorrect decision, *regardless of the unknown truth of nature*. Definition of Type I error rate in Lehmann (1986), p. 69, and Casella and Berger (1990), p. 361, takes the *supremum* of this probability over all parameter configurations. From a decision-theoretic point of view, controlling risk over independent but not necessarily identically distributed studies, such a supremum should be taken as well (see Chapter 1 of Berger (1985), p. 44). The risk of making gravely incorrect decisions while controlling the Type I error rate under a too restrictive null is very real, see Section 13.8 of Chapter 13 of this Handbook for an example with the log-rank test.

For this reason, family-wise error rate (FWER) has come to be defined as the *supremum* of the probability of rejecting at least one true null hypotheses, over all parameter configurations.[1] Methods in Chapters 2–5, 7, 10–13 all control FWER.

[1]Terminologically, to differentiate from the experimentwise error rate, this form of control is called strong FWER control, while experimentwise error rate control is termed weak control.

Controlling FWER is rather simplistic. As mentioned in Section 3.5 of Chapter 3 of this Handbook, in the 1990s, the concept of FWER was useful in getting statistical practice away from experimentwise Type I error rate control in general, and specifically in dose-response studies (which is the subject of Chapter 11 of this Handbook). But, as Table 1.1 shows, FWER does not care *how many* Type I errors are committed, two or five or seven false rejections all count as the same. In problems such as multiple endpoints (the subject of Chapter 12 of this Handbook), perhaps two false rejections should be considered worse than one? Tukey's *per family* error takes the number of false rejections into account, but there are other possibilities such as the following.

1.1.2 The False Discovery Rate (FDR)

Controlling the false discovery rate (FDR) is popular in discovery studies. While controlling Tukey's *per family* error rate controls the expected *number* of incorrect rejections, controlling the false discovery rate (FDR) controls the expected *proportion* of rejections that are false.

$$\begin{aligned} \text{FDR} &= \text{E}\left[\frac{\text{number of false rejections}}{\text{total number of rejections}}\right] \\ &= \text{E}\left[\frac{\text{number of rejected } H_{0i} \text{ that are true}}{\text{total number of rejected } H_{0i}}\right], \; i \in \{1, \ldots, k\}. \qquad (1.1) \end{aligned}$$

In Chapter 3 of this Handbook, some FDR-controlling multiple tests will be presented, including the famous linear step-up test by Benjamini and Hochberg (1995).

FDR control is not used in clinical trials, however, because it can be manipulated. As pointed out in Section 6 of Finner and Roters (2001), if the real null hypothesis of interest is H_{01} and the error rate to control is FDR, then one can increase the chance of rejecting H_{01} by artificially adding extremely false null hypotheses that would surely be rejected, inflating the denominator of (2.9). Supposing FDR is to be controlled at level-α,

- if H_{01} is true and one very false null hypothesis H_{02} is added, then Pr{reject the true null hypothesis H_{01}} $= 2\alpha$;
- if H_{01} is true and two very false null hypothesis H_{02} and H_{03} are added, then Pr{reject the true null hypothesis H_{01}} $= 3\alpha$;
- and so forth.

Such possibilities would be unacceptable in clinical trials. In general, FDR control becomes hard to interpret if proportionally there are lots of false null hypotheses, inflating the denominator in (2.9). This appears to be the case in clinical settings comparing treatments. If there is a causal single nucleotide polymorphism (SNP) for a targeted therapy, then Chapter 15 of this Handbook points out almost all other SNPs will pick up some of the causal effects inadvertently so that the chapter takes a confidence intervals formulation, controlling the per family error rate.

1.1.3 Conditional Error Rate and Selective Inference

Sometimes results of large-scale multiple testing controlling FDR are misinterpreted. For example, if multiple testing in a genome-wide association study (GWAS) controlling FDR at 5% results in 1,000 rejection, then it is not uncommon to see the interpretation

"50 out these 1,000 rejections are expected to be false rejections"

This interpretation is incorrect because it is a conditional statement, conditional not only on the number of rejections being 1,000 but also on which specific 1,000 null hypotheses are rejected.[2] This demonstrates a desire for *selective inference.* To this end, Chapter 7 of this Handbook points to future directions of conditional inference for variable selection strategies with better FDR control. Chapter 9 focuses on post-hoc bounds for the number or the proportion of true/false null hypotheses, with important applications in selective inference. Chapter 14 provides an example of selective inference applied to neuroimaging.

1.2 Other Early Multiple Comparison Developments

Whereas Tukey (1953) was on *confidence intervals* inference, there was a line of multiple *testing* development before the establishment of the closed testing principle by Marcus et al. (1976). Instead of making a list of the tests, we describe one that played a significant role in subsequent thinking of multiple comparisons: the Newman–Keuls' multiple range test (Newman (1939) and Keuls (1952)), which is as follows.

Suppose that under the ith treatment a random sample $Y_{i1}, Y_{i2}, \ldots, Y_{in}$ of size n is taken, where the observations between the treatments are independent. Then under the usual normality and equality of variances assumptions, we have the balanced one-way model (4.4)

$$Y_{ia} = \mu_i + \epsilon_{ia}, \quad i = 1, \ldots, k, \quad a = 1, \ldots, n, \tag{1.2}$$

where μ_i is the effect of the ith treatment, $i = 1, \ldots, k$, and $\epsilon_{11}, \ldots, \epsilon_{kn}$ are i.i.d. normal errors with mean 0 and unknown variance σ^2. We use the notation

$$\hat{\mu}_i = \bar{Y}_i = \sum_{a=1}^{n} Y_{ia}/n,$$

$$\hat{\sigma}^2 = MSE = \sum_{i=1}^{k} \sum_{a=1}^{n} (Y_{ia} - \bar{Y}_i)^2/[k(n-1)]$$

for the sample means and the pooled sample variance.

Let $c_2, \ldots, c_k$ denote the $100(1-\alpha)$ quantile of the (Studentized) range of m of the $\hat{\mu}_i$s

$$\frac{\max_{1 \le i,j \le m} |\hat{\mu}_i - \hat{\mu}_j|}{\hat{\sigma}}, \quad m = 2, \ldots, k$$

under the null hypothesis that their means are equal, $H_0 : \mu_1 = \cdots = \mu_m$.

Let $[1], \ldots, [k]$ denote the random indices such that

$$\hat{\mu}_{[1]} \le \cdots \le \hat{\mu}_{[k]}.$$

(Since $\hat{\mu}_1, \ldots, \hat{\mu}_k$ are continuous random variables, ties occur among them with probability zero.)

As a first step, compare the k range $\hat{\mu}_{[k]} - \hat{\mu}_{[1]}$ with $c_k \hat{\sigma} \sqrt{2/n}$. If

$$\hat{\mu}_{[k]} - \hat{\mu}_{[1]} \le c_k \hat{\sigma} \sqrt{2/n}$$

then stop; else assert

$$\mu_{[1]} \ne \mu_{[k]}$$

[2]There are versions of FDR that are more conditional in spirit, such as Efron's Fdr (cf. Efron (2007)) (note the lower case "d" and "r").

and the two $(k-1)$ ranges $\hat{\mu}_{[k-1]} - \hat{\mu}_{[1]}$ and $\hat{\mu}_{[k]} - \hat{\mu}_{[2]}$ are compared with $c_{k-1}\hat{\sigma}\sqrt{2/n}$ (which reduces multiplicity adjustment by one). If both ranges are less than or equal to $c_{k-1}\hat{\sigma}\sqrt{2/n}$, then stop. Otherwise, assert

$$\mu_{[1]} \neq \mu_{[k-1]}$$

if

$$\hat{\mu}_{[k-1]} - \hat{\mu}_{[1]} > c_{k-1}\hat{\sigma}\sqrt{2/n}$$

and/or assert

$$\mu_{[2]} \neq \mu_{[k]}$$

if

$$\hat{\mu}_{[k]} - \hat{\mu}_{[2]} > c_{k-1}\hat{\sigma}\sqrt{2/n},$$

and the three or appropriate one $(k-2)$ range(s) are compared with $c_{k-2}\hat{\sigma}\sqrt{2/n}$ (which reduces multiplicity adjustment by one more), and so on. Once a range has been found to be less than or equal to its scaled critical value, its subranges are no longer tested.

For example, suppose $k = 4$, and $\hat{\mu}_2 \leq \hat{\mu}_1 \leq \hat{\mu}_3 \leq \hat{\mu}_4$. Then, at Step 1, Newman–Keuls' multiple range test would compare $\hat{\mu}_4 - \hat{\mu}_2$ with $c_4\hat{\sigma}\sqrt{2/n}$ which adjusts for a multiplicity of four. If $\hat{\mu}_4 - \hat{\mu}_2 > c_4\hat{\sigma}\sqrt{2/n}$, then $\mu_4 > \mu_2$ is inferred, and go to Step 2; else stop.

At Step 2, compare $\hat{\mu}_3 - \hat{\mu}_2$ with $c_3\hat{\sigma}\sqrt{2/n}$ which adjusts for a multiplicity of three, as well as compare $\hat{\mu}_4 - \hat{\mu}_1$ with $c_3\hat{\sigma}\sqrt{2/n}$. If both ranges are less than or equal to $c_3\hat{\sigma}\sqrt{2/n}$, then stop. Otherwise, assert $\mu_3 > \mu_2$ if $\hat{\mu}_3 - \hat{\mu}_2 > c_3\hat{\sigma}\sqrt{2/n}$, and/or assert $\mu_4 > \mu_1$ if $\hat{\mu}_4 - \hat{\mu}_1 > c_3\hat{\sigma}\sqrt{2/n}$, and the three or appropriate one pairwise differences are compared with $c_2\hat{\sigma}\sqrt{2/n}$ which is like the critical value of a two-sample t-test.

Newman–Keuls' multiple range test appealed to the (vague) notion that

> *Multiplicity needs to be adjusted for (only) to the extent that multiple (ordered adjacent) groups are equal to each other.*

However, Newman–Keuls' multiple range test is not a confident inequalities method in the sense that it does not control the maximum probability of inferring $\mu_i \neq \mu_j$ for some pair when in fact $\mu_i = \mu_j$, unless $k = 3$. This can be seen as follows. Suppose $k = 4, \alpha = 0.05$, and $\mu_1 = \mu_2 \ll \mu_3 = \mu_4$ where $\ll$ means *much smaller*. Then all ranges except $\hat{\mu}_2 - \hat{\mu}_1$ and $\hat{\mu}_4 - \hat{\mu}_3$ will exceed their scaled critical values with virtual certainty. Thus,

$$\begin{aligned} & P_{\boldsymbol{\mu},\sigma^2}\{\text{at least one incorrect assertion}\} \\ = \; & P_{\boldsymbol{\mu},\sigma^2}\{\text{assert } \mu_1 \neq \mu_2 \text{ or } \mu_3 \neq \mu_4\} \\ \approx \; & 1 - (0.95)^2 \\ = \; & 0.0975. \end{aligned}$$

Therefore, the Newman–Keuls multiple range test is not a confident inequalities method and cannot be recommended. The culprit is that there are multiple decision paths, and the worst-case scenario is not under a single homogeneity null hypothesis.[3]

Nevertheless, the intriguing step-down flavor of Newman–Keuls' and other multiple range tests played a part in spurring a flurry of research activities on step-wise multiple comparison methods. The closed testing principle of Marcus et al. (1976) then established how FWER can be controlled in testing multiple hypotheses in general, with various stepwise methods that do control FWER turning out to be special cases. For testing multiple

[3]Therefore, it is best to avoid using Newman–Keuls' multiple range test which remains in the MEANS statement in Proc GLM of SAS.

hypotheses such as $H_0 : \mu_i = \mu_j$, $1 \leq i < j \leq k$, the closed testing principle considers all possible *intersections* of them that may be true, and a more correct phrasing of the previous notion would be

> *Multiplicity needs to be adjusted for (only) to the extent that multiple null hypotheses may be true simultaneously.*

1.3 Closed Testing Principle

The closed testing principle (cf. Marcus et al. (1976)) or closure principle is a "general solution to multiple testing problems" (Sonnemann (2008)) for finite systems of null hypotheses in the case that control of the FWER at a prespecified level α is desired. The general idea behind this method is to add to the system $\mathcal{H}$ of the null hypotheses $H_1, \ldots, H_m$ of interest all their possible (non-empty) intersections, where a null hypothesis is identified with a non-empty subset of the parameter space Θ of the statistical model under consideration. The extended system $\bar{\mathcal{H}}$ is called the closure of $\mathcal{H}$ (with respect to intersection). Let us denote such an intersection hypothesis by H_S, where $S \subseteq \{1, \ldots, m\}$ contains the indices of those null hypotheses which are used in the intersection. Even if these intersection hypotheses are not of scientific interest, they are tested in an auxiliary manner in order to provide a multiplicity correction. Namely, a closed testing procedure tests every such intersection hypothesis (including $H_i = H_{\{i\}}$ for $1 \leq i \leq m$) at full level α by an arbitrarily chosen level α test. The adjustment for multiplicity is then performed via the decision rule that only those null hypotheses H_ℓ are rejected for which all intersection hypotheses H_S with $\ell \in S$ have been rejected. Thus, the price to pay for the multiplicity of the problem is that one has to perform up to $2^m - 1$ tests.

A formal mathematical proof for the fact that this principle leads to a strongly FWER-controlling multiple tests can be found, e. g., in Section 3.3 of Dickhaus (2014). Here, we only explain the main idea: we may without loss of generality assume that at least one of the null hypotheses $H_1, \ldots, H_m$ is true (i. e., contains the true parameter value ϑ) because otherwise the FWER equals zero by definition. Hence, there exists a non-empty subset $S^* = S^*(\vartheta)$ of $\{1, \ldots, m\}$ which contains exactly those indices corresponding to true null hypotheses (under ϑ). The closed testing procedure as described before tests S^* at level α because it tests *all* non-empty subsets of $\{1, \ldots, m\}$ at level α. Thus, the probability of falsely rejecting S^* is upper-bounded by α. Due to the structure of the decision rule of the closed testing procedure, the event that at least one type I error occurs is a subset of the event that S^* is falsely rejected. By monotonicity of probability measures, we can conclude that the FWER is controlled at level α.

Assume that $H_0 = \bigcap_{i=1}^m H_i = H_{\{1,\ldots,m\}}$ is non-empty. We call H_0 the "global null hypothesis" or the "complete null hypothesis" of the system $\mathcal{H}$. In practice, one then starts with testing H_0. If H_0 cannot be rejected at level α, then the closed testing procedure cannot reject any of the H_i $(1 \leq i \leq m)$. Hence, it is in such cases not necessary to carry out any additional level α tests for intersection hypotheses H_S with $S \subset \{1, \ldots, m\}$. In the case that H_0 is rejected at level α one proceeds with testing (non-empty) intersection hypotheses H_S with $|S| = m - 1$, and in the case that rejections occur one continues in the same manner by iteratively decreasing the cardinality of the sets S. The following scheme illustrates this strategy for the case of $m = 3$.

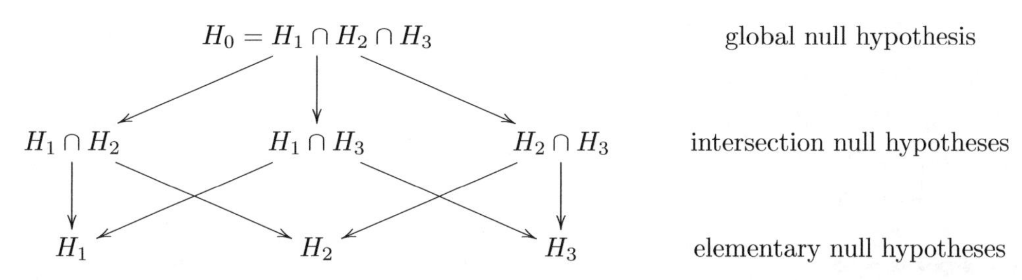

The arrows indicate which (intersection) null hypotheses to test next if rejections have taken place. The null hypotheses H_1, H_2, and H_3 at the bottom layer of the hierarchy are called elementary null hypotheses because their corresponding index set $S = \{i\}$ contains only one element $i \in \{1, 2, 3\}$.

Remark 1.1 *The number $2^m - 1$ of individual tests grows rapidly with the number m of elementary null hypotheses. Therefore, it is in practice often necessary to find a way to avoid carrying out all of these tests explicitly. Computational methods which serve that purpose are called "shortcuts". Plainly phrased, a shortcut of a closed testing procedure avoids explicit testing of all intersection hypotheses in $\bar{\mathcal{H}}$ by traversing $\bar{\mathcal{H}}$ a group-wise manner. Finding shortcuts and establishing general conditions for the existence of shortcuts in closed testing procedures is a vivid field of multiple test theory; see, among many others, Hommel et al. (2007).*

1.4 Overview of the Rest of the Material

The remaining chapters of Part I of this handbook contain general methodological concepts and results pertaining to multiple comparisons and simultaneous inference. Chapter 2 presents generic multiple tests based on marginal p-values, whereas Chapter 3 deals with methods utilizing the *joint* null distribution of p-values or test statistics, respectively, or approximations thereof. In Chapters 4 and 5, important construction principles for multiple tests and simultaneous confidence regions are presented. Namely, Chapter 4 is devoted to partitioning decision theory, and Chapter 5 to graphical approaches. Chapter 6 takes a decision-theoretic view on multiple tests and discusses concepts like admissibility in this context. In Chapter 7, the authors draw connections to the field of model selection. Chapter 8 discusses data-adaptive multiple tests that rely on an estimate of the proportion π_0 of true null hypotheses and a wide variety of estimators for π_0. Part I of the handbook is concluded with Chapter 9, dealing with post-hoc bounds for the number or the proportion of true/false null hypotheses. These bounds have important applications in selective inference.

Part II of the handbook is devoted to applications of multiple testing and simultaneous inference in medicine. Chapter 10 deals with adaptive and group-sequential designs of (clinical) trials, and with related data analysis issues. In Chapter 11, methods for dose-finding studies are described. In Chapter 12, the authors are concerned with studies involving multiple endpoints. Chapter 13, entitled "Personalized and precision medicine", contains methodological contributions to subgroup analyses. Finally, Chapters 14 and 15 refer to specific applications in brain imaging (Chapter 14) and genetics (Chapter 15, dealing with the analysis of SNP data), respectively.

1.5 Supplementary Material

There is a companion website for this handbook, which can be accessed here:

https://www.math.uni-bremen.de/~dickhaus/handbook-multiple-comparisons.html

This website will provide supplementary material such as computer code, video lectures, etc. referring to chapters of this handbook.

Bibliography

Benjamini, Y. and Y. Hochberg (1995). Controlling the false discovery rate: A practical and powerful approach to multiple testing. *J. R. Stat. Soc. Ser. B Stat. Methodol. 57*(1), 289–300.

Berger, J. O. (1985). *Statistical Decision Theory and Bayesian Analysis.* Second Edition. New York: Springer.

Casella, G. and R. L. Berger (1990). *Statistical Inference.* Duxbury Press, Belmont, CA.

Dickhaus, T. (2014). *Simultaneous Statistical Inference with Applications in the Life Sciences.* Berlin, Heidelberg: Springer.

Efron, B. (2007). Correlation and large-scale simultaneous significance testing. *J. Amer. Statist. Assoc. 102*(477), 93–103.

Finner, H. and M. Roters (2001). On the false discovery rate and expected type I errors. *Biom. J. 43*(8), 985–1005.

Hommel, G., F. Bretz, and W. Maurer (2007). Powerful short-cuts for multiple testing procedures with special reference to gatekeeping strategies. *Stat Med 26*(22), 4063–4073.

Keuls, M. (1952). The use of the 'Studentized range' in connection with an analysis of variance. *Euphytica 1*, 112–122.

Lehmann, E. L. (1986). *Testing Statistical Hypotheses.* Second Edition. Wiley Series in Probability and Mathematical Statistics: Probability and Mathematical Statistics. New York: John Wiley & Sons, Inc.

Marcus, R., E. Peritz, and K. R. Gabriel (1976). On closed test procedures with special reference to ordered analysis of variance. *Biometrika 63*(3), 655–660.

Newman, D. (1939). The distribution of the range in samples from a normal population, expressed in terms of an independent estimate of standard deviation. *Biometrika 35*, 16–31.

Sonnemann, E. (2008). General solutions to multiple testing problems. Translation of "Sonnemann, E. (1982). Allgemeine Lösungen multipler Testprobleme. EDV in Medizin und Biologie 13(4), 120–128". *Biom. J. 50*, 641–656.

Tukey, J. W. (1953). The Problem of Multiple Comparisons. Dittoed manuscript of 396 pages, Department of Statistics, Princeton University.

Tukey, J. W. (1994). The problem of multiple comparisons. In Henry I. Braun, editor, *The Collected Works of John W. Tukey*, Volume VIII, Chapter 1, pp. 1–300. New York, London: Chapman & Hall.

2

Multiple Test Procedures Based on p-Values

Ajit C. Tamhane
Northwestern University

Jiangtao Gou
Villanova University

CONTENTS

DOI: 10.1201/9780429030888-3

2.1 Introduction

This chapter gives an extensive overview of multiple test procedures (MTPs) based on marginal p-values. These MTPs are popular in practice because of their ease of application and ready availability of marginal p-values. Joint p-values are generally not readily available because the joint distributions of the underlying test statistics are complicated especially when the test statistics are of different types, e.g., t, χ^2, log-rank or Wilcoxon. Even when they are of the same type, the correlations between them are often unknown, e.g., t-statistics for multiple endpoints. Because of the popularity of the MTPs based on marginal p-values, the FDA Multiplicity Draft Guidance Document (U.S. Department of Health and Human Services, Food and Drug Administration, 2017) has devoted a lengthy subsection IV-C to their discussion.

We mainly focus on the MTPs that strongly control the *family-wise error rate (FWER)* (Hochberg and Tamhane, 1987), defined as follows. Let $H_1, \ldots, H_m$ be $m \geq 2$ null hypotheses treated together as a *family*. Then the strong FWER control requires that

$$\text{FWER} = P\{\text{Reject at least one true } H_i, i = 1, \ldots, m\} \leq \alpha \tag{2.1}$$

under all possible combinations of the true and false null hypotheses.

FWER is commonly used in confirmatory studies, e.g., Phase 3 clinical trials. These studies typically involve a relatively small number of hypotheses. Section 2.8 discusses the *false discovery rate (FDR)* approach of Benjamini and Hochberg (1995). FDR is useful in exploratory studies involving a large number of hypotheses.

The outline of the chapter is as follows. Section 2.2 introduces the Bonferroni and Simes tests, which are used to construct stepwise MTPs discussed in Section 2.3. The latter include the MTPs of Holm (1979), Hochberg (1988) and Hommel (1988) and their extensions and refinements. Section 2.5 introduces gatekeeping MTPs. Section 2.6 reviews the graphical approach proposed by Bretz et al. (2009) and Burman et al. (2009). Section 2.7 discusses group sequential versions of the Holm and Hochberg MTPs. Section 2.8 gives a brief review of the Benjamini and Hochberg (1995) MTP to control the FDR. Finally, Section 2.9 gives a few concluding remarks. Numerical examples with clinical trial contexts are given to illustrate all the procedures.

This chapter is an expanded and revised version of Tamhane and Gou (2018).

2.2 Two Basic Test Procedures

2.2.1 Bonferroni MTP

The *Bonferroni MTP* is historically the first and still the most widely used MTP because of its simplicity and omnibus applicability. It tests all $m \geq 2$ specified null hypotheses, each at level α/m, rejecting those H_i with $p_i \leq \alpha/m$.

We may choose to divide α unequally among the hypotheses by assigning weights $w_i > 0$ to the H_i such that $\sum_{i=1}^{m} w_i = 1$ with higher weights to the more important H_i and lower weights to the less important ones. Then we can reject any H_i if $p_i \leq w_i\alpha$. This is known as the *weighted Bonferroni MTP*. If all w_i are equal to $1/m$ then it reduces to the unweighted Bonferroni MTP.

The Bonferroni MTP is based on the well-known Bonferroni inequality that the probability of a union of events is bounded above by the sum of their probabilities. Let $\{H_i, i \in I\}$ be any nonempty subset of the true null hypotheses where $I \subseteq \{1, \ldots, m\}$ and let $H_I = \cap_{i \in I} H_i$. Further, let $\mathcal{E}_i$ denote the event that H_i is rejected, e.g., $\mathcal{E}_i = \{p_i \leq w_i \alpha\}$ for the weighted Bonferroni MTP. Then

$$\text{FWER} = P_{H_I}\{\cup_{i \in I} \mathcal{E}_i\} \leq \sum_{i \in I} P_{H_i}(\mathcal{E}_i) = \alpha \sum_{i \in I} w_i \leq \alpha. \tag{2.2}$$

Under positive dependence and when m is large, the Bonferroni inequality is not very sharp. Therefore, the Bonferroni MTP is too conservative and lacks power under these conditions.

The additive Bonferroni upper bound on the probability of a union of events can be sharpened by replacing it with the following multiplicative upper bound:

$$P_{H_I}\{\cup_{i \in I} \mathcal{E}_i\} \leq 1 - \prod_{i \in I} [1 - P_{H_i}(\mathcal{E}_i)]. \tag{2.3}$$

The above is an equality if the $\mathcal{E}_i$ are independent and a conservative inequality if the $\mathcal{E}_i$ are positively dependent. It is maximized when $I = \{1, \ldots, m\}$. Equating this maximum upper bound to α, we can set the critical constant for each H_i to $\alpha' = 1 - (1 - \alpha)^{1/m}$, which is slightly sharper than the critical constant α/m used by the Bonferroni MTP. The resulting MTP is known as the Šidák (1967) MTP or prospective alpha allocation scheme (PAAS) (Moyé, 2003).

The Bonferroni MTP provides an α-level union-intersection (UI) test (Gabriel, 1969) of the intersection hypothesis $H_0 = \cap_{i=1}^m H_i$ by rejecting it if at least one $p_i \leq \alpha/m$ or equivalently $p_{\min} \leq \alpha/m$. In the case of weighted Bonferroni MTP, this test is modified to $\min_{1 \leq i \leq m}(p_i/w_i) \leq \alpha$. These UI tests are useful for constructing stepwise MTPs as we shall see in the sequel.

For a given MTP, an *adjusted p-value* of an hypothesis H_i, denoted by $\widetilde{p}_i$, is defined as the smallest α at which H_i can be rejected using that MTP. If $\widetilde{p}_i \leq \alpha$, then H_i can be rejected. For the weighted Bonferroni MTP,

$$\widetilde{p}_i = \min(p_i/w_i, 1), \quad i = 1, \ldots, m.$$

2.2.2 Simes Test

Let $p_{(1)} \leq \cdots \leq p_{(m)}$ denote the ordered p-values and let $H_{(1)}, \ldots, H_{(m)}$ denote the corresponding hypotheses. The Simes (1986) test rejects the intersection hypothesis $H_0 = \cap_{i=1}^m H_i$ if at least one $p_{(i)} \leq i\alpha/m$ $(1 \leq i \leq m)$. It is an exact α-level test under independence of the p-values, which follows from the following identity: If the p-values are i.i.d. $U[0, 1]$ under H_0, then

$$P_{H_0}\left(\bigcup_{i=1}^{m} \left\{p_{(i)} \leq \frac{i\alpha}{m}\right\}\right) = \alpha. \tag{2.4}$$

Note that the Simes test is a test of a single hypothesis H_0 and cannot be used as an MTP to reject any individual $H_{(i)}$ for which $p_{(i)} \leq i\alpha/m$ $(1 \leq i \leq m)$ since such an MTP does not control the FWER.

The Simes test is more powerful than the Bonferroni test of H_0. However, it controls the type I error rate only under independence and under certain positive dependence conditions as shown by Sarkar and Chang (1997) and Sarkar (1998). Other contributions in the same vein include Sarkar (2008), Benjamini and Yekutieli (2001), and Sarkar (2002). Block et al. (2008) have shown that the Simes test is anti-conservative if the test statistics are negatively dependent. On the other hand, the Bonferroni test is valid under any type of dependence.

Some extensions of the Simes test include weighted Simes test (Benjamini and Hochberg, 1997; Hochberg and Liberman, 1994) and generalized Simes test based on critical constants of the form $c_i\alpha$ where $1 = c_1 \geq c_2 \geq \cdots \geq c_m$ (Cai and Sarkar, 2008; Gou and Tamhane, 2014). The latter test rejects H_0 if at least one $p_{(m-i+1)} \leq c_i\alpha$. For the basic Simes test, the critical constants are $c_i = (m - i + 1)/m$.

2.3 Stepwise MTPs

Well-known stepwise MTPs of Holm (1979) and Hochberg (1988) can be derived by applying the closure method proposed by Marcus et al. (1976) with the two basic tests described in the previous section as the local tests for all intersection hypotheses.

2.3.1 Closure Method

The closure method is best explained through a simple example. Consider three null hypotheses: H_1, H_2 and H_3. The closure method hierarchically orders their intersections: (a) the overall intersection: $H_1 \cap H_2 \cap H_3$, (b) the three paiwise intersections: $H_1 \cap H_2$, $H_1 \cap H_3$ and $H_2 \cap H_3$, and (c) the three singletons: H_1, H_2 and H_3. It tests each intersection hypothesis using a local α-level test and rejects it subject to the restriction that all intersection hypotheses implying it have been rejected by their individual local α-level tests. In the present example, the closure MTP begins by testing the overall intersection hypothesis $H_1 \cap H_2 \cap H_3$ at level α. If it is not rejected then it stops testing by accepting (throughout this chapter we use the term "accept" instead of "do not reject" for the sake of brevity) all hypotheses; otherwise, it tests all three pairwise intersection hypotheses by their individual local α-level tests. If $H_1 \cap H_2$ and $H_1 \cap H_3$ are rejected then since H_1 is contained in both, it next tests H_1. If only one of these two pairwise intersections is rejected, then H_1 is accepted without a test.

Since the closure MTP tests all intersection hypotheses at level α, it strongly controls the FWER. In addition, it satisfies the *coherence* requirement (Gabriel, 1969), namely if a given intersection hypothesis is not rejected then all intersection hypotheses implied by it are also not rejected. This MTP is shown graphically in Figure 2.1.

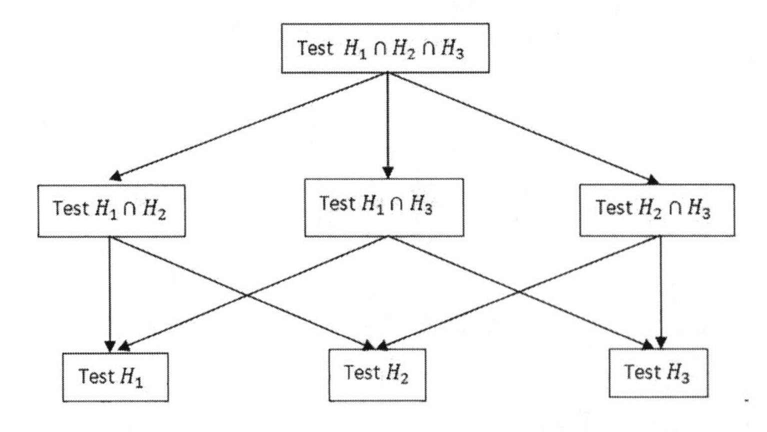

FIGURE 2.1
Graphical representation of the closure MTP for $m = 3$.

In general, the maximum number of tests required by this algorithm is $2^m - 1$. However, if the local tests are *consonant* (Gabriel, 1969), then a shortcut to the closed MTP exists that requires at most m tests. The Holm MTP is an example of such a shortcut.

Suppose the critical values used by the local tests of the intersection hypotheses are arranged in the form of a lower triangular matrix, called the *critical matrix* with the ith row consisting of the i critical values used to test the intersection hypothesis H_I of cardinality $|I| = i$. Liu (1996) showed that if the entries of the critical matrix are constant across each row then the shortcut to the closure MTP is a step-down MTP with its critical values given by the first column (from the bottom up) and if the entries are constant across each column then the shortcut to the closure MTP is a step-up MTP with its critical values given by the last row (from left to right). The examples of the critical matrices associated with the Holm (1979) and Hochberg (1988) stepwise MTPs are given in the following sections.

2.3.2 Holm Step-Down MTP

The Holm (1979) MTP can be derived as a shortcut to the closure MTP that uses the α-level Bonferroni test as the local test for every intersection hypothesis. The Bonferroni test of H_I rejects if $\min_{i \in I} p_i \leq \alpha/|I|$. These critical values are constant across each row (namely, α/i along the ith row) of the critical matrix shown below.

$$\begin{bmatrix} \frac{\alpha}{1} & & & & \\ \frac{\alpha}{2} & \frac{\alpha}{2} & & & \\ \frac{\alpha}{3} & \frac{\alpha}{3} & \frac{\alpha}{3} & & \\ \vdots & \vdots & \vdots & \ddots & \\ \frac{\alpha}{m} & \frac{\alpha}{m} & \frac{\alpha}{m} & \cdots & \frac{\alpha}{m} \end{bmatrix}$$

Hence, the shortcut is a step-down MTP, which is Holm's step-down MTP. The critical values of the Holm MTP are given by the first column (from the bottom up) of the critical matrix whose entries are $\alpha/m, \alpha/(m-1), \ldots, \alpha/1$.

The Holm MTP can be represented as shown below

$$\begin{array}{ccccccccc} H_{(1)} & & H_{(2)} & & \cdots & & H_{(m-1)} & & H_{(m)} \\ p_{(1)} & \leq & p_{(2)} & \leq & \cdots & \leq & p_{(m-1)} & \leq & p_{(m)} \\ \frac{\alpha}{m} & & \frac{\alpha}{m-1} & & \cdots & & \frac{\alpha}{2} & & \frac{\alpha}{1} \end{array}$$

Direction of testing: $\longrightarrow$

It operates as follows. Begin by testing $H_{(1)}$. If $p_{(1)} \leq \alpha/m$ then reject $H_{(1)}$ and continue to test $H_{(2)}$; otherwise accept all hypotheses and stop testing. In general, at the ith step test $H_{(i)}$. If $p_{(i)} \leq \frac{\alpha}{m-i+1}$, then reject $H_{(i)}$ and continue to test $H_{(i+1)}$; otherwise stop testing and accept all remaining hypotheses, $H_{(i)}, \ldots, H_{(m)}$. The adjusted p-values of the hypotheses $H_{(i)}$ are given by

$$\widetilde{p}_{(i)} = \min\left\{\max\left\{\widetilde{p}_{(i-1)}, (m-i+1)p_{(i)}\right\}, 1\right\}, \quad i = 1, \ldots, m.$$

Since the Holm MTP is based on the Bonferroni test, it is also free of any dependence assumptions among the p-values. Weighted extensions of the Holm MTP were studied by Holm (1979) and Benjamini and Hochberg (1997).

2.3.3 Hommel MTP

Since the Simes test is more powerful than the Bonferroni test, it is natural to use it as the local test in the closure MTP. The Simes test of the intersection hypothesis H_I rejects if at least one $p_{(i)} \leq i\alpha/|I|$ for $i \in I$. Note that, here for convenience, we have used a simplifying notation, namely, that although $i \in I$ ranges over a subset I of the index set $\{1, 2, \ldots, m\}$, the indexes (i) are assumed to range over $\{(1), (2), \ldots, (|I|)\}$ corresponding to the ordered values of $p_{(i)}$ for $i \in I$.

The Simes critical values to test intersection hypotheses H_I of cardinality $|I| = i$ can be arranged in the form of a critical matrix shown below:

$$\begin{bmatrix} \frac{\alpha}{1} & & & & \\ \frac{\alpha}{1} & \frac{\alpha}{2} & & & \\ \frac{\alpha}{1} & \frac{2\alpha}{3} & \frac{\alpha}{3} & & \\ \vdots & \vdots & \vdots & \ddots & \\ \frac{\alpha}{1} & \frac{(m-1)\alpha}{m} & \frac{(m-2)\alpha}{m} & \cdots & \frac{\alpha}{m} \end{bmatrix}.$$

Note that the entries of this critical matrix are not constant either across rows or across columns. As a result, the Hommel MTP does not have a simple step-down or step-up structure. It operates as follows: at Step 1, if $p_{(m)} \leq \alpha$ then reject all hypotheses and stop testing; otherwise accept $H_{(m)}$ and go to Step 2. In general, at Step $i = 2, \ldots, m-1$, if $p_{(m-j+1)} \leq [(i-j+1)/i]\alpha$ for at least one $j = 1, \ldots, i$ then reject all hypotheses with p-values $\leq \alpha/(i-1)$; otherwise accept $H_{(m-i+1)}$ and go to step $i+1$. At the final step, if $p_{(m-j+1)} \leq [(m-j+1)/m]\alpha$ for at least one $j = 1, \ldots, m$ then reject $H_{(1)}$ if $p_{(1)} \leq \alpha/(m-1)$ and stop testing.

2.3.4 Hochberg Step-Up MTP

The Hochberg (1988) MTP is a shortcut to the closure MTP, which uses a conservative Simes test as the local test for intersection hypotheses. The critical values of this conservative Simes test are $\alpha/(|I|-i+1)$, i.e., it rejects $H_I = \cap_{i\in I} H_i$ if at least one of the $p_{(i)} \leq \alpha/(|I|-i+1)$ for $i \in I$. These critical values are conservative compared to the exact Simes critical values $i\alpha/|I|$ since $\alpha/(|I|-i+1) \leq i\alpha/|I|$ with equality holding iff $i = 1$ or $i = |I|$. The critical values $\alpha/(|I|-i+1)$ are constant across each column of the critical matrix as shown below:

$$\begin{bmatrix} \frac{\alpha}{1} & & & & \\ \frac{\alpha}{1} & \frac{\alpha}{2} & & & \\ \frac{\alpha}{1} & \frac{\alpha}{2} & \frac{\alpha}{3} & & \\ \vdots & \vdots & \vdots & \ddots & \\ \frac{\alpha}{1} & \frac{\alpha}{2} & \frac{\alpha}{3} & \cdots & \frac{\alpha}{m} \end{bmatrix}.$$

Hence, the shortcut to this closure test is a step-up MTP with the critical values given by the last row of the critical matrix, viz., $\alpha/1, \alpha/2, \ldots, \alpha/m$, which are the same as those for the Holm MTP. This step-up MTP is the Hochberg MTP, which is conservative compared to the Hommel MTP but is simpler.

The Hochberg MTP can be represented using the same diagram as for the Holm MTP, but it operates in the reverse order as shown below.

$$\begin{array}{ccccccccc} H_{(1)} & & H_{(2)} & & \cdots & & H_{(m-1)} & & H_{(m)} \\ p_{(1)} & \leq & p_{(2)} & \leq & \cdots & \leq & p_{(m-1)} & \leq & p_{(m)} \\ \frac{\alpha}{m} & & \frac{\alpha}{m-1} & & \cdots & & \frac{\alpha}{2} & & \frac{\alpha}{1} \end{array}$$

Direction of testing: $\longleftarrow$

It operates as follows. Begin by testing $H_{(m)}$. If $p_{(m)} \leq \alpha$ then reject all hypotheses and stop testing; otherwise accept $H_{(m)}$ and continue to test $H_{(m-1)}$. In general, at the ith step test $H_{(m-i+1)}$. If $p_{(m-i+1)} \leq \frac{\alpha}{i}$ then stop testing and reject all remaining hypotheses, $H_{(m-i+1)}, \ldots, H_{(1)}$; otherwise accept $H_{(m-i+1)}$ and continue to test $H_{(m-i)}$.

It is easy to see that the Hochberg MTP is uniformly more powerful than the Holm MTP since it uses the same critical values but tests the hypotheses in the reverse order and rejects them by implication rather than accepting them as in the Holm MTP. However, it requires the same dependence assumptions as required by the Simes test, whereas the Holm MTP is assumption-free.

The adjusted p-values of the hypotheses $H_{(i)}$ are given by

$$\widetilde{p}_{(i)} = \min\left\{\min\left\{\widetilde{p}_{(i+1)}, (m-i+1)p_{(i)}\right\}, 1\right\}, \quad i = m, \ldots, 1.$$

Various ways of constructing weighted versions of the Hochberg MTP were studied by Tamhane and Liu (2008).

2.3.4.1 Hochberg MTP Under Negative Dependence

The Holm MTP is often preferred in practice to the Hochberg MTP even though it is less powerful because it does not require any distributional assumptions, whereas the Hochberg MTP requires positive dependence assumption among the test statistics. However, we saw in the previous section that the Hochberg MTP is based on the conservative Simes test, which may be conjectured to remain conservative under some types of negative dependence. However, it can be shown to be anti-conservative for $n = 2$ and $n = 3$ under negative dependence. These two cases can be handled by modifying its critical constants c_2 and c_3 as follows:

$$c_2 = \frac{1}{2} - \frac{\alpha}{2} \quad \text{and} \quad c_3 = \frac{1}{3} - \frac{\alpha}{36}.$$

The other critical constants remain the same, namely, $c_i = 1/i$. Gou and Tamhane (2018a) showed that this modified conservative Simes test is conservative when the joint distribution of the p-values is a mixture of their bivariate distributions where each bivariate distribution is negatively quadrant dependent (NQD) (Lehmann, 1966). Hence the Hochberg MTP based on it is also conservative under this type of negative dependence.

Of course, the Hochberg MTP may remain conservative under some other types of negative dependence as well. To verify this conjecture, Gou and Tamhane (2018a) carried out extensive simulations. They showed that the Hochberg MTP controls FWER under multivariate normal distributions with two types of negative correlation structures: (a) common negative correlation: $\rho_{ij} = \rho$ where $-1/(m-1) < \rho < 0$ and (b) product correlations: $\rho_{ij} = \lambda_i\lambda_j$, where $\lambda_i = \lambda > 0$ for $i = 1, \ldots, [m/2]$ and $\lambda_i = -\lambda < 0$ for $i = [m/2]+1, \ldots, m$ (where $[m/2]$ is the largest integer $\leq m/2$). Thus $\rho_{ij} = \lambda^2 > 0$ if $i, j \leq [m/2]$ or $i, j \geq [m/2]+1$ and $\rho_{ij} = -\lambda^2 < 0$ if $i \leq [m/2]$ and $j \geq [m/2]+1$. Thus, roughly half of the ρ_{ij} are positive and the other half are negative, all of them being of equal magnitude.

2.3.4.2 Hommel-Hochberg (HH) Hybrid MTP

Gou et al. (2014) improved upon the Hochberg MTP by combining it with a test similar to that used in the Hommel MTP. This Hochberg-Hommel (HH) hybrid MTP is applied using the same step-up algorithm as the Hochberg MTP as seen in the following: Fix critical values $c_1 \geq c_2 \geq \cdots \geq c_m$ and $d_1 \geq d_2 \geq \cdots \geq d_m$ where $c_i \geq d_i$, $c_1 = d_1 = 1$ and $c_m = d_m$. At Step 1, if $p_{(m)} > c_1\alpha$ then accept $H_{(m)}$ and go on to Step 2; otherwise reject all hypotheses and stop testing. At the ith step, if $p_{(m-i+1)} > c_i\alpha$ then accept $H_{(m-i+1)}$ and go onto Step $i+1$; otherwise reject all hypotheses H_j with $p_j \leq d_i\alpha$ and stop testing (this latter test is similar to the one used by the Hommel MTP). At $i = m$ if $p_{(1)} > c_m\alpha$ then accept $H_{(1)}$; otherwise reject $H_{(1)}$ and stop testing.

Gou et al. (2014) suggested the following choice of the critical constants:

$$c_i = \frac{i+1}{2i}, d_i = \frac{1}{i} \quad (1 \leq i \leq m-1), c_m = d_m = \frac{1}{m}. \tag{2.5}$$

This choice is the zeroth-order approximation to the solution of the equations for determining the c_i and d_i. The resulting HH MTP is therefore denoted by HH_0. It is identical to the Hochberg MTP for $m = 2$, but is uniformly more powerful for $m \geq 3$. Gou and Tamhane (2018b) suggested a more flexible choice of critical values for HH, but HH_0 is generally the best choice.

We now give a numerical example to illustrate all the MTPs introduced in this section.

Example 2.1 *Consider a dose-finding trial with five doses to be compared with a zero dose (placebo) based on a single primary endpoint. The FWER is to be controlled at* $\alpha = .05$. *The ordered one-sided p-values are as follows:*

$$p_{(1)} = .011, p_{(2)} = .012, p_{(3)} = .021, p_{(4)} = .032, p_{(5)} = .056.$$

For the sake of convenience, suppose that the p-values are ordered from the highest dose to the lowest dose. In other words, the highest dose is associated with the lowest p-value (most effective compared to placebo) and the lowest dose is associated with the highest p-value (least effective compared to placebo).

The critical constants $c_i\alpha$ *used by the Holm and Hochberg MTPs are as follows:*

$$c_1\alpha = .0500, c_2\alpha = .0250, c_3\alpha = .0167, c_4\alpha = .0125, c_5\alpha = .0100.$$

The Holm MTP begins by comparing $p_{(1)}$ *with* $c_5\alpha = .010$. *Since* $p_{(1)} > .0100$, *it accepts* $H_{(1)}$ *and hence accepts all the hypotheses (as does the Bonferroni MTP) and stops testing.*

The Hochberg MTP begins by comparing $p_{(5)}$ *with* $c_1\alpha = .0500$. *Since* $p_{(5)} > .0500$, *it accepts* $H_{(5)}$ *and goes on to test* $H_{(4)}$. *Since* $p_{(4)} > c_2\alpha = .0250$, *it accepts* $H_{(4)}$ *and goes on to test* $H_{(3)}$. *Since* $p_{(3)} > c_3\alpha = .0167$, *it accepts* $H_{(3)}$ *and goes on to test* $H_{(2)}$. *Since* $p_{(2)} < c_2\alpha = .0125$, *it rejects* $H_{(2)}$ *and by implication also* $H_{(1)}$ *and stops testing.*

Next consider the HH_0 *MTP. Its critical constants are given in the table below, where the* c_i *and* d_i *are calculated from* (2.5).

i	1	2	3	4	5
$c_i\alpha$	*.0500*	*.0375*	*.0333*	*.0313*	*.0100*
$d_i\alpha$	*.0500*	*.0250*	*.0167*	*.0125*	*.0100*

HH_0 *begins by comparing* $p_{(5)}$ *with* $c_1\alpha = .0500$. *Since* $p_{(5)} > .0500$, *it accepts* $H_{(5)}$ *and compares* $p_{(4)}$ *with* $c_2\alpha = .0375$. *Since* $p_{(4)} < .0375$, *it rejects all* H_i *with* $p_i \leq d_2\alpha = .0250$. *Thus it rejects* $H_{(1)}, H_{(2)}, H_{(3)}$ *and stops testing. Note that it does not reject* $H_{(4)}$.

Next consider the Hommel MTP. It accepts $H_{(5)}$ and $H_{(4)}$ since $p_{(5)} > .05$ and $p_{(4)} > .025$. At the third step, we have $p_{(5)} > .05, p_{(4)} \leq (2/3).05 = .033, p_{(3)} > (1/3).05 = .0167$. Since one of the inequalities is violated, we stop testing and reject those $H_{(i)}$ with $p_{(i)} \leq .05/2 = .025$; hence we reject $H_{(1)}, H_{(2)}$ and $H_{(3)}$.

Thus, in this example, the Holm MTP did not reject any hypothesis (did not find any dose effective compared to the placebo), the Hochberg MTP rejected $H_{(1)}$ and $H_{(2)}$ (found the highest and the second highest dose effective), while HH_0 and the Hommel MTP rejected $H_{(1)}, H_{(2)}$ and $H_{(3)}$ (found the three highest doses effective).

2.3.5 Sequentially Rejective Weighted Bonferroni MTPs

Hommel et al. (2007) studied a class of closed MTPs, which use weighted Bonferroni tests as local tests for all intersection hypotheses. Specifically, suppose that a collection of non-negative weights $w_i(I)$ is defined for the hypotheses H_i, $i \in I$ for all nonempty $I \subseteq \{1, \ldots, m\}$. The local test of any intersection hypothesis $H_I = \cap_{i \in I} H_i$ at level α rejects if $\min_{i \in I} p_i^*(I) \leq \alpha$ where $p_i^*(I) = \min(p_i/w_i(I), 1)$ is the weighted p-value. Hommel et al. (2007) showed that if the weights satisfy the monotonicity condition:

$$w_i(J) \leq w_i(I) \quad \text{for all } i \in I \subseteq J \subseteq \{1, \ldots, m\},$$

then the closed MTP is consonant (Gabriel, 1969) and admits a step-down shortcut. It is immediately clear that the unweighted Bonferroni test, which uses $w_i(I) = 1/|I|$, satisfies this monotonicity condition and the resulting step-down shortcut is the Holm MTP. Graphical implementation of these MTPs is discussed in Section 2.6.

2.4 Fixed Sequence and Fallback MTPs

In the stepwise MTPs studied in the previous section, the hypotheses are ordered based on their p-values, i.e., based on the observed data, while in some applications they are ordered *a priori*. Two examples are as follows: ordered testing of the doses of a drug and hypotheses ordered by their clinical importance. Let the hypotheses be ordered as $H_1 \to H_2 \to \cdots \to H_m$. The ordering of the hypotheses induces a nested structure on them: $H_1 \subseteq H_2 \subseteq \cdots \subseteq H_m$, where $H_i \subseteq H_j$ means that H_i is implied by H_j. Maurer et al. (1995) showed that a step-down shortcut results in the following *fixed sequence MTP*: begin by testing H_1 at level α. If H_1 is rejected, then test H_2 at the same level α; otherwise accept all hypotheses and stop testing. In general, if H_i is rejected, then test H_{i+1} at level α; otherwise accept all the remaining hypotheses and stop testing. This MTP illustrates the so-called *α-recycling principle*: the α used to test a rejected hypothesis is carried over to the next unrejected hypothesis in the sequence. Therefore, no "α-adjustment" is required if rejection occurs. However, the power of this MTP depends crucially on the "correct" ordering of the hypotheses in terms of their true effects. If an earlier hypothesis in the sequence has a small non-null effect, then testing can stop because of its acceptance. As a result, the later hypotheses with possibly larger effects and smaller p-values will be accepted without being tested.

The *fallback MTP* (Wiens, 2003; Wiens and Dmitrienko, 2005) addresses this problem of early stopping of the fixed sequence MTP. First weights $w_i > 0$ are preassigned to all hypotheses H_i such that $\sum_{i=1}^{m} w_i = 1$. The fallback MTP begins by testing H_1 at level $\alpha_1 = w_1\alpha$. If H_1 is rejected, then it tests H_2 at level $\alpha_2 = \alpha_1 + w_2\alpha$. If H_1 is accepted, then it tests H_2 at level $\alpha_2 = w_2\alpha$. In general, if H_{i-1} is rejected, then at Step i, it tests H_i at level $\alpha_i = \alpha_{i-1} + w_i\alpha$. If H_{i-1} is accepted, then it tests H_i at level $\alpha_i = w_i\alpha$.

Note that the fixed sequence MTP is a special case of the fallback MTP for $w_1 = 1$, $w_2 = \cdots = w_m = 0$. The fallback MTP is another example of α-recycling in action. If a given hypothesis H_i in the sequence is rejected, then the α_i used to test it is carried forward to H_{i+1}; on the other hand, if H_i is accepted, then the α_i used to test it is lost and H_{i+1} is tested at level $\alpha_{i+1} = w_{i+1}\alpha$.

2.5 Gatekeeping MTPs

Gatekeeping MTPs are used when the hypotheses under test are hierarchically ordered into several families with logical relations between successive families. For illustration purposes, consider two ordered families: primary family ($\mathcal{F}_1$) and secondary family ($\mathcal{F}_2$), but the stated results can be extended to any number of ordered families. Two types of gatekeeping MTPs are common: serial and parallel. *Serial gatekeeping MTPs* test a secondary hypothesis only if *all* primary hypotheses are rejected. This is the case when we have co-primary endpoints on all of which efficacy of the drug must be demonstrated. On the other hand, *parallel gatekeeping MTPs* test a secondary hypothesis only if *at least one* of the primary hypotheses is rejected. This is the case when the efficacy on any one of the primary endpoints is sufficient to demonstrate the efficacy of the drug. The fixed sequence MTP is an example of serial gatekeeping where each ordered hypothesis constitutes an ordered family.

Assume that there are m_1 hypotheses in $\mathcal{F}_1$, denoted by $H_1, \ldots, H_{m_1}$, and m_2 hypotheses in $\mathcal{F}_2$, denoted by $H_{m_1+1}, \ldots, H_{m_1+m_2}$. The total number of hypotheses is denoted by $m = m_1 + m_2$. We want to control the FWER at level α for the joint family $\mathcal{F} = \mathcal{F}_1 \cup \mathcal{F}_2$ subject to a specified gatekeeping condition.

2.5.1 Serial Gatekeeping MTPs

A serial gatekeeping MTP can be implemented by using the fixed sequence MTP at level α for $\mathcal{F}_1$ if the hypotheses can be "correctly" ordered. It allows each H_i, $i \in \mathcal{F}_1$ to be tested at full α level if the previous H_i in the sequence are rejected. Of course, if the hypotheses are not "correctly" ordered then fixed sequence MTP may not be a good choice and other MTPs may be preferable. If all hypotheses in the primary family are rejected, then one should use the most powerful MTP available at level α to test the hypotheses in the secondary family.

2.5.2 Parallel Gatekeeping MTPs

Dmitrienko et al. (2003) were the first to introduce parallel gatekeeping MTPs. Whereas in serial gatekeeping either the full α is propagated to $\mathcal{F}_2$ or none, in parallel gatekeeping a fraction α_2 of α $(0 \leq \alpha_2 \leq \alpha)$ is allowed to be propagated to $\mathcal{F}_2$ depending on the number of primary hypotheses rejected. Denote this number by $r_1 \leq m_1$. If we use the Bonferroni MTP to test primary hypotheses, then it is clear that $\alpha_2 = (r_1/m_1)\alpha$. In the following, we generalize this idea to stepwise MTPs such as the Holm and Hochberg MTPs. These stepwise MTPs need to be modified in order to use them to test primary hypotheses so that $\alpha_2 > 0$ when $r_1 > 0$, i.e., for the secondary hypotheses to be *testable* when the parallel gatekeeping condition is satisfied. Toward this end, we define some new concepts introduced in Dmitrienko et al. (2008).

2.5.2.1 Error Rate Function

Let $I \subseteq M_1 = \{1, 2, \ldots, m_1\}$ and let $H_I = \cap_{i \in I} H_i$. Then the *error rate function* under H_I of an α-level MTP is defined as

$$e(I|\alpha) = \sup_{H_I} P\{\text{Reject at least one } H_i,\ i \in I\}. \tag{2.6}$$

Further note that $e(\emptyset|\alpha) = 0, e(M_1|\alpha) = \alpha$ and $e(I|\alpha) \leq e(J|\alpha)$ if $I \subseteq J$.

It can be shown that if A_1 is the index set of the accepted hypotheses in $\mathcal{F}_1$ then the amount of α lost from those accepted hypotheses equals $e(A_1|\alpha)$. Hence, the amount of α propagated to the secondary family equals $\alpha_2 = \alpha - e(A_1|\alpha)$. For the Bonferroni MTP, we have $e(A_1|\alpha) = \frac{|A_1|}{m_1}\alpha = \frac{a_1}{m_1}\alpha$, where a_1 is the number of accepted hypotheses. Therefore,

$$\alpha_2 = \alpha - \frac{a_1}{m_1}\alpha = \frac{r_1}{m_1}\alpha.$$

2.5.2.2 Separable and Truncated MTPs

An MTP $\mathcal{P}_1$ used to test $H_i \in \mathcal{F}_1$ at level α is called *separable* if for all proper subsets $A_1 \subseteq M_1$, its error rate function $e(A_1|\alpha) \leq \alpha$ with an equality iff $A_1 = M_1$, i.e., iff all primary hypotheses are accepted and so all secondary hypotheses are accepted without tests. On the other hand, if at least one primary hypothesis is rejected, then $\alpha_2 > 0$ and so secondary hypotheses are tested at level α_2. If $\mathcal{P}_1$ is non-separable then $\alpha_2 = 0$ even if A_1 is a proper subset M_1. As a result, no secondary hypothesis can be tested even though $r_1 > 0$. Therefore, a nonseparable MTP cannot be used for the primary family as a part of parallel gatekeeping MTP. Note that the MTP $\mathcal{P}_2$ for $\mathcal{F}_2$ need not be separable; one should use the most powerful MTP available.

Single-step MTPs such as the Bonferroni and the Dunnett (1955) are separable, but their stepwise extensions such as the Holm and Hochberg, and step-up and step-down Dunnett (Dunnett and Tamhane, 1993; Naik, 1975), are not separable. Therefore, these stepwise MTPs cannot be used to test the primary family unless suitably modified to make them separable. This can be achieved by using a convex combination of the critical constants of the stepwise MTP and an appropriate single-step MTP. We call such hybrid stepwise procedures *truncated MTPs*. As we see below, making a nonseparable stepwise MTP separable comes at some loss in power.

As an example, the truncated Holm MTP compares ordered $p_{(i)}$ with the critical values,

$$\left[\frac{\gamma}{m_1 - i + 1} + \frac{1-\gamma}{m_1}\right]\alpha, \quad i = 1, 2, \ldots, m_1,$$

where $\gamma \in [0, 1)$ is called the *truncation fraction*. For $\gamma = 0$ we get the Bonferroni MTP and for $\gamma = 1$ we get the Holm MTP. The truncated Hochberg MTP uses the same critical values but tests the ordered hypotheses in the reverse order. Note that for $0 \leq \gamma < 1$, the truncated Holm MTP is less powerful than the Holm MTP. A smaller γ results in a lower primary power but a larger α_2 and hence higher secondary power. The choice of γ may be determined via simulation to balance the primary and secondary powers. Note that although $\gamma = 1$ gives the maximum primary power, it is not an acceptable choice since it corresponds to the Holm MTP, which is nonseparable.

The error rate function of the truncated Holm MTP is given by

$$e(I|\alpha) = \begin{cases} \left[\gamma + (1-\gamma)\frac{|I|}{m_1}\right]\alpha & \text{if } |I| > 0 \\ 0 & \text{if } |I| = 0. \end{cases}$$

Example 2.2 *We use the EPHESUS cardiovascular trial example from Dmitrienko and Tamhane (2009). The trial compared a new drug with placebo on two primary and two secondary endpoints:*

1. *P1: All-cause mortality, P2: Cardiovascular mortality + cardiovascular hospitalizations.*
2. *S1: Cardiovascular mortality, S2: All-cause mortality + all-cause hospitalizations.*

At least one of P1 or P2 must be found significant before testing either S1 or S2. Thus this is a parallel gatekeeping problem. Let H_1 *and* H_2 *denote the hypotheses associated with P1 and P2 and* H_3 *and* H_4 *denote the hypotheses associated with S1 and S2, respectively. Thus, the primary family is* $\mathcal{F}_1 = \{H_1, H_2\}$*, and the secondary family is* $\mathcal{F}_2 = \{H_3, H_4\}$*. Suppose* $\mathcal{P}_1$ *is the Bonferroni MTP (which is separable) and* $\mathcal{P}_2$ *is the Hochberg MTP (which is non-separable). The p-values (which have been altered for this example) and the critical constants used to test the primary and secondary hypotheses along with the test results are shown in the following table.*

$\mathcal{F}_i$	Hypothesis	p_i	α_i	Crit. Const.	Result
$\mathcal{F}_1$	H_1	.030	.05	.025	*Accept*
	H_2	.020	.05	.025	*Reject*
$\mathcal{F}_2$	H_3	.040	.025	.025	*Accept*
	H_4	.030	.025	.0125	*Accept*

Since the Bonferroni MTP rejects only one hypothesis (H_2) *from* $\mathcal{F}_1$, $\alpha_2 = (1/2)\alpha_1 = .025$. *The Hochberg MTP is unable to reject either* H_3 *or* H_4 *at* $\alpha_2 = .025$.

Next we will consider the same data but use the truncated Holm MTP with the truncation parameter $\gamma = 0.25$ *for* $\mathcal{F}_1$*. The results are summarized in the table below.*

$\mathcal{F}_i$	Hypothesis	p	α_i	Crit. Const.	Result
$\mathcal{F}_1$	H_1	.030	.05	.03125	*Reject*
	H_2	.020	.05	.025	*Reject*
$\mathcal{F}_2$	H_3	.040	.05	.05	*Reject*
	H_4	.030	.05	.025	*Reject*

The truncated Holm MTP rejects H_2 *since* $p_{(1)} = p_2 = .020 < .05/2$. *Next, it rejects* H_1 *since*

$$p_{(2)} = p_1 = .030 \le \left[\frac{0.25}{1} + \frac{0.75}{2}\right](.05) = .03125.$$

Note that the truncated Holm MTP uses a larger critical value $= .03125$ *at the second step than that used by the Bonferroni MTP, and so it rejects both* H_1 *and* H_2*. Thus the full* $\alpha = .05$ *is propagated to* $\mathcal{F}_2$*. The Hochberg MTP rejects* H_3 *(and hence also* H_4*) at* $\alpha = .05$.

2.5.3 General Gatekeeping

In practice, often more general gatekeeping conditions are encountered. Xi and Tamhane (2014) developed a gatekeeping MTP, which requires that at least $k \ge 1$ out of m_1 primary null hypotheses must be rejected to test any secondary hypotheses. This condition generalizes to parallel gatekeeping for $k = 1$ and to serial gatekeeping for $k = m_1$.

Dmitrienko et al. (2007) proposed tree gatekeeping MTPs in which testing of each hypothesis is conditional on rejection or acceptance of subsets of previously tested hypotheses. The hypotheses that are assigned a positive α initially can be tested unconditionally at the first step. Tree gatekeeping is restricted to Bonferroni-type MTPs. Graphical MTPs discussed in the next section follow naturally from the tree-structured MTPs.

2.6 Graphical Approach

The graphical approach was proposed by Bretz et al. (2009) and Burman et al. (2009) for testing hierarchically ordered hypotheses using sequentially rejective weighted Bonferroni tests. It provides a simple visual representation of the testing algorithm, which is convenient for communicating with clinicians. Bretz et al. (2009) and Bretz et al. (2011) have developed an R library package, called gMCP with a user-friendly GUI for designing and implementing graphical MTPs.

The hierarchical (gatekeeping) relations between the hypotheses are represented by a directed graph with hypotheses as nodes. A directed edge from node H_i to node H_j denotes that if H_i is rejected then a fraction of the α used to test H_i is propagated to H_j. This fraction is called the *transition fraction* and denoted by g_{ij} where $\sum_{j=1}^{m} g_{ij} \leq 1$ for all i. It is represented as a weight on the directed edge. If H_i is tested and rejected at level α_i then a fraction $\alpha_{ij} = g_{ij}\alpha_i$ is propagated to H_j.

Initially, significance levels α_i are assigned to hypotheses H_i such that $\sum_{i=1}^{m} \alpha_i = \alpha$, where some of the α_i may be set equal to 0, e.g., secondary hypotheses which cannot be tested unless some primary hypotheses are rejected. Any hypothesis H_i for which $p_i \leq \alpha_i$ can be rejected and the order of testing does not matter.

After rejection of each hypothesis H_i, the graph is updated according to the following algorithm.

1. Set $I = \{1, 2, \ldots, m\}$.
2. Reject any H_i with $p_i \leq \alpha_i$ and go to the next step. If all $p_i > \alpha_i$ then stop testing and accept all remaining hypotheses.
3. If H_i is rejected then update the graph

$$I \leftarrow I \setminus \{i\}$$

$$\alpha_j \leftarrow \begin{cases} \alpha_j + g_{ij}\alpha_i & j \in I \\ 0 & \text{otherwise} \end{cases}$$

$$g_{jk} \leftarrow \begin{cases} \frac{g_{jk}+g_{ji}g_{ik}}{1-g_{ji}g_{ij}} & j,k \in I, j \neq k, g_{ji}g_{ij} < 1 \\ 0 & \text{otherwise} \end{cases}$$

4. If $|I| \geq 1$, go to Step 2; otherwise stop testing.

The graph shows how the significance levels are transferred from rejected hypotheses to unrejected ones. The tests used for the hypotheses can be quite arbitrary as long as they yield valid p-values which can be compared against the assigned significance levels. The graphical approach also makes it easier to recycle the significance levels back to previously unrejected hypotheses, which helps boost the power of the MTP.

The following example gives a graphical representation of the truncated Holm MTP used in a primary family for parallel gatekeeping. The example helps to understand why the regular Holm MTP cannot be used for parallel gatekeeping and why the truncated Holm MTP must be used.

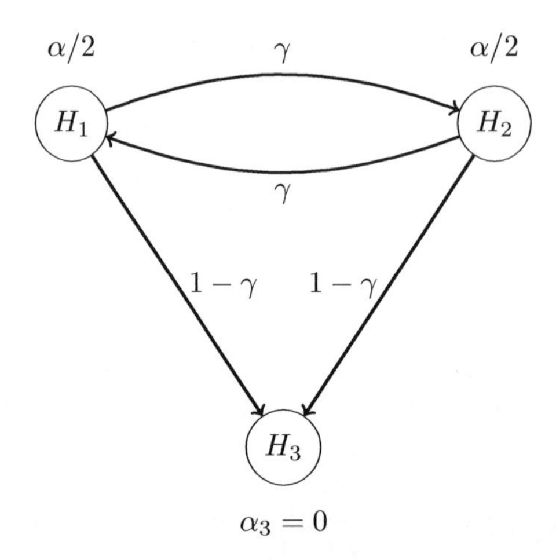

FIGURE 2.2
Graphical representation of the truncated Holm MTP for parallel gatekeeping.

Example 2.3 *Consider a clinical trial with two primary hypotheses (H_1 and H_2) and one secondary hypothesis (H_3). There is a parallel gatekeeping condition that at least one of H_1 and H_2 must be rejected before testing H_3. As shown in Figure 2.2, each primary hypothesis is tested at level $\alpha/2$ and upon rejection of either hypothesis a γ fraction of its $\alpha/2$ level is transferred to the other primary hypothesis and a $1-\gamma$ fraction is transferred to H_3 ($0 < \gamma < 1$). Thus, the MTP used to test H_1 and H_2 is the truncated Holm MTP with truncation fraction γ.*

Suppose H_1 is rejected first. Then, H_2 is tested at level $(1+\gamma)(\alpha/2)$. Also, since the node H_1 is deleted, the transition fraction γ on the edge from H_2 to H_1 is reallocated to the edge from H_2 to H_3 making its fraction equal to $\gamma + 1 - \gamma = 1$. If only H_1 is rejected, then H_3 is tested at level $(1-\gamma)(\alpha/2)$. If both H_1 and H_2 are rejected then H_3 is tested at the full $(1-\gamma)(\alpha/2) + (1+\gamma)(\alpha/2) = \alpha$ level.

It may be noted that if we set $\gamma = 1$ then H_1 and H_2 are tested using the regular Holm MTP, but then no α level can be transferred to H_3 upon their rejections since $1-\gamma = 0$. Therefore, a separable truncated MTP must be used for parallel gatekeeping.

2.7 Group Sequential MTPs

2.7.1 Review of Group Sequential Procedures

Group sequential procedures (GSPs) have become popular in clinical trials since originally proposed by Pocock (1977) and O'Brien and Fleming (1979). For ease of exposition, assume a simple setting where the patient outcomes are normally distributed with an unknown mean μ and a known variance σ^2. We want to test $H_0 : \mu = 0$ versus $H_1 : \mu > 0$. The total number of stages (looks) is denoted by $K \geq 2$, and the group sample sizes, $n_1, \ldots, n_K$, are assumed to be fixed. Let $N_j = \sum_{i=1}^{j} n_i$ denote the cumulative sample sizes and let $t_j = N_j/N_K$ denote the *information fractions* or *information times* ($j = 1, \ldots, K$). Note that $0 = t_0 < t_1 < \cdots < t_K = 1$. Further let $Z_1, \ldots, Z_K$ denote the standardized sample mean test statistics at the successive looks. Under H_0, $Z_1, \ldots, Z_K$ are jointly normally distributed with 0 means, unit variances and $\text{corr}(Z_i, Z_j) = \sqrt{N_i/N_j} = \sqrt{t_i/t_j}$ for $i < j$. H_0 is tested using an α-level GSP using a critical boundary $\{c_1, \ldots, c_K\}$, which satisfies

$$P_{H_0}\{\cup_{j=1}^K (Z_j \geq c_j)\} = \alpha.$$

It rejects H_0 at the first look j when $Z_j \geq c_j$. If $Z_j < c_j$ for all $j = 1, \ldots, K$, then H_0 is accepted.

The Pocock (1977) GSP (denoted by POC) uses $c_j = c$ for all j while the O'Brien and Fleming (1979) GSP (denoted by OBF) uses $c_j = c/\sqrt{t_j}$, where the critical constant $c > 0$ is different for POC and OBF and is determined from the above equation for any given α, K and the information fractions t_j. Wang and Tsiatis (1987) generalized the POC and OBF critical boundaries by defining $c_j = ct_j^{\lambda}$ where $\lambda = 0$ corresponds to the POC boundary and $\lambda = -0.5$ corresponds to the OBF boundary. The above-described GSP approach to testing H_0 is referred to as the *fixed boundary approach.*

Lan and DeMets (1983) proposed a *flexible boundary approach* in which neither the number of stages K nor the information fractions t_j are fixed in advance. Instead, an *error spending function (e.s.f.)*, denoted by $f(t, \alpha)$, is specified. For fixed α, $f(t, \alpha)$ is a monotonically nondecreasing continuous function of the information fraction $t \in [0, 1]$ with $f(0, \alpha) = 0$ and $f(1, \alpha) = \alpha$. Approximate e.s.f.s for the POC and OBF boundaries are given by

$$\text{POC: } f(t, \alpha) = \alpha \ln[1 + (e - 1)t], \quad \text{OBF: } f(t, \alpha) = 2\Phi\left(-z_{\alpha/2}/\sqrt{t}\right), \tag{2.7}$$

where $\Phi(\cdot)$ is the standard normal c.d.f.

Let the p_j denote the sequential p-values for testing H_0 at the information times t_j, $j = 1, \ldots, K$. In the normal mean testing setting $p_j = 1 - \Phi(Z_j)$, but the p-values may also come from other types of test statistics, e.g., χ^2 or log-rank. Denote by $\{\alpha_1, \ldots, \alpha_K\}$ the critical boundary induced by the e.s.f. $f(t, \alpha)$, which satisfies

$$P_{H_0}\left\{\cup_{i=1}^j (p_i \leq \alpha_i)\right\} = f(t_j, \alpha), \quad j = 1, \ldots, K.$$

Refer to the literature on GSPs (see, e.g., Jennison and Turnbull (2000)).

In the following three subsections, we describe the group sequential versions of the Simes test, the Holm MTP and the Hochberg MTP, respectively. For the sake of simplicity, throughout we assume that all $m \geq 2$ null hypotheses, $H_1, \ldots, H_m$, use the same error spending function $f(t, \alpha)$ and are tested at the same information times t_j $(j = 1, \ldots, K)$.

2.7.2 Group Sequential Simes Test (GSST)

The group sequential Simes test discussed below is different from the one given in Tamhane et al. (2020) although both are abbreviated as GSST. The one given in Tamhane et al. (2020) tests a *single* null hypothesis based on p-values from the sequential looks of a GSP. On the other hand, the one discussed below tests the intersection of *multiple* null hypotheses; thus, at each look, multiple p-values are available from multiple null hypotheses.

Fu (2018) proposed the following GSST for testing the overall intersection null hypothesis $H_0 = \cap_{i=1}^m H_i$. Let $p_{1j}, \ldots, p_{mj}$ denote the p-values of the m null hypotheses $H_1, \ldots, H_m$ at Stage j and let $p_{(1),j} \leq \cdots \leq p_{(m),j}$ denote their ordered values. Let $\{\alpha_1^{(i)}, \ldots, \alpha_K^{(i)}\}$ denote the critical boundary of level $i\alpha/m$ induced by the e.s.f. $f(t, i\alpha/m)$ so that

$$P_{H_*}\left\{\cup_{j=1}^K \left(p_{*j} \leq \alpha_j^{(i)}\right)\right\} = \frac{i\alpha}{m}, \quad i = 1, \ldots, m,$$

where the subscript * stands for the index $i = 1, \ldots, m$ of any fixed hypothesis. Thus the boundary $\{\alpha_1^{(i)}, \ldots, \alpha_K^{(i)}\}$ is the same for all m hypotheses.

GSST rejects H_0 if at least one $p_{(i),j}$ exits the $(i\alpha/m)$-level boundary, i.e., $p_{(i),j} \leq \alpha_j^{(i)}$ for some $i = 1, \ldots, m$ and $j = 1, \ldots, K$. This is a group sequential version of the fixed-sample Simes test, which rejects H_0 if $p_{(i)} \leq i\alpha/m$.

Example 2.4 *Let $K = m = 2$ and consider testing $H_0 = H_1 \cap H_2$ at $\alpha = 0.05$ using a two-stage GSP with one interim look at $t_1 = 0.5$. Suppose that the p-values at the two stages for H_1 and H_2 are*

$$\underline{\textit{Stage 1:}}\; p_{11} = 0.018, p_{21} = 0.034, \; \underline{\textit{Stage 2:}}\; p_{12} = 0.022, p_{22} = 0.016.$$

First suppose that POC is used as the reference GSP. Its critical boundaries at the two stages for $i = 1$ (or $i\alpha/2 = 0.025$) and for $i = 2$ (or $i\alpha/2 = 0.05$) are

$$\underline{\textit{Stage 1:}}\; \alpha_1^{(1)} = 0.0155, \alpha_1^{(2)} = 0.0310, \; \underline{\textit{Stage 2:}}\; \alpha_2^{(1)} = 0.0139, \alpha_2^{(2)} = 0.0297.$$

Note that these critical values are computed using the approximate e.s.f. of POC in (2.7). Alternatively we could have computed them using the fixed boundary approach with $c_1 = c_2 = c$. In that case $\alpha_1^{(1)} = \alpha_2^{(1)} = 0.0147$ and $\alpha_1^{(2)} = \alpha_2^{(2)} = 0.0304$, which are not too different from those given above.

The ordered p-values at the two stages are

$$\underline{\textit{Stage 1:}}\; p_{(1),1} = 0.018, p_{(2),1} = 0.034, \; \underline{\textit{Stage 2:}}\; p_{(1),2} = 0.016, p_{(2),2} = 0.022.$$

We see that at Stage 1, $p_{(1),1} = 0.018 > \alpha_1^{(1)} = 0.0155$ and $p_{(2),1} = 0.034 > \alpha_1^{(2)} = 0.0310$. Therefore the critical boundaries for $i = 1$ and $i = 2$ are not crossed by $\{p_{(1),1}, p_{(2),1}\}$ at Stage 1 and so H_0 is not rejected. At Stage 2. $p_{(1),2} = 0.016 > \alpha_1^{(1)} = 0.0139$ but $p_{(2),2} = 0.022 < \alpha_2^{(2)} = 0.0297$. Thus the critical boundary for $i = 2$ is crossed at Stage 2 and so H_0 is rejected.

Next suppose that OBF is used as the reference GSP. Its critical boundaries at the two stages for $i = 1$ (or $i\alpha/2 = 0.025$) and for $i = 2$ (or $i\alpha/2 = 0.05$) are

$$\underline{\textit{Stage 1:}}\; \alpha_1^{(1)} = 0.0015, \alpha_1^{(2)} = 0.0056, \; \underline{\textit{Stage 2:}}\; \alpha_2^{(1)} = 0.0245, \alpha_2^{(2)} = 0.0482.$$

As in the case of POC, these critical values are computed using the approximate e.s.f. of OBF in (2.7). Alternatively we could have computed them using the fixed boundary approach with $c_1 = c/\sqrt{t_1}, c_2 = c$. In that case $\alpha_1^{(1)} = 0.0026, \alpha_2^{(1)} = 0.0240$ and $\alpha_1^{(2)} = 0.0088, \alpha_2^{(2)} = 0.0467$. Once again, these critical values are not too different from those given above.

Since the GSST based on POC did not reject H_0 at Stage 1, we know that the GSST based on OBF will also not reject H_0 at Stage 1. On the other hand, since the former rejected H_0 at Stage 2, we know that the latter will also reject H_0 at Stage 2. This can be confirmed by checking that $p_{(1),2} = 0.016 < \alpha_2^{(1)} = 0.0245$ and $p_{(2),2} = 0.022 < \alpha_2^{(2)} = 0.0482$.

2.7.3 Group Sequential Holm MTP (GSHM)

Ye et al. (2013) used the principle of α recycling (Bretz et al., 2009; Burman et al., 2009) to derive a group sequential Holm MTP. They proposed two versions of it, which they referred to as group sequential Holm variable (GSHv) and group sequential Holm fixed (GSHf). For simplicity, consider two hypotheses, H_1 and H_2, tested by their individual GSPs at levels α_1 and α_2, respectively, with $\alpha_1 + \alpha_2 = \alpha$. If H_1 is rejected, then its α_1 level is recycled to H_2 and it is tested using a GSP at level α. Similarly, if H_2 is rejected, then its α_2 level is recycled to H_1, and it is tested using a GSP at level α. In GSHv, the α_i from a rejected hypothesis H_i is recycled to all stages of the GSP of the unrejected hypothesis H_j. In GSHf the α_i is recycled only to the final stage of the GSP of H_j. Xi and Tamhane (2015) generalized this idea to recycling α to all stages $k' \geq k$ starting at an arbitrary stage k $(1 \leq k \leq K)$ of the GSP of the unrejected hypothesis. GSHv corresponds to $k = 1$ and GSHf corresponds to $k = K$. Here we restrict to GSHv, which we refer to as GSHM. This MTP can also be presented using the graphical approach of Maurer and Bretz (2013).

Ye et al. (2013) and Fu (2018) extended this MTP to $K > 2$ and $m > 2$. We present the extension proposed by Fu (2018) since the group sequential Hochberg MTP readily follows from it by applying it in the reverse order using the same critical values. The extension proposed by Ye et al. (2013) is for the general weighted case but it is based on the idea of recycling α, which does not apply to the Hochberg MTP. The two extensions can be shown to be equivalent in the unweighted case.

At Stage j, let $\{\alpha_{1\ell}^{(j)}, \ldots, \alpha_{K\ell}^{(j)}\}$ denote the critical boundary for testing any hypothesis, induced by the e.s.f. $f(t, \alpha/(m_j - \ell + 1))$, i.e.,

$$P_{H_*}\{\cup_{k=1}^{i}(p_{*k} \leq \alpha_{k\ell}^{(j)})\} = f(t_i, \alpha/(m_j - \ell + 1)),\ i = 1, \ldots, K, \ell = 1, \ldots, m_j, \tag{2.8}$$

where m_j is the number of hypotheses under test at Stage j and the subscript * stands for the index of any fixed hypothesis. Note that m_j is updated not only at stages $j = 1, \ldots, K$ but also within each stage j as additional hypotheses are rejected. Thus, m_j denotes the current number of unrejected hypotheses still under test at Stage j.

Stage 1: Let $p_{(1),1} \leq \cdots \leq p_{(m_1),1}$ be the ordered p-values and denote the corresponding hypotheses by $H_{(1),1}, \ldots, H_{(m_1),1}$, respectively. Begin by testing $H_{(1),1}$. If $p_{(1),1} > \alpha_{11}^{(1)}$ then retain all hypotheses and go on to test them in Stage 2; otherwise reject $H_{(1),1}$ and go on to test $H_{(2),1}$. In general, reject $H_{(k),1}$ iff $H_{(1),1}, \ldots, H_{(k-1),1}$ are rejected and

$$p_{(k),1} \leq \alpha_{1k}^{(1)},$$

where $\alpha_{1k}^{(1)}$ is the first component of the critical boundary for Stage 1 computed from (2.8) $(k = 1, \ldots, m_1)$. Let r_1 be the last value of k when $p_{(k),1} \leq \alpha_{1k}^{(1)}$ $(0 \leq r_1 \leq m_1)$. If $r_1 = m_1$, i.e. if all hypotheses are rejected then stop testing; otherwise retain the $m_2 = m_1 - r_1$ hypotheses, $H_{(r_1+1),1}, \ldots, H_{(m_1),1}$, for testing in Stage 2.

Stage j: At Stage j, let m_j be the number of the hypotheses still under test and relabel them as $1, \ldots . m_j$. Let $p_{(1),j} \leq \cdots \leq p_{(m_j),j}$ denote their ordered p-values and let $H_{(1),j}, \ldots, H_{(m_j),j}$ denote the corresponding hypotheses. Finally let $\{\alpha_{1k}^{(j)}, \ldots, \alpha_{Kk}^{(j)}\}$ denote the critical boundary induced by the e.s.f. $f(t, \alpha/(m_j - k + 1))$ for testing any hypothesis. Reject $H_{(k),j}$ iff $H_{(1),j}, \ldots, H_{(k-1),j}$ are rejected and

$$p_{(k),j} \leq \alpha_{jk}^{(j)},$$

where $\alpha_{jk}^{(j)}$ is the jth component of the critical boundary for Stage j computed from (2.8) $(k = 1, \ldots, m_j)$. Let r_j denote the last value of k when $p_{(k),j} \leq \alpha_{jk}^{(j)}$ $(0 \leq r_j \leq m_j)$. If $r_j = m_j$, i.e., if all remaining hypotheses are rejected at Stage j or if $j = K$ then accept the remaining hypotheses, $H_{(r_j+1),j}, \ldots, H_{(m_j),j}$, and stop testing. Otherwise retain the remaining hypotheses for testing in Stage $j+1$. Let $j \leftarrow j+1$ and return to the beginning of this step.

2.7.4 Group Sequential Hochberg MTP (GSHC)

GSHC applies GSHM in the reverse order of the p-values at each stage, exactly like the Hochberg MTP applies the Holm MTP in the reverse order, using the same critical constants. We use the same notation for GSHC as for GSHM and the critical constants $\alpha_{ik}^{(j)}$ are defined in the same way.

Stage 1: Begin by testing $H_{(m_1),1}$. If $p_{(m_1),1} \leq \alpha_{1m_1}^{(1)}$ then reject all hypotheses and stop testing; otherwise retain $H_{(m_1),1}$ and go on to test $H_{(m_1-1),1}$. In general, retain $H_{(k),1}$ iff $H_{(m_1),1}, \ldots, H_{(k+1),1}$ have been retained and

$$p_{(k),1} > \alpha_{1k}^{(1)}.$$

Let r_1 be the first value of k when $p_{(k),1} \leq \alpha_{1k}^{(1)}$ $(0 \leq r_1 < m_1)$. Then reject $H_{(r_1),1}, \ldots, H_{(1),1}$. Retain the $m_2 = m_1 - r_1$ hypotheses, $H_{(r_1+1),1}, \ldots, H_{(m_1),1}$, for testing at Stage 2.

Stage j: At Stage j let m_j be the number of the hypotheses still under test and relabel them as $1, \ldots .m_j$. Let $p_{(1),j} \leq \cdots \leq p_{(m_j),j}$ denote their ordered p-values and $H_{(1),j}, \ldots, H_{(m_j),j}$ denote the corresponding hypotheses. Retain $H_{(k),j}$ iff $H_{(m_j),j}, \ldots, H_{(k+1),j}$ are retained and

$$p_{(k),j} > \alpha_{jk}^{(j)}.$$

Let r_j be the first value of k when $p_{(k),j} \leq \alpha_{jk}^{(j)}$ $(0 \leq r_j \leq m_j)$. Then reject $H_{(r_j),j}, \ldots, H_{(1),j}$. If $r_j = m_j$, i.e., if all hypotheses are rejected or if $j = K$ then stop testing and accept the remaining hypotheses, $H_{(r_j+1),j}, \ldots, H_{(m_j),j}$. Otherwise retain the remaining hypotheses for testing in Stage $j+1$, let $j \leftarrow j+1$ and return to the beginning of this step.

We next illustrate GSHM and GSHC by a numerical example.

Example 2.5 *Consider testing three hypotheses, H_1, H_2 and H_3 using a three-stage GSP (POC or OBF). Assume that all tests are one-sided and overall $\alpha = 0.05$. The critical values $\alpha_{ik}^{(j)}$ for GSHM and GSHC based on POC and OBF are listed in Table 2.1; these values are obtained by solving (2.8) using the approximate e.s.f.s of POC and OBF from (2.7). Note that $m_j - k + 1$ only takes values 1, 2 or 3 for $k = 1, \ldots, m_j$ and $m_j = 1, 2, 3$. Accordingly, the critical values are given only for $\alpha/1, \alpha/2$ and $\alpha/3$. The p-values of the three hypotheses are shown in Table 2.2.*

Let us first apply GSHM to these data using both POC and OBF critical boundaries.

Stage 1: The ordered p-values at Stage 1 are

$$p_{(1),1} = p_{31} = .008, p_{(2),1} = p_{21} = .010, p_{(3),1} = p_{11} = .026.$$

POC Boundary: Since $p_{(1),1} = .008 > .0076$, retain all three hypotheses for testing in Stage 2.

TABLE 2.1
Critical Values for GSHM and GSHC

$\frac{\alpha}{m_j-k+1}$	POC			OBF		
	Stage 1	Stage 2	Stage 3	Stage 1	Stage 2	Stage 3
$\frac{\alpha}{1}$	.0226	.0231	.0238	.0007	.0162	.0451
$\frac{\alpha}{2}$	.0113	.0109	.0108	.0001	.0060	.0231
$\frac{\alpha}{3}$	.0076	.0070	.0069	.0000	.0036	.0156

TABLE 2.2
p-Values p_{ij}

H_i	Stage 1	Stage 2	Stage 3
H_1	.026	.029	.022
H_2	.010	.010	.006
H_3	.008	.009	.014

OBF Boundary: Since $p_{(1),1} = .008 > .0000$, *retain all three hypotheses for testing in Stage 2.*

Stage 2: The ordered p-values at Stage 2 are

$$p_{(1),2} = p_{32} = .009, p_{(2),2} = p_{22} = .010, p_{(3),2} = p_{12} = .029.$$

POC Boundary: Since $p_{(1),2} = .009 > .0070$, *retain all three hypotheses for testing in Stage 3.*
OBF Boundary: Since $p_{(1),2} = .009 > .0036$, *retain all three hypotheses for testing in Stage 3.*

Stage 3: The ordered p-values at Stage 3 are

$$p_{(1),3} = p_{23} = .006, p_{(2),3} = p_{33} = .014, p_{(3),3} = p_{13} = .022.$$

POC Boundary: Since $p_{(1),3} = p_{23} = .006 < .0069$, *reject* H_2. *Next* $p_{(2),3} = p_{33} = .014 > .0108$, *so accept* H_3 *and hence also* H_1 *and stop testing.*
OBF Boundary: Since $p_{(1),3} = p_{23} = .006 < .0156$, *reject* H_2. *Next* $p_{(2),3} = p_{33} = .014 < .0231$, *so reject* H_3. *Finally* $p_{(3),3} = p_{13} = .022 < .0451$, *so reject* H_1. *Thus all three hypotheses are rejected.*

Next let us apply GSHC to these data using both POC and OBF critical boundaries. We use the ordered p-values and the critical constants (in the reverse order) given above.

Stage 1:
POC Boundary: Since $p_{(3),1} = p_{1,1} = .026 > .0226$, *retain* H_1. *Next since* $p_{(2),1} = p_{2,1} = .010 < .0113$, *reject* H_2 *and hence also* H_3.
OBF Boundary: Since $p_{(3),1} = .026 > .007$, $p_{(2),1} = .010 > .0001$ *and* $p_{(1),1} = .008 > .0000$, *retain all three hypotheses for testing in Stage 2.*

Stage 2:
POC Boundary: There is only one hypothesis H_1 left. Since $p_{(1),2} = p_{(1,2} = .029 > .0231$, retain H_1 for testing in Stage 3.
OBF Boundary: Since $p_{(3),2} = .029 > .0162, p_{(2),2} = .010 > .006, p_{(1),2} = .009 > .006$, retain all three hypotheses for testing in Stage 3.

Stage 3:
POC Boundary: Since $p_{(1),3} = p_{1,3} = .022 > .0228$, accept H_1.
OBF Boundary: Since $p_{(3),3} = p_{1,3} = .022 < .0451$, reject H_1 and hence also H_2 and H_3.

To summarize, GSHM rejects only H_2 using the POC boundary and rejects all three hypotheses using the OBF boundary. On the other hand, GSHC rejects H_2 and H_3 using the POC boundary and rejects all three hypotheses using the OBF boundary. These results are in agreement with the general rule that the Hochberg MTP rejects all those hypotheses rejected by the Holm MTP and possibly more. Also, the OBF boundary is uniformly more powerful than the POC boundary as shown in Tamhane et al. (2018). However, because the OBF boundary is more stringent in the earlier stages than the POC boundary, rejections of the hypotheses take place at later stages compared to the POC boundary, thus requiring more samples. This may be a consideration in some trials.

2.8 Benjamini-Hochberg MTP for False Discovery Rate Control

False discovery rate (FDR) is defined as follows: let R denote the number of rejected hypotheses of which V are true null hypotheses (and so are falsely rejected). Then

$$\text{FDR} = \begin{cases} E\left(\frac{V}{R}\right) & \text{if} \quad R > 0 \\ 0 & \text{if} \quad R = 0 \end{cases} \tag{2.9}$$

It is easy to show that FDR $\leq$ FWER with equality holding if all hypotheses are true. Therefore, FWER $\leq \alpha$ implies FDR $\leq \alpha$. Thus, FDR control is less stringent than FWER control. FDR control is generally used in exploratory studies involving a very large number of hypotheses, e.g., genomic and microarray studies. Because these studies are not confirmatory, FWER control is too stringent and is unnecessary since their goal is only to highlight interesting findings.

Benjamini and Hochberg (1995) proposed the following MTP. They showed that this MTP strongly controls FDR $\leq \alpha$. Let $p_{(1)} \leq p_{(2)} \leq \cdots \leq p_{(m)}$ be the ordered p-values of the m null hypotheses and let

$$i^* = \max\left\{i : p_{(i)} \leq \frac{i\alpha}{m}\right\}.$$

Then reject hypotheses $H_{(1)}, \ldots, H_{(i^*)}$ associated with the p-values $p_{(1)} \leq p_{(2)} \leq \cdots \leq p_{(i^*)}$. The Benjamini-Hochberg (BH) MTP can be applied in a stepwise manner by starting with $H_{(m)}$. If $p_{(m)} \leq \alpha$ then reject all hypotheses; otherwise accept $H_{(m)}$ and continue to the next step. In general, if $p_{(i)} \leq i\alpha/m$ then reject $H_{(1)}, \ldots, H_{(i)}$ and stop testing; otherwise accept $H_{(i)}$ and continue to the next step. Note that the BH MTP is very similar to the Hochberg MTP except that instead of the critical constants $\alpha/(m - i + 1)$ used by the Hochberg MTP, it uses the critical constants $i\alpha/m$ to test the ordered hypotheses $H_{(i)}$. Since $i\alpha/m \geq \alpha/(m-i+1)$ with equality holding iff $i = 1$ or $i = m$, the BH MTP is liberal compared to the Hochberg MTP as it controls FDR while the Hochberg MTP controls FWER. The BH MTP can also be applied graphically by plotting the ordered p-values, $p_{(i)}$, versus i and drawing the straight line $i\alpha/m$ versus i and finding the largest $i = i^*$ where the plot of the $p_{(i)}$ falls below the straight line.

2.9 Concluding Remarks

In this chapter, we have given an extensive overview of the MTPs based on marginal p-values. They do not consider correlations between the p-values but are easy to apply in practice. Correlations can be considered by applying Westfall and Young (1993)'s resampling approach to multiple testing.

This chapter did not focus on power and sample size calculations. Analytical studies of powers of MTPs are generally very difficult, and one must resort to simulations. East software by Cytel Inc. (`http://www.cytel.com/software/east`) is a good choice for this purpose.

Bibliography

Benjamini, Y. and Y. Hochberg (1995). Controlling the false discovery rate: a practical and powerful approach to multiple testing. *Journal of the Royal Statistical Society. Series B (Methodological) 57*(1), 289–300.

Benjamini, Y. and Y. Hochberg (1997). Multiple hypotheses testing with weights. *Scandinavian Journal of Statistics 24*(3), 407–418.

Benjamini, Y. and D. Yekutieli (2001). The control of the false discovery rate in multiple testing under dependency. *The Annals of Statistics 29*(4), 1165–1188.

Block, H. W., T. H. Savits, and J. Wang (2008). Negative dependence and the Simes inequality. *Journal of Statistical Planning and Inference 138*, 4107–4110.

Bretz, F., W. Maurer, W. Brannath, and M. Posch (2009). A graphical approach to sequentially rejective multiple test procedures. *Statistics in Medicine 28*(4), 586–604.

Bretz, F., M. Posch, E. Glimm, F. Klinglmueller, W. Maurer, and K. Rohmeyer (2011). Graphical approaches for multiple comparison procedures using weighted Bonferroni, Simes, or parametric tests. *Biometrical Journal 53*(6), 894–913.

Burman, C.-F., C. Sonesson, and O. Guilbaud (2009). A recycling framework for the construction of Bonferroni-based multiple tests. *Statistics in Medicine 28*(5), 739–761.

Cai, G. and S. K. Sarkar (2008). Modified Simes' critical values under independence. *Statistics & Probability Letters 78*(12), 1362–1368.

Dmitrienko, A., W. W. Offen, and P. H. Westfall (2003). Gatekeeping strategies for clinical trials that do not require all primary effects to be significant. *Statistics in Medicine 22*(15), 2387–2400.

Dmitrienko, A. and A. C. Tamhane (2009). Gatekeeping procedures in clinical trials. In A. Dmitrienko, A. C. Tamhane, and F. Bretz (Eds.), *Multiple Testing Problems in Pharmaceutical Statistics*, Chapter 5, pp. 165–192. Boca Raton, FL: Chapman and Hall/CRC Press.

Dmitrienko, A., A. C. Tamhane, and B. L. Wiens (2008). General multistage gatekeeping procedures. *Biometrical Journal 50*(5), 667–677.

Dmitrienko, A., B. L. Wiens, A. C. Tamhane, and X. Wang (2007). Tree-structured gatekeeping tests in clinical trials with hierarchically ordered multiple objectives. *Statistics in Medicine 26*(12), 2465–2478.

Dunnett, C. W. (1955). A multiple comparison procedure for comparing several treatments with a control. *Journal of the American Statistical Association 50*(272), 1096–1121.

Dunnett, C. W. and A. C. Tamhane (1993). Power comparisons of some step-up multiple test procedures. *Statistics & Probability Letters 16*, 55–58.

Fu, Y. (2018). A group sequential Hochberg step-up procedure for testing multiple hypotheses. unpublished manuscript.

Gabriel, K. R. (1969). Simultaneous test procedures–some theory of multiple comparisons. *The Annals of Mathematical Statistics 40*(1), 224–250.

Gou, J. and A. C. Tamhane (2014). On generalized Simes critical constants. *Biometrical Journal 56*(6), 1035–1054.

Gou, J. and A. C. Tamhane (2018a). Hochberg procedure under negative dependence. *Statistica Sinica 28*, 339–362.

Gou, J. and A. C. Tamhane (2018b). A flexible choice of critical constants for the improved hybrid Hochberg-Hommel procedure. *Sankhya B 80*, 85–97.

Gou, J., A. C. Tamhane, D. Xi, and D. Rom (2014). A class of improved hybrid Hochberg-Hommel type step-up multiple test procedures. *Biometrika 101*(4), 899–911.

Hochberg, Y. (1988). A sharper Bonferroni procedure for multiple tests of significance. *Biometrika 75*(4), 800–802.

Hochberg, Y. and U. Liberman (1994). An extended Simes' test. *Statistics & Probability Letters 21*(2), 101–105.

Hochberg, Y. and A. C. Tamhane (1987). *Multiple Comparison Procedures.* New York: John Wiley and Sons.

Holm, S. (1979). A simple sequentially rejective multiple test procedure. *Scandinavian Journal of Statistics 6*, 65–70.

Hommel, G. (1988). A stagewise rejective multiple test procedure based on a modified Bonferroni test. *Biometrika 75*(2), 383–386.

Hommel, G., F. Bretz, and W. Maurer (2007). Powerful short-cuts for multiple testing procedures with special reference to gatekeeping strategies. *Statistics in Medicine 26*(22), 4063–4073.

Jennison, C. and B. W. Turnbull (2000). *Group Sequential Methods with Applications to Clinical Trials.* New York: Chapman and Hall/CRC.

Lan, K. K. G. and D. L. DeMets (1983). Discrete sequential boundaries for clinical trials. *Biometrika 70*(3), 659–663.

Lehmann, E. L. (1966). Some concepts of dependence. *The Annals of Mathematical Statistics 37*, 1137–1153.

Liu, W. (1996). On multiple tests of a non-hierarchical finite family of hypotheses. *Journal of the Royal Statistical Society, Series B 58*(2), 455–461.

Marcus, R., E. Peritz, and K. R. Gabriel (1976). On closed testing procedures with special reference to ordered analysis of variance. *Biometrika 63*, 655–660.

Maurer, W. and F. Bretz (2013). Multiple testing in group sequential trials using graphical approaches. *Statistics in Biopharmaceutical Research 5*(4), 311–320.

Maurer, W., L. Hothorn, and W. Lehmacher (1995). Multiple comparisons in drug clinical trials and preclinical assays: A-priori ordered hypotheses. In J. Vollmar (Ed.), *Testing Principles in Clinical and Preclinical Trials.* Stuttgartv: Gustav Fischer Verlag.

Moyé, L. A. (2003). *Multiple Analyses in Clinical Trials: Fundamentals for Investigators.* New York: Springer-Verlag.

Naik, U. D. (1975). Some selection rules for comparing p processes with a standard. *Communications in Statistics. Series A 4*, 519–535.

O'Brien, P. C. and T. R. Fleming (1979). A multiple testing procedure for clinical trials. *Biometrics 35*(3), 549–556.

Pocock, S. J. (1977). Group sequential methods in the design and analysis of clinical trials. *Biometrika 64*(2), 191–199.

Sarkar, S. K. (1998). Some probability inequalities for ordered MTP2 random variables: a proof of the Simes conjecture. *The Annals of Statistics 26*(2), 494–504.

Sarkar, S. K. (2002). Some results on false discovery rate in stepwise multiple testing procedures. *The Annals of Statistics 30*(1), 239–257.

Sarkar, S. K. (2008). On the simes inequality and its generalization. In N. Balakrishnan, E. A. Peña, and M. J. Silvapulle (Eds.), *Beyond Parametrics in Interdisciplinary Research: Festschrift in Honor of Professor Pranab K. Sen*, Volume 1 of *IMS Collection: Beyond Parametrics in Interdisciplinary Research: Festschrift in Honor of Professor Pranab K. Sen*, pp. 231–242. Beachwood, OH: Institute of Mathematical Statistics.

Sarkar, S. K. and C. K. Chang (1997). The Simes method for multiple hypothesis testing with positively dependent test statistics. *Journal of the American Statistical Asociation 92*, 1601–1608.

Šidák, Z. (1967). Rectangular confidence regions for the means of multivariate normal distributions. *Journal of the American Statistical Association 62*(318), 626–633.

Simes, R. J. (1986). An improved Bonferroni procedure for multiple tests of significance. *Biometrika 73*, 751–754.

Tamhane, A. C. and J. Gou (2018). Advances in p-value based multiple test procedures. *Journal of Biopharmaceutical Statistics 28*(1), 10–27.

Tamhane, A. C., J. Gou, and A. Dmitrienko (2020). Some drawbacks of the Simes test in the group sequential setting. *Statistics in Biopharmaceutical Research 12*(3), 390-393.

Tamhane, A. C., J. Gou, C. Jennison, C. R. Mehta, and T. Curto (2018). A gatekeeping procedure to test a primary and a secondary endpoint in a group sequential design with multiple interim looks. *Biometrics 74*(1), 40–48.

Tamhane, A. C. and L. Liu (2008). On weighted Hochberg procedures. *Biometrika 95*(2), 279–294.

U.S. Department of Health and Human Services, Food and Drug Administration (2017). Guidance for industry: multiple endpoints in clinical trials, draft guidance.

Wang, S. K. and A. A. Tsiatis (1987). Approximately optimal one-parameter boundaries for group sequential trials. *Biometrics 43*, 193–199.

Westfall, P. H. and S. S. Young (1993). *Resampling-Based Multiple Testing: Examples and Methods for p-Value Adjustment.* New York: John Wiley & Sons.

Wiens, B. L. (2003). A fixed sequence Bonferroni procedure for testing multiple endpoints. *Pharmaceutical Statistics 2*(3), 211–215.

Wiens, B. L. and A. Dmitrienko (2005). The fallback procedure for evaluating a single family of hypotheses. *Journal of Biopharmaceutical Statistics 15*(6), 929–942.

Xi, D. and A. C. Tamhane (2014). A general multistage procedure for k-out-of-n gatekeeping. *Statistics in Medicine 33*(8), 1321–1335.

Xi, D. and A. C. Tamhane (2015). Allocating recycled significance levels in group sequential procedures for multiple endpoints. *Biometrical Journal 57*(1), 90–107.

Ye, Y., A. Li, L. Liu, and B. Yao (2013). A group sequential Holm procedure with multiple primary endpoints. *Statistics in Medicine 32*(7), 1112–1124.

3

Multivariate Multiple Test Procedures

Thorsten Dickhaus
University of Bremen

André Neumann
University of Bremen

Taras Bodnar
Stockholm University

CONTENTS

3.1 Introduction and preliminaries

Dependencies among data points are present in virtually all modern statistical applications. This holds especially true for studies with multiple endpoints which are all measured for the same observational units. For example, consider the case of a gene expression study. In that context, expression levels of m genes are measured for n individuals. The goal of the study typically is to detect statistically significant expression differences, either in a two-groups model or in a one-group model under different experimental conditions. Due to biological and technological reasons, the expression levels will typically exhibit strong dependencies, at least for genes which are functionally related; cf. Yona et al. (2006). In this chapter, we will discuss multiple tests which take such dependencies explicitly into account. Such multiple tests are called multivariate multiple tests, because they rely on joint distributions or on approximations thereof.

DOI: 10.1201/9780429030888-4

3.1.1 Motivation

Example 3.1 *As a simple motivating example for utilizing a multivariate multiple test, consider the case of* $m = 2$ *simultaneous tests for Gaussian means. Let* $Z = (Z_1, Z_2)^\top$ *denote an observable* $\mathbb{R}^2$*-valued random vector which follows the bivariate normal distribution with an unknown mean vector* $\mu = (\mu_1, \mu_2)^\top$*, but a known covariance (and correlation) matrix* $\Sigma = \begin{pmatrix} 1 & \rho \\ \rho & 1 \end{pmatrix}$*, where* $|\rho| < 1$*. Consider the two (one-sided) null hypotheses* $H_j = \{\mu_j \leq \mu_j^*\}$*,* $j = 1, 2$*, for a given vector* $\mu^* = (\mu_1^*, \mu_2^*)^\top \in \mathbb{R}^2$*. The corresponding alternative hypotheses are given by* $K_j = \{\mu_j > \mu_j^*\}$*,* $j = 1, 2$*. This is a typical setup for a multiple test for superiority in a clinical study concerning two endpoints. Let* α_{loc} *denote a local significance level, meaning that the two tests* φ_1 *and* φ_2 *for testing* H_1 *and* H_2 *are carried out at level* α_{loc} *each. Throughout the remainder of this chapter, we will use the term "local significance level" in the same manner. Namely,* α_{loc} *denotes the nominal (multiplicity-adjusted) significance level at which each "local test"* φ_j *for* $1 \leq j \leq m$ *is carried out. Here, it is canonical to use* $\varphi_j = \mathbf{1}_{(c,\infty)}(Z_j - \mu_j^*)$*,* $j = 1, 2$*, with* $c = \Phi^{-1}(1 - \alpha_{loc})$ *denoting the* $(1 - \alpha_{loc})$*-quantile of the univariate standard normal distribution. The Bonferroni correction for family-wise error rate (FWER) control at level* α *would advise us to take* $\alpha_{loc} = \alpha/2$*.*

Figure 3.1 displays the effect of choosing $\alpha_{loc} = 5\%$ *for varying values of the correlation coefficient* ρ *in the case that* $\mu = \mu^*$*, implying that both null hypotheses* H_1 *and* H_2 *are true. For strong negative correlations among* Z_1 *and* Z_2 $(\rho \to -1)$*, the realized FWER tends to* $\alpha = 10\% = 2\alpha_{loc}$*, meaning that the Bonferroni inequality becomes an equality for* $\rho \to -1$*. On the other hand, for positive* ρ *the realized FWER is below* α*, and it even monotonically decreases to* $\alpha_{loc} = 5\% = \alpha/2$ *for* $\rho \to +1$*, meaning that in the extreme case of perfect positive correlation among* Z_1 *and* Z_2 *no adjustment for multiplicity would be necessary at all.*

Since ρ *is assumed to be known here, the exhaustion of the FWER level* α *and, hence, the power of the multiple test could be improved by choosing* α_{loc} *by means of a quantile of the bivariate joint distribution of* $T = (T_1, T_2)^\top$ *under* $\mu = \mu^*$*, where* $T_j = Z_j - \mu_j^*$*,* $j = 1, 2$*. Denoting the joint cumulative distribution function (cdf) of* T *under* μ^* *by* F_2*, one possibility is to search for the constant* c_α *such that* $F_2(c_\alpha, c_\alpha) = 1 - \alpha$*. The corresponding (exact) local significance level is then given by* $\alpha_{loc} = 1 - \Phi(c_\alpha)$*. If an unequal weighting (for importance) of the two null hypotheses is desired, one can search for a solution of the form* $F_2(c_\alpha^{(1)}, c_\alpha^{(2)}) = 1 - \alpha$ *for* $c_\alpha^{(1)} \neq c_\alpha^{(2)}$*.*

Section 3.2 will deal with straightforward generalizations of Example 3.1 to arbitrary dimensions $m \geq 2$, to more general types of null hypotheses, and to more general test statistics which are assumed to be (asymptotically) jointly normally distributed, at least under the global null hypothesis.

3.1.2 Preliminaries

Throughout the chapter, we will assume a finite family $\mathcal{H}_m = (H_1, \ldots, H_m)$ of $m \in \mathbb{N}$ null hypotheses which are to be tested simultaneously under one and the same statistical model $(\mathcal{Y}, \mathcal{F}, (\mathbb{P}_\vartheta : \vartheta \in \Theta))$ with sample space $\mathcal{Y}$ and parameter ϑ taking values in the parameter space Θ. It is convenient to interpret each H_j as a subset of Θ, i. e., we will write $H_j \subset \Theta$, $1 \leq j \leq m$. The corresponding alternative hypotheses will be denoted by $K_j = \Theta \setminus H_j$, $1 \leq j \leq m$. Then we call $(\mathcal{Y}, \mathcal{F}, (\mathbb{P}_\vartheta : \vartheta \in \Theta), \mathcal{H}_m)$ a multiple test problem. A (non-randomized) multiple test $\varphi = (\varphi_1, \ldots, \varphi_m)^\top$ for $\mathcal{H}_m$ is a (measurable) mapping from the sample space $\mathcal{Y}$ to $\{0, 1\}^m$, where $\varphi_j(y) = 1$ means that we reject H_j in favor of K_j on the basis of the observed data $y \in \mathcal{Y}$, for $1 \leq j \leq m$. For given $\vartheta \in \Theta$, let $I_0(\vartheta) \subseteq \{1, \ldots, m\}$

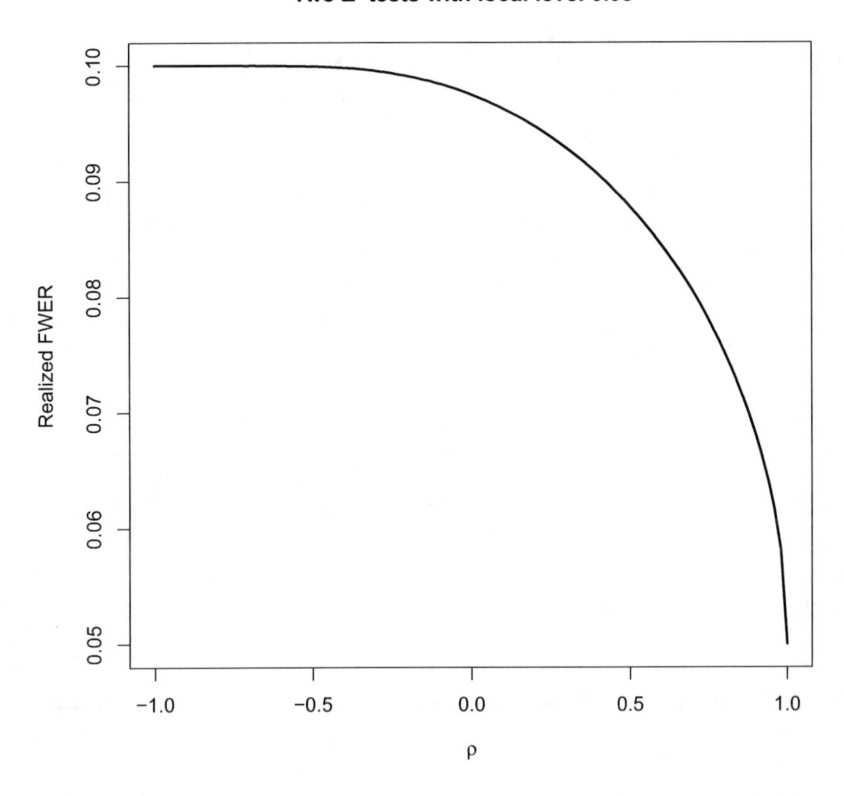

FIGURE 3.1
Realized FWER in the case of two simultaneous Z-tests, where the two test statistics are jointly normally distributed with correlation coefficient ρ. The nominal FWER level has been set to $\alpha = 0.1$. The resulting formula for the realized FWER is given by $1 - F_2\left(\Phi^{-1}(0.95), \Phi^{-1}(0.95)\right)$, where F_2 denotes the cdf of the bivariate centered normal distribution with unit variances and correlation coefficient ρ, and Φ denotes the cdf of the univariate standard normal distribution. Details are provided in Example 3.1.

denote the index set of true null hypotheses under ϑ, meaning that exactly those H_j are true for which $j \in I_0(\vartheta)$. For type I error control of φ, we will consider the two random variables $V_m = \sum_{j \in I_0(\vartheta)} \varphi_j$ and $R_m = \sum_{j=1}^{m} \varphi_j$. The random variable V_m denotes the (random) number of type I errors of φ (under ϑ) and the random variable R_m denotes the total number of rejections of φ (under ϑ). Two important type I error measures, which we will consider in this chapter, are the FWER and the false discovery rate (FDR). For a given value of $\vartheta \in \Theta$, they are given by

$$\text{FWER}_\vartheta(\varphi) = \mathbb{P}_\vartheta(V_m > 0) = \mathbb{P}_\vartheta\left(\bigcup_{j \in I_0(\vartheta)} \{\varphi_j = 1\}\right), \quad (3.1)$$

$$\text{FDR}_\vartheta(\varphi) = \mathbb{E}_\vartheta[\text{FDP}_\vartheta(\varphi)] = \mathbb{E}_\vartheta\left[\frac{V_m}{\max(R_m, 1)}\right]. \quad (3.2)$$

The random variable $\text{FDP}_\vartheta(\varphi) = V_m / \max(R_m, 1)$ appearing in (3.2) is called the false discovery proportion (FDP) of φ (under ϑ), where the max in the denominator is to avoid an expression of the form $0/0$.

Let $H_0 = \bigcap_{j=1}^{m} H_j$ denote the global (null) hypothesis in $\mathcal{H}_m$, meaning that all m null hypotheses $H_1, \ldots, H_m$ are true for any $\vartheta \in H_0$. The multiple test φ is said to control the FWER at level $\alpha \in (0, 1)$ in the weak sense, if $\text{FWER}_\vartheta(\varphi) \leq \alpha$ for all $\vartheta \in H_0$. It is said to control the FWER at level α in the strong sense, if $\text{FWER}_\vartheta(\varphi) \leq \alpha$ for all $\vartheta \in \Theta$. Clearly, strong FWER control implies weak FWER control. The reverse implication (i. e., equivalence of weak FWER control and strong FWER control) holds true, if the statistical model, the system $\mathcal{H}_m$ of null hypotheses, and the multiple test φ are such, that the FWER of φ becomes largest for parameter values in H_0. In Sections 3.2–3.5 below, we will mainly study the FWER behavior of certain multiple tests under H_0, meaning that we design procedures for weak FWER control (in the first place). However, the application of the closed test principle (see Chapter 1 of this handbook for details) allows one to utilize multiple tests with weak FWER control also for the purpose of strong FWER control. Namely, such tests have to be applied to every intersection hypothesis appearing in the closure of $\mathcal{H}_m$. We will not elaborate further on this strategy in this chapter. It has been explained in detail, among many others, by Stange et al. (2016), Chung and Romano (2016), and Schildknecht et al. (2015), where the latter article also contains applications to real data in a medical context. We note in passing that the concept of weak FDR control is not of independent interest, because FDR and FWER coincide if H_0 holds true. Hence, FDR control always has to be considered in the strict sense.

We assume, that every marginal test φ_j is carried out in terms of a real-valued test statistic T_j or a (random) p-value P_j taking values in $[0, 1]$, respectively, where $1 \leq j \leq m$. Our goal is to utilize the joint null distribution (or an approximation thereof) of the random vector $T = (T_1, \ldots, T_m)^\top$ or $P = (P_1, \ldots, P_m)^\top$, respectively, in the calibration of φ for type I error control.

3.1.3 Overview of the remainder of the chapter

The rest of the material in this chapter is structured as follows. In Section 3.2, we will consider linear hypotheses testing in the presence of (asymptotically) jointly normally distributed test statistics $T_1, \ldots, T_m$. Here, a quantile of the full m-variate (asymptotic) distribution of T under H_0 is taken as the critical value for FWER control of the multiple test. Sections 3.3 and 3.4 deal with the case, that only k-th order marginal distributions of T under H_0 are available, for some $k < m$. This leads to the utilization of probability bounds/ approximations (see Section 3.3), which can be expressed in terms of the "effective number of tests" (see Section 3.4). The methods in these two sections can be used under (asymptotic) joint normality of T under H_0, but they are not restricted to this distributional assumption. In Section 3.5, non-Gaussian dependencies are considered, which can conveniently be expressed in terms of so-called copula functions. Here, a point on the contour line of the copula of the test statistics (or their distributional transforms, respectively) determines the local significance level(s) for FWER control. In Section 3.6, we review some multivariate approaches to FDR control, mainly in terms of an adjustment of the nominal FDR level in the very popular Benjamini-Hochberg procedure. Finally, Section 3.7 provides some conclusions and practical recommendations.

3.1.4 How large can m be for the different methods under consideration?

For practical purposes, let us comment here up front about the values of m which can be handled with the methods that we are going to describe in the subsequent sections.

Our `R` implementations in Section 3.2 rely on the `R` package `mvtnorm`. The functions in that package can cope with dimensionalities of $m \leq 1{,}000$. The methods that we are going to present in Sections 3.3 and 3.4 are computationally intensive, so that we recommend their usage only for $m \leq 100$. However, as explained by Stange et al. (2016) and Steffen and Dickhaus (2020), these methods can be applied to different blocks of test statistics separately, where every such block may have a dimensionality of up to 100. The aggregation of the results from the different blocks is computationally inexpensive. Explicit copula modeling, as proposed in Section 3.5, is typically only feasible for m at most in the dozens. On the other hand, the FDR criterion considered in Section 3.6 is typically applied to large systems of hypotheses, where m is at least in the hundreds. The methods that we are going to present in Section 3.6 rely on the joint distribution of order statistics. There exist numerically stable and efficient computational methods to evaluate this distribution even for larger values of m; see, e. g., von Schroeder and Dickhaus (2020) and the references therein.

3.2 Methods based on multivariate normal distributions

In this section, we will discuss simultaneous test procedures in the sense of Gabriel (1969) and Hothorn et al. (2008).

Let ϑ with values in $\mathbb{R}^k$ for $k \in \mathbb{N}$ denote the statistical parameter of interest. In many applications, multiple test problems concerning ϑ can be formalized as systems of so-called linear hypotheses. To this end, let $C \in \mathbb{R}^{m \times k}$ denote a given matrix (the so-called contrast matrix) and $d \in \mathbb{R}^m$ a given vector, where $m \in \mathbb{N}$ denotes the number of null hypotheses to be tested simultaneously, as outlined in Section 3.1.2. Then, the system $\mathcal{H}_m = (H_1, \ldots, H_m)$ of (two-sided) linear hypotheses regarding ϑ, which is defined by C and d, can be written as

$$C\vartheta = d, \tag{3.3}$$

where we interpret (3.3) as a system of m null hypotheses. This means, that we define H_i by line i of the system of equations in (3.3), where $1 \leq i \leq m$. Each H_i encodes one linear restriction concerning (components of) ϑ. It is also possible to consider inequality relations in (3.3), leading to one-sided null hypotheses as in Example 3.1.

Example 3.2

(a) *Assume that we want to test, which of the components of ϑ are different from zero. We let $m = k$, $C = I_k$ (the identity matrix in $\mathbb{R}^{k \times k}$), and $d = 0 \in \mathbb{R}^k$. Then, line i of* (3.3) *encodes the i-th null hypothesis $H_i = \{\vartheta_i = 0\}$, for $1 \leq i \leq k$.*

(b) *Assume that we want to compare all components ϑ_i for $1 \leq i \leq k-1$ with the component ϑ_k. This has the interpretation, that the component ϑ_k corresponds to a "control group/treatment" against which all other groups/treatments shall be compared. We let $m = k-1$, $d = 0 \in \mathbb{R}^{k-1}$, and $C = C_{\text{Dunnett}} \in \mathbb{R}^{k-1 \times k}$. The contrast matrix C_{Dunnett} is Dunnett's contrast matrix with $k-1$ rows and k columns, where in each row j the j-th entry equals $+1$, the k-th entry equals -1 and all other entries are equal to zero.*

Now, assume for the moment that an (at least asymptotically for large sample sizes) unbiased and normally distributed estimator $\hat{\vartheta}$ of ϑ is at hand. Then, it is near at hand to employ the vector $T = C\hat{\vartheta} - d$ of test statistics for testing $\mathcal{H}_m$ defined by (3.3). If $\hat{\vartheta}$ (approximately) follows a normal distribution with mean ϑ and covariance matrix $\Sigma \in \mathbb{R}^{k \times k}$ (where Σ is functionally independent of ϑ), then (under standard regularity assumptions) T (approximately) follows a centered normal distribution with covariance matrix $C\Sigma C^\top \in \mathbb{R}^{m\times m}$ under the global hypothesis $H_0 = \bigcap_{i=1}^m H_i$ of $\mathcal{H}_m$. A simultaneous test procedure (STP) based on the vector T chooses a suitable quantile of the (approximate) distribution $\mathcal{N}_m(0, C\Sigma C^\top)$ of T under H_0 as the rejection threshold c_α for each individual test statistic T_i (which is the i-th component of T or its absolute value, respectively). The i-th null hypothesis H_i gets rejected at FWER level α if and only if (iff) T_i exceeds c_α.

Remark 3.1

(a) *Exemplary model classes under which asymptotically unbiased and normally distributed estimators are available (under certain regularity conditions) have been discussed, among others, by Hothorn et al. (2008), Bretz et al. (2010), Hochberg and Tamhane (1987), Miller (1981), and in Section 4.2 of Dickhaus (2014). They comprise, for example, analysis of variance models, multiple linear regression models, generalized linear models, survival models, and various time series models.*

(b) *Strong FWER control at (asymptotic) level α of the multiple contrast test defined by T and c_α can be established under the (asymptotic) "subset pivotality" condition (see Section 2.2.3 in Westfall and Young (1993)). Detailed calculations can be found, for example, in Section III of Dickhaus and Gierl (2013) and in Lemma 3.1 of Dickhaus and Stange (2013).*

(c) *The multivariate normal distribution can be replaced by a suitable multivariate Student's t distribution, if there is uncertainty about the marginal variances of $T_1, \ldots, T_m$ under H_0 and Studentization techniques are applied; see Example 3.3 for an application in the context of the analysis of variance.*

Example 3.3 (ANOVA1) *Under the balanced, homoscedastic one-factorial analysis of variance (ANOVA) model with $k \geq 3$ groups, two important multiple test problems are "all pairwise multiple comparisons" (MCA) and "all multiple comparisons with a control group" (MCC). Here, $\vartheta \in \mathbb{R}^k$ is the vector of the group-specific population means, which is estimated by the vector $\hat{\vartheta}$ of the group-specific sample means. The error variance is assumed to be unknown. In the context of multiple contrast tests, MCA leads to the so-called Tukey contrast matrix and the Tukey test, respectively (see Tukey (1953)), while MCC leads to the so-called Dunnett contrast matrix and the Dunnett test, respectively (see Dunnett (1955) and Dunnett (1964)). These are classical multiple tests which have been treated, for instance, by Hochberg and Tamhane (1987). The following source code in* R *demonstrates how the critical values for the Tukey test can be derived (i) based on the general theory of multiple contrast tests, and (ii) based on the built-in* R *routine* `qtukey()`. *One has to notice, that the* R *routine* `qtukey()` *actually computes quantiles of the closely related Studentized range distribution. To obtain these quantiles, the critical values of the Tukey test have to be multiplied by $\sqrt{2}$. Letting $Z_1, \ldots, Z_k$ denote stochastically independent and identically standard normally distributed random variables and $S > 0$ a further random variable which is stochastically independent of $(Z_1, \ldots, Z_k)^\top$ such that νS^2 is chi-square distributed with ν degrees of freedom for some $\nu \in \mathbb{N}$, the Studentized range distribution with parameters k and ν is the distribution of $(\max_{1\leq i\leq k} Z_i - \min_{1\leq \ell\leq k} Z_\ell)/S$. In the case of the Tukey test, $\nu = k(n-1)$.*

```
library("mvtnorm");
library("multcomp");

##################################################
# ANOVA1 with k groups, "all pairs" contrasts #
##################################################
n <- c(11,11,11,11);      #group-specific sample sizes
```

```
k <- length(n);          #number of groups
C <- contrMat(n, type = "Tukey");
M <- diag(1/n);
combis <- combn(1:k, 2);
D <- diag(sqrt((n[combis[1, ]] * n[combis[2, ]]) /
          ( n[combis[1, ]] + n[combis[2, ]])));

#correlation matrix of the test statistics
R <- D %*% C %*% M %*% t(C) %*% D;
alpha <- 0.1;
my_df <- sum(n) - k;

my_Tukey_quantile <- qmvt(p=1-alpha, tail="both.tails",
                          df=my_df, corr=R)$quantile;
my_StudRangeQuantile <- my_Tukey_quantile*sqrt(2);
my_Tukey_quantile;
my_StudRangeQuantile; my_df; k; alpha;
# In agreement with Table 8 on page 408 of
# Hochberg and Tamhane (1987)!

R_Tukey_quantile <- qtukey(p=1-alpha, nmeans=k, df=my_df);
R_Tukey_quantile;
```

Table 3.1 tabulates some numerical values of quantiles of the Studentized range distribution. All values in Table 3.1 are in very good agreement with Table 8 on pages 408 - 409 in Hochberg and Tamhane (1987).

TABLE 3.1
Some quantiles of the Studentized range distribution. The FWER level is denoted by α, k denotes the number of groups, n denotes the sample size per group, and $\nu = k(n-1)$ denotes the resulting degrees of freedom. The column "Contrast" contains the solution based on the general methodology of multiple contrast tests, and the column "qtukey" the one obtained by the built-in R routine qtukey().

α	k	n	ν	**Contrast**	**qtukey**
0.05	3	10	27	3.506	3.506
0.05	3	20	57	3.403	3.403
0.05	3	50	147	3.348	3.348
0.05	5	10	45	4.018	4.018
0.05	5	20	95	3.932	3.933
0.05	5	50	245	3.887	3.887
0.05	10	10	90	4.592	4.588
0.05	10	20	190	4.528	4.528
0.05	10	50	490	4.493	4.495
0.1	3	10	27	3.030	3.030
0.1	3	20	57	2.962	2.962
0.1	3	50	147	2.925	2.925
0.1	5	10	45	3.591	3.590
0.1	5	20	95	3.531	3.531
0.1	5	50	245	3.498	3.499
0.1	10	10	90	4.215	4.212
0.1	10	20	190	4.170	4.168
0.1	10	50	490	4.141	4.145

Remark 3.2

(a) *Explicit formulas and* R *code for multiple contrast tests under many other model classes can be found in Bretz et al. (2010). Applications in nonparametric multiple comparisons, together with* R *code, have been worked out by Konietschke et al. (2012) and Konietschke et al. (2015).*

(b) *While in the specific case of the Tukey test the utilization of the* R *routine* `qtukey()` *appears more convenient than the more involved code referring to the general multiple contrast test methodology, one has to keep in mind that the setup of the latter has a much broader scope. As a very simple example, consider the case of unequal group-specific sample sizes under Example 3.3 (i. e., an unbalanced design). In that case, the (scaled) Studentized range distribution is not the correct null distribution for the maximum of the test statistics anymore. However, the code referring to the general multiple contrast test methodology still delivers the correct quantile, if the group-specific sample sizes at hand are entered in its first line.*

(c) *One further important generalization of the presented methodology is to extend the scope of multiple contrast tests to the case of flexible study designs with several stages; see, e. g., Magirr et al. (2012) and the references therein.*

3.3 Methods based on higher-order probability bounds

Let $m \in \mathbb{N}$ denote the number of null hypotheses to be tested simultaneously, and assume that real-valued test statistics $T_1, \ldots, T_m$ are at hand, which tend to larger values under alternatives. For calibrating a multivariate STP φ based on $T = (T_1, \ldots, T_m)^\top$ or for calculating corresponding multiplicity-adjusted p-values, respectively, we have to evaluate expressions of the following form:

$$F_m(x) = \mathbb{P}_0 \left(\bigcap_{j=1}^{m} \{T_j \leq x\} \right), \tag{3.4}$$

or equivalently

$$\bar{F}_m(x) = 1 - F_m(x) = \mathbb{P}_0 \left(\bigcup_{j=1}^{m} \{T_j > x\} \right), \quad x \in \mathbb{R}, \tag{3.5}$$

where $\mathbb{P}_0$ denotes some probability measure under the global null hypothesis $H_0 = \bigcap_{j=1}^m H_j$. The quantities in (3.4) or (3.5), respectively, can often not be evaluated exactly. Reasons for this can be (i) lacking information about the full m-variate null distribution of T, or (ii) computational infeasibility. For example, the R package `mvtnorm` (computation of multivariate normal/Student's t probabilities and quantiles) which is based on Genz and Bretz (2009) gives an error message whenever m exceeds 1000. Therefore, two basic ideas for approximating $F_m(x)$, which only require the computation of lower-dimensional marginal distributions, i. e., $F_k(x)$ for some $k < m$, are given by sum-type and product-type probability bounds/approximations.

Lemma 3.1

a) *(Bonferroni inequalities, sum-type probability bounds (STPBs).)*
Let $A_1, \ldots, A_m$ *be arbitrary events, and let* $\mathbb{P}$ *denote any probability measure. Then*

$$\forall p \geq 1: \sum_{k=1}^{2p}(-1)^{k-1}S_k \leq \mathbb{P}\left(\bigcup_{j=1}^{m} A_j\right) \leq b_{2p-1} := \sum_{k=1}^{2p-1}(-1)^{k-1}S_k, \tag{3.6}$$

where

$$S_k = \sum_{1\leq j_1<j_2<\ldots<j_k\leq m} \mathbb{P}(A_{j_1} \cap A_{j_2} \cap \ldots \cap A_{j_k}) \tag{3.7}$$

for $1 \leq k \leq m$, *and* $S_k = 0$ *for* $k > m$. *See Section 4.7 of Comtet (1974) and Galambos and Simonelli (1996) for proofs, related results, and further references. A bivariate variant of the aforementioned upper Bonferroni bounds is due to Worsley (1982) and is given by*

$$\mathbb{P}\left(\bigcup_{j=1}^{m} A_j\right) \leq b_2 := \sum_{j=1}^{m} \mathbb{P}(A_j) - \sum_{j=1}^{m-1} \mathbb{P}(A_j \cap A_{j+1}). \tag{3.8}$$

For our purposes we have to consider the events $A_j = \{T_j > x\}$ *and the probability measure* $\mathbb{P} = \mathbb{P}_0$, *so that the probability expression in* (3.6) *and on the left-hand side of* (3.8) *equals* $\bar{F}_m(x)$.

b) *(Product-type probability bounds (PTPBs).)*
Define the events $O_j := \{T_j \leq x\} = A_j^c$ *for* $1 \leq j \leq m$. *Due to chain factorization, it holds for any* $1 \leq k \leq m-1$ *that*

$$F_m(x) = \mathbb{P}_0(O_1, \ldots, O_m) = \mathbb{P}_0(O_1, \ldots, O_k) \prod_{j=k+1}^{m} \mathbb{P}_0(O_j | O_{j-1}, \ldots, O_1).$$

Now assume that T *is sub-Markovian of order* $k \geq 2$ *(*SM_k*) in the sense of Definition 2.2 in Dickhaus and Stange (2013) under* H_0. *Then it holds for all* $k \leq j \leq m$ *that*

$$\mathbb{P}_0(O_j | O_{j-1}, \ldots, O_1) \geq \mathbb{P}_0(O_j | O_{j-1}, \ldots, O_{j-k+1}) \tag{3.9}$$

and, consequently,

$$F_m(x) \geq \beta_k := \mathbb{P}_0(O_1, \ldots, O_k) \prod_{j=k+1}^{m} \mathbb{P}_0(O_j | O_{j-1}, \ldots, O_{j-k+1}). \tag{3.10}$$

Occasionally, we will write $b_\ell(x)$ *or* $\beta_k(x)$, *respectively, instead of* b_ℓ *or* β_k, *respectively, to indicate the argument* x *at which the approximations are evaluated. Furthermore, we refer to* ℓ *and* k, *respectively, as the order of these (sum- or product-type) approximations.*

Remark 3.3

a) *We note that the complexity of computing the sums* S_k *in* (3.7) *is high, because* $\binom{m}{k}$ k*-dimensional marginal probabilities have to be evaluated. On the other hand,* (3.6) *always holds true, regardless of the dependency structure among* $T_1, \ldots, T_m$. *A computationally inexpensive alternative is the utilization of* b_2 *from* (3.8). *Under certain structural assumptions, sum-type bounds of higher order can be improved. For example, the derivations of Naiman and Wynn (2005, 1992) are based on geometric or topological arguments.*

b) *In the general case the inequality relation in* (3.9) *is not fulfilled; cf., among others, Block et al. (1992) and Glaz and Johnson (1984). However, β_k often yields a good approximation of $F_m(x)$ already for $k \in \{2, 3\}$; see, for example, Section 4 of Stange et al. (2016) for numerical results pertaining to multivariate chi-square probabilities. In the remainder, we refer to β_k as the product-type probability approximation (PTPA) of order k to $F_m(x)$. The word "approximation" instead of "bound" indicates, that the inequality in* (3.10) *may fail if T is not SM_k under H_0.*

c) *The vector T is called positive lower orthant dependent (PLOD) under H_0, if for all $t = (t_1, \ldots, t_m)^\top \in \mathbb{R}^m$, it holds that*

$$\mathbb{P}_0(T_1 \leq t_1, \ldots, T_m \leq t_m) \geq \prod_{j=1}^{m} \mathbb{P}_0(T_j \leq t_j).$$

This entails, in particular, that

$$F_m(x) \geq \prod_{j=1}^{m} \mathbb{P}_0(T_j \leq x) =: \beta_1. \tag{3.11}$$

Calibrating a multivariate multiple test by means of (3.11) *leads to a so-called Šidák test, see Šidák (1967). If $T_1, \ldots, T_m$ are jointly stochastically independent under H_0, we obtain equality in* (3.11).

Based on the aforementioned considerations, the following multiplicity- and dependency-adjusted p-values have been proposed by Stange et al. (2016).

Definition 3.1 *For a given order ℓ or k, respectively, the procedure MADAM (standing for "multiplicity- and dependency-adjustment method") from Stange et al. (2016) transforms the observed values $t_1, \ldots, t_m$ of the test statistics $T_1, \ldots, T_m$ into one of the following multiplicity- and dependency-adjusted p-values:*

$$p_{\Sigma,j} = b_\ell(t_j), \tag{3.12}$$
$$p_{\Pi,j} = 1 - \beta_k(t_j), \tag{3.13}$$

for all $1 \leq j \leq m$. The subscript Σ in (3.12) *indicates, that a STPB is utilized, and the subscript Π in* (3.13) *indicates, that a PTPA is utilized.*

For (approximate) FWER control at level α, the p-values from (3.12) or (3.13), respectively, may simply be thresholded at α.

3.4 Effective numbers of tests

The STPBs b_ℓ for $\ell \geq 2$ as well as the PTPAs β_k for $k \geq 2$ utilize information about the dependency structure among $T_1, \ldots, T_m$ under H_0 by incorporating ℓ-variate or k-variate marginal distributions, respectively, of $T = (T_1, \ldots, T_m)^\top$ under H_0. The bounds b_1 and β_1 only utilize univariate marginal distributions. One question of practical interest is, how much gain in FWER exhaustion and, consequently, power can be achieved by the exploitation of higher-order marginal distributions. This may also aid in selecting the appropriate order ℓ or k, respectively. On the one hand, the order should be chosen as large as possible in order to exhaust α as tightly as possible. On the other hand, as mentioned in Remark 3.3, the computational complexity of computing the bounds/approximations increases with increasing order.

One way to quantify the aforementioned gain is to compute the effective number of tests (see Dickhaus and Stange (2013) and the references therein) corresponding to b_ℓ or β_k, respectively. The concept of effective numbers of tests is very popular in the context of genetic association studies; see Dickhaus et al. (2012), Chapter 9 of Dickhaus (2014), Section 4.1 of Dickhaus and Stange (2013), Section 5 of Stange et al. (2016), and Section 5.1 of Dickhaus (2015) for details and many references. To describe this concept formally, assume that FWER control at a given level α is targeted and that the α_{loc}-quantile of T_j under H_0 is chosen as the critical value for the marginal test φ_j, where $1 \leq j \leq m$. Due to multiplicity, it is immediately clear that $\alpha_{loc} \leq \alpha$. For b_1 or β_1, respectively, the value of α_{loc} can be calculated straightforwardly. Namely, we obtain the (first-order) Bonferroni correction $\alpha_{loc} \equiv \alpha_{loc}(b_1) = \alpha/m$ in the case of b_1 and the Šidák correction $\alpha_{loc} \equiv \alpha_{loc}(\beta_1) = 1 - (1-\alpha)^{1/m}$ in the case of β_1.

By equating b_ℓ or β_k, respectively, for ℓ or k larger than one with α (where $x = x_j$ is chosen as the α_{loc}-quantile of the distribution of T_j under H_0), we can (numerically) determine $\alpha_{loc}(b_\ell)$ or $\alpha_{loc}(\beta_k)$, respectively. The effective number of tests of order ℓ or k, respectively, which we will denote by $M_{\text{eff}}^{(\ell)}$ or $M_{\text{eff}}^{(k)}$, is now found by (numerically) solving

$$M_{\text{eff}}^{(\ell)} \alpha_{loc}(b_\ell) \quad = \quad \alpha \quad \text{or} \tag{3.14}$$

$$(1 - \alpha_{loc}(\beta_k))^{M_{\text{eff}}^{(k)}} \quad = \quad 1 - \alpha, \tag{3.15}$$

respectively. This means, that we write $\alpha_{loc}(b_\ell)$ in the form of a univariate Bonferroni correction with m replaced by $M_{\text{eff}}^{(\ell)}$ or that we write $\alpha_{loc}(\beta_k)$ in the form of a Šidák correction with m replaced by $M_{\text{eff}}^{(k)}$, respectively. Now, if the effective number of tests is smaller than m, we interpret this as "effectively" having to correct for only $M_{\text{eff}}^{(\ell)}$ or $M_{\text{eff}}^{(k)}$ comparisons instead of m ones. This reduction/relaxation of the "effective" multiplicity correction is due to the fact that we exploit dependencies among the tests (or test statistics), such that not every marginal test "fully counts" in $M_{\text{eff}}^{(\ell)}$ or $M_{\text{eff}}^{(k)}$, because of certain similarities between them.

Example 3.4 *Let $m = 50$ and assume that the vector $T = (T_1, \ldots, T_{50})^\top$ follows under the global hypothesis H_0 a centered m-variate normal distribution. The correlation matrix of T is assumed to be an equi-correlation matrix, such that all non-diagonal elements of it are identical and equal to ρ, where ρ ranges from 0 to 0.9 in steps of 0.1. In Table 3.2, we display the local significance levels and the effective numbers of tests resulting from three different calibration methods for a target FWER level of $\alpha = 5\%$: (i) Exact calibration (up to numerical inaccuracies) by means of the full 50-variate null distribution of T. We have used the* R*-routine* `qmvnorm()` *for this purpose. (ii) Approximate calibration by means of the STPB b_2 from (3.8). (ii) Approximate calibration by means of the PTPA β_3 from (3.10). For the marginal variances of each T_j, $1 \leq j \leq m$, we have chosen the value $1/n$ for $n = 30$. This mimics the case of a multiple test for Gaussian means (in the case of a unit error variance), where the sample size for each marginal test problem equals $n = 30$.*

The numerical values displayed in Table 3.2 are very much in line with the FWER behavior of the multiple test discussed in our simple motivating Example 3.1: The stronger the positive correlation among the test statistics, the larger α_{loc} and, consequently, the smaller M_{eff}. In the setting studied here, the PTPA β_3 delivers a conservative approximation, which is in most cases much closer to the exact value (both in terms of α_{loc} and in terms of M_{eff}) than the STPB b_2.

TABLE 3.2

Local significance levels and effective numbers of tests corresponding to `qmvnorm()`, b_2 from (3.8), and β_3 from (3.10), respectively, assuming jointly normally distributed test statistics under the global null hypothesis. Their covariance matrix equals $\Sigma = n^{-1}(\rho J + (1-\rho) I)$, where J denotes the matrix with every entry equal to one and I is the identity matrix. The parameter ρ is the equi-correlation coefficient. The values of $\alpha_{loc}^{\text{qmvnorm}}$ and $M_{\text{eff}}^{\text{qmvnorm}}$ are based on the full m-variate joint distribution of the test statistics. The global FWER level was chosen as $\alpha = 0.05$, the number of hypotheses equals $m = 50$, and the marginal variances are all equal to $1/n$, for $n = 30$. The values of $M_{\text{eff}}^{\text{qmvnorm}}$ have been computed according to (3.14).

ρ	$\alpha_{loc}^{\text{qmvnorm}}$	$M_{\text{eff}}^{\text{qmvnorm}}$	$\alpha_{loc}^{b_2}$	$M_{\text{eff}}^{b_2}$	$\alpha_{loc}^{\beta_3}$	$M_{\text{eff}}^{\beta_3}$
0	0.001025	48.78	0.001001	49.95	0.001026	49.97
0.1	0.001070	46.73	0.001003	49.85	0.001027	49.92
0.2	0.001161	43.07	0.001007	49.65	0.001038	49.39
0.3	0.001321	37.85	0.001015	49.26	0.001067	48.05
0.4	0.001573	31.79	0.001030	48.54	0.001140	44.97
0.5	0.001991	25.11	0.001057	47.30	0.001288	39.80
0.6	0.002609	19.16	0.001106	45.21	0.001575	32.54
0.7	0.003756	13.31	0.001194	41.88	0.002137	23.98
0.8	0.005900	8.47	0.001371	36.47	0.003259	15.71
0.9	0.011086	4.51	0.001836	27.23	0.006695	7.64

Remark 3.4 *In the context of genetic association studies, m contingency tables (one per considered genetic locus) have to be analyzed simultaneously with respect to association of a (typically binary) phenotype and the genotype at the respective locus. In this, m can be a very large number of an order of magnitude of up to 10^5 or 10^6 in the case of a genome-wide association study. Typically, at least in the presence of large sample sizes, a chi-square test statistic T_j is computed for each contingency table j, $1 \leq j \leq m$. Due to the biological mechanism of inheritance (and due to technical aspects of the measurements), there exist pronounced dependencies among the T_j's, and the vector $T = (T_1, \ldots, T_m)^\top$ follows a multivariate (central) chi-square distribution under the global hypothesis of independence of the phenotype of interest and the genotype at all m loci under investigation. The computation of multivariate chi-square probabilities is rather involved (see Dickhaus and Royen (2015), Stange et al. (2016), and the references therein), and it seems that up to date explicit analytical formulas only exist for two-, three- and four-variate chi-square probabilities. Therefore, the utilization of PTPAs of order 2 – 4 has been proposed in this context by Dickhaus and Stange (2013) and Stange et al. (2016). Another multivariate approach to addressing the multiplicity problem in genetic association analyses is to combine multiple test procedures with (inherently multivariate) statistical learning methods like support vector machines; see Mieth et al. (2016) and Dickhaus (2018) for details.*

3.5 Copula-based methods

Again, we assume that $m \in \mathbb{N}$ null hypotheses are to be tested simultaneously, and that real-valued test statistics $T_1, \ldots, T_m$ are at hand, each tending to larger values under the alternative. As discussed around (3.4), the joint cdf F_m under a fixed parameter value $\vartheta^* \in H_0$ is needed in order to calibrate a multivariate STP based on $T = (T_1, \ldots, T_m)^\top$ for

FWER control (under ϑ^*). The idea in this section is, to decompose F_m into the marginal cdfs of the T_j's and the dependency structure among the T_j's. To this end, denote by G_j the marginal cdf of T_j under ϑ^*, for $1 \leq j \leq m$. Then, we have the following result.

Theorem 3.1 (Sklar's Theorem, see Sklar (1959, 1996).) *There exists a function $C_T : [0,1]^m \to [0,1]$, called the copula (function) of T, such that for all $t = (t_1, \ldots, t_m)^\top \in \bar{\mathbb{R}}^m$, it holds*

$$F_m(t) = C_T(G_1(t_1), \ldots, G_m(t_m)). \tag{3.16}$$

If G_j is a continuous function for all $1 \leq j \leq m$, then the copula C_T is unique.

Equation (3.16) formalizes the decomposition of F_m into the marginal cdfs $G_1, \ldots, G_m$ and the dependency structure among $T_1, \ldots, T_m$, which is mathematically described by the copula (or: dependence function) C_T. The advantages of working with (3.16) are threefold: (i) The transformation with the marginal cdfs leads to a distributional standardization under H_0, meaning that by the principle of quantile/probability integral transformation, we have that $G_j(T_j)$ is uniformly distributed on $[0,1]$ under ϑ^* in the case of a continuous G_j, $1 \leq j \leq m$. Of course, the same holds true for the corresponding (random) p-value $P_j = 1 - G_j(T_j)$. The random variable $G_j(T_j)$ is often referred to as the distributional transform of T_j, cf. Rüschendorf (2009). If all G_j's are strictly increasing on their supports, then the copula of the distributional transforms coincides with C_T. (ii) By means of (3.16), we obtain a very high degree of modeling flexibility. Namely, the marginal models referring to $G_1, \ldots, G_m$ can be coupled with any copula C_T. For example, in the case of marginal tests for means, univariate normal cdfs $G_1, \ldots, G_m$ can be coupled with a non-Gaussian copula, to describe situations where univariate normality of marginal arithmetic means can (approximately) be assumed, but multivariate normality of the vector T (under ϑ^*) may be questionable. (iii) Modeling dependencies by means of copula functions has meanwhile become a standard tool in applied multivariate statistics and quantitative risk management; see, e. g., Nelsen (2006), Joe (2014), Härdle and Okhrin (2010), Embrechts et al. (2003), and Chapter 5 of McNeil et al. (2005). There exists a rich and ever growing body of literature on appropriate copula models for many applications. By means of (3.16), these models are available for multiple testing.

The following result connects Sklar's Theorem with FWER control under $\vartheta^* \in H_0$.

Lemma 3.2 Theorem 2 in Dickhaus and Gierl (2013), Lemma 3.5 in Neumann et al. (2019).

Under the assumptions of Theorem 3.1, assume that $G_1, \ldots, G_m$ are known and fixed, at least asymptotically for large sample sizes. Furthermore, assume that C_T does not depend on ϑ. Let critical values $c_j(\alpha)$ for $1 \leq j \leq m$ be given, such that the j-th null hypothesis H_j is rejected at FWER level α by the STP φ iff T_j exceeds $c_j(\alpha)$. Finally, let $\alpha_{loc}^{(j)} = 1 - G_j(c_j(\alpha))$ denote a local significance level for the j-th marginal test problem, $1 \leq j \leq m$. Then we have that

$$\text{FWER}_{\vartheta^*, C_T}(\varphi) = 1 - C_T\left(1 - \alpha_{loc}^{(1)}, \ldots, 1 - \alpha_{loc}^{(m)}\right).$$

Lemma 3.2 shows, that the multiplicity-adjusted local significance levels $\alpha_{loc}^{(1)}, \ldots, \alpha_{loc}^{(m)}$ or, equivalently, the multiplicity-adjusted critical values $c_1(\alpha), \ldots, c_m(\alpha)$ can be determined by means of the contour line of the copula C_T at contour level $1-\alpha$. In practice, it is convenient to carry out the resulting multiple test φ in terms of the (realized) p-values $p_j = 1 - G_j(t_j)$, where t_j denotes the observed value of T_j, $1 \leq j \leq m$. We reject H_j at FWER level α, iff p_j is smaller than $\alpha_{loc}^{(j)}$. Figure 3.2 is an adapted and extended version of Figure 2 in Dickhaus and Gierl (2013) and depicts this construction in the bivariate case ($m = 2$). Figure 3.2 also shows how a possible weighting for importance of the null hypotheses is automatically incorporated in the methodology.

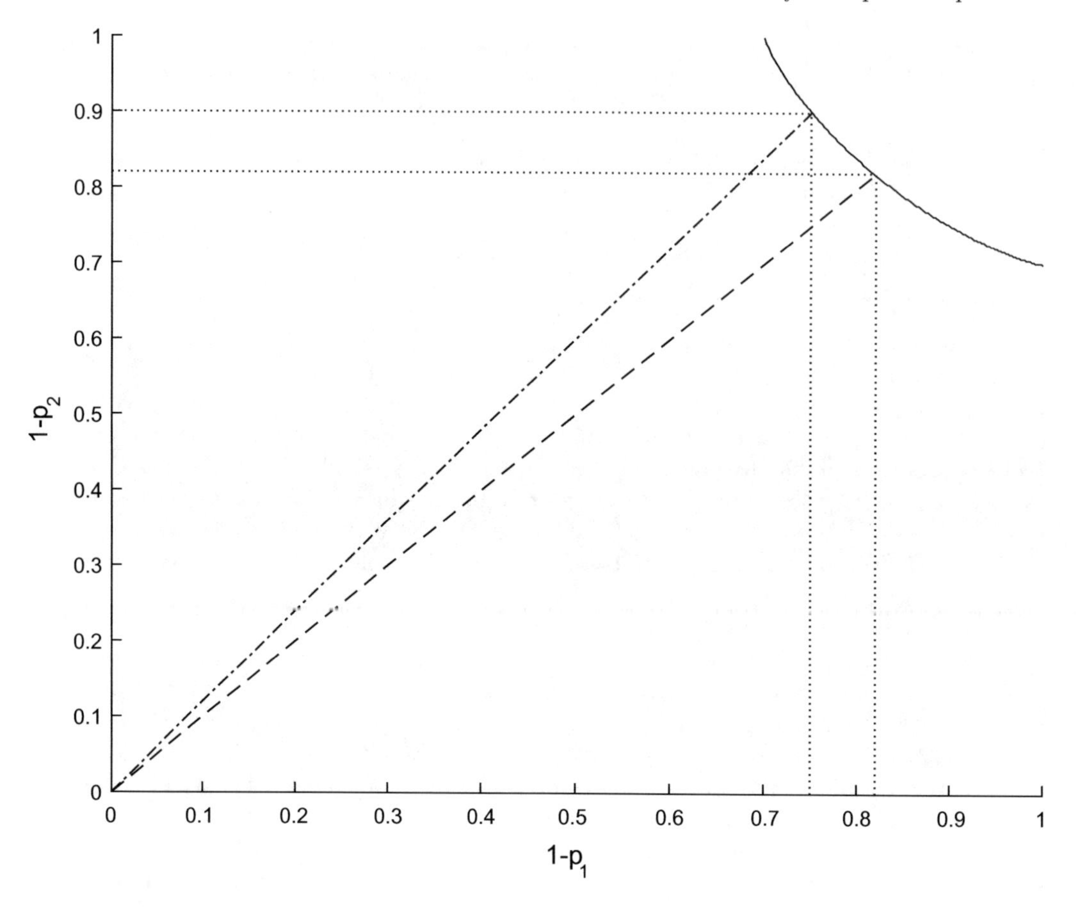

FIGURE 3.2
Copula-based calibration of a local significance level. The solid line displays the contour line of a specific bivariate copula at contour level 0.7, corresponding to an FWER level of $\alpha = 0.3$. The dashed line corresponds to an equal weighting of the two null hypotheses to which the p-values p_1 and p_2 refer. By projecting the point of intersection of the dashed line and the contour line of the copula onto the coordinate axes, we obtain an equi-coordinate local significance level of $1 - 0.82 = 0.18 > 0.15 = \alpha/2$. The dash-dotted line corresponds to a case in which the first null hypothesis gets a higher weight than the second one. We obtain a local significance level of $1 - 0.75 = 0.25$ for the first null hypothesis and a local significance level of $1 - 0.9 = 0.1$ for the second null hypothesis.

Remark 3.5 *Lemma 3.2 refers to a given, fixed copula C_T. In practice, the dependency structure among the test statistics $T_1, \ldots, T_m$ will however often not be known exactly. In such cases, C_T has to be (pre-)estimated. Stange et al. (2015) studied parametric estimation methods for copula functions and their impact on the FWER behavior of multivariate STPs. In particular, they provide a method to derive confidence regions for the realized FWER in the case that there is estimation uncertainty about C_T . Neumann et al. (2019) have obtained analogous results for certain nonparametric copula estimators, in particular so-called Bernstein copula estimators.*

Example 3.5 *Let us assume that the test statistics or their distributional transforms, respectively, follow under the global hypothesis H_0 an m-variate Frank copula with parameter*

TABLE 3.3
Comparison of the local significance levels $\alpha_{loc}^{\text{Sidak}} = 1 - (1-\alpha)^{1/m}$ and $\alpha_{loc}^{\text{Frank}(\eta)}$ calibrated under the Frank copula with parameter η. The global FWER level equals $\alpha = 0.05$ and the number of null hypotheses equals $m = 50$. The line $\eta = 0$ corresponds to joint independence.

η	$\alpha_{loc}^{\text{Sidak}}$	$\alpha_{loc}^{\text{Frank}(\eta)}$
0	0.00103	0.00101
2	0.00103	0.00107
4	0.00103	0.00113
6	0.00103	0.00116
8	0.00103	0.00122
10	0.00103	0.00129
12	0.00103	0.00135
14	0.00103	0.00142

η. *Denoting this copula by* C_η, *it holds that*

$$C_\eta(u_1, \ldots, u_m) = \psi_\eta\left(\sum_{j=1}^{m} \psi_\eta^{-1}(u_j)\right),$$

where $u_j \in [0,1]$ *for all* $1 \leq j \leq m$. *The function* ψ_η *is called the generator function of* C_η, *and it is given by*

$$\psi_\eta(t) = -\frac{1}{\eta}\log\left(1 - \frac{1-\exp(-\eta)}{\exp(t)}\right), \; t \in [0,1].$$

For $\eta \to 0$, *the model tends to the case of joint independence, while the degree of positive dependency increases with increasing* $\eta > 0$.

Table 3.3 compares the equi-coordinate local significance levels resulting from Lemma 3.2 (applied to $C_T = C_\eta$ *with varying values of* η*) with the local significance level of the Šidák test (which is exact under* H_0 *when assuming joint independence of the test statistics or p-values, respectively). In the case of* $\eta = 0$ *and* $\eta = 2$, *the two local significance levels are very similar. For larger values of* η, *however, the copula-based (multivariate) calibration method exploits the positive dependencies among the test statistics, and it leads to a markedly larger local significance level than the Šidák method. All values in Table 3.3 refer to* $m = 50$ *and an FWER level of* $\alpha = 0.05$.

3.6 Multivariate multiple tests for control of the false discovery rate

Up to date, the still by far most popular multiple test procedure for control of the FDR is the linear step-up (LSU) test φ^{LSU}, which has been proposed in the seminal paper by Benjamini and Hochberg (1995). The multiple test φ^{LSU} is also often referred to as the Benjamini-Hochberg procedure or simply *the* FDR procedure. In our notation, the decision rule of φ^{LSU} may be written as follows.

(i) Let $p_{1:m}, \ldots, p_{m:m}$ denote the ordered values of the m (random) p-values $P_1, \ldots, P_m$.

(ii) Determine

$$k = \max\{1 \le j \le m : p_{j:m} \le j\alpha/m\}, \tag{3.17}$$

where α denotes the target FDR level.

(iii) If the maximum in (3.17) does not exist, retain all m null hypotheses $H_1, \ldots, H_m$. Otherwise, reject exactly $H_{1:m}, \ldots, H_{k:m}$, where $H_{1:m}, \ldots, H_{m:m}$ denote the re-ordered null hypotheses in $\mathcal{H}_m$, according to the ordering of the corresponding p-values.

Early results on FDR control of φ^{LSU} dealt with the case of jointly independent p-values (or test statistics); see Benjamini and Hochberg (1995) and Finner and Roters (2001). However, since the beginning of the 21st century several authors have analyzed the FDR behavior of φ^{LSU} and related stepwise rejective multiple tests under (positive) dependency; cf., e. g., Benjamini and Yekutieli (2001), Sarkar (2002), Finner et al. (2007), Sarkar (2008), Block et al. (2013), Bodnar and Dickhaus (2014), Bodnar and Dickhaus (2017), Finner et al. (2017), and references therein.

Let ϑ denote the parameter of the statistical model under consideration, denote by $\mathbb{P}_\vartheta$ the distribution of the data sample under ϑ, and let $P_1, \ldots, P_m$ denote the (random) p-values on which φ^{LSU} operates. Benjamini and Yekutieli (2001) proved the following result (see also Section 4 of Finner et al. (2009) for slightly more general calculations).

Lemma 3.3 (see Equation (10) in Benjamini and Yekutieli (2001).) *Assume that exactly $m_0 \equiv m_0(\vartheta)$ out of the m null hypotheses that are to be tested simultaneously are true under ϑ. Without loss of generality, assume that H_i is true for $1 \le i \le m_0$, and that H_i is false for $m_0 + 1 \le i \le m$. Then it holds for the FDR of φ^{LSU} under ϑ, that*

$$FDR_\vartheta\left(\varphi^{LSU}\right) = \sum_{i=1}^{m_0} \sum_{k=1}^{m} \frac{1}{k} \mathbb{P}_\vartheta(P_i \le q_k \cap C_k^{(i)}), \tag{3.18}$$

where $q_j = j\alpha/m$, $1 \le j \le m$, and $C_k^{(i)}$ denotes the event that exactly $k-1$ null hypotheses additionally to H_i are rejected by φ^{LSU}.

Clearly, the probability expression on the right-hand side of (3.18) refers to the joint distribution of $P_1, \ldots, P_m$ under ϑ. Hence, it is possible to employ the multivariate approaches from our earlier sections to calculate/bound the FDR of φ^{LSU} under well-defined dependency structures among $P_1, \ldots, P_m$. This can for instance be used to adjust the nominal FDR level α. Namely, if the FDR of φ^{LSU} (or an upper bound for it) under the assumed dependency structure exceeds α, we may replace the nominal value of α by some smaller value such that the FDR is controlled. For example, Benjamini and Yekutieli (2001) showed that replacing α by $\alpha / \left(\sum_{i=1}^m i^{-1}\right)$ in the definition of φ^{LSU} always controls the FDR, no matter the dependency structure among $P_1, \ldots, P_m$. On the other hand, Bodnar and Dickhaus (2014) have shown that the FDR of φ^{LSU} is typically strictly smaller than α if the copula of $P_1, \ldots, P_m$ is an Archimedean copula with a completely monotone generator function (which does not depend on ϑ), and they derived an adjustment factor by which the nominal level α can be increased in order to exhaust the FDR level and, hence, optimize the power of φ^{LSU} by exploiting the (positive) dependencies among $P_1, \ldots, P_m$.

Both of these proposals follow the general construction method (exploitation of FDR bounds) which has been mentioned on page 81 of Dickhaus (2014). However, it has to be mentioned that the calculations referring to the right-hand side of (3.18) can become rather tedious, due to the complicated structure of the event $\{P_i \le q_k \cap C_k^{(i)}\}$. Furthermore, as indicated for instance by Finner et al. (2007) and Blanchard et al. (2014), the false discovery proportion (FDP) is typically not well concentrated around its expectation (the FDR) under dependency. Hence, many authors consider it more appropriate to control exceedance probabilities (over some given threshold) of the FDP under dependency rather than its mean; cf. also Delattre and Roquain (2015) and the references therein. The distribution of the FDP relies on the joint distribution of all m (random) p-values, too; cf. Blanchard et al. (2014), von Schroeder and Dickhaus (2020), and the references therein.

3.7 Conclusions and practical recommendations

We have presented multivariate approaches to the calibration of multiple tests for control of the FWER and the FDR, respectively. Multivariate statistical models and resulting multiple tests have several advantages over generic procedures which only take into account univariate marginal distributions of test statistics or p-values: (i) Multivariate statistical models are often more realistic, because they take into account the dependencies in the data. Such dependencies are ubiquitous in nowadays' (high-throughput) measurements, because of the underlying (neuro-)biological or technological mechanisms. Hence, data from such experiments typically exhibit strong temporal, spatial, or spatio-temporal dependencies. (ii) Especially in the presence of positive dependencies, the power of the multiple test can be enhanced by explicitly modeling and incorporating marginal distributions of higher order into the decision rule of the multiple test. This has been demonstrated by numerical examples in Sections 3.4 and 3.5. Disadvantages of the presented methods are a potentially high computational effort, and the need for information about the kind of dependencies among the test statistics or p-values, leading to a higher model complexity than in the case of univariate marginal approaches. Therefore, it is recommendable in practice to apply multivariate techniques whenever computationally feasible, given that the type of dependency structure among the test statistics is known. In the case of a totally unknown dependency structure among $T_1, \ldots, T_m$, nonparametric copula (pre-)estimation methods may be applied, but the methodology of Neumann et al. (2019) at least requires that the copula of T is a nuisance parameter in the sense that it does not depend on the parameter ϑ to which the null hypotheses $H_1, \ldots, H_m$ refer.

One different class of multivariate multiple tests, which has not been treated in this chapter, is constituted by resampling-based procedures. Such procedures implicitly take into account the aformentioned dependencies by employing appropriate resampling schemes which approximate the (joint) null distribution of the entire vector of test statistics or p-values, respectively. Resampling-based multiple tests for FWER control have been worked out for instance by Westfall and Young (1993), Troendle (1995), Romano and Wolf (2005a), Romano and Wolf (2005b), and Chung and Romano (2016). Resampling-based methods for FDR control have been derived, among others, by Yekutieli and Benjamini (1999), Troendle (2000), Romano et al. (2008), and Dudoit and van der Laan (2008).

Of course, it is also possible to combine explicit multivariate modeling of the data and resampling-based calibration of a multiple test for control of the FWER or FDR, respectively. For example, a model-based bootstrap procedure for multiple specification tests in dynamic factor models has been theoretically derived by Dickhaus and Pauly (2016) and implemented and applied by Dickhaus and Sirotko-Sibirskaya (2019).

Bibliography

Benjamini, Y. and Y. Hochberg (1995). Controlling the false discovery rate: A practical and powerful approach to multiple testing. *J. R. Stat. Soc., Ser. B 57*(1), 289–300.

Benjamini, Y. and D. Yekutieli (2001). The control of the false discovery rate in multiple testing under dependency. *Ann. Stat. 29*(4), 1165–1188.

Blanchard, G., T. Dickhaus, E. Roquain, and F. Villers (2014). On least favorable configurations for step-up-down tests. *Statistica Sinica 24*(1), 1–23.

Block, H. W., T. Costigan, and A. R. Sampson (1992). Product-type probability bounds of higher order. *Probab. Eng. Inf. Sci. 6*(3), 349–370.

Block, H. W., T. H. Savits, J. Wang, and S. K. Sarkar (2013). The multivariate-t distribution and the Simes inequality. *Stat. Probab. Lett. 83*(1), 227–232.

Bodnar, T. and T. Dickhaus (2014). False discovery rate control under Archimedean copula. *Electron. J. Stat. 8*(2), 2207–2241.

Bodnar, T. and T. Dickhaus (2017). On the Simes inequality in elliptical models. *Ann. Inst. Statist. Math. 69*(1), 215–230.

Bretz, F., T. Hothorn, and P. Westfall (2010). *Multiple Comparisons Using R.* Boca Raton, FL: Chapman and Hall/CRC.

Chung, E. and J. P. Romano (2016). Multivariate and multiple permutation tests. *Journal of Econometrics 193*(1), 76 – 91.

Comtet, L. (1974). *Advanced combinatorics. The art of finite and infinite expansions* (enlarged ed.). Dordrecht: D. Reidel Publishing Co..

Delattre, S. and E. Roquain (2015). New procedures controlling the false discovery proportion via Romano-Wolf's heuristic. *Ann. Statist. 43*(3), 1141–1177.

Dickhaus, T. (2014). *Simultaneous Statistical Inference with Applications in the Life Sciences.* Berlin, Heidelberg: Springer-Verlag.

Dickhaus, T. (2015). Simultaneous Bayesian analysis of contingency tables in genetic association studies. *Stat. Appl. Genet. Mol. Biol. 14*(4), 347–360.

Dickhaus, T. (2018). Combining high-dimensional classification and multiple hypotheses testing for the analysis of big data in genetics. In *Statistics and its applications*, Volume 244 of *Springer Proc. Math. Stat.*, pp. 47–50. Springer, Singapore.

Dickhaus, T. and J. Gierl (2013). Simultaneous test procedures in terms of p-value copulae. In *Proceedings on the 2nd Annual International Conference on Computational Mathematics, Computational Geometry & Statistics (CMCGS 2013)*, pp. 75–80. Global Science and Technology Forum (GSTF).

Dickhaus, T. and M. Pauly (2016). Simultaneous statistical inference in dynamic factor models. In I. Rojas and H. Pomares (Eds.), *Time Series Analysis and Forecasting*, pp. 27–45. Springer, New York, NY.

Dickhaus, T. and T. Royen (2015). A survey on multivariate chi-square distributions and their applications in testing multiple hypotheses. *Statistics 49*(2), 427–454.

Dickhaus, T. and N. Sirotko-Sibirskaya (2019). Simultaneous statistical inference in dynamic factor models: chi-square approximation and model-based bootstrap. *Comput. Statist. Data Anal. 129*, 30–46.

Dickhaus, T. and J. Stange (2013). Multiple point hypothesis test problems and effective numbers of tests for control of the family-wise error rate. *Calcutta Statistical Association Bulletin 65*(257–260), 123–144.

Dickhaus, T., K. Strassburger, D. Schunk, C. Morcillo-Suarez, T. Illig, and A. Navarro (2012). How to analyze many contingency tables simultaneously in genetic association studies. *Stat. Appl. Genet. Mol. Biol. 11*(4), Article 12.

Dudoit, S. and M. J. van der Laan (2008). *Multiple testing procedures with applications to genomics.* Springer Series in Statistics. Springer, New York.

Dunnett, C. W. (1955). A multiple comparison procedure for comparing several treatments with a control. *J. Am. Stat. Assoc. 50*, 1096–1121.

Dunnett, C. W. (1964). New tables for multiple comparisons with a control. *Biometrics 20*, 482–491.

Embrechts, P., F. Lindskog, and A. McNeil (2003). Modelling dependence with copulas and applications to risk management. In S. Rachev (Ed.), *Handbook of Heavy Tailed Distributions in Finance*, pp. 329–384. Amsterdam, Boston, MA: Elsevier Science B.V.

Finner, H., T. Dickhaus, and M. Roters (2007). Dependency and false discovery rate: Asymptotics. *Ann. Stat. 35*(4), 1432–1455.

Finner, H., T. Dickhaus, and M. Roters (2009). On the false discovery rate and an asymptotically optimal rejection curve. *Ann. Stat. 37*(2), 596–618.

Finner, H. and M. Roters (2001). On the false discovery rate and expected type I errors. *Biom. J. 43*(8), 985–1005.

Finner, H., M. Roters, and K. Strassburger (2017). On the Simes test under dependence. *Statist. Papers 58*(3), 775–789.

Gabriel, K. R. (1969). Simultaneous test procedures - some theory of multiple comparisons. *Ann. Math. Stat. 40*, 224–250.

Galambos, J. and I. Simonelli (1996). *Bonferroni-type inequalities with applications.* Probability and its Applications (New York). Springer-Verlag, New York, NY.

Genz, A. and F. Bretz (2009). *Computation of multivariate normal and t probabilities.* Lecture Notes in Statistics 195. Springer, Berlin.

Glaz, J. and B. M. Johnson (1984). Probability inequalities for multivariate distributions with dependence structures. *J. Am. Stat. Assoc. 79*, 436–440.

Härdle, W. K. and O. Okhrin (2010). De copulis non est disputandum - Copulae: an overview. *AStA Adv. Stat. Anal. 94*(1), 1–31.

Hochberg, Y. and A. C. Tamhane (1987). *Multiple comparison procedures.* Wiley Series in Probability and Mathematical Statistics. Applied Probability and Statistics. John Wiley & Sons, Inc, New York, NY.

Hothorn, T., F. Bretz, and P. Westfall (2008, Jun). Simultaneous inference in general parametric models. *Biom. J. 50*(3), 346–363.

Joe, H. (2014). *Dependence modeling with copulas.* Chapman and Hall/CRC Press, New York, NY.

Konietschke, F., L. A. Hothorn, and E. Brunner (2012). Rank-based multiple test procedures and simultaneous confidence intervals. *Electron. J. Stat. 6*, 738–759.

Konietschke, F., M. Placzek, F. Schaarschmidt, and L. Hothorn (2015). nparcomp: An R Software Package for Nonparametric Multiple Comparisons and Simultaneous Confidence Intervals. *Journal of Statistical Software 64*(9), 1–17.

Magirr, D., T. Jaki, and J. Whitehead (2012). A generalized Dunnett test for multi-arm multi-stage clinical studies with treatment selection. *Biometrika 99*(2), 494–501.

McNeil, A. J., R. Frey, and P. Embrechts (2005). *Quantitative risk management. Concepts, techniques, and tools.* Princeton, NJ: Princeton University Press.

Mieth, B., M. Kloft, J. A. Rodriguez, S. Sonnenburg, R. Vobruba, C. Morcillo-Suarez, X. Farre, U. M. Marigorta, E. Fehr, T. Dickhaus, G. Blanchard, D. Schunk, A. Navarro, and K. R. Müller (2016, Nov). Combining Multiple Hypothesis Testing with Machine Learning Increases the Statistical Power of Genome-wide Association Studies. *Scientific Reports 6*, Article 36671.

Miller, Jr., R. G. (1981). *Simultaneous statistical inference* (Second ed.). Springer-Verlag, New York-Berlin. Springer Series in Statistics.

Naiman, D. and H. Wynn (2005). The algebra of Bonferroni bounds: discrete tubes and extensions. *Metrika 62*(2–3), 139–147.

Naiman, D. Q. and H. P. Wynn (1992). Inclusion-exclusion-Bonferroni identities and inequalities for discrete tube-like problems via Euler characteristics. *Ann. Stat. 20*(1), 43–76.

Nelsen, R. B. (2006). *An introduction to copulas. 2nd ed.* Springer Series in Statistics. Springer, New York, NY.

Neumann, A., T. Bodnar, D. Pfeifer, and T. Dickhaus (2019). Multivariate multiple test procedures based on nonparametric copula estimation. *Biom. J. 61*(1), 40–61.

Romano, J. P., A. M. Shaikh, and M. Wolf (2008). Control of the false discovery rate under dependence using the bootstrap and subsampling. *TEST 17*(3), 417–442.

Romano, J. P. and M. Wolf (2005a). Exact and approximate stepdown methods for multiple hypothesis testing. *J. Am. Stat. Assoc. 100*(469), 94–108.

Romano, J. P. and M. Wolf (2005b). Stepwise multiple testing as formalized data snooping. *Econometrica 73*(4), 1237–1282.

Rüschendorf, L. (2009). On the distributional transform, Sklar's theorem, and the empirical copula process. *J. Stat. Plann. Inference 139*(11), 3921–3927.

Sarkar, S. K. (2002). Some results on false discovery rate in stepwise multiple testing procedures. *Ann. Stat. 30*(1), 239–257.

Sarkar, S. K. (2008). On the Simes inequality and its generalization. *IMS Collections Beyond Parametrics in Interdisciplinary Research: Festschrift in Honor of Professor Pranab K. Sen 1*, 231–242.

Schildknecht, K., S. Olek, and T. Dickhaus (2015). Simultaneous Statistical Inference for Epigenetic Data. *PLOS ONE 10*(5), Article e0125587.

Šidák, Z. (1967). Rectangular confidence regions for the means of multivariate normal distributions. *J. Am. Stat. Assoc. 62*, 626–633.

Sklar, A. (1959). Fonctions de répartition à n dimensions et leurs marges. *Publ. Inst. Statist. Univ. Paris 8*, 229–231.

Sklar, A. (1996). Random variables, distribution functions, and copulas - a personal look backward and forward. In *Distributions with Fixed Marginals and Related Topics.*, pp. 1–14. Institute of Mathematical Statistics, Hayward, CA.

Stange, J., T. Bodnar, and T. Dickhaus (2015). Uncertainty quantification for the family-wise error rate in multivariate copula models. *AStA Adv. Stat. Anal. 99*(3), 281–310.

Stange, J., T. Dickhaus, A. Navarro, and D. Schunk (2016). Multiplicity- and dependency-adjusted p-values for control of the family-wise error rate. *Stat. Probab. Lett. 111*, 32–40.

Stange, J., N. Loginova, and T. Dickhaus (2016). Computing and approximating multivariate chi-square probabilities. *Journal of Statistical Computation and Simulation 86*(6), 1233–1247.

Steffen, N. and T. Dickhaus (2020). Optimizing effective numbers of tests by vine copula modeling. *Depend. Model. 8*, 172–185.

Troendle, J. F. (1995). A stepwise resampling method of multiple hypothesis testing. *J. Am. Stat. Assoc. 90*(429), 370–378.

Troendle, J. F. (2000). Stepwise normal theory multiple test procedures controlling the false discovery rate. *J. Stat. Plann. Inference 84*(1–2), 139–158.

Tukey, J. W. (1953). The problem of multiple comparisons. In *The collected works of John W. Tukey. Volume VIII: Multiple comparisons: 1948–1983. Ed. by Henry I. Braun, with the assistance of Bruce Kaplan, Kathleen M. Sheehan, Min-Hwei Wang.*, pp. 1–300. Chapman & Hall, New York, NY.

von Schroeder, J. and T. Dickhaus (2020). Efficient calculation of the joint distribution of order statistics. *Comput. Statist. Data Anal. 144*, Article 106899.

Westfall, P. H. and S. S. Young (1993). *Resampling-based multiple testing: examples and methods for p-value adjustment.* Wiley Series in Probability and Mathematical Statistics. Applied Probability and Statistics. Wiley, New York, NY.

Worsley, K. (1982). An improved Bonferroni inequality and applications. *Biometrika 69*, 297–302.

Yekutieli, D. and Y. Benjamini (1999). Resampling-based false discovery rate controlling multiple test procedures for correlated test statistics. *J. Stat. Plann. Inference 82*(1–2), 171–196.

Yona, G., W. Dirks, S. Rahman, and D. M. Lin (2006, Jul). Effective similarity measures for expression profiles. *Bioinformatics 22*(13), 1616–1622.

4

Partitioning for Confidence Sets, Confident Directions, and Decision Paths

Helmut Finner

University of Dusseldorf

Szu-Yu Tang

Pfizer

Xinping Cui

University of California, Riverside

Jason C. Hsu

The Ohio State University

CONTENTS

DOI: 10.1201/9780429030888-5

Partitioning is a fundamental principle in multiple comparisons. In this chapter, we discuss and illustrate three applications of the partitioning principle corresponding to three motivations.

4.1 Motivations for Partitioning

Partitioning, as the name implies, refers to a partitioning of the entire *parameter* space.

Some scientific problems naturally partition the parameter space. In comparing seven treatments for a disease, for example, if treatment one is the best, then treatment two is not; if treatment three is the best, then treatment seven is not, and so forth. Such natural partitioning provides the first of the three motivations listed below for developing the partitioning principle (PP).

Section 4.2 Some scientific problems naturally partition.

Section 4.3 By providing associated confidence sets, partitioning can reduce multiplicity adjustment while guaranteeing control of the directional error rate.

Section 4.4 Partitioning can formulate multiple testing problems so that decision-making automatically follow desirable paths.

Both the PP and the closed testing principle can control the family-wise error rate (FWER) in testing multiple hypotheses while keeping multiplicity adjustment only to the extent that it is needed. (See Chapter 1 of this Handbook for descriptions of the Closed Testing Principle and FWER.) A second motivation for PP, stated in both Takeuchi (1973, 2010) and in Stefansson et al. (1988), is to be able to derive confidence sets associated with multiple tests. A motivation for that, in turn, is making decisions based on confidence sets naturally controls the directional error rate. In contrast, we will cite examples in Section 4.3.5 of multiple tests that control the FWER in testing *equality* null hypotheses but do not control the directional error rate, tests without clearly associated confidence sets because the union of their null hypotheses make up only a (small) part of the entire parameter space.

The third motivation for PP is that some decision-making processes naturally have paths. To assess the efficacy of a medicine, in some (but not all) therapeutic areas, it is natural to test doses from high to low in that order. As another example, efficacy in the primary endpoint would be tested before the secondary endpoint because efficacy in the secondary endpoint is relevant only if there is efficacy in the primary endpoint. While gatekeeping methods impose rules on closed tests to keep decision-making on paths, the PP can transparently partition the parameter space to channel decision-making onto desirable decision paths.

4.2 Multiple Comparisons with the Best: A Scientific Problem Which Naturally Partitions

Whether a drug starts as an active compound and gets metabolized and eliminated from the body or starts as an inactive compound and gets metabolized to an active form, patients in subgroups separated by polymorphism of a gene metabolizing the drug might derive differential benefit from that drug. Therefore, one might wonder which subgroup or subgroups of patients derive maximum benefit or practically maximum benefit from the drug, and which other subgroups do not.

Most drugs are "soft drugs," active compounds that, after performing their activity, are metabolized into an inactive form that is then excreted from the body. Other drugs, such as tamoxifen and clopidogrel, are "pro-drugs," inactive compounds needing to be metabolized to their active form.

The cytochrome P450 family of enzymes (abbreviated as CYP) is associated with the metabolism of many drugs (Nebert 2002). Perhaps the two most prominent genes in the P450 family are 2D6 and 2C19. Efficacy of some high profile drugs has been reported to be impacted by polymorphism in 2D6 and 2C19. For example, Schroth et al. (2009), Hoskins, Carey and McLeod (2009), and Abraham et al. (2010) discuss differential efficacy of tamoxifen for patients with variants of the CYP2D6 gene. Mega et al. (2010), Paré et al. (2010), and Holmes et al. (2011) compare the efficacy of Plavix (clopidogrel) for patients with variants of the CYP2C19 gene. This is not surprising because patients in different subpopulations defined by such polymorphisms will not metabolize the drug at the same rate and therefore will not derive the same benefit from that drug. As John W. Tukey (1992) said:

> Our experience with the real world teaches us – if we are willing learners – that, provided we measure to enough decimal places, no two 'treatments' ever have identically the same long-run value.

Polymorphisms in P450 genes are annotated by the so-called *star-allele* nomenclature. CYP2D6 has more than 90 alleles. CYP2C19 is somewhat less polymorphic. Its major alleles are *1, *2, *3, and *17, with *1 being normal (wild-type), *2 and *3 being loss-of-function, and *17 being gain-of-function alleles.

Paré et al. (2010) obtained 5,059 samples from a randomized, double-blind, placebo-controlled trial with 12562 patients and studied the effect of CYP2C19 genotype on the efficacy of clopidogrel as measured by cardiovascular outcomes. Paré et al. (2010) classified the population into five metabolizer subgroups according to their CYP2C19 genotype, as shown in Table 4.1. However, one can easily imagine classifying the population into finer subgroups, separating *1/*17 from *17/*17 for example, and separating *2/*17 from *3/*17.

Therefore, let us say we have k patient subpopulations. Identifying the following three kinds of subgroups will be very useful.

$S^{>}$ The subgroup deriving *the maximum* efficacy from the drug

$S_{<}$ Subgroups deriving *less than* maximum efficacy from the drug

S^{δ} Subgroups deriving *practically the maximum* efficacy

TABLE 4.1
Metabolizer Subgroups Defined by CYP2C19 Polymorphism

Metabolizer	Alleles
Poor	*2/*2 or *2/*3 or *3/*3
Intermediate	*1/*2 or *1/*3
Extensive	*1/*1
Ultra	*1/*17 or *17/*17
Unknown	*2/*17 or *3/*17

With finite amount of data, we cannot identify $S^>$, $S^<$, and S^δ with 100% confidence. However, we can certainly identify these subgroups with $100(1-\alpha)\%$ confidence (in a confidence sets sense). One possibility is to compare every subgroup with every other subgroup, i.e., do all-pairwise comparisons, and deduce information about $S^>, S^<, S^\delta$. However, since comparisons among bad treatments are not of primary interest, one might ask whether confident $S^>, S^<, S^\delta$ subgroup identifications is possible without deducing the information from all pairwise comparisons. Surprisingly, the answer is "yes", to a large extent.

Let us call the subgroup that receives the most efficacy the "best" subgroup and think about testing the hypotheses

H_{01}: The $1st$ subgroup is the best

H_{02}: The $2nd$ subgroup is the best

$\vdots$

H_{0k}: The kth subgroup is the best

There is only one best subgroup. If the $1st$ group receives the most efficacy, then the $2nd$ group does not, and so forth. Therefore, exactly one of the hypotheses is true, all others are false.

Therefore, one cannot make more than one type-I error in testing these k null hypotheses, and no multiplicity adjustment is needed for testing these k hypotheses simultaneously. That is, if each of $H_{0i}, i = 1, \ldots, k$, is tested at level-α, FWER is controlled at level-α.

However, testing the ith hypothesis involves comparing the ith subgroup with the other $k-1$ subgroups. The $k-1$ comparisons are one-sided because no subgroup can be better than the "best". Therefore, there is a one-sided multiplicity adjustment of $k-1$ within each of the k tests. It is less than the $k(k-1)$ multiplicity adjustment for one-sided all-pairwise comparisons or the $k(k-1)/2$ multiplicity adjustment for two-sided all-pairwise comparisons.

Note that the null hypotheses $H_{0i}, i = 1, \ldots, k$, essentially partition the parameter space. This formulation of multiple comparisons is called multiple comparisons with the best (MCB). The MCB formulation is convenient for us to explain how confidence sets for multiple comparisons can be constructed.

4.2.1 Confidence Sets Associated with Multiple Tests

Let Θ denote the parameter space. The connection between a family of tests for a parameter and a confidence set for that parameter is given by the following theorem.

Theorem 4.1 (Lehmann 1986, p. 90, Casella and Berger 2001, p. 421) *Let Θ denote the parameter space and let $\hat{\boldsymbol{\theta}}$ be a random vector whose distribution depends on $\boldsymbol{\theta} \in \Theta$. If $\{\phi_{\boldsymbol{\theta}}(\hat{\boldsymbol{\theta}}) : \boldsymbol{\theta} \in \Theta\}$ is a family of tests such that*

$$P_{\boldsymbol{\theta}}\{\phi_{\boldsymbol{\theta}}(\hat{\boldsymbol{\theta}}) = 0\} \geq 1 - \alpha$$

for each $\boldsymbol{\theta} \in \Theta$, then

$$C(\hat{\boldsymbol{\theta}}) = \{\boldsymbol{\theta} : \phi_{\boldsymbol{\theta}}(\hat{\boldsymbol{\theta}}) = 0, \boldsymbol{\theta} \in \Theta\}$$

is a level $100(1-\alpha)\%$ confidence set for $\boldsymbol{\theta}$.

One of the earliest uses of this correspondence is by Fieller (1964), to get a confidence interval for the *ratio* of two normal means. His size-α test for each hypothesized value of the true ratio of means is a clever linear combination of the estimated means, and these tests are then pivoted to obtain the confidence set.

The key point to note is that in order to obtain a confidence set, there needs to be a test for each parameter value of the parameter space, that is, the *family* of tests should partition the parameter space. Note that if each test is actually of *size* α, $P_{\boldsymbol{\theta}}\{\phi_{\boldsymbol{\theta}}(\hat{\boldsymbol{\theta}}) = 0\} = 1 - \alpha$, then the confidence set $C(\hat{\boldsymbol{\theta}})$ is *exact*, that is, it has confidence level exactly equal to $1 - \alpha$. Depending on the choice of the family of tests, the confidence set may or may not be convex.

With Θ denoting the parameter space, let $\mathcal{X}$ denote the sample space. Given $x \in \mathcal{X}$, once the family of tests have been executed, confidence bounds for each parameter can be deduced by calculating its minimum and maximum values in the confidence set.

Lemma 4.1 (Projection-Lemma) *Suppose $C(x)$ is a level $1 - \alpha$ confidence set for the parameter $\boldsymbol{\theta} = (\theta_1, \ldots, \theta_p) \in \Theta$. Given $x \in \mathcal{X}$, define, for $i \in \{1, \ldots, p\}$,*

$$\begin{aligned} U_i(x) &= \sup\{\eta : \exists\, \boldsymbol{\theta} \in C(x) : \theta_i = \eta\}, \\ L_i(x) &= \inf\{\eta : \exists\, \boldsymbol{\theta} \in C(x) : \theta_i = \eta\}. \end{aligned}$$

Then $D_i(x) = [L_i(x), U_i(x)], i = 1, \ldots, p$, constitute level $(1 - \alpha)$ simultaneous confidence intervals for $\boldsymbol{\theta} = (\theta_1, \ldots, \theta_p)$.

Note that even if the confidence set $C(x)$ is *exact*, the simultaneous confidence intervals $(D_1, \ldots, D_p)$ may be conservative, that is, their confidence level may be greater than $1 - \alpha$. The confidence set $D = D_1 \times D_2 \times \cdots \times D_p$ is of course convex.

Now consider a partition $\Theta_1, \ldots, \Theta_M$ of the parameter space, that is,

$$\bigcup_{m=1}^{M} \Theta_m = \Theta$$

and

$$\Theta_i \cap \Theta_j = \emptyset \text{ for all } i \neq j.$$

If parameters in each Θ_m is tested by a *different* family of tests, then these families of tests can be pivoted separately in eacn Θ_m and then combined to yield a confidence set for $\boldsymbol{\theta}$, leading naturally to a PP for multiple comparisons confidence set construction. Letting $\mathcal{C}_{1-\alpha}(\Theta)$ denote all possible level $1 - \alpha$ confidence sets $C(x)$ for the parameter $\boldsymbol{\theta} = (\theta_1, \ldots, \theta_p) \in \Theta$, the following Partition-Projection corollary gives a formal guideline for calculating the final confidence bounds.

Corollary 4.1 (Partition-Projection Corollary) *Let $\{\Theta_1, \ldots, \Theta_M\}$ be a partition of the parameter space Θ. For each $m \in \{1, \ldots, M\}$, let $\tilde{C}_m(x)$ be a level $1 - \alpha$ confidence set for $\boldsymbol{\theta}$. Given $x \in \mathcal{X}$, $i \in \{1, \ldots, p\}$, $m \in M_+(x) = \{m \in M : \tilde{C}_m(x) \neq \emptyset\}$, define*

$$\begin{aligned} U_{im}(x) &= \sup\{\eta : \exists\, \boldsymbol{\theta} \in \tilde{C}_m(x) : \theta_i = \eta\}, \\ L_{im}(x) &= \inf\{\eta : \exists\, \boldsymbol{\theta} \in \tilde{C}_m(x) : \theta_i = \eta\}, \\ U_i(x) &= \sup_{m \in M_+(x)} U_{im}(x), \\ L_i(x) &= \inf_{m \in M_+(x)} L_{im}(x), \end{aligned}$$

then $(D_1, \ldots, D_p)$ with $D_i(x) = [L_i(x), U_i(x)]$ constitute level $(1 - \alpha)$ simultaneous confidence intervals for $\boldsymbol{\theta} = (\theta_1, \ldots, \theta_p)$.

When the family of distributions is a location family of distributions, one can start with one or more tests for a particular hypothesized parameter value and employ equivariance to generate the family of tests for the entire parameter space. In the presence of an unknown (nuisance) scale parameter, usually this hypothesized parameter value is chosen so that a statistic whose distribution depends on neither the location parameters nor the scale parameter (i.e., a pivotal quantity) is available. Pivoting within each subspace then taking their union gives the confidence set.

Theorem 4.2 (Pivoting-Partitioning Confidence Set Construction) *Suppose the distribution of* $\hat{\boldsymbol{\theta}} - \boldsymbol{\theta}$ *does not depend on* $\boldsymbol{\theta} = (\theta_1, \ldots, \theta_p)$*. Consider a partition* $\Theta_1, \ldots, \Theta_M$ *of the parameter space. If each* $\phi_m(\hat{\boldsymbol{\theta}}) = \phi_m(\hat{\theta}_1, \ldots, \hat{\theta}_p)$ *is a level-*α *test for*

$$H_0 : \theta_1 = \cdots = \theta_p = 0$$

with acceptance region $A_m, m = 1, \ldots, M$*, then a level* $100(1-\alpha)\%$ *confidence region for* $\boldsymbol{\theta} = (\theta_1, \ldots, \theta_p)$ *is*

$$C(\hat{\theta}_1, \ldots, \hat{\theta}_p) = \bigcup_{m=1}^{M} \Big(\{-\boldsymbol{\theta} + \hat{\boldsymbol{\theta}} : \boldsymbol{\theta} \in A_m\} \cap \Theta_m \Big).$$

Proof 1 *The test*

$$\phi(\hat{\boldsymbol{\theta}} - \boldsymbol{\theta}^0) = \phi_m(\hat{\theta}_1 - \theta_1^0, \ldots, \hat{\theta}_p - \theta_p^0)$$

is a level-α *test for*

$$H_{\boldsymbol{\theta}^0} : \theta_1 = \theta_1^0, \ldots, \theta_p = \theta_p^0.$$

Therefore, by Corollary 4.1, a level $100(1-\alpha)\%$ *confidence set for* $\boldsymbol{\theta}$ *is*

$$C(\hat{\boldsymbol{\theta}}) = \{\boldsymbol{\theta}^0 : H_{\boldsymbol{\theta}^0} \textit{ is accepted } \} \tag{4.1}$$

$$= \bigcup_{m=1}^{M} \{\boldsymbol{\theta}^0 : \hat{\boldsymbol{\theta}} - \boldsymbol{\theta}^0 \in A_m \textit{ and } \boldsymbol{\theta}^0 \in \Theta_m\} \tag{4.2}$$

$$= \bigcup_{m=1}^{M} \Big(\{-\boldsymbol{\theta} + \hat{\boldsymbol{\theta}} : \boldsymbol{\theta} \in A_m\} \cap \Theta_m \Big). \tag{4.3}$$

Stefansson et al. (1988) used this PP to construct a confidence set for the step-down version of Dunnett's method as well as MCB confidence sets which we will show in Section 4.2.2. (See the introductory Chapter 1 for a general description of step-wise methods.) Other examples of uses of the PP to construct multiple comparison confidence sets include Hayter and Hsu (1994), Finner (1994), Finner and Strassburger (2007), and Strassburger and Bretz (2008).

4.2.2 Partition Confidence Set for MCB

Suppose that under the ith treatment a random sample $Y_{i1}, Y_{i2}, \ldots, Y_{in}$ of size n is taken, where the observations between the treatments are independent. Then under the usual normality and equality of variances assumptions, we have the balanced one-way model (4.4)

$$Y_{ia} = \mu_i + \epsilon_{ia}, \quad i = 1, \ldots, k, \quad a = 1, \ldots, n, \tag{4.4}$$

where μ_i is the effect of the ith treatment, $i = 1, \ldots, k$, and $\epsilon_{11}, \ldots, \epsilon_{kn}$ are i.i.d. normal errors with mean 0 and unknown variance σ^2. We use the notation

$$\hat{\mu}_i = \bar{Y}_i = \sum_{a=1}^{n} Y_{ia}/n,$$

$$\hat{\sigma}^2 = MSE = \sum_{i=1}^{k}\sum_{a=1}^{n}(Y_{ia} - \bar{Y}_i)^2/[k(n-1)]$$

for the sample means and the pooled sample variance.

Suppose we partition of the parameter space by $\Theta_1, \ldots, \Theta_m$ where $\Theta_i = \{\mu_i > \max_{j \neq i} \mu_j\}$, i.e., Θ_i is the part of the parameter space where the *ith* subgroup is the best.

Within Θ_i we want to test the null hypothesis

$$H_{0i} : \text{the } i^{th} \text{ subgroup is the best}$$

If a larger treatment effect is better, then that null hypothesis becomes

$$H_{0i} : \quad \mu_i > \max_{j \neq i} \mu_j$$

Figure 4.1 shows, for the case of $k = 3$, H_{0i} : the *ith* subgroup is the best for $i = 2, 3$. The shaded area in Figure 4.1b is Θ_2, while the shaded area in Figure 4.1b is Θ_3.

For every parameter value $(\mu_1, \ldots, \mu_k)$ in Θ_i, Dunnett's size-α test for that parameter value being true has acceptance region

$$A_i = \left\{\hat{\mu}_i - \mu_i > \hat{\mu}_j - \mu_j - d\hat{\sigma}\sqrt{2/n} \text{ for all } j \neq i\right\}$$

where d is the quantile that makes the test size-α.

Following the pivoting Theorem 4.2, the parameters within each Θ_i that are not rejected are

$$\left\{\mu_i - \mu_j > \hat{\mu}_i - \hat{\mu}_j - d\hat{\sigma}\sqrt{2/n} \text{ for all } j \neq i\right\} \bigcap \Theta_i$$

Therefore, an exact $100(1-\alpha)$ confidence set is

$$C(\hat{\mu}_1, \ldots, \hat{\mu}_k, \hat{\sigma}) = \bigcup_{i=1}^{k} \left(\left\{\mu_i - \mu_j > \hat{\mu}_i - \hat{\mu}_j - d\hat{\sigma}\sqrt{2/n} \text{ for all } j \neq i\right\} \cap \Theta_i\right).$$

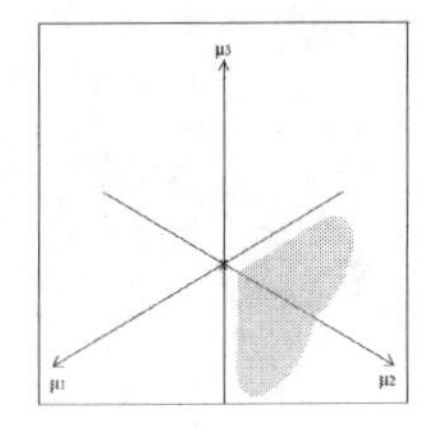

(a) Null space for $\mu_2 > \max_{j=1,3} \mu_j$

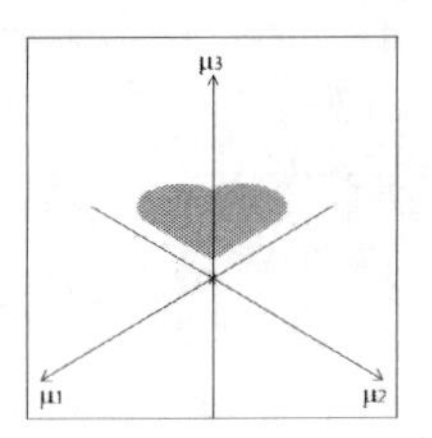

(b) Null space for $\mu_3 > \max_{j=1,2} \mu_j$

FIGURE 4.1
Examples of partitioned MCB null hypotheses.

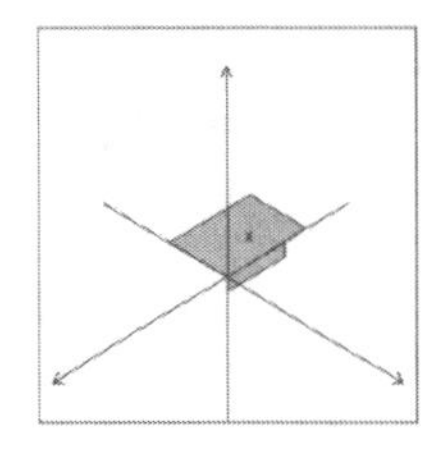

(a) Exact MCB confidence set

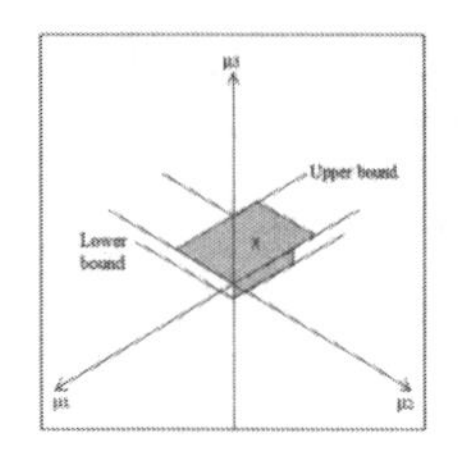

(b) Bounds for $\mu_3 - \max_{j=1,2} \mu_j$

FIGURE 4.2
Deducing MCB confidence bounds from its exact confidence set.

Figure 4.2 shows, for the case of $k = 3$, a particular example of an exact MCB confidence set. Location of "×" is the point estimate of $(\hat{\mu}_1, \hat{\mu}_2, \hat{\mu}_3)$. It is such that $\hat{\mu}_3$ is somewhat larger than $\hat{\mu}_2$ but much larger than $\hat{\mu}_1$. Thus H_{01} : the 1st subgroup is the best is rejected so that

$$\left\{\mu_i - \mu_j > \hat{\mu}_i - \hat{\mu}_j - d\hat{\sigma}\sqrt{2/n} \text{ for all } j \neq i\right\} \cap \Theta_i = \emptyset,$$

but neither H_{02} nor H_{03} is rejected.

As we discuss in more detail in the next section, the natural parameters for MCB are $\mu_i - \max_{j \neq i} \mu_j, i = 1, \ldots, k$. Contour of constant $\mu_i - \max_{j \neq i} \mu_j$ is a "V", with the tip of the V on the ith axis. Figure 4.2b shows, given an exact MCB confidence set such as the one in Figure 4.2a, how one would deduce lower and upper confidence bound for $\mu_3 - \max_{j \neq 3} \mu_j$.

In the next section, we relate MCB confidence sets derived algebraically in Hsu (1981) and (1984) with the ones obtained by using the geometrical pivoting technique above.

4.2.3 Multiple Comparisons with the Best

Earlier parameterization of MCB was in terms of "comparisons with the best". Thus, if a larger treatment effect is better, then even though which treatment is best is unknown, one could define the parameters of primary interest to be

$$\max_{j=1,\ldots,k} \mu_j - \mu_i, i = 1, \ldots, k, \tag{4.5}$$

the difference between the (unknown) true best treatment effect and each of the k treatment effects. This was the parametrization in Hsu (1981, 1982) and Edwards and Hsu (1983).

However, it turns out to be advantageous to parameterize MCB as "comparison with the best of the *others*". Suppose a larger treatment effect (e.g., survival time) implies a better treatment. Then the parameters

$$\mu_i - \max_{j \neq i} \mu_j, \; i = 1, \ldots, k \tag{4.6}$$

contain all the information that the parameters (4.5) contain, for if

$$\mu_i - \max_{j \neq i} \mu_j > 0,$$

then treatment i is the best treatment. On the other hand, if

$$\mu_i - \max_{j \neq i} \mu_j < 0,$$

then treatment i is not the best treatment. Further, even if the ith treatment is not the best, but nevertheless

$$\mu_i - \max_{j \neq i} \mu_j > -\delta$$

where δ is a small positive number, then the ith treatment is at least close to the best.

Note that whereas the range of the parameters (4.5) is $[0, \infty)$, the range of the parameters (4.6) is $(-\infty, \infty)$. Thus, the reference point to which confidence intervals for the parameters (4.6) should be compared is the usual 0. This is one advantage of the parametrization (4.6) over the parametrization (4.5). Another advantage is, as will be shown in Section 4.2.4, lower and upper confidence bounds on the (4.6) parameters correspond to Indifference Zone Selection and Subset Selection respectively. Starting with Hsu (1984), MCB parameterization switched to (4.6) from (4.5).

Contrasts such as those for all-pairwise comparisons (MCA)

$$\mu_i - \mu_j,\ i \neq j$$

and for multiple comparisons with a control (MCC)

$$\mu_i - \mu_k,\ i \neq k,$$

would be straight lines in Figure 4.2b, but multiple comparisons with the best (MCB) parameters (4.6) have "V" shaped contours, as shown in that figure.

Hsu (1984) showed that the closed intervals

$$[-(\hat{\mu}_i - \max_{j \neq i} \hat{\mu}_j - d\hat{\sigma}\sqrt{2/n})^-, (\hat{\mu}_i - \max_{j \neq i} \hat{\mu}_j + d\hat{\sigma}\sqrt{2/n})^+], i = 1, \ldots, k \tag{4.7}$$

form a set of $100(1-\alpha)\%$ simultaneous confidence intervals for $\mu_i - \max_{j \neq i} \mu_j$. Here $-x^- = \min\{0, x\}$ and $x^+ = \max\{0, x\}$.

While the derivation in Hsu (1984) was algebraic, if one were to deduce from the exact MCB confidence set not just confidence bounds for $\mu_3 - \max_{j \neq 3} \mu_j$ as in Figure 4.2b but for $\mu_i - \max_{j \neq i} \mu_j$ for all $i = 1, \ldots, k$, then indeed one would get the simultaneous confidence intervals (4.7).

4.2.4 Connection with Ranking and Selection

Multiple comparison with the best, an early example of what is called *selective inference*, has its roots in ranking and selection, which has two principal formulations: *subset selection*, and *indifference zone selection.*

Let $(1), \ldots, (k)$ denote the unknown indices such that

$$\mu_{(1)} \leq \mu_{(2)} \leq \cdots \leq \mu_{(k)}.$$

In other words, (i) is the *anti-rank* of μ_i among $\mu_1, \ldots, \mu_k$. For example, suppose $k = 3$ and $\mu_2 < \mu_3 < \mu_1$; then $(1) = 2$, $(2) = 3$, $(3) = 1$. For the balanced one-way model (4.4), Subset Selection, due to Gupta (1956; 1965), gives the set

$$G = \left\{ i : \hat{\mu}_i - \max_{j \neq i} \hat{\mu}_j + d\hat{\sigma}\sqrt{2/n} > 0 \right\}$$

as a $100(1-\alpha)\%$ confidence set for (k).[1] Subset selection infers treatments with indices not in G are not the best treatment. Assuming $\mu_{(k)} > \mu_{(k-1)}$, the subset selection confidence

---

[1]In traditional Subset Selection literature, when there are multiple indices for $\operatorname{argmax}_i \mu_i$, arbitrarily one such index is "tagged" to be (k).

statement

$$\inf_{\boldsymbol{\mu},\sigma^2} P_{\boldsymbol{\mu},\sigma^2}\{(k) \in G\} \geq 1 - \alpha$$

is implied by the confidence statement associated with constrained MCB upper bounds

$$\inf_{\boldsymbol{\mu},\sigma^2} P_{\boldsymbol{\mu},\sigma^2}\left\{\mu_i - \max_{j\neq i}\mu_j \leq \left(\hat{\mu}_i - \max_{j\neq i}\hat{\mu}_j + d\hat{\sigma}\sqrt{2/n}\right)^+\right\} \geq 1 - \alpha,$$

since a non-positive upper bound on $\mu_i - \max_{j\neq i}\mu_j$ implies $i \neq (k)$.

Indifference zone selection, due to Bechhofer (1954), has a *design* aspect and an *inference* aspect.

Traditional ranking and selection inferences were on *indices* that correspond to different rankings of the means of the populations; they were not directly on the values of means themselves. Let $[1], \ldots, [k]$ denote the random indices such that

$$\hat{\mu}_{[1]} < \hat{\mu}_{[2]} < \ldots < \hat{\mu}_{[k]}.$$

(Since $\hat{\mu}_i$ are continuous random variables, ties occur in them with probability zero.) In other words, $[i]$ is the anti-rank of $\hat{\mu}_i$ among $\hat{\mu}_1, \ldots, \hat{\mu}_k$. For example, suppose $k = 3$ and $\hat{\mu}_2 < \hat{\mu}_1 < \hat{\mu}_3$; then $[1] = 2$, $[2] = 1$, $[3] = 3$. To understand the explanation of the connection between traditional ranking and selection inferences and modern MCB inference below, it is important to keep in mind that $[1], \ldots, [k]$ are *random* variables.

For the balanced one-way model (4.4) with σ known, the design aspect of indifference zone selection sets the common sample size n to be the smallest integer such that

$$d\sigma\sqrt{2/n} \leq \delta^\star, \tag{4.8}$$

where $\delta^\star(> 0)$ represents the quantity of *indifference* to the user, that is, treatment means within $\delta^\star$ of the best are considered to be equivalent to the best.

Once data have been collected, the inference aspect of indifference zone selection then "selects" the $[k]$th treatment as the best treatment. The indifference zone confidence statement is that if $\boldsymbol{\mu} = (\mu_1, \ldots, \mu_k)$ is in the so-called *preference zone*

$$\{\mu_{(k)} - \mu_{(k-1)} > \delta^\star\},$$

then with a probability of at least $1 - \alpha$ the true best treatment will be selected. In other words, the indifference zone confidence statement is

$$\inf_{\mu_{(k)}-\mu_{(k-1)}>\delta^\star} P_{\boldsymbol{\mu},\sigma^2}\{\mu_{[k]} = \mu_{(k)}\} = 1 - \alpha. \tag{4.9}$$

This confidence statement is implied by the confidence statement

$$\inf_{\boldsymbol{\mu},\sigma^2} P_{\boldsymbol{\mu},\sigma^2}\{-\delta^\star \leq \mu_{[k]} - \max_{j\neq[k]}\mu_j\} \geq 1 - \alpha \tag{4.10}$$

because, for $\boldsymbol{\mu}$ in the preference zone, the only treatment mean μ_i with

$$-\delta^\star \leq \mu_i - \max_{j\neq i}\mu_j$$

is $\mu_{(k)}$. While Fabian (1962) gave a direct proof of (4.10), we can see that the confidence statement (4.10) is implied by the confidence statement

$$\inf_{\boldsymbol{\mu},\sigma^2} P_{\boldsymbol{\mu},\sigma^2}\{-d\sigma\sqrt{2/n} \leq \mu_{[k]} - \max_{j\neq[k]}\mu_j\} \geq 1 - \alpha$$

because

$$d\sigma\sqrt{2/n} \leq \delta^\star$$

by indifference zone sample size design (4.8). This last confidence statement, in turn, is implied by

$$\inf_{\boldsymbol{\mu},\sigma^2} P_{\boldsymbol{\mu},\sigma^2}\{-(\hat{\mu}_{[k]} - \max_{j\neq[k]} \hat{\mu}_j - d\sigma\sqrt{2/n})^- \leq \mu_{[k]} - \max_{j\neq[k]} \mu_j\} \geq 1-\alpha,$$

because

$$\hat{\mu}_{[k]} - \max_{j\neq[k]} \hat{\mu}_j \geq 0$$

and

$$-d\sigma\sqrt{2/n} < 0.$$

The last confidence bound now looks familiar. It is one of the constrained lower MCB confidence bounds on $\mu_i - \max_{j\neq i}\mu_j$, $i = 1,\ldots,k$, namely, the one on $\mu_{[k]} - \max_{j\neq[k]}\mu_j$, in the special case of σ known, which can be thought of as the case where the degree of freedom of $\hat{\sigma}$ equals infinity.

In essence, the design aspect of indifference zone selection guarantees a desired accuracy of the MCB lower bound for $\mu_{[k]} - \max_{j\neq[k]}\mu_j$, so that after experimentation, the useful level $1-\alpha$ confidence statement

$$-\delta^\star \leq \mu_{[k]} - \max_{j\neq[k]} \mu_j \tag{4.11}$$

can be made with probability one. For single-stage experiments, this can be achieved only when σ is known. When σ is unknown and must be estimated, there is a positive probability that the lower confidence bound

$$\hat{\mu}_{[k]} - \max_{j\neq[k]} \hat{\mu}_j - d\hat{\sigma}\sqrt{2/n}$$

on $\mu_{[k]} - \max_{j\neq[k]}\mu_j$ is less than $-\delta^\star$. However, in analogy with sample size computation based on power of tests, one can design a single-stage experiment so that, with a probability $1-\beta < 1-\alpha$, the lower bound on $\mu_{[k]} - \max_{j\neq[k]}\mu_j$ will be greater than $-\delta^\star$.

Book-length discussions of ranking and selection include Gibbons, Olkin and Sobel (1977), Gupta and Panchapakesan (1979), and Bechhofer, Santner and Goldsman (1995).

Having shown subset selection and indifference zone selection correspond to upper and lower MCB confidence bounds, the most important observation to make at this point is that, since the MCB confidence intervals are guaranteed to cover the parameters $\mu_i - \max_{j\neq i}\mu_j$, $i = 1,\ldots,k$, simultaneously with a probability of at least $1-\alpha$, subset selection inference and indifference zone selection inference can be given simultaneously with the guarantee that both inferences are correct with a probability of at least $1-\alpha$.

4.3 Partitioning for Confidence Sets and Confident Directions

Both the PP and the closed testing principle can be applied to reduce multiplicity adjustment in multiple testing. The idea behind both principles is that in order to control FWER, multiplicity need only be adjusted to the extent that multiple null hypotheses may be simultaneously true. (See Chapter 1 of this Handbook for descriptions of the closed testing principle and FWER.)

If a multiple test is a partition test, then (so long as it partitions the entire parameter space) generally it can be pivoted to have an associated confidence set (using Theorem 4.1 and Corollary 4.1). Why this is important is if multiple testing decision-making is "compatible" with a confidence set, then the directional error rate is controlled.

In testing a new treatment Rx against a control C, merely rejecting the null hypothesis that there is no difference is not useful. One has to make a decision: either Rx is better than C or Rx is worse than C. In practice, this decision is made according to what the data indicates, upon a rejection of an equality null. If one infers Rx is better than C but in fact Rx is worse than C, then this *directional error* should be counted as an error, and the rate of this incorrect decision should be controlled.

The problem of comparing Rx versus C, for two-sided inference, is often formulated as a test of equality. There are two issues with that, the first is, as Tukey (1991) said, "Statisticians classically asked the wrong question – and were willing to answer with a lie, one that was often a downright lie ... All we know about the world teaches us that the effects of A and B are always different – in some decimal place – for any A and B. Thus, asking 'Are the effects different' is foolish. What we should be answering first is 'Can we tell the direction in which the effects of A differ from the effects of B?' " Therefore, an equality null is almost surely false. Ding et al. (2018) in fact reported their observation that all equality nulls are statistically false in genome-wide association studies (GWAS), see the discussions in Sections 6.4 and 7.1 of Chapter 15 in this Handbook on testing for single nucleotide polymorphisms (SNPs) testing as well. Both in Ding et al. (2018) and in Chapter 15 of Handbook, instead of tests of equalities, confidence intervals are given instead. Chapter 14 on bioinfomatics and genomics in this Handbook also find exact equalities nulls in medical imaging to be false. Making extensive use of the PP, inferences given in Chapter 14, are also in the form of confidence intervals.

The other issue is that, in general, we cannot assume a multiple test that controls the FWER for testing

$$\begin{aligned} H_1^= &: \eta_1 = 0 \\ H_2^= &: \eta_2 = 0 \\ H_3^= &: \eta_3 = 0 \end{aligned} \tag{4.12}$$

automatically controls the directional error rate because directional errors are not counted in Type I error definition for testing equality null hypotheses. To control the directional error rate in two-sided testing, instead of testing equality nulls (4.12), the formulation by Finner (1999) is to test *pairs* of one-sided nulls:

$$\begin{aligned} H_1^{\leq} &: \eta_1 \leq 0, \quad H_1^{\geq} : \eta_1 \geq 0 \\ H_2^{\leq} &: \eta_2 \leq 0, \quad H_2^{\geq} : \eta_2 \geq 0 \\ H_3^{\leq} &: \eta_3 \leq 0, \quad H_3^{\geq} : \eta_3 \geq 0. \end{aligned} \tag{4.13}$$

Controlling the FWER of testing the null hypotheses (4.13) would indeed control the directional error rate because direction errors are counted as Type I errors.

Shaffer (1980) proved that, if the test statistics are independent and certain distributional assumptions are satisfied, then some step-down methods for testing equality nulls do control the directional error rate. Providing a unifying framework, Finner (1999) gave conditions under which tests controlling the FWER of testing equality nulls control the FWER of testing the paired one-sided nulls. Intricacy of their proofs and the distributional assumptions required make clear that control of directional error rate should not be taken for granted in general.

However, to control the FWER of testing, a set of null hypotheses using a $100(1-\alpha)\%$ confidence set is simple: reject a null hypothesis H_{0i} if the confidence set[2] does not contain any parameter point in H_{0i}. This is true regardless of whether the null hypotheses are equalities like (4.12), or one-sided inequalities, or pairs of one-sided inequalities like (4.13). Therefore, multiple tests based on confidence sets have the advantage that they can control the directional error rate when the problem is properly formulated to do so.

The multiple tests that we recommend in this chapter, such as the step-down version of one-sided Dunnett's test, are tests with associated confidence sets.

4.3.1 A Dose-Response Motivating Example

Schnell et al. (2017) described an Alzheimer Disease (AD) study which compared three doses of a new treatment (doses 1, 2, and 3) with a negative control (dose 0, a placebo). There was an active control (a Standard of Care or SoC, denoted as dose 4 for convenience) in the study as well. Response in this study is *improvement* in ADAS-Cog11 (baseline ADAS-Cog11 minus final ADAS-Cog11) after 24 weeks (a relatively short duration for an AD study). For illustration purpose, we look at the male subset of the data (Sex = 1).

We can model the n_i observed *improvements* from baseline $Y_{i1}, Y_{i2}, \ldots, Y_{in_i}$ under the *ith* dose, including dose 0, as a one-way model:

$$Y_{ir} = \mu_i + \epsilon_{ir}, \quad i = 0, \ldots, k, \quad r = 1, \ldots, n_i, \tag{4.14}$$

where μ_i is the mean improvement given the *ith* dose, $i = 0, \ldots, k$ with $k = 4$, and $\epsilon_{01}, \ldots, \epsilon_{kn_k}$ are i.i.d. normal errors with mean 0 and variance σ^2 unknown. This model differs from (4.4) in that sample sizes may be different. We use the notation

$$\hat{\mu}_i = \bar{Y}_i = \sum_{r=1}^{n_i} Y_{ir}/n_i,$$

$$\hat{\sigma}^2 = MSE = \sum_{i=0}^{k}\sum_{r=1}^{n_i}(Y_{ir} - \bar{Y}_i)^2 / \sum_{i=0}^{k}(n_i - 1)$$

for the sample means and the pooled sample variance.

Of interests are:

- Is there verification that the active control (Dose 4) is better than the placebo (Dose 0)?
- Which of Doses 1, 2, and 3 are better than the placebo (Dose 0)?

Let $\boldsymbol{\mu} = (\mu_0, \ldots, \mu_k)$, and for $i = 1, \ldots, 4$, let

$$H_{0i} : \mu_i \leq \mu_0$$

be the null hypotheses that Dose i is not better than the placebo, and let

$$\Theta_i = \{\boldsymbol{\mu} : \mu_i - \mu_0 \leq 0\}$$

be the corresponding subspace of the parameter space.

Each of the null hypotheses can be tested using a one-sided t-test, or a lower confidence bound on $\mu_i - \mu_0$. Dunnett's (1955) method adjusts for the multiplicity of testing four null hypotheses, providing four simultaneous confidence bounds, taking correlation of the point estimates into account, producing the results in Table 4.2. From the lower confidence

[2] which can be two-sided simultaneous confidence intervals for two-sided testing or one-sided simultaneous confidence bounds for one-sided testing

TABLE 4.2
Analysis of the Alzheimer Data Set from Single-Step Dunnett's Method with Dose 0 as the Control

Dose	Estimated Improvement	$\lvert t \rvert$ Adjusted p-value	95% One-Sided Lower Confidence Bound	t Adjusted p-Value
4	5.667	0.0061	1.806	0.0030
3	3.605	0.2076	−0.675	0.1040
2	3.710	0.0991	0.006	0.0496
1	1.757	0.6911	−1.914	0.3663

bounds, or the t adjusted p-value, we can infer that Dose 4 (the active control) and Dose 2 are better than the placebo.

However, there are two reasons for thinking beyond Dunnett's method:

1. In parts of the parameter space where not all four null hypotheses are true, adjustment for multiplicity conceptually can be less than four;

2. In testing to make decision about which doses are better than the placebo, it may seem natural to follow a path of testing in the order of active control → high dose → medium dose → low dose.

Both reasons lead to partitioning of the parameter space, different partitioning. The remainder of this section describes how to partition to reduce multiplicity, whereas in Section 4.4 we will describe how to partition to channel decision-making to follow paths.

4.3.2 Partitioning One-Sided Tests without Paths

Following Finner and Strassburger (2002), any family of hypotheses $\mathcal{H} = \{H_i : i \in I\}$ generates a natural partition which is the coarsest partition with the property that each H_i can be represented as a disjoint union of the members of the partition.

If $\boldsymbol{\theta} \in \Theta$ is the 'true' parameter, then H_i is said to be *true* if $\boldsymbol{\theta} \in H_i$. The index set $I(\boldsymbol{\theta}) = \{i \in I : H_i \ni \boldsymbol{\theta}\}$ will denote the set of all indices of true null hypotheses if $\boldsymbol{\theta}$ is the true parameter.

Let $\mathcal{J} = \{J : J \subseteq I\}$ and $\Theta_J = \{\boldsymbol{\theta} \in \Theta : I(\boldsymbol{\theta}) = J\}$, $J \subseteq I$. The natural partition is defined by

$$\Theta_{\mathcal{J}} = \{\Theta_J : J \in \mathcal{J}\}.$$

In words, Θ_J consists of parameter points for which $H_i, i \in J$, are true but $H_j, j \notin J$, are false.

Note that $\Theta_\emptyset$ is one of the partitioning subspaces. That is, one of the J is the empty set $\emptyset$, so $\Theta_\emptyset$ consists of parameter points for which all the null hypotheses are false. While these parameter points are not involved in any test of the original null hypotheses $\mathcal{H} = \{H_i : i \in I\}$, it is better for each parameter in $\Theta_\emptyset$ to be formally tested as well since the construction of confidence set associated with testing $\mathcal{H}$ by Theorem 4.1 expects every parameter point in the parameter space be tested.

Thereby, the closure $\overline{\mathcal{H}}$ of a family $\mathcal{H}$ generates the same natural partition as $\mathcal{H}$. Supposing that $\mathcal{H} = \overline{\mathcal{H}}$, one may set $\Theta_i = H_i \cap (\bigcup_{j:H_j \subset H_i} H_j)^c$ for $i \in I$ and $J_p = \{i \in I : \Theta_i \neq \emptyset\}$. Then the natural partition generated by $\mathcal{H}$ is given by

$$\Theta(J_p) = \{\Theta_i : i \in J_p\},$$

TABLE 4.3
Partition Testing of Four Null Hypotheses

Θ_1	Θ_2	Θ_3	Θ_4	H_{01}	H_{02}	H_{03}	H_{04}
◎	◎	◎	◎	✓	✓	✓	✓
◎	◎	◎	⊗	✓	✓	✓	✗
◎	◎	⊗	◎	✓	✓	✗	✓
◎	⊗	◎	◎	✓	✗	✓	✓
⊗	◎	◎	◎	✗	✓	✓	✓
◎	◎	⊗	⊗	✓	✓	✗	✗
◎	⊗	◎	⊗	✓	✗	✓	✗
◎	⊗	⊗	◎	✓	✗	✗	✓
⊗	◎	◎	⊗	✗	✓	✓	✗
⊗	◎	⊗	◎	✗	✓	✗	✓
⊗	⊗	◎	◎	✗	✗	✓	✓
◎	⊗	⊗	⊗	✓	✗	✗	✗
⊗	◎	⊗	⊗	✗	✓	✗	✗
⊗	⊗	◎	⊗	✗	✗	✓	✗
⊗	⊗	⊗	◎	✗	✗	✗	✓
⊗	⊗	⊗	⊗	✗	✗	✗	✗

i. e., $\Theta(J_p) = \Theta_{\mathcal{J}}$. If $J_p = I$, then each hypothesis H_i can be identified with Θ_i and vice versa. Note that J_p may be much smaller than I.

So in testing k null hypotheses

$$H_{0i} : \theta_i \leq 0, \; i = 1, \ldots, k, \tag{4.15}$$

for each $I \subseteq \{1, \ldots, k\}, I \neq \emptyset$, form $H^{\star}_{0I} : \theta_i \leq 0$ for all $i \in I$ and $\theta_j > 0$ for $j \notin I$. There are 2^k parameter subspaces and $2^k - 1$ hypotheses to be tested.

In the AD study, $\theta_i = \mu_i - \mu_0, \; i = 1, \ldots, 4$. So with $k = 4$ in the Alzheimer study, there are 16 combinations of these four null hypotheses being true or false. Partitioning divides the parameter space $\Theta = \{\theta_1, \theta_2, \theta_3, \theta_4\}$ into sixteen disjoint subspaces as depicted in Table 4.3, with a ✓representing a null hypothesis being true and an ✗ representing that null hypothesis being false.

To see explicitly the 16 parameter subspaces, for each of the 16 rows in Table 4.3, we take intersection of Θ_i and their complements where a ◎ represents Θ_i and a $\otimes$ represents Θ_i^c. For example, the second row, with H_{01} and H_{02} and H_{03} being true while H_{04} being false, that combination corresponds to $\boldsymbol{\mu} \in \Theta_1 \cap \Theta_2 \cap \Theta_3 \cap \Theta_4^c$ of the parameter space. Therefore, the 16 partitioning parameter subspaces are

$$\begin{aligned}
\Theta_{\{1,2,3,4\}} &= \{\theta_1 \leq 0 \text{ and } \theta_2 \leq 0 \text{ and } \theta_3 \leq 0 \text{ and } \theta_4 \leq 0\} \\
\Theta_{\{1,2,3\}} &= \{\theta_1 \leq 0 \text{ and } \theta_2 \leq 0 \text{ and } \theta_3 \leq 0 \text{ and } \theta_4 > 0\} \\
&\cdots \\
\Theta_{\{1,2\}} &= \{\theta_1 \leq 0 \text{ and } \theta_2 \leq 0 \text{ and } \theta_3 > 0 \text{ and } \theta_4 > 0\} \\
\Theta_{\{1,3\}} &= \{\theta_1 \leq 0 \text{ and } \theta_2 > 0 \text{ and } \theta_3 \leq 0 \text{ and } \theta_4 > 0\} \\
&\cdots \\
\Theta_{\{3\}} &= \{\theta_1 > 0 \text{ and } \theta_2 > 0 \text{ and } \theta_3 \leq 0 \text{ and } \theta_4 > 0\} \\
\Theta_{\{4\}} &= \{\theta_1 > 0 \text{ and } \theta_2 > 0 \text{ and } \theta_3 > 0 \text{ and } \theta_4 \leq 0\} \\
\Theta_{\emptyset} &= \{\theta_1 > 0 \text{ and } \theta_2 > 0 \text{ and } \theta_3 > 0 \text{ and } \theta_4 > 0\}
\end{aligned}$$

with the corresponding partitioning null hypotheses being

$$\begin{array}{rl}
H^{\star}_{\{1,2,3,4\}}: & \theta_1 \leq 0 \text{ and } \theta_2 \leq 0 \text{ and } \theta_3 \leq 0 \text{ and } \theta_4 \leq 0 \\
H^{\star}_{\{1,2,3\}}: & \theta_1 \leq 0 \text{ and } \theta_2 \leq 0 \text{ and } \theta_3 \leq 0 \text{ and } \theta_4 > 0 \\
 & \ldots \\
H^{\star}_{\{1,2\}}: & \theta_1 \leq 0 \text{ and } \theta_2 \leq 0 \text{ and } \theta_3 > 0 \text{ and } \theta_4 > 0 \\
H^{\star}_{\{1,3\}}: & \theta_1 \leq 0 \text{ and } \theta_2 > 0 \text{ and } \theta_3 \leq 0 \text{ and } \theta_4 > 0 \\
 & \ldots \\
H^{\star}_{\{3\}}: & \theta_1 > 0 \text{ and } \theta_2 > 0 \text{ and } \theta_3 \leq 0 \text{ and } \theta_4 > 0 \\
H^{\star}_{\{4\}}: & \theta_1 > 0 \text{ and } \theta_2 > 0 \text{ and } \theta_3 > 0 \text{ and } \theta_4 \leq 0 \\
H^{\star}_{\emptyset}: & \theta_1 > 0 \text{ and } \theta_2 > 0 \text{ and } \theta_3 > 0 \text{ and } \theta_4 > 0
\end{array}$$

In general, for each i, partition testing would infer $\theta_i > 0$ if and only if all $H^{\star}_{0I}$ with $i \in I$ are rejected, because H_{0i} is the union of $H^{\star}_{0I}$ with $i \in I$. To infer Dose i to be better than the placebo is to rule out the possibility that $\mu_i - \mu_0 \leq 0$, which means $\boldsymbol{\mu}$ does not belong to any of the partition null hypotheses that contain $\mu_i - \mu_0 \leq 0$, which, in turn, means all rows that contain a ✓ for H_{0i} is rejected.

These 16 parameter subspaces partition the parameter space, that is, the true $\boldsymbol{\mu}$ is in exactly one of these 16 partition null hypotheses. It is impossible for Dose 2 to be better than the placebo and at the same time to be worse than the placebo, for example. Therefore, each of the partitioned null hypotheses can be tested at level-α while controlling FWER at level-α, no multiplicity adjustment is needed (even though there are 16 partitioning null hypotheses). There is multiplicity adjustment within the test for most rows in Table 4.3 though, because partitioning null hypothesis such as $H^{\star}_{\{1,2,3\}} : \boldsymbol{\mu} \in \Theta_1 \cap \Theta_2 \cap \Theta_3 \cap \Theta_4^{\mathbf{c}}$ implies the three null hypotheses $H_{0i} : \theta_i \leq 0,\ i = 1, \ldots, 3$ are simultaneously true. However, for this particular partitioning null hypothesis, the extent to which multiplicity needs to be adjusted is three, not four. For testing $H^{\star}_{\{2,3\}}$, multiplicity adjustment is two, not four, for example. So compared to Dunnett's (1955) single-step method, which in essence adjusts for a multiplicity of four for all $H^{\star}_{I}$, partition testing potentially reduces multiplicity adjustment.

How to test any null hypothesis is not unique (one can flip a coin, for example, but that would be silly). Regardless of how each partitioning null hypothesis is tested, we can invoke Theorem 4.1 to obtain a corresponding confidence set.

Weak partition testing makes use the fact that a level-α test for the *intersection* null hypothesis

$$H_{\{1,2,3\}} : \boldsymbol{\mu} \in \Theta_1 \cap \Theta_2 \cap \Theta_3 \tag{4.16}$$

(which does not specify whether $\boldsymbol{\mu} \in \Theta_4$ or not) actually is also a level-α test for the *partitioning* null hypothesis

$$H^{\star}_{\{1,2,3\}} : \boldsymbol{\mu} \in \Theta_1 \cap \Theta_2 \cap \Theta_3 \cap \Theta_4^{\mathbf{c}} \tag{4.17}$$

for example. This is because a level-α test for (4.16) would not reject with a probability greater than α when $\boldsymbol{\mu} \in \Theta_1 \cap \Theta_2 \cap \Theta_3$ regardless of the value of μ_4, so it would not reject with a probability greater than α when $\boldsymbol{\mu}$ is in the subset $\boldsymbol{\mu} \in \Theta_1 \cap \Theta_2 \cap \Theta_3 \cap \Theta_4^{\mathbf{c}}$ with $\mu_4 > \mu_0$ in particular. Weak partition testing tests each of the 15 partitioning null hypotheses such as (4.17) at level-α by testing its corresponding intersection null hypothesis (4.16) at level-α.

Still, tests for the intersection null hypotheses are not unique. They could be F-tests, or max-T/min-P tests, for example. Technically, it would not be wrong to use an F-test to test $H_{\{1,2,3,4\}}$ and use max-T/min-P tests for the remaining intersection null hypotheses, for example, *so long as all 15 tests are executed without taking shortcuts.* What has caused confusion was that two legacy multiple tests, Holm's step-down method and Hochberg's step-up method, appear to execute only k tests based on the ordered p-values. In reality, those k tests are shortcuts to all $2^k - 1$ tests (see Huang and Hsu 2007). Without that realization, there were some incorrect shortcutting early on (see Chapters 3 and 4 of Hsu 1996). Therefore, we will use the analysis of the Alzheimer study to illustrate when and how to take legitimate shortcuts in executing a partition test.

Holm's step-down method adjusts for multiplicity within the test for each partitioning null hypothesis $H_I^\star$ by the Bonferroni inequality, while Hochberg's step-up method adjusts for multiplicity using a conservative modification of Simes' equality (see Huang and Hsu 2007). Neither method takes the correlations among the test statistics into account. Under model (4.14) though, joint distribution of the test statistics is readily computable. We thus illustrate partition testing using Dunnett's method to test each $H_I^\star$, to take the joint distribution of the test statistics into account. See Chapter 3 on multivariate methods in this Handbook for a comprehensive discussion of multiple tests that take joint distribution into account.

Conditions for taking shortcuts

To take step-down shortcuts, pretty much the form of the test for each partitioning null hypothesis $H_I^\star$ needs to be a maximum T (maxT) or, equivalent in form, a minimum p-value (minP) test. F-tests, which are based on sums of squares, do not allow shortcuts.

Shortcut condition 1 For the individual null hypothesis $H_{0i}^\star$ that has the largest test statistic value or, equivalent in form, the smallest p-value in testing a partitioning null hypothesis $H_I^\star$ with $I \ni i$, it remains the null hypothesis having the largest test statistic and the smallest p-value in testing any other partitioning null hypothesis $H_J^\star$ with a smaller set $J, J \subset I$;

Shortcut condition 2 For this individual null hypothesis $H_{0i}^\star$, its adjusted p-value in testing $H_J^\star$ with any $J \subset I$ is no larger than its adjusted p-value in testing $H_I^\star$.

Even with a maxT/minP test for each $H_I^\star$, there is subtlety involved in executing a multiple test to meet the shortcut conditions, which we will illustrate in executing a step-down version of Dunnett's method. (See Chapter 1 of this Handbook for a description of what are called step-wise methods.)

4.3.3 Confident Decision-Making Based on Step-Down Dunnett's Method

Even with the null hypotheses being one-sided (4.15), that is, with the intention being to infer which doses are better than the placebo, current practice is still to execute the testing as two-sided.

There is a perception that controlling the FWER of two-sided testing of equality nulls at level-α controls the FWER of testing one-sided nulls at level-$\alpha/2$. This perception is slightly wrong if the equality nulls are tested by a confidence intervals method, but can be quite wrong if the equality nulls are tested by a method based on p-values without an associated confidence set. If the one-sided method on which two-sided testing is based has an associated confidence set, then the one-sided FWER (including the directional error rate) is (to a close approximation) $\alpha/2$. However, such is not necessarily the case if two-sided testing is based on p-values for testing equalities.

The single-step Dunnett's method produces confidence intervals. In general, if we use the lower confidence bounds of $100(1-\alpha)\%$ simultaneous two-sided confidence intervals to

TABLE 4.4
Adjusted Two-Sided $|t|$ $p-$Values Facilitating Execution of Step-Down Dunnett's Method for the Alzheimer Study, to Be Compared with 0.10 for One-Sided FWER $\approx 5\%$

Dose	Adjusted for $\{1,2,3,4\}$	Adjusted for $\{1,2,3\}$	Adjusted for $\{1,3\}$	For $\{1\}$
4	0.0061	-	-	-
3	0.2076	0.1673	0.1217	-
2	0.0991	0.0783	-	-
1	0.6911	0.6021	0.4802	0.2949

test one-sided nulls (in any particular $H^{\star}_{0I}$), the one-sided Type I error rate (for testing that $H^{\star}_{0I}$) is close to but not exactly $100(\alpha/2)\%$, as can be seen by the fact that the two-sided $|t|$ adjusted $p-$values are not exactly twice the one-sided t adjusted p-values in Table 4.2. With equal-tailed confidence intervals and $\pm$ symmetry in the joint distribution, it actually can be shown that this practice is slightly liberal. However, our experience has been that liberalism is mostly slight, of not a big concern. Therefore, to reflect current practice, we test the one-sided null hypotheses (4.15) using the appropriate "side" of a method which has an associated confidence set.

Based on the PP, Chapter 3 of Hsu (1996) derived simultaneous confidence bounds for a step-down version of one-sided Dunnett's method, using the Partition Projection corollary 4.1.[3] Therefore, using the lower bounds of 90% two-sided Dunnet simultaneous confidence intervals to test each H_{0I}, the one-sided FWER including the directional error rate for testing (4.15) will be (approximately) 5%. However, in Section 4.3.5, we will explain the danger of using a method based on p-values for tests of equality nulls without an associated confidence set.

We also note (in passing) that, if one were to view the execution of step-down Dunnett's method controlling FWER for testing the equality nulls (4.12) at level-α as intended for two-sided inference, then it would be nontrivial to prove that the two-sided directional error rate is controlled at level-α (i.e., it would be nontrivial to prove the FWER for testing the *paired* one-sided nulls (4.13) is controlled at level-α). The reason for that is such proofs typically assume there is balance in the design (such as equal sample sizes) or even independence among the test statistics. Neither is true is in our real life Alzheimer study example.

Table 4.4 facilitates the execution of partition testing using the two-sided Dunnett's method for testing each partitioning $H^{\star}_{I}$. To be clear, our intended inferences are one-sided so the problem is formulated as testing the one-sided null hypotheses (4.15). We use the lower confidence bounds of 90% two-sided confidence intervals to test each $H^{\star}_{0I}$ at (approximately) 5% so that the FWER of testing the one-sided nulls (4.15) is approximately 5%.

The trade-off between the single-step Dunnett's method and its step-down version is that, while the step-down version potentially can infer more doses to be better than the placebo (the negative control), it gives up the ability to give strictly positive lower bounds. That is, the step-down version infers Dose i to be better than the placebo by giving the inference $\mu_i - \mu_0 > 0$. Instead of going through details of the derivation, we give an intuitive explanation of why this is so.

In accordance with Theorem 4.1, to provide positive lower confidence bounds for $\mu_i - \mu_0$, one has to test for possible positive values of $\mu_i - \mu_0$, such as $\mu_i - \mu_0 = 0.01, 0.02, \ldots$. Dunnett's method does that, testing each parameter configuration ($\mu_i - \mu_0 = \mu^{\star}_i - \mu^{\star}_0, i = 1, \ldots, k$), for all possible values of $\mu^{\star}_i - \mu^{\star}_0$ positive and negative, and then applying the pivoting Theorem 4.1 to get the lower bounds.

[3]Technically, the derivation in Hsu (1996) was for a balanced one-way design, but the idea generalizes.

Both closed testing and partition testing can potentially infer more doses to be better than the placebo than Dunnett's method. (See the introductory Chapter 1 for a description of closed testing.) How they do that is by *not* testing for possible positive values of $(\mu_i - \mu_0, i \notin J)$, in parts of the parameter space where $(\mu_i - \mu_0, i \notin J)$ are positive (thus reducing multiplicity adjustment in testing $H^{\star}_J$ comparing to testing $H^{\star}_{\{1,\ldots,k\}}$). To wit, $(\mu_i - \mu_0, i \notin J) > 0$ in $H^{\star}_J$ with $J \subset I$, so closed testing and (weak) partition testing do not bother testing for $(\mu_i - \mu_0, i \notin J)$ in testing $H^{\star}_J$, thus reducing multiplicity adjustment but giving up the ability to provide positive lower bounds for them. However, the partitioning version of the step-down one-sided Dunnett's method is a confidence set method, so it controls the directional error rate. That is the important point.

4.3.4 Executing Step-Down Dunnett's Method for the Alzheimer Study

The subtlety in execution alluded to after stating the shortcut conditions is that tests for all $H^{\star}_I, I \subseteq \{1,\ldots,k\}$, should be executed by fitting the entire data to the model. The reason for this is if, instead, testing for $H^{\star}_J$ is done by fitting only data involved in $H^{\star}_J$, then even with point estimates for treatment effects remaining the same, estimates for σ^2 would differ for different J, and shortcut condition 2 can be violated.

Suppose, for example, $\hat{\sigma}^2$ for $H^{\star}_{\{1,2,3\}}$ is computed based on Doses 1 and 2 and 3 data, while $\hat{\sigma}^2$ for $H^{\star}_{\{1,2\}}$ is computed based on Doses 1 and 2 data. Then the two $\hat{\sigma}^2$ would differ in value and in degrees of freedom, and p-value of H_{02} adjusted for $H^{\star}_{\{1,2\}}$ may be larger than the p-value of H_{02} adjusted for $H^{\star}_{\{1,2,3\}}$, so shortcut condition 2 may not hold.

We thus fit the entire data set to the model (4.14) throughout our demonstration of how using Dunnett's method to test each $H^{\star}_I$ has some shortcuts, facilitated by the adjusted p-values displayed in Table 4.4, which are to be compared with 0.10 for one-sided FWER $\approx 5\%$.

Note that the SAS codes in Program 12.9 of Dmitrienko et al. (2007) and the codes in Program 14.5 of Westfall et al. (2011) are meant for studies with balanced designs only. If one so desires, one can follow the concept demonstrated below to write his/her own codes for studies that are not perfectly balanced and/or have covariates, fitting the entire data set to a model, and specifying contrasts for each $H^{\star}_I$ that needs to be tested.

Step 1 Dose 4 has the smallest p-value. Its p-value adjusted for testing $H^{\star}_{\{1,2,3,4\}}$ is 0.0061, so $H^{\star}_{\{1,2,3,4\}}$ is rejected at the two-sided $\alpha = .10$ level. The adjusted p-value for Dose 4 in testing any $H^{\star}_J$ with $J \subset \{1,2,3,4\}$ would be smaller than 0.0030, so all $H^{\star}_J$ with $J \subset \{1,2,3,4\}$ would be rejected as well. We thus know all eight of the partitioning null hypotheses $H^{\star}_I$ with $4 \in I$ are rejected. Therefore, we can infer Dose 4 (the active control) to be better than the placebo, with a confidence bound of $\mu_4 - \mu_0 > 0$.

Step 2 Dose 2 has the second smallest p-value. Its p-value adjusted for testing $H^{\star}_{\{1,2,3\}}$ is 0.0783, so $H^{\star}_{\{1,2,3\}}$ is rejected at the two-sided $\alpha = .10$ level. The adjusted p-value for Dose 2 in testing any $H^{\star}_J$ with $J \subset \{1,2,3\}$ would be smaller than 0.0783, so all $H^{\star}_J$ with $J \subset \{1,2,3\}$ would be rejected as well. We know from earlier that all $H^{\star}_I$ with $4 \in I$ are rejected, and we now know that among the remaining $H^{\star}_I$, those with $2 \in I$ are rejected. Therefore, we can infer Dose 2 to be better than the placebo, with a confidence bound of $\mu_2 - \mu_0 > 0$.

Step 3 Dose 3 has the third smallest p-value. Its p-value adjusted for testing $H^{\star}_{\{1,3\}}$ is 0.1217, so $H^{\star}_{\{1,3\}}$ fails to be rejected at the two-sided $\alpha = .10$ level. Thus, we are unable to infer Dose 3 to be better than the placebo.[4] At this point, we might as well stop, not bother testing $H^{\star}_{\{1\}}$ or $H^{\star}_{\{3\}}$, because even if either is rejected, we cannot infer either dose to be better than the placebo because $H^{\star}_{\{1,3\}}$ fails to be rejected.

Therefore, for the Alzheimer study, one-sided Dunnett's method and its step-down version come to the same inference, both inferring Doses 4 and 2 to be better than the placebo. The single-step Dunnett's method provides more information, in giving positive lower bounds of 1.806 and 0.006 for $\mu_4 - \mu_0$ and $\mu_2 - \mu_0$ (instead of the lower bounds of zero by its step-down version). However, one can see that there is the possibility that the Dose 3 p-value adjusted for $H_{\{1,3\}}$ (instead of adjusting for $H_{\{1,2,3,4\}}$ by the single-step Dunnett's method) can potentially be small enough ($< .10$ instead of being 0.1217) to allow the step-down version to infer Dose 3 to be better than the placebo, had the data turned out a bit differently. Such is the trade-off between single-step and step-down, the potential of more doses inferred to be better than the placebo versus an inability to give strictly positive lower confidence bounds.

4.3.5 Testing Equality Null Hypotheses May Not Control the Directional Error Rate

Besides reminding readers that multiple tests that have associated confidence sets automatically control the directional error rate, below we describe some real-life situations in which (not confidence set-based) multiple tests that control the FWER of testing equality nulls may *not* control the directional error rate.

An earlier realization of this danger was documented in Hsu and Berger (1999). In the setting of dose-response studies, a Type I incorrect decision is erroneously inferring a minimum effective dose (MED) that is lower than the true MED. Hsu and Berger (1999) showed that most of the so-called contrasts tests (that were popular then) that technically control the FWER of testing equality null hypotheses have inflated Type I incorrect decision rates.

Let us say we test for the efficacy of Rx versus C, and there is a biomarker dividing the patients into a g^+ and a g^- subgroup. We are interested in answering the questions

Q^+ : Is efficacy η_{g^+} in the g^+ subgroup > 0?

Q^- : Is efficacy η_{g^-} in the g^- subgroup > 0?

$Q^{\pm}$: Is efficacy $\eta_{g^{\pm}}$ in the overall population $\{g^+, g^-\} > 0$?

If one formulates these questions properly as testing the three one-sided null hypotheses

$$\begin{aligned} H^+_{\leq} : \eta_{g^+} \leq 0 \quad vs. \quad K^+_{>} : \eta_{g^+} > 0 \\ H^-_{\leq} : \eta_{g^-} \leq 0 \quad vs. \quad K^-_{>} : \eta_{g^-} > 0 \\ H^{\pm}_{\leq} : \eta_{g^{\pm}} \leq 0 \quad vs. \quad K^{\pm}_{>} : \eta_{g^{\pm}} > 0 \end{aligned} \tag{4.18}$$

then it is possible that

3 true All three nulls are true;

2 true Two out of three are true (e.g., $H^-_{\leq}$ and $H^{\pm}_{\leq}$)

1 true Only one of the nulls is true (e.g. $H^-_{\leq}$)

0 true None is true.

[4]Technically, one could compute a lower confidence bound for $\mu_3 - \mu_0$ by projection, but it would be negative (< 0), so is not reported here.

For example, it is certainly possible that Rx is better than C by an amount δ (> 0) in g^+ but worse in g^- by an amount more than δ, in which case not all three null hypotheses are true ($H^+_\leq$ is false) but two out of the three nulls ($H^-_\leq$ and $H^\pm_\leq$) are true. Therefore, multiplicity adjustment in step-down testing would go from three for [3 true] to two for [2 true] to one for [1 true].

On the other hand, if one formulated these questions as testing three equality null hypotheses,

$$\begin{aligned} H^+_= &: \eta_{g^+} = 0 \\ H^-_= &: \eta_{g^-} = 0 \\ H^\pm_= &: \eta_{g^\pm} = 0 \end{aligned} \tag{4.19}$$

then any two of the null hypotheses being true implies the third is true[5]. Therefore, *if it is not the case that all three null hypotheses are true, then at most one of the null hypothesis is true.* Therefore, if a test for all three equality null hypotheses (4.19) being true is rejected, then one can go straight to testing the individual nulls with no multiplicity adjustment. In other words, multiplicity adjustment in testing would decrease from three directly to one, too drastic a jump to control the directional error rate. A calculation in Han et al. (2020) shows a multiple test which controls the FWER of testing the two-sided null (4.19) at 10% has a one-sided (directional) error rate at least 6.4%.

What Shaffer (1980) and Finner (1999) showed was it is nontrivial to prove a multiple test controlling the FWER of testing the equality nulls (4.12) at α actually controls the directional error rate testing the *pairs* of one-sided nulls (4.13). What we are cautioning here is a counterexample exists that a two-sided multiple test controlling the FWER for testing the equality nulls (4.12) at α does not control the directional error rate testing the one-sided nulls (4.15) at level-$\alpha/2$.

It may also be useful to mention how perspective on taking advantage of logical relationships among equality nulls has evolved. A simple example of such logical relationships similar to but different from our example above is in traditional all-pairwise comparisons. If the comparison of three means μ_1, μ_2, μ_3 is formulated as testing the three equality nulls:

$$\begin{aligned} H_{12} &: \mu_1 = \mu_2 \\ H_{23} &: \mu_2 = \mu_3 \\ H_{31} &: \mu_3 = \mu_1 \end{aligned}$$

then clearly any two nulls being true implies the third is true. Shaffer (1986) proposed to take advantage of such relationships to reduce multiplicity adjustment. Westfall (1997) and Westfall and Tobias (2007) followed up with computer algorithms for implementation, as `TYPE=LOGICAL` in the `STEPDOWN` option of the `LSMEANS` and `MSMESTIMATE` statements of SAS. However, further follow-up by Westfall et al. (2013) revealed that making use of logical relationships *in testing equality nulls* may cause the directional error rate to not be controlled. Therefore, the perspective has turned from being positive toward the negative. For two-sided inference when there are logical relationships among the parameters, a safe approach is to use confidence set methods such as Tukey's (1953) method for all-pairwise comparisons and the methods in Ding et al. (2016) and Lin et al. (2019) for inference on efficacy in subgroups and their mixtures (methods which are described in Chapter 13).

Finally, we point out that an extreme example of a test that is not capable of controlling the directional error rate is the log-rank test used in survival analysis everyday. It can be

[5]provided efficacy measure is logic-respecting as defined in Section 4 of Chapter 13 on Subgroups Analysis in this Handbook, so that $\eta_{g^\pm}$ is a weighted average of η_{g^+} and η_{g^-} say

thought of as testing infinitely many equality nulls (4.12) between Rx and C, that the survival probabilities are exactly equal at all time points or, equivalently, that the expected survival times are exactly equal for all quantiles. In multiple comparisons, such a null hypothesis is called a *Complete* null, where *all* the null hypotheses are true.[6] Controlling the Type I error rate of testing a *complete* null is termed *weak* control, which may be insufficient to control the incorrect decision rate. Section 8 of Chapter 13 on Subgroups Analysis in this Handbook contains a realistic example showing that a level-5% log-Rank test can have an incorrect (directional) decision rate exceeding 15%, so the danger of the log-rank test testing a very restrictive equality null can harm patients is real. Instead of reporting p-values based on the log-rank test while reporting confidence interval based on the Wald test, we suggest reporting confidence sets (which can of course be used to test hypotheses) or employ tests with compatible confidence sets.

4.4 Partition to Follow Decision Paths

In therapeutic areas such as diabetes and hypertension, higher doses generally give larger effects.[7] However, in psychiatric areas like schizophrenia, true response to increasing dose of a drug, as measured by reduction in Positive and Negative Syndrome Scale (PANSS) for example, may first increase then decrease. See Arvanitis et al. (1997) for an example.

In our Alzheimer study example, Dose 2 (medium dose) seems to be a bit more effective than Dose 3 (high dose). Whether that is real, or due to variability in a finite sample, is hard to tell.

In any case, whether the thinking is higher doses correspond to larger effects, or it is awkward to state there is evidence of efficacy at the medium dose but not at the high dose, it is not uncommon for the analysis plan of a clinical study to have a *decision path*, testing for efficacy from high dose to low dose. Testing for efficacy proceeds along the path, stopping as soon as efficacy fails to be established.

How this differs from testing without a path is, testing along a *single* path, FWER is controlled without multiplicity adjustment. A common misconception is this validity depends on an assumption that the true response is monotonically nondecreasing as dose increases. It is valid without any assumption on the response curve. The simplest proof of this validity is to "ask the questions," by partitioning.

4.4.1 The Decision Path Principle: Asking the Right Questions

A sequence of potential inferences is a *decision path*.

Decision Path Principle: Null hypotheses should be formulated so that decision-making naturally follows decision paths.

Implicitly used in Hsu and Berger (1999), this principle was explicitly stated in Liu and Hsu (2009). Applying this principle changes how the null hypotheses are formulated, by *asking the right questions*.

Suppose dose i is considered effective if $\mu_i > \mu_0 + \Delta$. To logically infer dose k is effective by the rejection of a null hypothesis, the null hypothesis tested has to be

[6]The *complete* null is also called the *global* null. See Chapter 1 of this Handbook.

[7]But too high a dose can be dangerous. For diabetic patients, injecting too much insulin can cause blood sugar level to drop too low and result in hypoglycemia. Too many diuretics for treating hypertension may cause the blood pressure to be too low resulting in syncope (fainting).

H_{0k} : Dose k is ineffective. To logically infer doses k and $k-1$ are effective by the rejection of the null hypotheses H_{0k} and $H_{0(k-1)}$, the union of the null hypotheses H_{0k} and $H_{0(k-1)}$ needs to include the possibilities dose k is ineffective and/or dose $k-1$ is ineffective.

Consider testing the null hypotheses

$$H_{0k}^{\downarrow} : \text{Dose } k \text{ is ineffective}$$
$$H_{0(k-1)}^{\downarrow} : \text{Dose } k \text{ is effective but dose } k-1 \text{ is ineffective}$$
$$\vdots$$
$$H_{0i}^{\downarrow} : \text{Doses } i+1, \ldots, k \text{ are effective but dose } i \text{ is ineffective}$$
$$\vdots$$
$$H_{01}^{\downarrow} : \text{Doses } 2, \ldots, k \text{ are effective but dose 1 is ineffective}$$

Statistically, the null hypotheses are as follows:

$$\begin{aligned} H_{0k}^{\downarrow} &: \mu_k \leq \mu_0 + \Delta \\ H_{0(k-1)}^{\downarrow} &: \mu_{k-1} \leq \mu_0 + \Delta < \mu_k \\ &\vdots \\ H_{0i}^{\downarrow} &: \mu_i \leq \mu_0 + \Delta < \min\{\mu_{i+1}, \ldots, \mu_k\} \\ &\vdots \\ H_{01}^{\downarrow} &: \mu_1 \leq \mu_0 + \Delta < \min\{\mu_2, \ldots, \mu_k\} \end{aligned} \tag{4.20}$$

Together with

$$H_{00}^{\downarrow} : \mu_0 + \Delta < \min\{\mu_1, \ldots, \mu_k\} \tag{4.21}$$

the null hypotheses (4.20) partition the parameter space, so no multiplicity adjustment is needed in testing them.

For any integer i, if $H_{0j}^{\downarrow}$, $j = i, \ldots, k$, are all rejected, then the logical inference is doses $i, \ldots, k$ are all efficacious: $\mu_j > \mu_1 + \Delta$, $j = i, \ldots, k$.

For example, suppose

$$H_{04}^{\downarrow} : \text{Dose 4 is not efficacious}$$

is rejected. Then, one can infer Dose 4 is efficacious.

Suppose

$$H_{04}^{\downarrow} : \text{Dose 4 is not efficacious}$$

and

$$H_{03}^{\downarrow} : \text{Dose 4 is efficacious but Dose 3 is not efficacious}$$

are both rejected, then since the union of $H_{04}^{\downarrow}$ and $H_{03}^{\downarrow}$ is "either Dose 4 or Dose 3 is not efficacious," the rejection of $H_{04}^{\downarrow}$ and $H_{03}^{\downarrow}$ implies "both Doses 4 and 3 are efficacious."

On the other hand, if

$$H_{04}^{\downarrow} : \text{Dose 4 is not efficacious}$$

is rejected,

$$H_{03}^{\downarrow} : \text{Dose 4 is efficacious but Dose 3 is not efficacious}$$

fails to be rejected, but

$$H_{02}^{\downarrow} : \text{Doses 4 and 3 are efficacious but Dose 2 is not efficacious}$$

is rejected, then still the only useful inference remains Dose 4 is efficacious. Therefore, one might as well stop testing when $H_{03}^{\downarrow}$ fails to be rejected. By asking the right questions (4.20), partition testing automatically follows the decision path.

Level-α tests for each $H_{0i}^{\downarrow}$, $i = 1, \ldots, k$, are of course not unique. Note, however, a level-α test for

$$H_{0i} : \mu_i \leq \mu_0 + \Delta \tag{4.22}$$

is also a level-α test for

$$H_{0i}^{\downarrow} : \mu_i \leq \mu_0 + \Delta < \min\{\mu_{i+1}, \ldots, \mu_k\}$$

For example, a test that rejects no more than 5% of the time when Dose 3 is ineffective, regardless of whether Dose 4 is effective, will reject no more than 5% of the time in particular when Dose 3 is ineffective and Dose 4 is effective. Therefore, the simplest level-α test for $H_{0i}^{\downarrow}$ is to use a one-sided two-sample size-α t-test comparing μ_i with μ_0 for each $H_{0i}^{\downarrow}$.

With this choice of test for $H_{0i}^{\downarrow}$, $i = 1, \ldots, k$, since the null hypotheses partition the parameter space, Hsu and Berger (1999) could apply the Partition-Project corollary 4.1 to give the confidence bounds version of the inference:

Step 1

If $\hat{\mu}_k - \hat{\mu}_0 - t_{\alpha,\nu}\hat{\sigma}\sqrt{1/n_k + 1/n_0} > \Delta$,
then infer $\mu_k - \mu_0 > \Delta$ and go to Step 2;
else infer $\mu_k - \mu_0 > \hat{\mu}_k - \hat{\mu}_0 - t_{\alpha,\nu}\hat{\sigma}\sqrt{1/n_k + 1/n_0}$ and stop.

Step 2

If $\hat{\mu}_{k-1} - \hat{\mu}_0 - t_{\alpha,\nu}\hat{\sigma}\sqrt{1/n_{k-1} + 1/n_0} > \Delta$,
then infer $\mu_{k-1} - \mu_0 > \Delta$ and go to Step 3;
else infer $\mu_{k-1} - \mu_0 > \hat{\mu}_{k-1} - \hat{\mu}_0 - t_{\alpha,\nu}\hat{\sigma}\sqrt{1/n_{k-1} + 1/n_0}$ and stop.

$\vdots$

Step k

If $\hat{\mu}_1 - \hat{\mu}_0 - t_{\alpha,\nu}\hat{\sigma}\sqrt{1/n_1 + 1/n_0} > \Delta$
then infer $\mu_1 - \mu_0 > \Delta$ and go to Step $k+1$;
else infer $\mu_1 - \mu_0 > \hat{\mu}_1 - \hat{\mu}_0 - t_{\alpha,\nu}\hat{\sigma}\sqrt{1/n_1 + 1/n_0}$ and stop.

Step $k+1$

Infer $\min_{i=1,\ldots,k} \mu_i - \mu_0 > \min_{i=1,\ldots,k}\{\hat{\mu}_i - \hat{\mu}_0 - t_{\alpha,\nu}\hat{\sigma}\sqrt{1/n_i + 1/n_0}\}$ and stop.

Note the Step $k+1$ confidence bound is from pivoting an Intersection-Union Tests (IUT) for (4.21).

Closed duplicate testing to stay on a decision path

Whether there are decision paths or not, closed testing would test all intersections of the null hypotheses in (4.15). To stay on a (single) decision path, what closed testing does (including the graphical approach) is to test all the intersection null hypotheses that make up each of $H_{0i}^{\downarrow}$ by one and the same pair-wise t test, rejecting if $\hat{\mu}_i - \hat{\mu}_0 - t_{\alpha,\nu}\hat{\sigma}\sqrt{1/n_i + 1/n_0} > \Delta$. (Chapter 5 of this Handbook is on the graphical approach.) In this scheme, testing $H_{04}^{\downarrow}$ corresponds to testing all eight rows with ⊚ for Θ_4 in Table 4.3 by the same test which rejects when $\hat{\mu}_4 - \hat{\mu}_0 - t_{\alpha,\nu}\hat{\sigma}\sqrt{1/n_4 + 1/n_0} > \Delta$. Testing $H_{03}^{\downarrow}$ corresponds to testing the four rows with ⊚ for Θ_3 but ⊗ for Θ_4 in Table 4.3 by the same test which rejects when $\hat{\mu}_3 - \hat{\mu}_0 - t_{\alpha,\nu}\hat{\sigma}\sqrt{1/n_3 + 1/n_0} > \Delta$, and so forth. Bauer et al. (1998) explains this (redundant) closed testing scheme as:

> "Now for strictly ordered null hypotheses, every level α-test can be formally considered as a level α-test for all intersections with null hypotheses at a lower hierarchical order."

4.4.2 Making Decisions along a Path for the Alzheimer Study

We take $\Delta = 0$ and fit the entire data to the model. Unlike the multivariate case, in the univariate case, comparing two-sided $|t|$ p-values to 10% corresponds exactly to comparing one-sided t p-values to 5%. So, using the two-sided $|t|$ p-values in Table 4.5 which are computed without multiplicity adjustment, we have

Step 1

Is $\hat{\mu}_4 - \hat{\mu}_0 - t_{.05,\nu}\hat{\sigma}\sqrt{1/n_4 + 1/n_0} > 0$?
Yes since the two-sided $|t|$ p-value for Dose 4 $= 0.0016$
So infer $\mu_4 - \mu_0 > 0$ and go to Step 2;

Step 2

Is $\hat{\mu}_3 - \hat{\mu}_0 - t_{.05,\nu}\hat{\sigma}\sqrt{1/n_3 + 1/n_0} > 0$?
No since the two-sided $|t|$ p-value for Dose 3 $= 0.0665$
So stop.

Therefore, for the particular case of this Alzheimer study, at the one-sided 5% level, making decisions along the Dose 4 → 3 → 2 → 1 path infers only Dose 4 to be better than the placebo, while the single-step and the step-down Dunnett's method infer Doses 4 and 2 to be better than the control.

TABLE 4.5
Unadjusted Two-Sided $|t|$ $p-$Values Facilitating Execution of Decision-Path Method for the Alzheimer Study, to Be Compared with .10 for One-Sided FWER = 5%

Dose	**$\|t\|$ p-Value for {4}**	**$\|t\|$ p-Value for {3}**	**$\|t\|$ p-Valued for {2}**	**$\|t\|$ p-Value for {1}**
4	0.0016	-	-	-
3	-	0.0665	-	-
2	-	-	0.0295	-
1	-	-	-	0.2949

The methods we have presented do not assume response has any particular form as a function of dose. Though, for the minimum effective dose (MED) problem to be meaningful, there is the tacit assumption that if a dose is efficacious, then all higher doses are efficacious as well. In the presence of a $\Delta > 0$ defining a clinically meaningful difference, this assumption is not as strong as the one that efficacy is monotonically increasing in dose (see Dmitrienko et al. 2007). In addition, see Chapter 11 on dose-finding in this Handbook for the MCP-Mod approach which models response as a continuously valued function of dose.

4.4.3 Partitioning When There Are Multiple Decision Paths

Let μ_{ij} denote the mean response in dose group i for endpoint j, $i = 0, 1, ..., k$, $j = 1, ..., m$, where $i = 0$ denotes the placebo group, while $i = 1, \ldots, k$ are additional doses. And $j = 1$ denotes the primary endpoint, $j = 2$ is the secondary endpoint, $j = 3$ is the tertiary endpoint, and so forth. Define $\theta_{ij} = \mu_{ij} - \mu_{0j}$ to be the difference in mean efficacy measurement between dose group i and the placebo group for endpoint j.

Assuming a larger measurement indicates a better treatment, the statistical inference of interest is to test, for each dose endpoint combination,

$$H_{0ij} : \theta_{ij} \leq \delta_j, \quad i = 1, ..., k, \ j - 1, ..., m. \tag{4.23}$$

The efficacy claim of the new experimental drug is based on the primary endpoint alone, additional claims on secondary endpoints are of interest only if the primary endpoint has shown efficacy.

Liu and Hsu (2009) showed how to apply the decision path principle by using partition testing to situations in which there is an ordering among the endpoints (in terms of a sequence of potential inferences), but there is no such ordering among the doses. The lower-ordered secondary endpoints, in this situation, are required to be tested only if higher-ordered ones are proven efficacious.

Decision paths are thus within, but not between, doses. Figure 4.3 illustrates such paths with k doses and m endpoints.

If a single secondary endpoint is involved, the decision path in Figure 4.3 becomes Figure 4.4 shown below. For notation conveniences, instead of using the second subscript to index the endpoint, we use superscripts ^{P} and ^{S} to denote primary and secondary endpoints.

Given decision paths in Figure 4.4, the parameter space is partitioned in two stages:

Path partition: Partition within each path.

Disjointness partition: Further partition by taking intersections to make hypotheses between paths disjoint.

Path partition is within each dose i. Starting with the primary endpoint, we test H_{0i}^P : $\theta_i^P \leq \delta^P$. If it is rejected, then efficacy in the primary endpoints has been established.

Then follow the path to the secondary endpoint. However, instead of testing $H_{0i}^S : \theta_i^S \leq \delta^S$, we make it disjoint with H_{0i}^P and test $H_{0i}^{\star S} : \theta_i^S \leq \delta^S$ and $\theta_i^P > \delta^P$. If both H_{0i}^P and $H_{0i}^{\star S}$ are rejected, then we logically conclude efficacy in both the primary and the secondary endpoints.

Whereas the reason for path partitioning is inference in the secondary endpoint is irrelevant unless efficacy in the primary endpoint is established, disjointness partitioning is for proper multiplicity adjustment.

Multiplicity adjustment is needed (only) to the extent that two or more hypotheses can be true simultaneously. Ask the question, "is it possible that high dose primary lacks efficacy and simultaneously there is efficacy in high dose primary but not in high dose secondary?" The answer is "no," so there is no need to adjust for multiplicity in testing H_{0i}^P and $H_{0i}^{\star S}$.

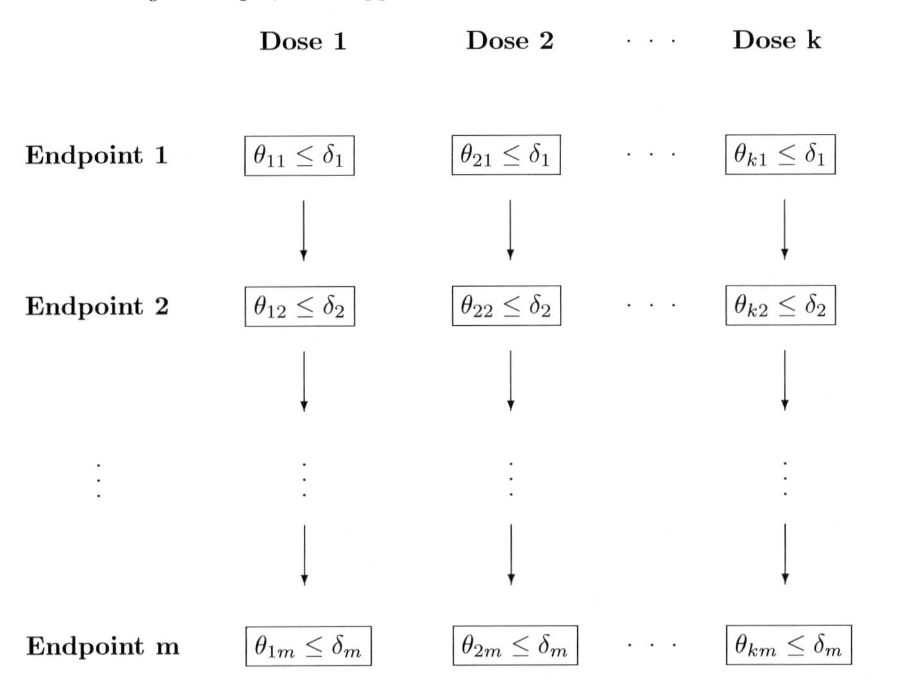

FIGURE 4.3
Decision paths for k doses m endpoints, with one path for each dose going from endpoint 1 to endpoint m.

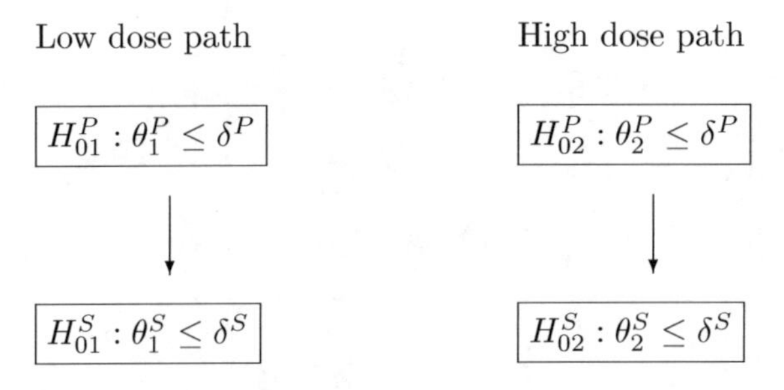

FIGURE 4.4
Decision paths for low and high doses.

However, "is it possible that high dose primary and low dose secondary lack efficacy?" "Yes it is," so that particular multiplicity of two needs to be adjusted for.

To figure out which combination of the path-partitioned null hypotheses can be true simultaneously, between paths, connect an *edge* between subspaces of the parameter space that are not disjoint, as illustrated in Figure 4.5. Edges represent hypotheses that can be true simultaneously. (There is no edge between $\{\theta_1^S > \delta^S$ and $\theta_1^P > \delta^P\}$ and $\{\theta_2^S > \delta^S$ and $\theta_2^P > \delta^P\}$, since the intersection of these two hypotheses, the ideal situation of having efficacy in all doses and endpoints, need not be tested.) Take intersections of connected subspaces to form new hypotheses. The resulting set of hypotheses, as presented in Table 4.6, partitions the parameter space. Therefore, so long as each partition hypothesis is tested at level α, the FWER is controlled strongly at level α. Inferences on which of the $m \times k$ combinations of dose and endpoint are efficacious are then obtained by collating the results from the $(m+1)^k - 1$ tests.

TABLE 4.6
Partition Hypotheses Following Decision Paths in Figure 4.4

Index	Partition Hypothesis	Rejection Rule
1	$\boldsymbol{\theta_1^P \leq \delta^P}$ and $\boldsymbol{\theta_2^P \leq \delta^P}$	$t_1^P > c_1$ or $t_2^P > c_1$
2	$\boldsymbol{\theta_1^P \leq \delta^P}$ and $\theta_2^P > \delta^P$ and $\boldsymbol{\theta_2^S \leq \delta^S}$	$t_1^P > c_1$ or $t_2^S > c_2$
	$\boldsymbol{\theta_1^P \leq \delta^P}$ and $\theta_2^P > \delta^P$ and $\theta_2^S > \delta^S$	$t_1^P > c_3$
3	$\theta_1^P > \delta^P$ and $\boldsymbol{\theta_2^P \leq \delta^P}$ and $\boldsymbol{\theta_1^S \leq \delta^S}$	$t_2^P > c_1$ or $t_1^S > c_2$
	$\theta_1^P > \delta^P$ and $\boldsymbol{\theta_2^P \leq \delta^P}$ and $\theta_1^S > \delta^S$	$t_2^P > c_3$
	$\theta_1^P > \delta^P$ and $\theta_2^P > \delta^P$ and $\boldsymbol{\theta_1^S \leq \delta^S}$ and $\boldsymbol{\theta_2^S \leq \delta^S}$	$t_1^S > c_1$ or $t_2^S > c_1$
4	$\theta_1^P > \delta^P$ and $\theta_2^P > \delta^P$ and $\boldsymbol{\theta_1^S \leq \delta^S}$ and $\theta_2^S > \delta^S$	$t_1^S > c_3$
	$\theta_1^P > \delta^P$ and $\theta_2^P > \delta^P$ and $\theta_1^S > \delta^S$ and $\boldsymbol{\theta_2^S \leq \delta^S}$	$t_2^S > c_3$

The index column corresponds to the labels of edges in Figure 4.5.

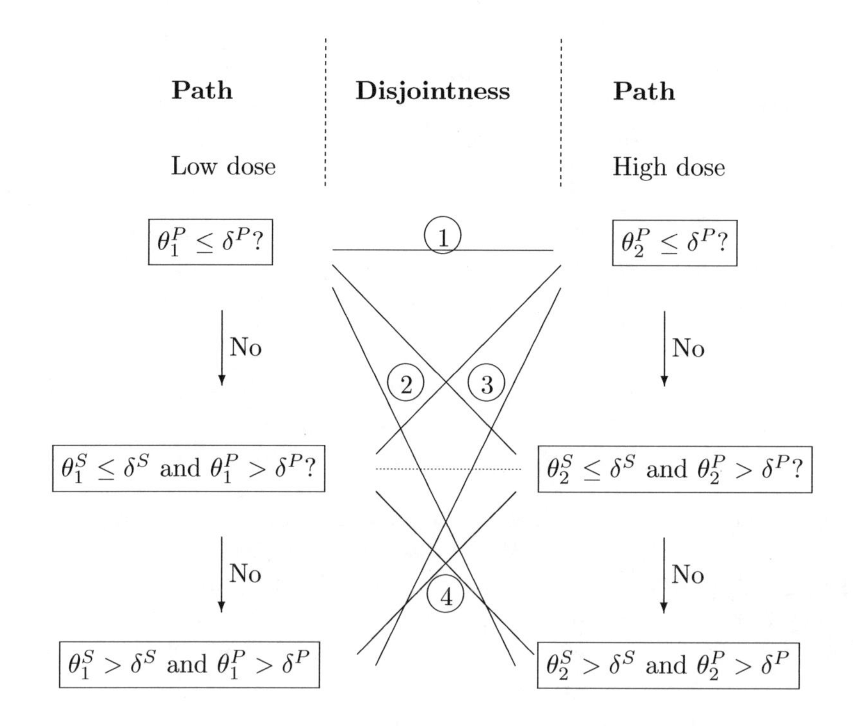

FIGURE 4.5
Graphical representation of two stages of partitioning in the setting of Figure 4.4.

Specifically, the inference $\theta_{ij} > \delta_j$ is made if all hypotheses implying that $\theta_{ij} \leq \delta_j$ could be true are rejected. This includes partitioning hypotheses which do not explicitly state an inequality for θ_{ij}. For example, the hypothesis that states $\theta_1^P \leq \delta^P$ and $\theta_2^P \leq \delta^P$ includes the possibility that $\theta_i^S \leq \delta^S$ (as well as the possibility $\theta_i^S > \delta^S$) and must be rejected before inference on any secondary endpoint is given.

4.4.3.1 Insights from the Path-Partitioning Principle

The path-partitioning principle is most useful in giving insights into the structure of multiple testing when there are multiple doses and decision paths. To execute multiple testing when there are decision paths, the graphical approach described in Chapter 5 is perhaps more convenient. We thus focus on explaining the *insights.*

Multiplicity adjustment: Inclusion of secondary endpoints in the analysis may necessitate multiplicity adjustment in inference on the primary endpoint. For example, in Table 4.6, one rejects the hypothesis H_{01}^P if

$$\{\theta_1^P \leq \delta^P \text{ and } \theta_2^P \leq \delta^P\}, \tag{4.24}$$

$$\{\theta_1^P \leq \delta^P \text{ and } \theta_2^P > \delta^P \text{ and } \theta_2^S \leq \delta^S\}, \tag{4.25}$$

$$\text{and } \{\theta_1^P \leq \delta^P \text{ and } \theta_2^P > \delta^P \text{ and } \theta_2^S > \delta^S\} \tag{4.26}$$

are all rejected. Thus, inference on a low dose of the primary endpoint (θ_1^P) may need multiplicity adjustment, to account for the possibility that high dose of the secondary endpoint may lack efficacy ($\theta_2^S \leq \delta^S$ in (4.25)). The mere presence of high-dose secondary endpoint necessitates multiplicity adjustment in testing for efficacy in low dose primary, at least initially.

Note, however, this multiplicity adjustment is removed if data shows efficacy in the secondary endpoint at high dose (i.e., if the partition hypothesis (4.25) is rejected). Partitioning makes transparent that, if efficacy at a dose has been established for all endpoints, then that dose needs no longer be included in multiplicity adjustment.

This last realization from Liu and Hsu (2009), which some have come to phrase as "there is no need to leave money on the table," explains a key difference from Figure 2 in Bretz et al. (2009), to Figure 1 in Bretz et al. (2011) and Figure 6.3(a) in Chapter 5 on Graphical Approach in this Handbook, that there is an arrow from the bottom node of each path to the top node of the other path in the latter two figures, as the graphical approach to this problem has evolved.

Appearance: Whether or not one readily sees it, inference on the primary endpoint may depend on observations on a secondary endpoint (at a different dose) because initially one must account for possibilities such as efficacy is lacking for the primary endpoint at high dose and for the secondary endpoint at low dose. However, this multiplicity can be removed if data indicates otherwise.

One can give the appearance that such dependence does not occur, by choosing not to remove the multiplicity adjustment, even if data indicates it can be. This is the approach taken in Dmitrienko et al. (2006), Xu et al. (2009), and Dmitrienko and Tamhane (2011), to ensure "inference made in primary endpoints not affected by the inference made in secondary endpoints." We feel there is no need for such loss of power, for the sake of appearance.

4.4.4 Controlling FWER May Be too Simplistic for Primary-Secondary Endpoint Problems

The original proof (in Hsu and Berger 1999) that no multiplicity adjustment is needed to control FWER in testing along a (single) path was in the setting of dose-response studies. In the setting of testing high dose first, with evidence of efficacy in high dose, one then tests low dose, there may be one or two Type I errors. If neither dose is effective, then inferring high dose is effective only commits one Type I error while inferring both high dose and low dose as effective commits two Type I errors. FWER, being the probability of making at least one Type I error, counts one or two Type I errors as the same. With the principal purpose of Hsu and Berger (1999) being to show that most of the contrast methods that pool information across doses popular then do not control incorrect decision rate, they viewed this oversimplification of FWER control as an acceptable first approximation, since those two Type I errors are of the same kind: "too low is too low."

However, the ordered endpoints setting is different. In testing primary and secondary endpoints in sequence, unconditional FWER refers to, over many studies each with a primary and a secondary endpoint, roughly how many percent of the studies have incorrect efficacy claim in either primary or secondary or both. However, since testing primary is for approval, while secondary testing is for additional labeling claim, we might consider conditional Type I error rate on testing for efficacy in the Secondary, conditional on inferring efficacy in the primary. This conditional error rate is, over many drug submissions which get approval (and therefore have drug labels), roughly how many percent of the labels have incorrect additional claims (beyond indication), and might offer useful additional information toward sound decision-making.

4.5 Key Messages of This Chapter

- In multiple comparisons, error rate controls are useful if they translate to controlling the probability of making *incorrect decisions.*
- One can be confident that the *directional* error rate is controlled if the null hypotheses of a multiple test partition the entire parameter space, but not if the null hypotheses are mere equalities.
- If the null hypotheses partition the entire parameter space, then that multiple tests can be pivoted to give a compatible confidence set.
- Using a compatible confidence set to execute multiple tests guarantees the directional error rate is controlled.
- Partitioning is also useful in formulating null hypotheses to channel multiple tests onto prespecified decision paths.

Bibliography

Abraham, J. E., M. J. Maranian, K. E. Driver, R. Platte, B. Kalmyrzaev, C. Baynes, C. Luccarini, M. Shah, S. Ingle, D. Greenberg, H. M. Earl, A. M. Dunning, P. D. Pharoah, and C. Caldas (2010). CYP2D6 gene variants: association with breast cancer specific survival in a cohort of breast cancer patients from the United Kingdom treated with adjuvant tamoxifen. *Breast Cancer Research 12*, R64.

Arvanitis, L. A., B. G. Miller, and the Seroquel Trial 13 Study Group (1997). Multiple fixed doses of "Seroquel" (Quetiapine) in patients with acute exacerbation of Schizophrenia: A comparison with Haloperidol and placebo. *Biological Psychiatry 42*, 233–246.

Bauer, P., J. Rohmel, W. Maurer, and L. Hothorn (1998). Testing strategies in multi-dose experiments including active control. *Statistics in Medicine 17*, 2133–2146.

Bechhofer, R. E. (1954). A single-sample multiple decision procedure for ranking means of normal populations with known variances. *Annals of Mathematical Statistics 25*, 16–39.

Bechhofer, R. E., T. J. Santner, and D. M. Goldsman (1995). *Design and Analysis of Experiments for Statistical Selection, Screening and Multiple Comparisons.* John Wiley & Sons, New York, NY.

Bretz, F., W. Maurer, W. Brannath, and M. Posch (2009). A graphical approach to sequentially rejective multiple test procedures. *Statistics in Medicine 28*, 586–604.

Bretz, F., M. Posch, E. Glimm, F. Klinglmueller, W. Maurer, and K. Rohmeyer (2011). Graphical approaches for multiple comparison procedures using weighted bonferroni, simes, or parametric tests. *Biometrical Journal 53*, 894–913.

Casella, G. and R. L. Berger (2001). *Statistical Inference* (2nd ed.). Thomson Learning, Pacific Grove, CA.

Ding, Y., Y. G. Li, Y. Liu, S. J. Ruberg, and J. C. Hsu (2018). Confident inference for snp effects on treatment efficacy. *Ann. Appl. Statist. 12*(3), 1727–1748.

Ding, Y., H.-M. Lin, and J. C. Hsu (2016). Subgroup mixable inference on treatment efficacy in mixture populations, with an application to time-to-event outcomes. *Statistics in Medicine 35*, 1580–1594.

Dmitrienko, A., K. Fritsch, J. Hsu, and S. Ruberg (2007). *Pharmaceutical Statistics Using SAS: A Practical Guide*, Chapter Design and Analysis of Dose-Ranging Clinical Studies, pp. 273–311. SAS Institute, Inc.

Dmitrienko, A., W. Offen, O. Wang, and D. Xiao (2006). Gatekeeping procedures in dose-response clinical trials based on the Dunnett test. *Pharmaceutical Statistics 5*, 19–28.

Dmitrienko, A. and A. C. Tamhane (2011). Mixtures of multiple testing procedures for gatekeeping applications in clinical trials. *Statistics in Medicine 30*, 1473–1488.

Edwards, D. G. and J. C. Hsu (1983). Multiple comparisons with the best treatment. *Journal of the American Statistical Association 78*, 965–971.

Fabian, V. (1962). On multiple decision methods for ranking population means. *Annals of Mathematical Statistics 33*, 248–254.

Fieller, E. C. (1954). Some problems in interval estimation. *Journal of the Royal Statistical Society. Series B (Methodological) 16*(2), 175–185.

Finner, H. (1994). Two-sided tests and one-sided confidence bounds. *Annals of Statistics 22*, 1502–1516.

Finner, H. (1999). Stepwise multiple test procedures and control of directional errors. *The Annals of Statistics 27*, 274–289.

Finner, H. and K. Strassburger (2002). The partitioning principle: a powerful tool in multiple decision theory. *Annals of Statistics 30*, 1194 –1213.

Finner, H. and K. Strassburger (2007). Step-up related simultaneous confidence intervals for MCC and MCB. *Biometrical Journal 49*(1), 40–51.

Gibbons, J. D., I. Olkin, and M. Sobel (1977). *Selecting and Ordering Populations: A New Statistical Methodology.* Wiley, New York, NY.

Gupta, S. S. (1956). On a decision rule for a problem in ranking means. Mimeo Series 150, Institute of Statistics, University of North Carolina, Chapel Hill, NC.

Gupta, S. S. (1965). On some multiple decision (selection and ranking) rules. *Technometrics 7*, 225–245.

Gupta, S. S. and S. Panchapakesan (1979). *Multiple Decision Procedures – Theory and Methodology of Selecting and Ranking Populations.* John Wiley, New York, NY.

Han, Y., S.-Y. Tang, H.-M. Lin, and J. C. Hsu (2020). Exact simultaneous confidence intervals for logical selection of a biomarker cut-point. Unpublished.

Hayter, A. J. and J. C. Hsu (1994). On the relationship between stepwise decision procedures and confidence sets. *Journal of the American Statistical Association 89*, 128–136.

Holmes, M. V., P. Perel, T. Shah, A. D. Hingorani, and J. P. Casas (2011). Cyp2c19 genotype, clopidogrelmetabolism, platelet function, and cardiovascular events: A systematic review and meta-analysis. *Journal of the American Medical Association 306*, 2704–2714.

Hoskins, J. M., L. A. Carey, and H. L. McLeod (2009). CYP2D6 and tamoxifen: DNA matters in breast cancer. *Nature Reviews: Cancer 9*, 576–586.

Hsu, J. C. (1981). Simultaneous confidence intervals for all distances from the 'best'. *Annals of Statistics 9*, 1026–1034.

Hsu, J. C. (1982). Simultaneous inference with respect to the best treatment in block designs. *Journal of the American Statistical Association 77*, 461–467.

Hsu, J. C. (1984). Constrained two-sided simultaneous confidence intervals for multiple comparisons with the best. *Annals of Statistics 12*, 1136–1144.

Hsu, J. C. and R. L. Berger (1999). Stepwise confidence intervals without multiplicity adjustment for dose response and toxicity studies. *Journal of the American Statistical Association 94*, 468–482.

Huang, Y. and J. C. Hsu (2007). Hochberg's step-up method: Cutting corners off Holm's step-down method. *Biometrika 22*, 2244–2248.

Kil, S., E. Kaizar, S.-Y. Tang, and J. C. Hsu (2020). *Principles and Practice of Clinical Trials*, Chapter Confident Statistical Inference with Multiple Outcomes, Subgroups, and Other Issues of Multiplicity. Cham: Springer International Publishing, Cham.

Lawrence, J. (2019). Familywise and per-family error rates of multiple comparison procedures. *Statistics in Medicine 38*(19), 3586–3598.

Lehmann, E. L. (1986). *Testing Statistical Hypotheses* (Second ed.). John Wiley, New York, NY.

Lin, H.-M., H. Xu, Y. Ding, and J. C. Hsu (2019). Correct and logical inference on efficacy in subgroups and their mixture for binary outcomes. *Biometrical Journal 61*, 8–26.

Liu, Y. and J. C. Hsu (2009). Testing for efficacy in primary and secondary endpoints by partitioning decision paths. *Journal of the American Statistical Association 104*, 1661–1670.

Mega, J. L., S. L. Close, S. D. Wiviott, L. Shen, J. R. Walker, T. Simon, E. M. Antman, E. Braunwald, and M. S. Sabatine (2010). Genetic variants in ABCB1 and CYP2C19 and cardiovascular outcomes after treatment with clopidogrel and prasugrel in the TRITON-TIMI 38 trial: a pharmacogenetic analysis. *The Lancet 376*, 1312–1319.

Mega, J. L., W. Hochholzer, A. L. F. III, M. J. Kluk, D. J. Angiolillo, D. J. Kereiakes, S. Isserman, W. J. Rogers, C. T. Ruff, C. Contant, M. J. Pencina, B. M. Scirica, J. A. Longtine, A. D. Michelson, and M. S. Sabatine (2011). Dosing clopidogrel based on cyp2c19 genotype and the effect on platelet reactivity in patients with stable cardiovascular disease. *Journal of the American Medical Association 306*, 2221–2228.

Nebert, D. and D. Russell (2002). Clinical importance of the cytochromes P450. *The Lancet 360*, 1155–1162.

Paré, G., S. R. Mehta, S. Yusuf, S. S. Anand, S. J. Connolly, J. Hirsh, K. Simonsen, D. L. Bhatt, K. A. Fox, and J. W. Eikelboom (2010). Effects of CYP2C19 genotype on outcomes of clopidogrel treatment. *New England Journal of Medicine 363*, 1704–1714.

Schnell, P., Q. Tang, P. Muller, and B. P. Carlin (2017). Subgroup inference for multiple treatments and multiple endpoints in an alzheimer's disease treatment trial. *Ann. Appl. Stat. 11*, 949–966.

Schroth, W. (2009). Association between CYP2D6 polymorphisms and outcomes among women with early stage breast cancer treated with tamoxifen. *Journal of the American Medical Association 302*, 1429–1436.

Shaffer, J. P. (1980). Control of directional errors with stagewise multiple test procedures. *Annals of Statistics 8*, 1342–1348.

Shaffer, J. P. (1986). Modified sequentially rejective multiple test procedures. *Journal of the American Statistical Association 81*, 826–831.

Stefansson, G., W. Kim, and J. C. Hsu (1988). On confidence sets in multiple comparisons. In S. S. Gupta and J. O. Berger (Eds.), *Statistical Decision Theory and Related Topics IV*, Volume 2, pp. 89–104. New York: Springer-Verlag.

Strassburger, K. and F. Bretz (2008). Compatible simultaneous lower confidence bounds for the Holm procedure and other Bonferroni-based closed tests. *Statistics in Medicine 27*(24), 4914–4927.

Takeuchi, K. (1973). *Studies in Some Aspects of Theoretical Foundations of Statistical Data Analysis (in Japanese).* Tokyo: Toyo Keizai Shinposha.

Takeuchi, K. (2010). Basic ideas and concepts for multiple comparison procedures. *Biometrical Journal 52*, 722–734.

Tukey, J. W. (1953). The Problem of Multiple Comparisons. Dittoed manuscript of 396 pages, Department of Statistics, Princeton University.

Tukey, J. W. (1991). The philosophy of multiple comparisons. *Statistical Science 6*, 100–116.

Tukey, J. W. (1992). Where should multiple comparisons go next? In F. M. Hoppe (Ed.), *Multiple Comparisons, Selection, and Applications in Biometry: A Festschrift in Honor of Charles W. Dunnett*, Chapter 12, pp. 187–208. New York: Marcel Dekker.

Westfall, P. H. (1997). Multiple testing of general contrasts using logical constraints and correlations. *Journal of the American Statistical Association 92*(437), 299–306.

Westfall, P. H., F. Bretz, and R. D. Tobias (2013). Directional error rates of closed testing procedures. *Statistics in Biopharmaceutical Research 5*, 345–355.

Westfall, P. H. and R. D. Tobias (2007). Multiple testing of general contrasts: Truncated closure and the extended shafferroyen method. *Journal of the American Statistical Association 102*, 487–494.

Westfall, P. H., R. D. Tobias, and R. D. Wolfinger (2011). *Multiple Comparisons and Multiple Tests Using SAS* (2nd ed.). SAS Publishing.

Xu, H., I. Nuamah, J. Liu, P. Lim, and A. Sampson (2009). A Dunnett-Bonferroni-based parallel gatekeeping procedure for dose–response clinical trials with multiple endpoints. *Pharmaceutical Statistics 8*(4), 301–316.

5

Graphical Approaches for Multiple Comparison Procedures

Dong Xi

Novartis Pharmaceuticals

Frank Bretz

Novartis AG

CONTENTS

5.1 Introduction

Regulatory guidelines (EMA, 2017; FDA, 2017) mandate the strong control of the family-wise error rate (FWER) for primary and secondary objectives in confirmatory clinical trials. That is, the probability to erroneously reject at least one true null hypothesis is controlled at a prespecified significance level $\alpha \in (0, 1)$ for any configuration of true and false null hypotheses. Many multiple comparison procedures are available, such as those by Bonferroni, Holm (1979), Hochberg (1988), Hommel (1988), and Dunnett (1991). However, confirmatory clinical trials are becoming increasingly more complex with simultaneously

DOI: 10.1201/9780429030888-6

investigating multiple doses or regimens of a new treatment, more than one endpoint, multiple populations and others. Thus, the clinical trial team needs to design a multiple comparison procedure that reflects a hierarchical structure among primary and secondary objectives, a comparison of paramount importance or a few comparisons sharing equal importance, etc. As a result, the aforementioned procedures are often not suitable for the structured hypothesis test problems coupled with different importance for different hypotheses, because they treat all hypotheses equally without addressing the underlying structures.

As clinical trials are becoming increasingly complex, a variety of multiple comparison procedures have been proposed such as fixed sequence (or hierarchical) (Maurer et al., 1995; Westfall and Krishen, 2001), fallback (Wiens, 2003; Wiens and Dmitrienko, 2005) and gatekeeping procedures (Dmitrienko et al., 2003, 2008, 2007), see Alosh et al. (2014) for an overview of recent advancement. In this chapter, we focus on the graphical approaches to construct, visualize and perform multiple comparison procedures that are tailored to the structured families of hypotheses (Bretz et al., 2009; Burman et al., 2009). In this framework, nodes with weights represent individual null hypotheses and their local significance levels. A directed edge with a weight from one hypothesis to another specifies how the significance level of the origin hypothesis is propagated to the end hypothesis when the origin hypothesis is rejected. Together with an updating algorithm, the graphical procedure controls the FWER in the strong sense at the prespecified level α across all hypotheses. Many commonly used multiple comparison procedures, including various gatekeeping procedures, can be visualized and performed intuitively using the graphical approach.

Since their introduction, graphical approaches have been applied to address different trial applications, such as to comparing multiple doses of an investigational treatment with a control (Bretz et al., 2009), combined non-inferiority and superiority testing (Guilbaud, 2011; Hung and Wang, 2010; Lawrence, 2011), testing of composite endpoints and their components (Huque et al., 2011; Rauch and Beyersmann, 2013) and subgroup analyses (Goteti et al., 2014). While these applications have mainly used Bonferroni-based tests, extensions have been proposed to include weighted Simes tests (Bretz et al., 2011; Lu, 2016; Maurer et al., 2011) or weighted parametric tests (Bretz et al., 2011; Millen and Dmitrienko, 2011; Xi et al., 2017). Extensions to group sequential trials (Maurer and Bretz, 2013b; Rosenblum et al., 2016; Xi and Tamhane, 2015) and adaptive designs (Klinglmueller et al., 2014; Sugitani et al., 2016) have also been discussed. More advanced graphs include symmetric graphs for equally weighted tests, graphs for families of hypotheses (Kordzakhia and Dmitrienko, 2013; Maurer and Bretz, 2014; Xi and Bretz, 2019) and entangled graphical procedures (Maurer and Bretz, 2013a). Practical considerations were also discussed for power and sample size calculation (Bretz et al., 2011) and software implementation in SAS (Bretz et al., 2011) and R (Bretz et al., 2011; Rohmeyer and Klinglmueller, 2014). Other overviews of the graphical approaches were provided with different focuses (Bretz et al., 2014; Glimm et al., 2019; Xi et al., 2016). In this chapter, we introduce the key ideas and highlight the connections between the graphical approach and other multiple comparison procedures.

5.1.1 Clinical Trial Example

Consider a confirmatory clinical trial to compare two doses of an investigational treatment (high and low) versus control in patients with chronic obstructive pulmonary disease (COPD). There are two endpoints included in the multiplicity adjustment which are the forced expiratory volume in 1 second (FEV1) and the number of exacerbations. While improvement in both FEV1 and exacerbation is important for COPD patients, we consider FEV1 as the primary endpoint because it measures the lung function and could be assessed earlier than exacerbation. Thus, there are four hypotheses: the primary hypotheses H_1 and H_2 on FEV1 comparing high and low dose against control, respectively, and the secondary hypotheses H_3 and H_4 on exacerbation for the two dose-control comparisons. For each dose-control comparison, the secondary hypothesis on exacerbation will only be tested after the benefit of the investigational treatment has been established for the primary endpoint FEV1. This hierarchical structure is useful to reflect different relationships among clinical and regulatory importance between the primary and the secondary hypotheses. In addition, high and low doses are considered to be equally important and to be investigated simultaneously.

The goal is to derive a multiple comparison procedure that controls the FWER at a prespecified significance level for all four hypotheses while considering the aformentioned clinical considerations. Common multiple comparison procedures, such as those by Bonferroni, Holm, Hochberg, or Dunnett, are not applicable because they treat all hypotheses equally important and do not address the hierarchical structure within each dose-control comparison. As illustrated in the rest of the chapter, the graphical procedure provides a convenient approach to construct, visualize and perform multiple comparison procedures that are tailored to structured families of hypotheses.

5.2 Bonferroni-Based Graphical Procedures

In this section, we introduce the graphical approach based on Bonferroni tests by Bretz et al. (2009), which serves as the basis to extend to other tests. Also, we provide step-by-step illustrations of how to update graphs after each rejection, which can be used to facilitate explanation and discussion with other clinical team members.

5.2.1 Heuristics

Assume that we are interested in testing m null hypotheses $H_1, \ldots, H_m$, which may include primary and/or secondary hypotheses. A multiple comparison procedure for $H_i, i \in I = \{1, \ldots, m\}$ should control the FWER strongly at a prespecified significance level $\alpha \in (0,1)$ (EMA, 2017; FDA, 2017). Consider that these hypotheses could be of different importance, represented by different weights. Hypothesis H_i has a local weight $0 \leq w_i \leq 1$ with $\sum_{i=1}^{m} w_i \leq 1$. Thus, the local significance level for H_i is $w_i\alpha, i \in I$. In other words, the overall significance level α is split among these m hypotheses, which is the weighted Bonferroni split. Let p_i denote the unadjusted p-value for hypothesis $H_i, i \in I$. Hypothesis H_i is rejected if $p_i \leq w_i\alpha$. Upon rejected, its local weight w_i will be propagated to the remaining not-yet-rejected hypotheses and added to their local weights, according to a prespecified rule. The graphical procedure continues testing the remaining hypotheses with the updated local significance levels for further rejections. The procedure updates and repeats until no further hypothesis could be rejected. With the formal notation introduced in Section 5.2.2, we show that this procedure leads to a multiple comparison procedure that controls the FWER strongly at level α.

The graphical approach visualizes a multiple comparison procedure using nodes and edges to represent trial objectives with different relationships for importance and hierarchical structures. The m hypotheses are visualized as weighted nodes, with hypothesis H_i being assigned the local weight w_i and thus the local significance level $w_i\alpha, i \in I$. The propagation of local weights is represented as weighted, directed edges. The direction indicates the propagation from a rejected hypothesis to a not-yet-rejected hypothesis. The weight associated with an edge determines the fraction of the local weight of the origin node

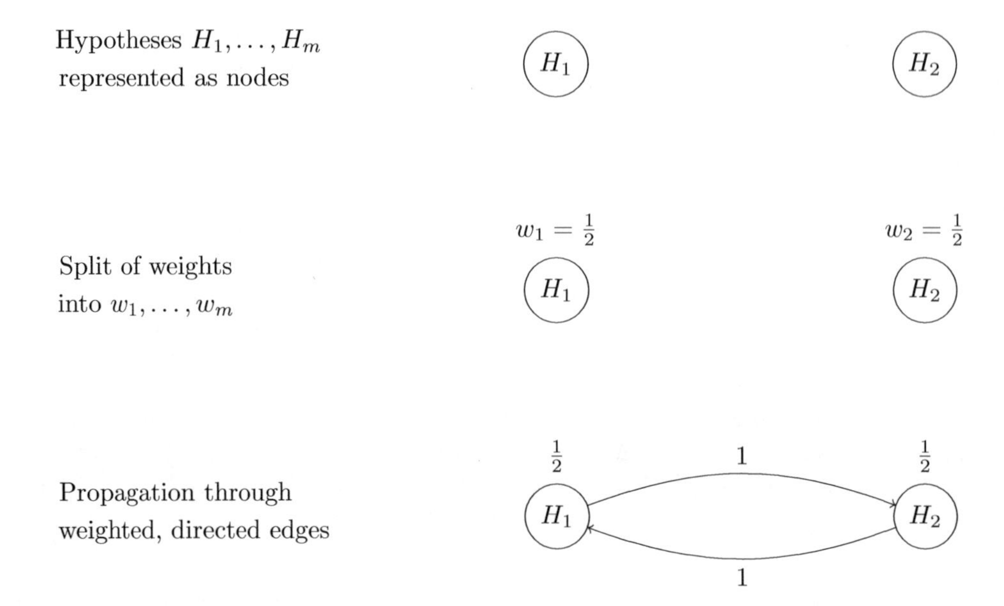

FIGURE 5.1
Conventions for the graphical approach illustrated with $m = 2$ hypotheses. (Adapted from Bretz et al. (2019).)

that is added to the local weight of the end node when the origin hypothesis is rejected. For an example with $m = 2$ hypotheses, we illustrate a graph in Figure 5.1 with conventions.

Figure 5.1 already illustrates two common multiple comparison procedures for two hypotheses. The middle graph visualizes the Bonferroni procedure that tests each hypothesis at level $w_i\alpha = \alpha/2$; the procedure rejects hypothesis H_i when $p_i \leq w_i\alpha = \alpha/2, i = 1, 2$. Because the rejection or non-rejection of H_1 does not affect H_2 and vice versa, there is no propagation and hence no edge between H_1 and H_2. If we add propagation and edges as in the bottom graph in Figure 5.1, this means that if H_1 is rejected with $p_1 \leq w_1\alpha = \alpha/2$, its local weight of $1/2$ will be propagated to H_2, which then has a local weight of $1/2+1/2 = 1$ and a local significance level of α. This process is the same as the Holm procedure with two hypotheses. First reject the hypothesis with the smaller p-value if it is not larger than $\alpha/2$. If this hypothesis is rejected, continue testing the other hypothesis at level α. Note if $w_1 \neq w_2$, the middle graph extends naturally to the weighted Bonferroni procedure and the bottom graph to the weighted Holm (1979) procedure.

Before formally describing the graphical approach, we illustrate graphically the above Holm procedure, the fixed sequence procedure, and the fallback procedure using a numerical example. Consider testing two hypotheses with the FWER controlled strongly at one-sided level of $\alpha = 0.025$. Let the unadjusted p-values be $p_1 = 0.02$ and $p_2 = 0.01$ for H_1 and H_2, respectively. Figure 5.2 provides the initial graph and the updated graph after a hypothesis is rejected. For the Holm procedure, H_2 is first rejected because $p_2 = 0.01 < \alpha/2 = 0.0125$. Then H_1 is tested at level $\alpha \cdot (1/2 + 1/2) = 0.025$ because of propagation of H_2's weight $1/2$ to H_1. The fixed sequence procedure proceeds in a predefined order (Maurer et al., 1995; Westfall and Krishen, 2001). For two hypotheses, it tests H_1 at level α, and if it is rejected, it tests H_2 at level α. Thus, H_1 is first rejected because $p_1 = 0.02 < \alpha \cdot 1 = 0.025$. Then H_2 is tested at level $\alpha \cdot (0 + 1) = 0.025$ because of propagation of H_1's weight 1 to H_2. The fallback procedure (Wiens, 2003) also proceeds in a predefined order but splits the local weights between H_1 and H_2. It tests H_1 at level $w_1\alpha$, and if it is rejected, it tests

(a) Holm procedure

Initial graph

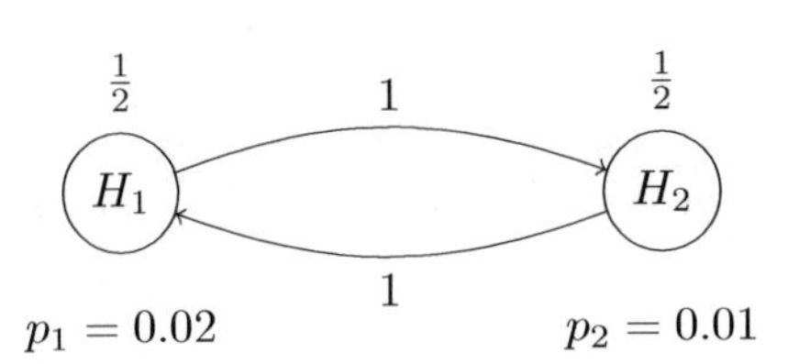

Updated graph after rejecting H_2
$p_2 = 0.01 < \alpha/2 = 0.0125$

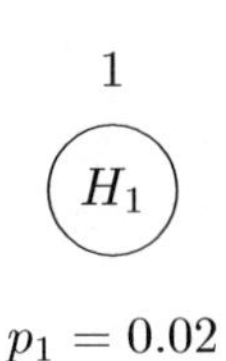

(b) Fixed sequence procedure

Initial graph

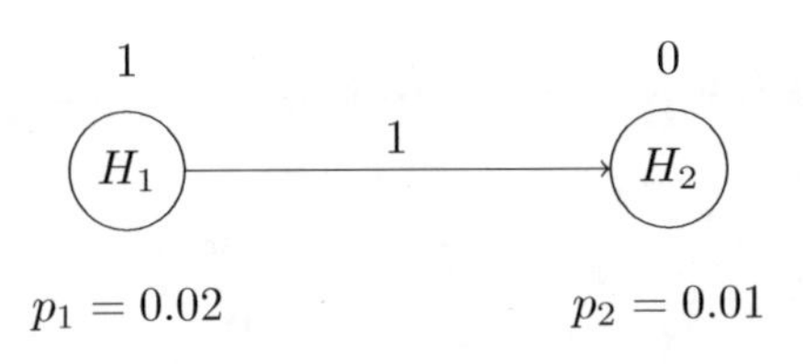

Updated graph after rejecting H_1
$p_1 = 0.02 < \alpha \cdot 1 = 0.025$

1
H_2
$p_2 = 0.01$

(c) Fallback procedure

Initial graph

1/2 1/2
H_1 —1→ H_2
$p_1 = 0.02$ $p_2 = 0.01$

Updated graph after rejecting H_2
$p_2 = 0.01 < \alpha/2 = 0.0125$

1/2
H_1
$p_1 = 0.02$

FIGURE 5.2
Graphical examples of (a) the Holm procedure, (b) the fixed sequence procedure and (c) the fallback procedure, with $m = 2$ hypotheses and $\alpha = 0.025$.

H_2 at level $(w_1 + w_2)\alpha = \alpha$ or otherwise at level $w_2\alpha$. Setting $w_1 = w_2 = 1/2$, H_1 is not rejected because $p_1 = 0.02 > \alpha/2 = 0.0125$. Then H_2 is tested at level $\alpha/2 = 0.0125$ because of no propagation from H_1. After H_2 is rejected, there is no propagation to H_1, which thus remains not rejected. We could add an edge from H_2 back to H_1 to allow the possibility of testing H_1 at level α if H_2 is rejected first. This leads a graph of an improved fallback procedure by Wiens and Dmitrienko (2005). In addition, Bretz et al. (2009) provided graphical visualizations for these procedures with three null hypotheses.

5.2.2 General Results

In this section, we introduce the necessary notation to describe the graphical approach more formally based on Bonferroni tests. For m hypotheses $H_1, \ldots, H_m$, we assume that they satisfy the free combination condition (Holm, 1979), which means that any subset of these hypotheses can be true while the rest can be false simultaneously. If this condition is not satisfied, the methods in this chapter still control the FWER at level α, although they can possibly be improved (Brannath and Bretz, 2010). Let the initial split of local weights be $\boldsymbol{w} = (w_1, \ldots, w_m)$ such that $\sum_{i=1}^{m} w_i \leq 1$. Then the local significance level for hypothesis H_i is $w_i\alpha, i \in I = \{1, \ldots, m\}$. Let $\boldsymbol{G} = (g_{ij})_{i,j=1,\ldots,m}$ denote the m by m transition matrix to determine propagation between hypotheses. The transition weight g_{ij} represents the fraction of the local weight w_i of H_i that is added to the local weight of $H_j, i \neq j \in I$, when H_i is rejected. While $\boldsymbol{w}$ determines the initial weight split, $\boldsymbol{G}$ specifies the weighted directed edges in a graph. We require the following regularity conditions on local weights and transition weights:

$$0 \leq w_i \leq 1, \ \sum_{i=1}^{m} w_i \leq 1 \text{ for all } i = 1, \ldots, m \text{ and,}$$

$$0 \leq g_{ij} \leq 1, \ g_{ii} = 0, \ \sum_{k=1}^{m} g_{ik} \leq 1 \text{ for all } i, j = 1, \ldots, m. \tag{5.1}$$

In other words, local weights should be non-negative and their sum should be bounded by 1. Transition weights should also be non-negative, with no loop going from a hypothesis to the same hypothesis, and the sum of all transition weights from edges going out of a single hypothesis should be bounded by 1. Given the unadjusted p-values $p_1, \ldots, p_m$ calculated for hypotheses $H_1, \ldots, H_m$, the following algorithm determines how to reject hypotheses and to update the graph (Bretz et al., 2009).

Algorithm 5.1 *Graphical approach based on weighted Bonferroni tests*

(0) Set $I = \{1, \ldots, m\}$.

(1) Select $j \in I$ such that $p_j \leq w_j\alpha$ and reject H_j; otherwise stop.

(2) Update the graph:

$$I \to I \setminus \{j\}$$

$$w_\ell(I) \to \begin{cases} w_\ell(I) + w_j(I)g_{j\ell}, & \ell \in I, \\ 0, & \text{otherwise,} \end{cases}$$

$$g_{\ell k} \to \begin{cases} \frac{g_{\ell k} + g_{\ell j}g_{jk}}{1 - g_{\ell j}g_{j\ell}}, & \ell \neq k \in I, g_{\ell j}g_{j\ell} < 1, \\ 0, & \text{otherwise.} \end{cases}$$

(3) If the cardinality of I satisfies $|I| \geq 1$, go to Step (1); otherwise stop.

The rationale for the updating algorithm is that the updated local weight is the sum of the original weight and the weight propagated from the rejected hypothesis. The updated transition weight is a fraction. For simplicity, we drop its dependency on I. Its numerator is the sum of the original transition weight and the transition weight of the indirect path through the rejected hypothesis; its denominator is a normalizing factor to remove the possible loop involving the rejected hypothesis. As shown in Algorithm 5.1, there are at most m steps to reject m hypotheses, if not stopping early with no further rejection. Thus, this is a shortcut or sequentially rejective procedure (Holm, 1979). The algorithm, together with the local weight vector $\boldsymbol{w}$ and the transition matrix $\boldsymbol{G}$, uniquely determines a multiple comparison procedure that controls the FWER strongly at level α (Bretz et al., 2009). The proof involves constructing a closed test procedure with weighted Bonferroni tests for intersection hypotheses, and some details are provided in Section 5.2.7. To illustrate the idea, we revisit the clinical trial example mentioned in Section 5.1.1 in the next section.

5.2.3 Application of the Graphical Approach

In the COPD trial example in Section 5.1.1, there are four hypotheses to account for two dose-control comparisons on a primary and a secondary endpoint. Hypotheses H_1 and H_2 are the primary hypotheses for the high-control and the low-control comparisons, respectively; hypotheses H_3 and H_4 are the secondary hypotheses for the high-control and the low-control comparisons, respectively. The two doses are considered equally important but in each dose-control comparison, the secondary hypothesis will be tested only if the corresponding primary hypothesis has been rejected.

One graph that incorporates these considerations is illustrated in the panel (a) of Figure 5.3. In this graph, there are two edges with transition weights 1/2 going out of each primary hypothesis. This indicates that once $H_1(H_2)$ is rejected, its local weight will be equally split. Then one half will be propagated to the other primary hypothesis for a higher chance to claim success on the primary endpoint for the other dose, and the other half will be propagated to the corresponding secondary hypothesis for a higher chance to claim success on both endpoints for the same dose-control comparison. In addition, there is an edge from $H_3(H_4)$ to $H_2(H_1)$, which means that once the secondary hypothesis is rejected for a dose-control comparison (it also means that the primary hypothesis has been rejected already), the local weight will be propagated to the primary hypothesis of the other dose-control comparison. It is important to note that we deliberately avoid the edge from H_3 to H_4 because H_4 should be tested only once H_2 has been rejected. As shown later, the propagation between H_3 and H_4 will appear after updating the graph once a primary hypothesis is rejected. The same argument holds for the edge from H_4 to H_3.

Assume the unadjusted p-Values of $p_1 = 0.01, p_2 = 0.012, p_3 = 0.01, p_4 = 0.02$ for an overall significance level $\alpha = 0.025$. From the initial graph, we see that H_1 could be rejected because $p_1 < \alpha/2 = 0.0125$. Then we remove the node and its associated edges from the graph, and update the graph as in panel (b) of Figure 5.3. The local weights are updated to split equally the local weight of H_1 and to propagate to H_2 and H_3. Thus, $w_2 = 1/2 + 1/2 \cdot 1/2 = 3/4$ and $w_3 = 0 + 1/2 \cdot 1/2 = 1/4$. As a result of updating, we can see that the local weights of the remaining hypotheses are increased because of receiving weights from the rejected hypothesis through propagation.

There are two outcomes in the updated edges: an existing edge may get a new weight, and a new edge may appear. For example, the edge of g_{24} is updated because there was a direct edge from H_2 to H_4 with a transition weight of 1/2 and thus $g_{24} = 1/2/(1-1/2 \cdot 1/2) = 2/3$. The denominator of $1 - 1/2 \cdot 1/2$ is due to the fact that removing H_1 leads a loop from H_2 and indirectly through H_1 back to H_2, and thus the weight associated with the loop $1/2 \cdot 1/2$ would be subtracted from the graph. In addition, a new edge appears from H_2 to

H_3 because there was an indirect path through H_1. After removing H_1, this indirect path becomes a direct edge with the transition weight $g_{23} = (0 + 1/2 \cdot 1/2)/(1 - 1/2 \cdot 1/2) = 1/3$.

After rejecting H_1 and updating the graph as in panel (b) of Figure 5.3, we rejected H_2 because $p_2 = 0.012 < 3\alpha/4 = 0.01875$. Then we update the graph for the remaining hypotheses H_3 and H_4. The resulting graph shown in panel (c) of Figure 5.3 is the same graph for the Holm procedure as in Figure 5.2. Then we reject H_3 because $p_3 = 0.01 < \alpha/2 = 0.0125$ and after this, we reject H_4 because $p_4 = 0.02 < \alpha/2 + \alpha/2 = 0.025$.

As one may notice, both H_1 and H_2 are "rejectable" in the first step because $p_1 < \alpha/2$ and $p_2 < \alpha/2$. In Step (1) of Algorithm 1 in Section 5.2.2, there is no specific rule as to how to select j if there are multiple hypotheses satisfying $p_j \leq w_j\alpha$. In this case, a natural question is whether rejecting H_2 first instead of H_1 would lead to a different set of decisions (rejections and non-rejections) for the remaining hypotheses. To answer this question, we also provide the updated graph after H_2 being rejected first in panel (d) of Figure 5.3. In this updated graph, the local weight of H_1 is increased to $w_1 = 1/2 + 1/2 \cdot 1/2 = 3/4$, which means that H_1 could still be rejected since $p_1 < \alpha/2 < 3\alpha/4$. In addition, if we further reject H_1, the updated graph will be the same as the one in panel (c) of Figure 5.3. This means that the decisions to reject H_3 and H_4, or not, remain the same, as long as both H_1 and H_2 are rejected. The rejection order does not matter, and this observation holds true in

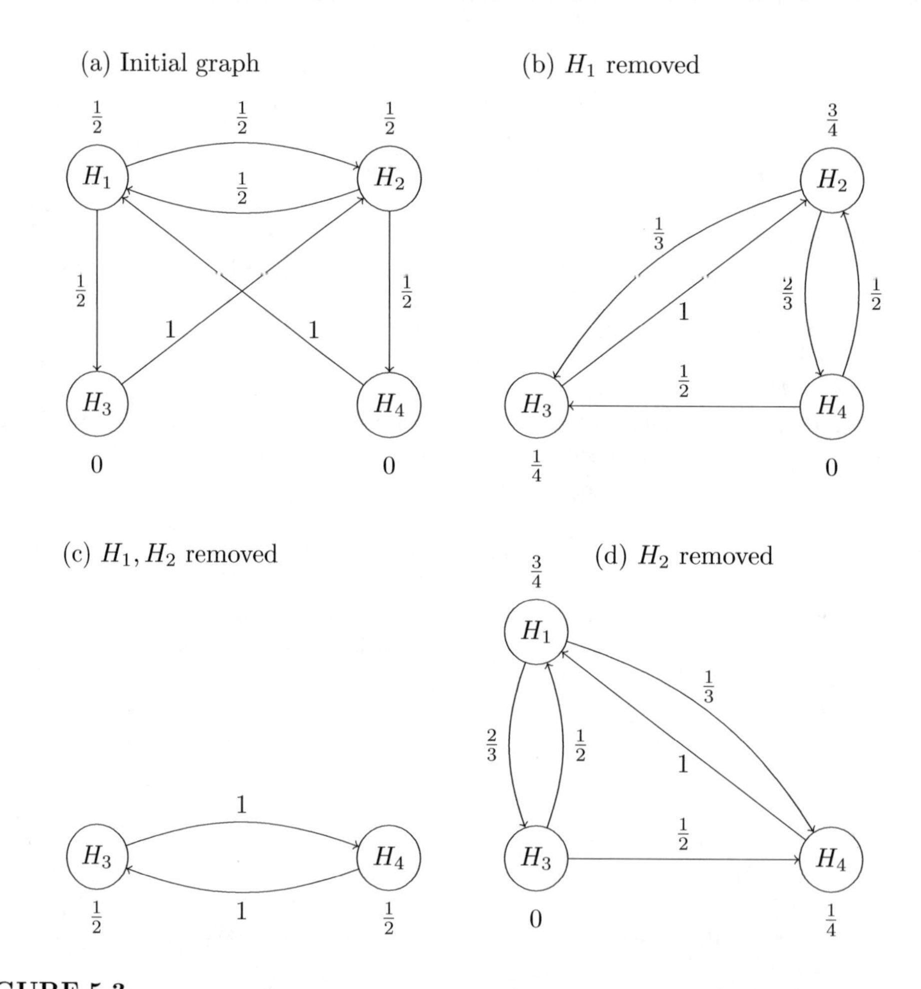

FIGURE 5.3
A graphical multiple comparison procedure for the COPD example in Section 5.1.1.

general for the graphical approach, i.e., the final decision to reject or not reject a hypothesis do not depend on the order of selecting hypotheses in Step (1) of Algorithm 5.1 in Section 5.2.2. We refer to Section 5.2.7 for more explanation and Bretz et al. (2009) for a detailed proof. In practice, a convenient and sensible choice would be to select the hypothesis with the smallest p-value relative to the local weight, i.e., $j = \arg\min_{i \in I}\{p_i/w_i\}$.

5.2.4 Adjusted p-Values

Adjusted p-values provide another way to describe a multiple comparison procedure. The construction of adjusted p-values incorporates the underlying multiplicity adjustment, and thus they could be compared directly with the overall significance level α for decision-making. By definition, an adjusted p-value of a hypothesis is the smallest significance level at which the hypothesis could be rejected by the multiple comparison procedure (Westfall et al., 1993). To derive an adjusted p-value of a hypothesis, the significance level is set to be a variable; by varying it, we could see what significance levels lead to the rejection of the hypothesis and denote the smallest level as its adjusted p-value; and finally, the hypothesis is rejected if its adjusted p-value is less than or equal to the prespecified significance level α.

To formally calculate adjusted p-values, let p_i^{adj} denote the adjusted p-value for $H_i, i \in I = \{1, \ldots, m\}$. We modify Algorithm 1 in Section 5.2.2 to derive adjusted p-values based on unadjusted p-values $p_1, \ldots, p_m$. Given initial local weights $\boldsymbol{w}$ and a transition matrix $\boldsymbol{G}$, adjusted p-values could be derived in the following algorithm (Bretz et al., 2009).

Algorithm 5.2 *Adjusted p-values based on the Bonferroni-based graphical approach*

(0) Set $I = \{1, \ldots, m\}$ and $p_{\max} = 0$.

(1) Let $j = \arg\min_{i \in I}\{\frac{p_i}{w_i(I)}\}$, $p_j^{\text{adj}} = \max\{\frac{p_j}{w_j(I)}, p_{\max}\}$, and $p_{\max} = p_j^{\text{adj}}$.

(2) Update the graph:

$$I \to I \setminus \{j\}$$

$$w_\ell(I) \to \begin{cases} w_\ell(I) + w_j(I)g_{j\ell}, & \ell \in I, \\ 0, & \text{otherwise}, \end{cases}$$

$$g_{\ell k} \to \begin{cases} \frac{g_{\ell k} + g_{\ell j}g_{jk}}{1 - g_{\ell j}g_{j\ell}}, & \ell \neq k \in I, g_{\ell j}g_{j\ell} < 1, \\ 0, & \text{otherwise}. \end{cases}$$

(3) If $|I| \geq 1$, go to Step (1); otherwise go to Step (4).

(4) Reject hypotheses H_i with $p_i^{\text{adj}} \leq \alpha$ for all $i = 1, \ldots, m$.

To illustrate using the COPD example in Section 5.2.3, consider the unadjusted p-values $p_1 = 0.01, p_2 = 0.012, p_3 = 0.01, p_4 = 0.02$. As shown in Figure 5.3, initial weights are $w_1 = 1/2, w_2 = 1/2, w_3 = 0, w_4 = 0$. In the first step, $j = 1$ because $p_1/w_1 < p_2/w_2$. Thus, $p_1^{\text{adj}} = p_1/w_1 = 0.01/(1/2) = 0.02$ and $p_{\max} = 0.02$. After removing H_1, the graph is updated with the local weights $w_2 = 3/4, w_3 = 1/4, w_4 = 0$. Then $j = 2$ because $p_2/w_2 < p_3/w_3$. Thus, $p_2^{\text{adj}} = \max\{p_2/w_2, p_{\max}\} = \max\{0.012/(3/4), 0.02\} = 0.02$ and $p_{\max} = 0.02$. After removing H_1 and H_2, the graph is updated with the local weights $w_3 = 1/2, w_4 = 1/2$. Then $j = 3$ because $p_3/w_3 < p_4/w_4$. Thus, $p_3^{\text{adj}} = \max\{p_3/w_3, p_{\max}\} = \max\{0.01/(1/2), 0.02\} = 0.02$ and $p_{\max} = 0.02$. Finally after removing H_1, H_2 and H_3, there is only one hypothesis left, H_4, with a local weight of 1. Thus, $p_4^{\text{adj}} = p_4/w_4 = \max\{0.02/1, p_{\max}\} = 0.02$. In summary, $p_1^{\text{adj}} = p_2^{\text{adj}} = p_3^{\text{adj}} = p_4^{\text{adj}} = 0.02 < \alpha = 0.025$ and therefore, all four hypotheses are rejected. Note that this conclusion is the same as the one from Section 5.2.2 using unadjusted p-values for decision-making. In general, the conclusions are expected to be the same between using adjusted p-values against the overall significance level (Algorithm 5.2) and using unadjusted p-values against adjusted significance levels (Algorithm 5.1).

5.2.5 Simultaneous Confidence Intervals

Confidence intervals quantify uncertainty around the point estimate of the parameter of interest. Consider testing one-sided null hypotheses $H_i : \theta_i \leq \delta_i$, where θ_i is the parameter of interest and δ_i is the prespecified margin against which θ_i is hypothesized. Let $L_i(\gamma)$ denote the lower limit of the one-sided confidence interval for θ_i at level γ, which means that $p_i \leq \gamma$ is equivalent to $L_i(\gamma) \geq \delta_i$. Simultaneous confidence intervals also adjust for the Bonferroni-based multiple comparison procedure for a simultaneous coverage probability of at least $1-\alpha$ (Guilbaud, 2008; Strassburger and Bretz, 2008), i.e., $P\left[\cap_{i\in I}(\theta_i > L_i)\right] \geq 1-\alpha$.

For an initial graph with local weights $\boldsymbol{w}$ and transition matrix $\boldsymbol{G}$, we apply Algorithm 5.1 and assume that the set of rejected hypotheses is $R \subseteq I = \{1, \ldots, m\}$. With a simultaneous coverage probability of at least $1-\alpha$, Strassburger and Bretz (2008) provided the lower limits of the one-sided confidence intervals for $\theta_1, \ldots, \theta_m$ as

$$L_i = \begin{cases} \delta_i, & \text{for } i \in R \text{ if } R \neq I, \\ L_i(\alpha_i(I \setminus R)), & \text{for } i \notin R \text{ if } R \neq I, \\ \max\left(\delta_i, L_i(w_i\alpha)\right), & \text{if } R = I, \end{cases} \tag{5.2}$$

where $\alpha_i(I \setminus R) = \alpha w_i(I \setminus R)$ represents the significance level for hypothesis H_i after all hypotheses in R are removed from I according to Algorithm 5.1. For hypotheses that are not rejected, their lower limits merely reflect the significance level at which they are retained. If not all hypotheses are rejected, the lower limits of the rejected ones reflect the test decision $\theta > \delta$. If all hypotheses are rejected, the lower limits become larger, and thus more informative, because they reflect both the test decision and the initial significance level.

The proof for the simultaneous confidence intervals (5.2) relies on the partitioning principle (Finner and Strassburger, 2002). This approach refers to partitioning the entire parameter space and is closely related to the technique of acceptance set inversion. In general, the partitioning principle allows one to derive confidence sets associated with multiple comparisons. Finner et al. (2021) discuss and illustrate three applications of the partitioning principle corresponding to three motivations, including the transparent partition of the parameter space to channel decision-making onto desirable decision paths that are closely related to the graphical approaches discussed in this chapter.

To illustrate this with the COPD example in Section 5.2.3, consider the p-values $p_1 = 0.01,\ p_2 = 0.012,\ p_3 = 0.01, and\ p_4 = 0.02$. We know from the previous section that all hypotheses are rejected. Let the margin be $\delta_i = 0$ for all hypotheses. Thus, the lower limit of the one-sided confidence interval with a simultaneous coverage probability of at least $1 - \alpha = 0.975$ is $\max(\delta_i, L_i(w_i\alpha)) = \max\{0, L_i(w_i\alpha)\}$. For H_1 and H_2, L_1 and L_2 are thus between 0 and the marginal lower limit of the one-sided confidence interval at level $1 - 0.0125$, respectively. For H_3 and H_4, we have $L_3 = L_4 = 0$. Note that the lower bound could be further increased in some cases for the first and the last scenarios in (5.2), and we refer to the default graph (Guilbaud, 2018) for more details.

5.2.6 Power and Sample Size

Power and sample size are important aspects related to the design of a clinical trial. For a hypothesis on a single endpoint for a comparison between an investigational treatment and control, the sample size is usually calculated to achieve a target power of the test for

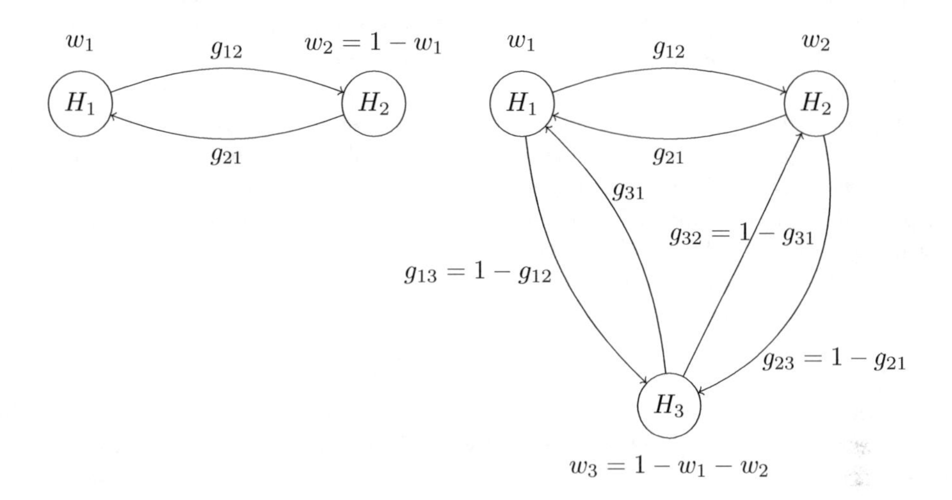

FIGURE 5.4
General graphs for two hypotheses and three hypotheses.

a parameter specification under the alternative hypothesis. When investigating multiple hypotheses simultaneously with a multiple comparison procedure, one needs to define the power in more detail because the trial success criteria may include winning not only on one or more primary endpoints but also on secondary endpoints. Various extensions to define success criteria with multiple hypotheses have been proposed (Chen et al., 2011; Julious and McIntyre, 2012; Maurer and Mellein, 1988; Senn and Bretz, 2007; Sozu et al., 2010; Xiong et al., 2005). In this section, we briefly discuss how to translate these power considerations into graphs.

For a single hypothesis, the power to reject it using a statistical test is usually related to the significance level and the sample size. The power increases if the significance level increases or the sample size increases. Thus, for a given sample size, an increased significance level usually leads to an increased power. Extending to two hypotheses, we consider two multiple comparison procedures previously presented in Figure 5.2. The Holm procedure splits the overall significance level α equally between H_1 and H_2. On the other hand, the fixed sequence procedure assigns the overall significance level α to H_1 initially. Thus, H_1 is expected to have a higher power in the fixed sequence procedure than in the Holm procedure. This power advantage for H_1 comes with a trade-off for H_2. In the Holm procedure, H_2 has an initial significance level of $\alpha/2$ and if H_1 is rejected, H_2 has an updated significance level of $\alpha/2 + \alpha/2 = \alpha$. For the fixed sequence procedure, however, H_2 has an initial significance level of 0 and only if H_1 is rejected, H_2 has an updated significance level of $0 + \alpha = \alpha$. Thus, H_2 is expected to have a lower power in the fixed sequence procedure than in the Holm procedure.

For two hypotheses, both graphs for the Holm and the fixed sequence procedures could be viewed as examples of a general graph with local weights and transition weights defined as in panel (a) of Figure 5.4. Local weights need to satisfy $w_1 + w_2 = 1$ and the exact choices of w_1, g_{12}, g_{21} depend on the power considerations above. For the Holm procedure, $w_1 = w_2 = 1/2$ and $g_{12} = g_{21} = 1$; for the fixed sequence procedure, $w_1 = 1, w_2 = 0$ and $g_{12} = 1, g_{21} = 0$.

Consider a clinical trial in which there is a secondary hypothesis H_3, in addition to the two primary hypotheses H_1, H_2. First, we need to split local weights such that

$w_1 + w_2 + w_3 = 1$. The clinical team is interested in testing H_3 if any primary hypothesis is rejected and thus $w_3 = 0$. For every hypothesis, there are two outgoing edges with transition weights such that $g_{ij} + g_{ik} = 1$ for $i = 1, 2, 3$. This general graph is illustrated in panel (b) of Figure 5.4. The power trade-off discussed above also applies here: to increase the power of a hypothesis, we could increase its local weight and the transition weights of edges that go into this hypothesis. As a result, however, the power of another hypothesis may decrease because of a reduced local weight and/or reduced transition weights of edges that go into it. If more information is available on the possible parameters of alternative hypotheses, we could perform a more quantitative assessment by simulating from the joint distribution of test statistics and checking the proportion of times one or more hypotheses are rejected. Bretz et al. (2011) provided more detailed discussions and illustrative examples.

5.2.7 Weighting Schemes and Connection to Closed Testing

In this section, we introduce weighting schemes as the basis of the graphical approach. They also provide the connection to closed testing by which strong FWER control can be formally established. In addition, they form the foundation to extend the Bonferroni-based graphical approach to other tests.

A weighting scheme essentially summarize the complete set of a closed test procedure. Consider an initial graph for m hypotheses $H_i, i \in I = \{1, \ldots, m\}$ with initial local weights $\boldsymbol{w}_I = (w_1, \ldots, w_m)$ and initial transition weights $\boldsymbol{G} = (g_{ij})_{i,j=1,\ldots,m}$. Suppose that a hypothesis H_k is removed from the graph and we update the local weights $\boldsymbol{w}_{I\setminus\{k\}}$ and the transition matrix $\boldsymbol{G}(I \setminus \{k\})$ according to Algorithm 5.1 in Section 5.2.2. Further, we remove a second hypothesis H_ℓ from the graph and update the local weights $\boldsymbol{w}_{I\setminus\{k,\ell\}}$ and the transition matrix $\boldsymbol{G}(I \setminus \{k, \ell\})$. We could continue this process until all hypotheses have been removed from the graph, and repeat this process for all permutations of indices in I. As a result, we obtain a collection of local weights $\boldsymbol{W} = \{\boldsymbol{w}(J) : \varnothing \neq J \subseteq I\}$. This collection is a weighting scheme of the initial graph, which includes local weights of all hypotheses in every proper subset of I (Bretz et al., 2011).

Formally, we have to tailor Algorithm 1 in order to determine a weighting scheme. For a given index set $\varnothing \neq J \subsetneq I$, let $J^c = I \setminus J$ denote the set of indices that are not contained in J. Then the following algorithm determines the weights $w_j(J), j \in J$. This algorithm has to be repeated for each $J \subseteq I$ to generate the complete set of weights for the underlying closed test procedure.

***Algorithm 5.3** Weighting scheme*

(0) Set $I = \{1, \ldots, m\}$, $\varnothing \neq J \subsetneq I$, and $J^c = I \setminus J$.

(1) Select $j \in J^c$ and remove H_j.

(2) Update the graph:

$$I \to I \setminus \{j\}, \ J^c \to J^c \setminus \{j\}$$

$$w_\ell(I) \to \begin{cases} w_\ell(I) + w_j(I) g_{j\ell}, & \ell \in I, \\ 0, & \text{otherwise,} \end{cases}$$

$$g_{\ell k} \to \begin{cases} \frac{g_{\ell k} + g_{\ell j} g_{jk}}{1 - g_{\ell j} g_{j\ell}}, & \ell \neq k \in I, g_{\ell j} g_{j\ell} < 1, \\ 0, & \text{otherwise.} \end{cases}$$

(3) If $|J^c| \geq 1$, go to Step (1); otherwise stop.

Similar to what we stated in Section 5.2.3, the test decisions do not depend on the order of rejections. That is, the weighting scheme $\boldsymbol{W}$ is uniquely defined for any initial graph, regardless of the order to select indices from J^c (Bretz et al., 2009). Note that the weighting scheme corresponds to the closed family for $H_1, \ldots, H_m$, which consists of intersection hypotheses $H_J = \cap_{j \in J} H_j$ for all $\varnothing \neq J \subseteq I$. The closed test procedure applies an intersection test to every intersection hypothesis at the significance level α and rejects H_i if all H_J with $i \in J \subseteq I$ are rejected by their intersection tests, thus leading to strong FWER control at level α (Marcus et al., 1976). More precisely, the weighting scheme specifies the local weights of hypotheses involved in every intersection hypothesis and thus guides the intersection test on how to split weights among hypotheses. If we apply the Bonferroni test to every intersection hypothesis, the resulting closed test procedure reduces to the one implemented in Algorithm 5.1 in Section 5.2.2. This leads to a sequentially rejective procedure that requires at most m steps to make rejections, whereas the full closed test procedure requires $2^m - 1$ steps to go through every intersection hypothesis (Bretz et al., 2009).

To illustrate the connection between the graphical approach and the closed test procedure, we return to the COPD example in Section 5.2.3. Figure 5.3 shows the initial graph for the four hypotheses under investigation. Figure 5.3 also visualizes the updated graphs when $H_1 and\ H_2$ are removed from the graph. These already provide the local weights for four intersection hypotheses in the weighting scheme: $H_1 \cap H_2 \cap H_3 \cap H_4$, $H_2 \cap H_3 \cap H_4$, $H_1 \cap H_3 \cap H_4$, and $H_3 \cap H_4$. We complete this process using Algorithm 3 and provide the complete weighting scheme in Table 5.1. For any $\varnothing \neq J \subseteq I$, the local weights are non-negative numbers for hypotheses involved in J and '$-$' for the remaining ones. Note that the sum of local weights is bounded by 1 for every J, which is a result of the regularity conditions in (5.1) and Algorithm 5.3.

Consider a closed test procedure for the weighting scheme in Table 5.1 and apply the Bonferroni test to every intersection hypothesis. Then $H_J = \cap_{j \in J} H_j$ is rejected if $p_j \leq w_j(J)\alpha$ for at least one $j \in J \subseteq I$. For example, $H_I = H_1 \cap H_2 \cap H_3 \cap H_4$ is rejected if $p_1 \leq \alpha/2$ or $p_2 \leq \alpha/2$. Assume $p_1 = 0.01, p_2 = 0.012, p_3 = 0.01, p_4 = 0.02$. We reject $H_I = H_1 \cap H_2 \cap H_3 \cap H_4$ because $p_1 \leq \alpha/2 = 0.0125$. We also reject every intersection hypothesis involving H_1 because $p_1 \leq \alpha/2 < 3\alpha/4 < \alpha = 0.025$. Thus, we reject H_1 by closed testing. Similarly, $p_2 \leq \alpha/2 < 3\alpha/4 < \alpha = 0.025$ and we also reject H_2 by closed testing. For the remaining intersection hypotheses, $p_3 < \alpha/2 < p_4 < \alpha$ and we reject H_3 and H_4 as well. As expected, the decisions from the closed test procedure are identical with the decisions from the graphical approach using Algorithm 5.1 in Section 5.2.3.

Formally, the local weight of a hypothesis is non-decreasing from a bigger set to a smaller one (i.e., from an intersection hypothesis involving more hypotheses to another involving fewer). For example, $w_1(\{1,2,3,4\}) = 1/2 = w_1(\{1,2,3\}) = 1/2 < w_1(\{1,3\}) = 1$. This

TABLE 5.1
Weighting Scheme of the COPD Example in Section 5.2.3

J	$w_1(J)$	$w_2(J)$	$w_3(J)$	$w_4(J)$	J	$w_1(J)$	$w_2(J)$	$w_3(J)$	$w_4(J)$
$\{1,2,3,4\}$	1/2	1/2	0	0	$\{2,3\}$	–	3/4	1/4	–
$\{1,2,3\}$	1/2	1/2	0	–	$\{2,4\}$	–	1	–	0
$\{1,2,4\}$	1/2	1/2	–	0	$\{3,4\}$	–	–	1/2	1/2
$\{1,3,4\}$	3/4	–	0	1/4	$\{1\}$	1	–	–	–
$\{2,3,4\}$	–	3/4	1/4	0	$\{2\}$	–	1	–	–
$\{1,2\}$	1/2	1/2	–	–	$\{3\}$	–	–	1	–
$\{1,3\}$	1	–	0	–	$\{4\}$	–	–	–	1
$\{1,4\}$	3/4	–	–	1/4					

monotonicity condition provided by Hommel et al. (2007) as

$$w_j(J) \leq w_j(J'), \text{ for all } j \in J' \subset J \subseteq I \tag{5.3}$$

enables the so-called consonance property (Gabriel, 1969), which means that the rejection of H_J always leads to the rejection of at least one $H_{J'}$ with $J' \subset J$. This condition, combined with the Bonferroni test, leads to a shortcut or sequentially rejective procedure as implemented by Algorithm 5.1, which has the same decision as the underlying closed test procedure but with at most m steps. The formal proof was provided by Bretz et al. (2009).

5.2.8 Graphical Representation of Gatekeeping Procedures

The graphical approach includes many common multiple comparison procedures, such as the Bonferroni, Holm, fixed sequence, and fallback procedures. It also extends to more recent gatekeeping procedures that handle the hierarchical structure among families of hypotheses. Consider, for example, two families of hypotheses: the primary family includes primary hypotheses and the secondary family includes secondary hypotheses. To control the FWER at level α, we first test the primary family at level α using a multiple comparison procedure, for example, the Holm procedure. In the following, we outline the conditions under which we can test the secondary family.

A serial gatekeeping procedure requires that the secondary family is tested only if all hypotheses in the primary family have been rejected (Maurer et al., 1995; Westfall and Krishen, 2001). For example, we visualize in panel (a) of Figure 5.5 such a serial gatekeeping procedure for two primary hypotheses H_1, H_2 and one secondary hypotheses H_3. Because the procedure tries to reject both primary hypotheses before testing the secondary hypothesis, it has an edge from H_1 to H_2 with a transition weight of 1 and an edge from H_2 to H_1 with a transition weight of $1-\varepsilon$. The remaining ε is propagated from H_2 to H_3 to facilitate testing H_3 when both H_1 and H_2 are rejected. Here, ε denotes an infinitesimally small transition weight that effectively ensures the serial gatekeeping condition (Bretz et al., 2009). When not both H_1 and H_2 are rejected, ε is so small that no weight of H_2 will be propagated to H_3. If H_1 is rejected, g_{23} is updated to $\varepsilon/\left[1 - 1 \cdot (1-\varepsilon)\right] = 1$ such that the local weight of H_2 will be propagated to H_3 once H_2 is rejected as well. Although there is no edge from H_1 to H_3, a similar effect happens if H_2 is rejected first and the transition weight from H_1 to H_3 becomes 1. Bretz et al. (2009) provided more detailed discussions about ε edges.

A parallel gatekeeping procedure requires that the secondary family is tested only if at least one primary hypothesis has been rejected (Dmitrienko et al., 2003). For example, we visualize in panel (b) of Figure 5.5 such a parallel gatekeeping procedure for two primary hypotheses H_1, H_2 and two secondary hypotheses H_3, H_4. The example procedure by Dmitrienko et al. (2003) uses a Bonferroni test for primary hypotheses. If any primary hypothesis is rejected, its local weight will be equally split and propagated to secondary hypotheses. For secondary hypotheses, a Holm procedure could be used and thus H_3 propagates to H_4 and vice versa. The absence of propagation from secondary hypotheses to primary hypotheses ensures that inferences on primary hypotheses are not affected by inferences on secondary hypotheses (Dmitrienko et al., 2007).

However, this may not be the only way to propagate from secondary hypotheses. For example, after rejecting the primary hypothesis, both secondary hypotheses could be tested. If one of them is rejected, its local weight will also be propagated back to the other not-yet-rejected primary hypothesis. This represents the trade-off in power between rejecting both primary hypotheses and rejecting both secondary hypotheses, see Bretz et al. (2009) for such an alternative graph.

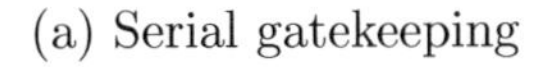

(b) Parallel gatekeeping

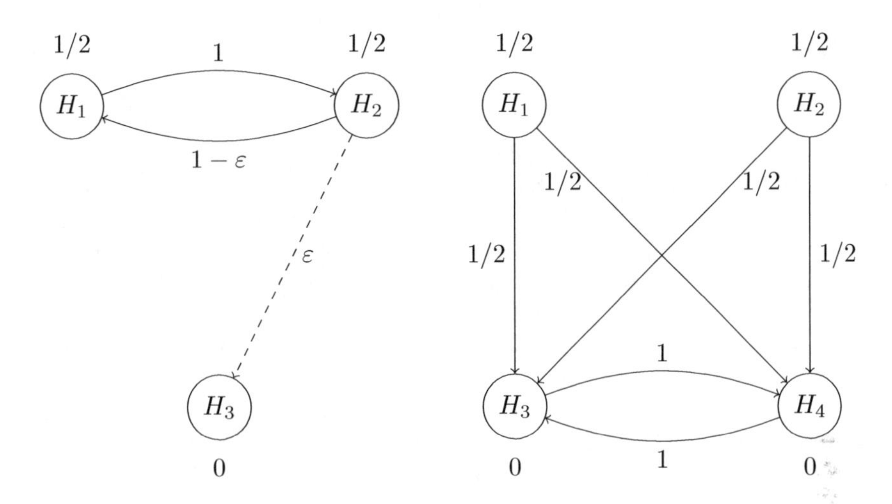

FIGURE 5.5
Graphical representation of two gatekeeping procedures.

5.3 Extensions

In this section, we extend the Bonferroni-based graphical approach from Section 5.2 to other types of multiple comparison procedures. We first introduce graphical procedures using weighted parametric and Simes tests. Then, we describe an extension to allow group sequential tests. Finally, we present symmetric graphs for equally weighted tests, graphs for families of hypotheses, and entangled graphs which have memory to account for more complex structures.

5.3.1 Parametric Graphical Procedures

Bonferroni-based graphical procedures are widely applicable but may suffer from low power when test statistics are highly correlated. In some cases, the joint distribution of test statistics is known (or partially known), and parametric tests could utilize better the correlation structure to increase power. We describe parametric graphical procedures (Bretz et al., 2011; Millen and Dmitrienko, 2011; Xi et al., 2017), which are based on the weighting scheme described in Section 5.2.7. The idea of parametric tests was previously discussed in various contexts (Dunnett, 1955; Dunnett and Tamhane, 1991; Westfall et al., 1998; Westfall and Young, 1993).

Consider an initial graph for m null hypotheses with local weights $\boldsymbol{w}$ and transition matrix $\boldsymbol{G}$. Its weighting scheme $\boldsymbol{W}$ is generated by Algorithm 5.3 in Section 5.2.7. This determines a closed test procedure based on the initial graph such that for each intersection hypothesis $H_J = \cap_{j \in J} H_j, J \subseteq I = \{1, \ldots, m\}$, its local weights are $w_j(J), j \in J$ from the weighting scheme. We know that H_j is rejected by the closed test procedure at level α if all $H_J, j \in J \subseteq I$ are rejected at level α. If we apply the Bonferroni test to H_J at level α, it is rejected if $p_j \leq w_j(J)\alpha$ for any $j \in J$, or equivalently, if $\min_{j \in J}\{p_j / w_j(J)\} \leq \alpha$. If instead we know the joint distribution of $p_j, j \in J$, we could use a parametric test to reject H_J if $\min_{j \in J}\{p_j / w_j(J)\} \leq c_J \alpha$ (or equivalently, if $p_j \leq c_J w_j(J)\alpha$ for any $i \in J$), where c_j is calculated such that

$$P\left[\min_{j\in J}\left\{\frac{p_j}{w_j(J)}\right\}\leq c_J\alpha\right]=\sum_{j\in J}w_j(J)\alpha. \tag{5.4}$$

If the Bonferroni test is applied, $c_J = 1$, and if a parametric test accounting for the correlation is applied, $c_J > 1$ leads to a more powerful test.

If we know that the joint distribution of test statistics is multivariate normal or t, probabilities can be calculated using the `mvtnorm` package in R (Genz et al., 2020) and we can solve c_J from (5.4). If the joint distribution of the test statistics is only partially known, i.e., for only a subset of statistics, we could still apply the parametric test to blocks of hypotheses and refer to Xi et al. (2017) for further discussions. For other distributions, resampling methods could be used to approximate the multivariate distribution of the p-values (Westfall and Young, 1993).

Because solving numerically for c_J in (5.4) involves evaluating multidimensional integration, its precision may be affected as the dimension increases. Instead, a computationally more efficient way is to derive the p-value for the intersection hypothesis. Let $q_J = \min_{j\in J}\{p_j/w_j(J)\}$ denote the smallest observed weighted p-value for H_J. Then the p-value $\tilde{p}_J$ for the intersection hypothesis H_J is

$$\tilde{p}_J=\min\left\{1,\frac{1}{\sum_{j\in J}w_j(J)}P\left[\min_{j\in J}\left\{\frac{p_j}{w_j(J)}\right\}\leq q_J\right]\right\}. \tag{5.5}$$

Therefore, we reject H_J if $p_j \leq c_J w_j(J)\alpha$ for any $j \in J$ with c_J determined in (5.4), or equivalently, if $\tilde{p}_J \leq \alpha$ (Xi et al., 2017).

The monotonicity condition for consonance in (5.3) is adapted such that a parametric graphical procedure is consonant if

$$c_J w_j(J) \leq c_{J'} w_j(J') \text{ for all } j \in J' \subset J \subseteq I. \tag{5.6}$$

If this condition is satisfied, we could perform a sequentially rejective procedure as follows: (Bretz et al., 2011).

Algorithm 5.4 *Graphical approach based on weighted parametric tests*

(0) Set $I = \{1,\ldots,m\}$.

(1) Select $j \in I$ such that $p_j \leq c_I w_j(I)\alpha$ and reject H_j; otherwise stop.

(2) Update the graph:

$$I \to I\setminus\{j\}$$

$$w_\ell(I)\to\begin{cases}w_\ell(I)+w_j(I)g_{j\ell}, & \ell\in I,\\ 0, & \text{otherwise,}\end{cases}$$

$$g_{\ell k}\to\begin{cases}\frac{g_{\ell k}+g_{\ell j}g_{jk}}{1-g_{\ell j}g_{j\ell}}, & \ell\neq k\in I, g_{\ell j}g_{j\ell}<1,\\ 0, & \text{otherwise.}\end{cases}$$

(3) If $|I| \geq 1$, go to Step (1); otherwise stop.

The monotonicity condition (5.6) is satisfied if $w_j(J) = c_J w_j$ for all j and all J (Xi et al., 2017). Both the single-step Dunnett (1955) and the step-down Dunnett (1991) procedures satisfy this condition. In many other practical situations, however, consonance is violated. In these cases, we could still perform the closed test following the weighting scheme defined by the graphical approach and use the p-value in (5.5) to test each intersection hypothesis.

Returning to the COPD example in Section 5.2.3, the initial graph in Figure 5.3 leads to the weighting scheme in Table 5.1. Since the primary hypotheses represent the high and low doses compared with the common control, their test statistics are correlated. This is the setting where the Dunnett (1955) test is applicable. Assume that there are equal number of patients in all three treatment groups and the test statistics follow the normal distribution. Under the null hypothesis of no treatment effect, the two test statistics for the primary hypotheses follow a standard bivariate normal distribution with mean 0 and a correlation of 0.5.

For $H_I = H_1 \cap H_2 \cap H_3 \cap H_4$, the local weights are $1/2, 1/2, 0, 0$. Using a Bonferroni test at level α, H_I is rejected if $p_1 \leq \alpha/2$ or $p_2 \leq \alpha/2$. If we use instead a parametric test utilizing the correlation of 0.5 between the two test statistics, we have $c_I = 1.078$ and thus H_I is rejected if $p_1 \leq 1.078 \cdot \alpha/2$ or $p_2 \leq 1.078 \cdot \alpha/2$. Alternatively, we could calculate the p-value of H_I as $\tilde{p}_I = P\left[(p_1 \leq q_I/2) \cup (p_2 \leq q_I/2)\right]$, where $q_I = \min\{2p_1, 2p_2\}$. If, for example, $p_1 = 0.01, p_2 = 0.012$, we can reject H_I because $p_1 < 1.078 \cdot 0.0125$ or equivalently $\tilde{p}_I = 0.0187 < \alpha = 0.025$. For other intersection hypotheses not involving H_1 and H_2 simultaneously, we could still apply the Bonferroni test. Finally, the closed test procedure rejects H_i if all H_J are rejected for $J \ni i$. In this case, the monotonicity condition in (5.6) is satisfied because $c_J/2 = 0.539 < 3/4 < 1$ for H_1 for all H_J with $J \ni 1$ and $c_J/2 = 0.539 < 3/4 < 1$ for H_2 for all H_J with $J \ni 2$. Thus, this graphical test procedure with a parametric test for H_1 and H_2 has a shortcut procedure as mentioned in Algorithm 5.4.

5.3.2 Simes-Based Graphical Procedures

Extending Bonferroni-based graphical procedures is also possible when the correlation between test statistics is not fully known but could be limited to a certain range. For example, the Simes (1986) test is a popular choice if the correlations are known to be non-negative. To incorporate the Simes test into the graphical framework, we consider the weighted Simes test (Benjamini and Hochberg, 1997). For the overall intersection hypothesis $H_I = \cap_{i \in I} H_i$, it is rejected if $p_j \leq \sum_{i \in I_j} w_i \alpha$ for at least one $j \in I$, where $I_j = \{k \in I : p_k \leq p_j\}$ includes all hypotheses whose p-values are not larger than p_j. When H_I is true, the Type I error rate is $P\left[\bigcup_{j \in I} \left\{p_j \leq \sum_{i \in J_j} w_i(J)\alpha\right\}\right]$. This weighted Simes test protects the Type I error rate at one-sided level α if, for example, the test statistics follow a multivariate normal distribution with non-negative correlations (Benjamini and Heller, 2007). Under this condition, it is obvious that the weighted Simes is more powerful than the Bonferroni test. For example if $p_1 < p_2$, the weighted Simes test rejects $H_1 \cap H_2$ if $p_1 \leq w_1\alpha$ or $p_2 \leq (w_1 + w_2)\alpha$, whereas the threshold of the Bonferroni test for p_2 is only $w_2\alpha$. When $w_i = 1/m$ for all $i = 1, \ldots, m$, the weighted Simes test reduces to the original Simes (1986) test.

Consider an initial graph for m hypotheses with local weights $\boldsymbol{w}$ and transition matrix $\boldsymbol{G}$. Its weighting scheme $\boldsymbol{W}$ can be generated by Algorithm 5.3 in Section 5.2.7. An intersection hypothesis $H_J = \cap_{j \in J} H_j, J \subseteq I = \{1, \ldots, m\}$ is rejected by the weighted Simes test if there is at least one $j \in J$ such that

$$p_j \leq \sum_{i \in J_j} w_i(J)\alpha, \tag{5.7}$$

where $J_j = \{k \in J : p_k \leq p_j\}$ (Bretz et al., 2011). Alternatively, the p-value for H_J is

$$\tilde{p}_J = \min_{j \in J} \left\{ \frac{p_j}{\sum_{i \in J_j} w_i(J)} \right\},$$

and H_J is rejected at level α if $\tilde{p}_J \leq \alpha$ (Lu, 2016). If $w_j(J)$ are equal for all $j \in J$ and all $J \subseteq I$, the resulting closed test procedure is equivalent to the Hommel (1988) procedure. When the condition for the Type I error rate control of the weighted Simes test is satisfied among some test statistics but not all, we could apply the weighted Simes test to blocks of hypotheses (Lu, 2016).

Although consonance may not always hold for Simes-based graphical procedures, it is possible to derive a partially sequentially rejective algorithm without going through the full closed test procedure. The weighted Simes test is uniformly more powerful than the Bonferroni test, which means that the former rejects all hypotheses rejected by the latter and possibly more. Then we could apply the Bonferroni-based graphical procedure first and only use the Simes-based procedure for the remaining not-yet-rejected hypotheses. The following algorithm adopts the one proposed by Bretz et al. (2011) and assumes that $\sum_{j\in J} w_j(J) = 1$ for all $J \subseteq I$.

Algorithm 5.5 *Graphical approach based on weighted Simes tests*

(1) If $p_i > \alpha$ for all $i \in I$, stop and retain all m hypotheses.

(2) If $p_i \leq \alpha$ for all $i \in I$, stop and reject all m hypotheses.

(3) Perform the Bonferroni-based graphical procedure based on Algorithm 5.1 in Section 5.2.2. Let $I = I^r \cup I^c$, where I^r denotes the index set of rejected hypotheses from Step (2) and I^c denotes its complement set. If $|I^c| < 3$, stop and retain the remaining hypotheses in I^c.

(4) If $|I^c| \geq 3$, calculate the local weights $w_i(I^c)$ and the update transition matrix for the set I^c using Algorithm 5.3 in Section 5.2.7.

(5) Reject $H_i, i \in I^c$ if for each $J \subseteq I^c$ with $i \in J$, there exists at least one $j \in J$ such that $p_j \leq \sum_{k\in J_j} w_k(J)\alpha$.

In Step (2), if all unadjusted p-values are not larger than α, every intersection hypothesis H_J is rejected because its largest p-value $p_j, j \in J$ is always compared with $\sum_{i\in J_j} w_i(J)\alpha = \sum_{i\in J} w_i(J)\alpha = \alpha$. In Step (3), if $|I^c| < 3$, there is a p-value larger than α; otherwise, all hypotheses would have been rejected in Step (2). In this case, the weighted Simes test is equivalent to the weighted Bonferroni test for one or two hypotheses.

Returning to the COPD example in Section 5.2.3, the initial graph in Figure 5.3 leads to the weighting scheme in Table 5.1. For $H_I = H_1 \cap H_2 \cap H_3 \cap H_4$, the local weights are $1/2, 1/2, 0, 0$. Let $p_1 = 0.01$ and $p_2 = 0.012$. Using a Bonferroni test at level α, H_I is rejected because $p_1 \leq \alpha/2 = 0.0125$ or $p_2 \leq \alpha/2 = 0.0125$. Since the primary hypotheses represent the high and low doses compared with the common control, their test statistics are positively correlated. Thus, the Simes test may be applicable and H_I is therefore rejected because $p_1 \leq \alpha/2 = 0.0125$ or $p_2 \leq \alpha = 0.025$. Equivalently, we also reject H_I since the p-value of H_I is $\tilde{p}_I = \min\{p_1/w_1, p_2/(w_1 + w_2)\} = \min\{2p_1, p_2\} = p_2 \leq \alpha = 0.025$. For other intersection hypotheses not involving H_1 and H_2 simultaneously, we could still apply the Bonferroni test. Note that for $H_J = H_1 \cap H_3 \cap H_4$, it can be rejected by the Bonferroni test if $p_1 \leq 3\alpha/4$ or $p_4 \leq \alpha/4$. Thus the rejection of H_I does not necessarily imply the rejection of H_J, which is the case if, for example, $p_1 > 3\alpha/4$, $p_2 \leq \alpha$ and $p_4 > \alpha/4$. Thus, we need to go through the closed test procedure and reject H_i if all H_J are rejected for $J \ni i$.

5.3.3 Graphical Approaches for Group Sequential Designs

In group sequential designs, hypotheses are tested repeatedly in time after groups of patients complete assessment (Jennison and Turnbull, 1999). Group sequential designs allow stopping the trial early to reject a null hypothesis at an interim analysis or otherwise testing the same hypothesis at the final analysis. Spending functions are often used to control the Type I error rate such that a boundary value is calculated to compare with the unadjusted p-value for the rejection decision. With a monotonicity condition, this design could be combined with the graphical approach to test multiple hypotheses at multiple time points, while controlling the FWER strongly at a prespecified level α.

Consider testing a null hypothesis $H: \theta \leq 0$ versus the upper alternative, at $h-1$ interim analyses and the final analysis. Using the canonical distribution (Jennison and Turnbull, 1999), we assume a multivariate normal distribution among test statistics $Z_t, t = 1, \ldots, h$ for analysis t, which has $\mathrm{E}(Z_t) = \theta\sqrt{I_t}$ and $\mathrm{Cov}(Z_t, Z_{t'}) = \sqrt{I_t/I_{t'}}$ for $t < t'$. Here, I_t is the statistical information available at analysis t, which is often proportional to the number of patients completing assessment at analysis t and inversely proportional to the standard deviation of the measurement. Define the information time as $y_t = I_t/I_h \in [0,1]$ which is the proportion of information at analysis t relative to that at the final analysis h. Then we define a spending function $\alpha(\gamma, y)$ with the total significance level $0 < \gamma < 1$ and the information time point y. This function determines the cumulative spending of γ up to the time y with $\alpha(\gamma, 0) = 0$, $\alpha(\gamma, 1) = \gamma$ and $\alpha(\gamma, y) \leq \alpha(\gamma, y')$ for $0 \leq y < y' \leq 1$. For analysis t with the information time y_t, the level spent at this analysis is $\alpha_t(\gamma) = \alpha(\gamma, y_t) - \alpha(\gamma, y_{t-1})$. The boundary value for the unadjusted p-value at analysis t could be obtained by solving $\alpha_t^*(\gamma)$ from

$$\alpha_t(\gamma) = P\left[\{p_t \leq \alpha_t^*(\gamma)\} \cap \bigcap_{s=1}^{t-1} \{p_s > \alpha_s^*(\gamma)\}\right],$$

where p_s is the unadjusted p-value of the test statistic at analysis s and $\alpha_s^*(\gamma)$ is the boundary value at analysis s.

To extend to multiple hypotheses, a monotonicity condition is needed on the spending function such that for $\gamma' \geq \gamma$ and for all $t = 1, \ldots, h$,

$$\alpha_t(\gamma') \geq \alpha_t(\gamma) \Rightarrow \alpha_t^*(\gamma') \geq \alpha_t^*(\gamma). \tag{5.8}$$

This condition ensures the validity of the Bonferroni-based graphical procedure for group sequential designs (Maurer and Bretz, 2013b). It has been verified that many commonly used spending functions satisfy this condition including La-DeMets O'Brien-Fleming and Pocock functions (Lan and DeMets, 1983).

Consider testing m null hypotheses $H_1, \ldots, H_m$ at analyses $t = 1, \ldots, h$ using a Bonferroni-based graphical procedure. Assign a spending function $\alpha_i(\gamma, y)$ to H_i, which has the spent level $\alpha_{i,t}(\gamma)$ at analysis t and the boundary value $\alpha_{i,t}^*(\gamma)$ for the unadjusted p-value. Based on the weighting scheme, an intersection hypothesis $H_J, J \subseteq I = \{1, \ldots, m\}$ is rejected at analysis t if there is an index $j \in J$ such that $p_{j,t} \leq \alpha_{j,t}^*(w_j(J)\alpha)$. It has been shown that under the monotonicity condition (5.8) the Bonferroni-based graphical procedure for group sequential designs could be implemented with a sequentially rejective algorithm as follows (Maurer and Bretz, 2013b).

Algorithm 5.6 *Graphical approach based on weighted Bonferroni tests, at $h-1$ interim analyses*

(0) Set $t = 1$ and $I = \{1, \ldots, m\}$.

(1) Calculate the unadjusted p-value $p_{i,t}$ and the boundary value $\alpha^*_{i,t}(w_i(I)\alpha)$ for $H_i, i \in I$.

(2) Select $j \in I$ such that $p_{j,t} \le \alpha^*_{j,t}(w_j(I)\alpha)$, reject H_j, and continue to Step (3); otherwise, if $t < h$, the trial continues to the next analysis $t \to t+1$ and go to Step (1).

(3) Update the graph:

$$I \to I \setminus \{j\}$$

$$w_\ell(I) \to \begin{cases} w_\ell(I) + w_j(I) g_{j\ell}, & \ell \in I, \\ 0, & \text{otherwise}, \end{cases}$$

$$g_{\ell k} \to \begin{cases} \frac{g_{\ell k} + g_{\ell j} g_{jk}}{1 - g_{\ell j} g_{j\ell}}, & \ell \neq k \in I, g_{\ell j} g_{j\ell} < 1, \\ 0, & \text{otherwise}. \end{cases}$$

(4) If $|I| \geq 1$, go to Step (1); otherwise stop.

We refer to Maurer and Bretz (2013b) for numerical examples illustrating Algorithm 5.6. Further extensions include choosing more flexible group sequential plans (Xi and Tamhane, 2015), incorporating the correlation between the test statistics (Rosenblum et al., 2016), and adaptive group sequential designs (Klinglmueller et al., 2014; Sugitani et al., 2016).

5.3.4 Symmetric Graphs for Equally Weighted Tests and the Hochberg Procedure

In the previous sections, we focused on tests for intersection hypotheses allowing different local weights for different hypotheses. These are readily compatible with the weighting scheme and thus the graphical approach. However, other tests that treat hypotheses with equal weights such as the Hochberg procedure and its variants (Gou et al., 2014; Hochberg, 1988). In addition, many omnibus tests do not rely on weighted individual p-values, including Fisher's combination (1922), O'Brien's (1984) and F tests. They combine in different ways the information from elementary hypotheses, but the common underlying assumption is that all elementary hypotheses are equally weighted. Thus, in order to use the Hochberg procedure and the aforementioned omnibus tests with the graphical approach, a weighting scheme needs to be developed that assigns equal weights to the elementary hypotheses.

Consider a weighting scheme $\boldsymbol{W}$ generated via Algorithm 5.3 in Section 5.2.7 by an initial graph with local weights $\boldsymbol{w}$ and transition matrix $\boldsymbol{G}$. The weighting scheme is symmetric if $w_i(J) = w_j(J)$ for all $i \neq j \in J$ and all $J \subseteq I = \{1, \ldots, m\}$. In this case, we could apply any of these tests at the significance level $w_{.}(J)\alpha$ without relying on the individual local weights, where $w_{.}(J) = \sum_{j \in J} w_j(J)$. To generate such weighting schemes, we introduce symmetric graphs (Xi and Bretz, 2019). An initial graph is symmetric if $w_i = w_j$ and $g_{ij} = g_{ik}$ for all $i \neq j \neq k \in I$, which means that all nodes have the same local weights and all outgoing edges from the same node have the same transition weights. It has been shown that updating a symmetric initial graph according to Algorithm 5.3 in Section 5.2.7 leads to symmetric reduced graphs that assign to each intersection hypothesis a set of equal local weights to the elementary hypotheses (Xi and Bretz, 2019). To further make the graph exhaustive in transition weights, we could require $\sum_{j \in I} g_{ij} = 1$ and thus $g_{ij} = 1/(m-1)$ for all $i \neq j \in I$.

(a) Initial graph

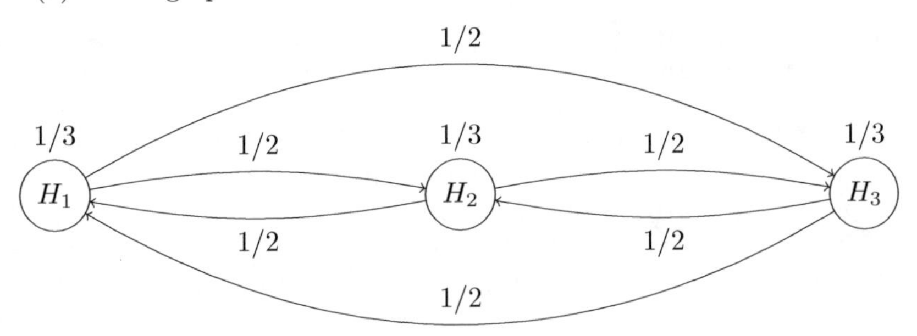

(b) Reduced graph after removing H_2

1/2 1 1/2

H_1 H_3

1

(c) Reduced graph after removing H_1, H_2

1

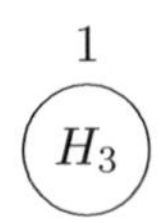

FIGURE 5.6
Symmetric graphs for three hypotheses using the Hochberg procedure.

To visualize a symmetric graph, consider three equally weighted null hypotheses H_i with $w_i = 1/3, i \in I = \{1, 2, 3\}$. Let the transition weights be 1/2 between any pair of hypotheses. Figure 5.6 visualizes the initial graph and the updated graphs, which are all symmetric. This is the same graph as the one for the Holm (1979) procedure as illustrated by Bretz et al. (2009). It could be verified that the resulting weighting scheme is also symmetric because the local weights are 1/3, 1/2 and 1 for intersection hypotheses with 3, 2 and 1 hypothesis, respectively.

To apply, for example, the Hochberg procedure, assume unadjusted p-values $p_1, \dots, p_m$ for $H_1, \dots, H_m$, respectively. Let $p_{(1)} \leq p_{(2)} \leq \dots \leq p_{(m)}$ denote the ordered p-values corresponding to $H_{(1)}, H_{(2)}, \dots, H_{(m)}$, respectively. The Hochberg procedure operates as follows:

1. If $p_{(m)} \leq \alpha$, reject all hypotheses and stop. Otherwise, retain $H_{(m)}$ and test $H_{(m-1)}$.

2. In general, if $p_{(j)} \leq \alpha/(m-j+1)$, reject $H_{(j)}, \dots, H_{(1)}$ and stop. Otherwise, retain $H_{(j)}$ and test $H_{(j-1)}$ until $j = 1$.

Liu (1996) showed that the Hochberg procedure performed at level α is equivalent to a closed test procedure which rejects an intersection hypothesis H_J at level α if

$$p_{(j)}(J) \leq \alpha/(|J| - j + 1) \text{ for at least one } j = 1, \dots, |J|, \tag{5.9}$$

where $p_{(1)}(J) \leq \dots \leq p_{(|J|)}(J)$ are the ordered p-values based on $p_j, j \in J$. Because the rejection rule (5.9) relies on the significance level of the intersection instead of the local weights of its elementary hypotheses, symmetric graphs in Figure 5.6 visualize the Hochberg procedure with three hypotheses. We can use this graph to visualize also the Hommel (1988) procedure as well as several omnibus tests, such as Fisher's combination (1922), O'Brien's (1984) and F tests. In addition, Xi and Bretz (2019) provided further discussions on conditions for consonance.

5.3.5 Graphical Approaches for Families of Hypotheses

We now extend the graphical approach to test multiple families of hypotheses. We represent each of such family by a single node and describe the relationship among families using the graphical approach. A multiple comparison procedure is applied to the hypotheses within each family. We remove a node from the graph if all hypotheses within a single family are rejected. Going back to the COPD example in Section 5.2.3, we consider adding a primary endpoint which is the symptom score describing a patient's quality of life. Along with the original primary endpoint of FEV1, we test the exacerbation endpoint only if the investigational drug is superior to control on both FEV1 and the symptom score. As a result, Figure 5.3 still applies but now H_1 represents two hypotheses: one for FEV1 and one for the symptom score. In this case, the local weight of H_1 is propagated only if both hypotheses within H_1 are rejected, followed by removing H_1 and updating the graph according to Algorithm 1. The same holds true for H_2.

Formally, let $\mathcal{H}_i, i \in I = \{1, \ldots, m\}$ denote family i with k_i hypotheses $H_{i,j}$ in this family and j denote the index in $I_i = \{(i,1), \ldots, (i,k_i)\}$. Assume that a multiple comparison procedure satisfying the α-consistency condition is applied to family $\mathcal{H}_i$, i.e., if any hypothesis in $\mathcal{H}_i$ is rejected at level α, it will also be rejected at level $\alpha' > \alpha$ (Hommel and Bretz, 2008). We can then calculate adjusted p-values for each hypothesis in this family as the smallest significance level at which this hypothesis can be rejected (see Section 5.2.4), denoted by $p^*_{i,j}$. Then we could define a sequentially rejective procedure as follows (Maurer and Bretz, 2014).

Algorithm 5.7 *Graphical approach based on weighted Bonferroni tests for families of hypotheses*

(1) Select $(i,j) \in I_i$ such that $p^*_{ij} \leq w_i\alpha$, reject H_{ij} and set I_i to $I_i \setminus \{(i,j)\}$; otherwise stop.

(2) If $|I_i| \geq 1$, go to Step (1); otherwise update the graph:

$$
\begin{aligned}
I &\rightarrow I \setminus \{i\} \\
w_\ell(I) &\rightarrow \begin{cases} w_\ell(I) + w_j(I) g_{j\ell}, & \ell \in I, \\ 0, & \text{otherwise}, \end{cases} \\
g_{\ell k} &\rightarrow \begin{cases} \frac{g_{\ell k} + g_{\ell j} g_{jk}}{1 - g_{\ell j} g_{j\ell}}, & \ell \neq k \in I, g_{\ell j} g_{j\ell} < 1, \\ 0, & \text{otherwise}. \end{cases}
\end{aligned}
$$

(3) If $|I| \geq 1$, go to Step (1); otherwise stop.

This framework includes several procedures mentioned previously such as serial gatekeeping (see Section 5.2.8) and symmetric graphs (see Section 5.3.4). A similar approach to allow for parallel gatekeeping among families was discussed by Kordzakhia and Dmitrienko (2013).

5.3.6 Entangled Graphs

All graphical test procedures described in the previous sections do not have memory. That is, the local weight propagated from two hypotheses to a third hypotheses are combined without being memorized where it came from. In some clinical trial applications, however, it is desirable to remember from where the local weight is propagated, to further reflect underlying structures among hypotheses. The entangled graphs allow such a more complex situation.

Returning to the COPD example in Section 5.2.3, we consider adding an endpoint on the symptom score between the primary endpoint on FEV1 and the secondary endpoint on exacerbation. For a dose-control comparison, if we reject the null hypothesis on FEV1, we will test the hypothesis on the symptom score, using data from both doses, i.e., we will pool data from the high and low dose groups and compare against data from control. If the hypothesis on the symptom score is rejected, we then test the hypothesis on exacerbation within each dose-control comparison, respectively, following the requirement that the primary hypothesis on FEV1 has to be rejected before testing the secondary hypothesis on exacerbation. A graphical visualization could be found in Figure 5.7 by adding H_5 to the graph in Figure 5.3 for the hypothesis on the symptom score. Suppose we first reject H_1. The updated graph assigns H_5 a local weight of 1/2. If we further reject H_5, both H_3 and H_4 have local weights of 1/4 in the updated graph. Here this graph does not remember that H_2 has not been rejected but gives H_4 a positive local weight without having rejected the primary hypothesis H_2 for that dose-control comparison.

To allow memory, we use one subgraph to memorize propagation from H_1 and one subgraph for H_2. In each subgraph, we have all nodes for all hypotheses. The local weights and the transition weights are zero if they are not associated with propagation from H_1. When testing a hypothesis, its local weight is the combination of the local weights in both subgraphs. For example, the local weight of H_1 is 1/2 in the first subgraph and 0 in the second subgraph. Thus, its local weight for testing is $1/2+0=1/2$. Figure 5.8 visualizes the two subgraphs, the entangled graph and the updated graphs after H_1 and H_5 are rejected. We could see that subgraphs are presented as separate dimensions in the entangled graph, and the updating is done within each dimension.

Formally, consider testing m hypotheses using n subgraphs to handle n underlying structures. Let $\boldsymbol{w}_h = (w_{h1}, \ldots, w_{hm})$ and $\boldsymbol{G}_h = (g_{hij})_{i,j=1,\ldots,m}$ denote the local weight vector and the transition matrix for subgraph $h = 1, \ldots, n$. To control the FWER at level α, we have $\sum_{h=1}^{n} \sum_{i=1}^{m} w_{hi} \leq 1$ and each transition matrix satisfies the regularity conditions in (5.1). Hypothesis H_i could be tested at level $\sum_{h=1}^{n} w_{hi}\alpha$ and if rejected, the updating of graph is done within each subgraph (or dimension) as follows:

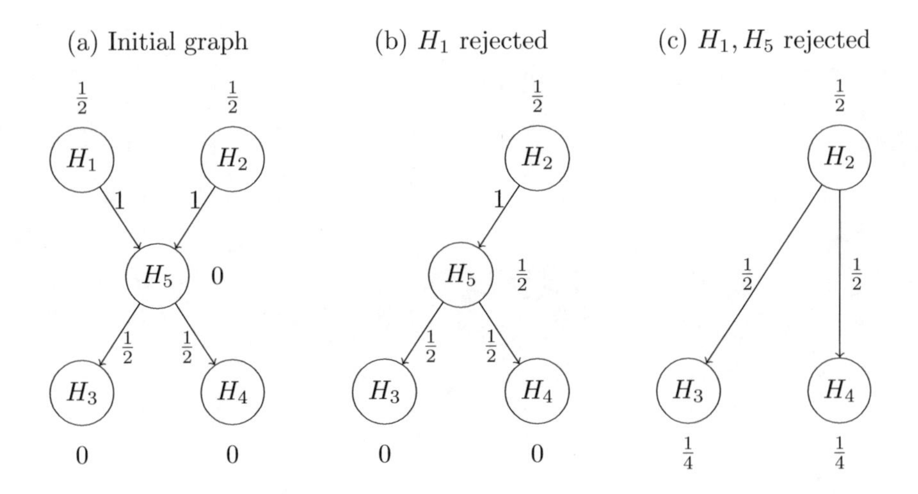

FIGURE 5.7
Graphical multiple comparison procedure without memory.

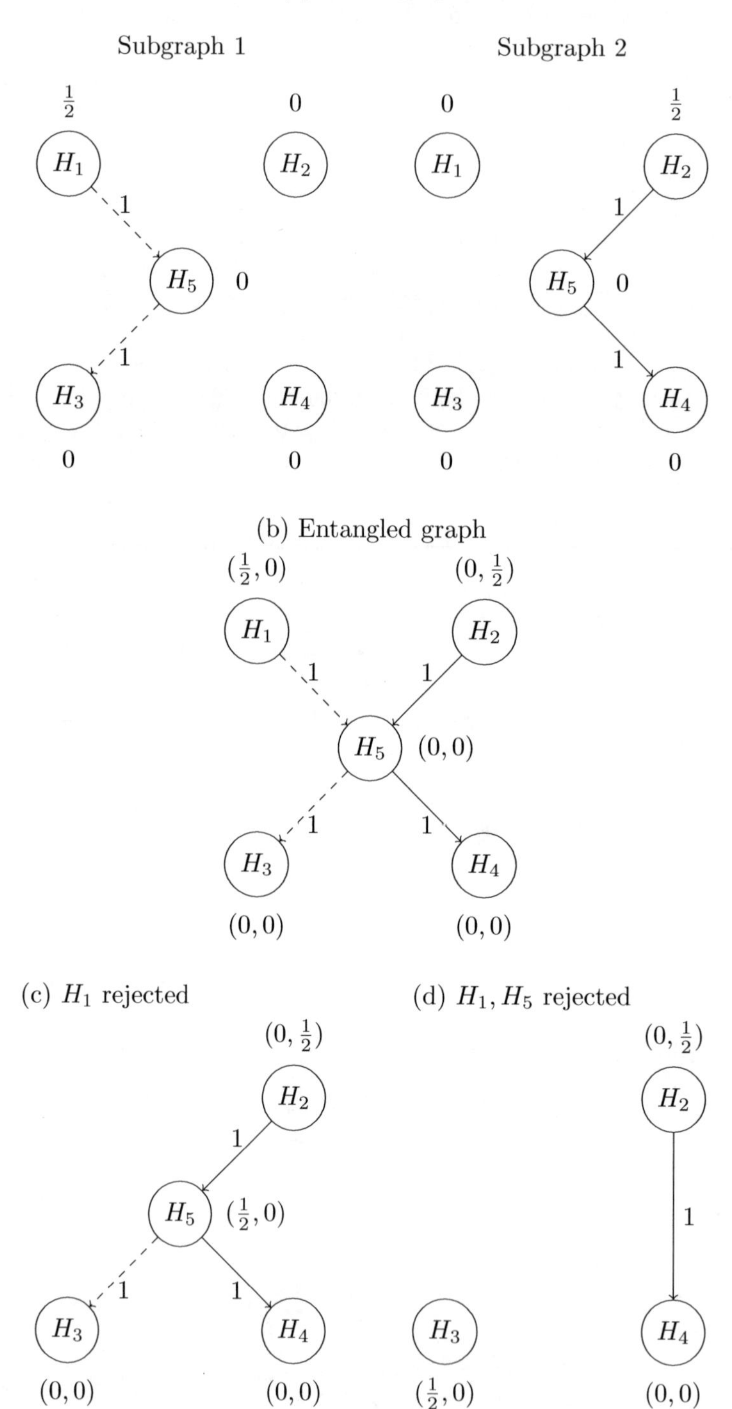

FIGURE 5.8
Graphical multiple comparison procedure with memory.

Algorithm 5.8 *Entangled graphical approach based on weighted Bonferroni tests*

(0) Set $I = \{1, \dots, m\}$.

(1) Select $j \in I$ such that $p_j \leq \sum_{h=1}^{n} w_{hj}\alpha$ and reject H_j; otherwise stop.

(2) Update the graph:

$$I \to I \setminus \{j\}$$

$$w_{h\ell}(I) \to \begin{cases} w_{h\ell}(I) + w_{hj}(I) g_{hj\ell}, & \ell \in I, h = 1, \dots, n, \\ 0, & \text{otherwise,} \end{cases}$$

$$g_{h\ell k} \to \begin{cases} \frac{g_{h\ell k} + g_{h\ell j} g_{hjk}}{1 - g_{h\ell j} g_{hj\ell}}, & \ell \neq k \in I, g_{h\ell j} g_{hj\ell} < 1, h = 1, \dots, n, \\ 0, & \text{otherwise.} \end{cases}$$

(3) If $|I| \geq 1$, go to Step (1); otherwise stop.

The validity of the entangled graph was established by Maurer and Bretz (2013a), and it was shown to be equivalent to the default graph by Burman et al. (2009). Following the idea of weighting schemes, we can also extend Algorithm 5.3 to derive the weighting scheme for entangled graphs. As a result, previously mentioned tests remain applicable, including parametric, Simes, Hochberg, and group sequential tests. In addition, entangled graphs can also be used to visualize the truncated Holm procedure (Dmitrienko et al., 2008; Strassburger and Bretz, 2008) and several k-out-of-n gatekeeping procedures (Xi and Tamhane, 2014). In addition, Maurer and Bretz (2013a) provided more detail and further extensions.

5.4 Conclusions

Graphical approaches offer the possibility and flexibility to tailor multiple comparison procedures to structured families of hypotheses and to visualize complex decision strategies in an efficient and easily communicable way, while controlling the FWER strongly at a prespecified significance level α. As we illustrated, this framework includes many commonly used multiple comparison procedures, and it opens the door to advanced multiple comparison procedures for prioritizing clinical trial objectives. While the discussions in this chapter are more focused on decision-making in clinical trials, it should be acknowledged that multiplicity has a much broader impact throughout many applications. Good practice to ensure sound decision-making and reproducibility should always be appreciated.

Bibliography

Alosh, M., F. Bretz, and M. Huque (2014). Advanced multiplicity adjustment methods in clinical trials. *Statistics in Medicine 33*(4), 693–713.

Benjamini, Y. and R. Heller (2007). False discovery rates for spatial signals. *Journal of the American Statistical Association 102*(480), 1272–1281.

Benjamini, Y. and Y. Hochberg (1997). Multiple hypotheses testing with weights. *Scandinavian Journal of Statistics 24*(3), 407–418.

Brannath, W. and F. Bretz (2010). Shortcuts for locally consonant closed test procedures. *Journal of the American Statistical Association 105*(490), 660–669.

Bretz, F., W. Maurer, W. Brannath, and M. Posch (2009). A graphical approach to sequentially rejective multiple test procedures. *Statistics in Medicine 28*, 586–604.

Bretz, F., W. Maurer, and G. Hommel (2011). Test and power considerations for multiple endpoint analyses using sequentially rejective graphical procedures. *Statistics in Medicine 30*(13), 1489–1501.

Bretz, F., W. Maurer, and J. Maca (2014). Graphical approaches to multiple testing. In W. R. Young and D.-G. D. Chen (Eds.), *Clinical Trial Biostatistics and Biopharmaceutical Applications*, Chapter 14, pp. 349–394. Chapman and Hall/CRC, Boca Raton, FL.

Bretz, F., W. Maurer, and D. Xi (2019). Replicability, reproducibility, and multiplicity in drug development. *Chance 32*(4), 4–11.

Bretz, F., M. Posch, E. Glimm, F. Klinglmueller, W. Maurer, and K. Rohmeyer (2011). Graphical approaches for multiple comparison procedures using weighted Bonferroni, Simes, or parametric tests. *Biometrical Journal 53*(6), 894–913.

Burman, C.-F., C. Sonesson, and O. Guilbaud (2009). A recycling framework for the construction of Bonferroni-based multiple tests. *Statistics in Medicine 28*, 739–761.

Chen, J., J. Luo, K. Liu, and D. V. Mehrotra (2011). On power and sample size computation for multiple testing procedures. *Computational Statistics & Data Analysis 55*(1), 110–122.

Dmitrienko, A., W. W. Offen, and P. H. Westfall (2003). Gatekeeping strategies for clinical trials that do not require all primary effects to be significant. *Statistics in Medicine 22*, 2387–2400.

Dmitrienko, A., A. C. Tamhane, and B. L. Wiens (2008). General multistage gatekeeping procedures. *Biometrical Journal 50*, 667–677.

Dmitrienko, A., B. L. Wiens, A. C. Tamhane, and X. Wang (2007). Tree-structured gatekeeping tests in clinical trials with hierarchically ordered multiple objectives. *Statistics in Medicine 26*(12), 2465–2478.

Dunnett, C. W. (1955). A multiple comparison procedure for comparing several treatments with a control. *Journal of the American Statistical Association 50*(272), 1096–1121.

Dunnett, C. W. and A. C. Tamhane (1991). Step-down multiple tests for comparing treatments with a control in unbalanced one-way layouts. *Statistics in Medicine 10*(6), 939–947.

EMA (2017). *Guideline on multiplicity issues in clinical trials (Draft)*. European Medicines Agency. Available at `http://www.ema.europa.eu/docs/en_GB/document_library/Scientific_guideline/2017/03/WC500224998.pdf`.

FDA (2017). *Guidance for Industry Multiple Endpoints in Clinical Trials (Draft)*. Food and Drug Administration. Available at `https://www.fda.gov/regulatory-information/search-fda-guidance-documents/multiple-endpoints-clinical-trials-guidance-industry`.

Finner, H. and K. Strassburger (2002). The partitioning principle: a powerful tool in multiple decision theory. *Annals of Statistics*, 1194–1213.

Finner, H., S.-Y. Tang, X. Cui, and J. C. Hsu (2021). Partitioning for confidence sets, confident directions, and decision paths. In X. Cui, T. Dickhaus, Y. Ding, and J. C. Hsu (Eds.), *Handbook of Multiple Comparisons*. Boca Raton, FL: Chapman and Hall/CRC.

Fisher, R. A. (1922). *Statistical Methods for Research Workers*. Oliver and Boyd, London.

Gabriel, K. R. (1969). Simultaneous test procedures–some theory of multiple comparisons. *The Annals of Mathematical Statistics 40*(1), 224–250.

Genz, A., F. Bretz, T. Miwa, X. Mi, F. Leisch, F. Scheipl, B. Bornkamp, M. Maechler, T. Hothorn, and M. T. Hothorn (2020). Package 'mvtnorm'. *Journal of Computational and Graphical Statistics 11*, 950–971.

Glimm, E., D. Xi, and P. Gallo (2019). Multiple comparisons, multiple primary endpoints and subpopulation analysis. In S. Halabi and S. Michiels (Eds.), *Textbook of Clinical Trials in Oncology: A Statistical Perspective*, Chapter 10, pp. 183–202. Boca Raton, FL: CRC Press.

Goteti, S., S. Hirawat, C. Massacesi, N. Fretault, F. Bretz, and B. Dharan (2014). Some practical considerations for phase iii studies with biomarker evaluations. *Journal of Clinical Oncology 32*(8), 854–855.

Gou, J., A. C. Tamhane, D. Xi, and D. Rom (2014). A class of improved hybrid hochberg–hommel type step-up multiple test procedures. *Biometrika 101*(4), 899–911.

Guilbaud, O. (2008). Simultaneous confidence regions corresponding to holm's step-down procedure and other closed-testing procedures. *Biometrical Journal 50*(5), 678–692.

Guilbaud, O. (2011). Note on simultaneous inferences about non-inferiority and superiority for a primary and a secondary endpoint. *Biometrical Journal 53*(6), 927–937.

Guilbaud, O. (2018). Simultaneous confidence intervals compatible with sequentially rejective graphical procedures. *Statistics in Biopharmaceutical Research 10*(3), 220–232.

Hochberg, Y. (1988). A sharper Bonferroni procedure for multiple tests of significance. *Biometrika 75*(4), 800–802.

Holm, S. (1979). A simple sequentially rejective multiple test procedure. *Scandinavian Journal of Statistics 6*(2), 65–70.

Hommel, G. (1988). A stagewise rejective multiple test procedure based on a modified Bonferroni test. *Biometrika 75*(2), 383–386.

Hommel, G. and F. Bretz (2008). Aesthetics and power considerations in multiple testing–a contradiction? *Biometrical Journal 50*(5), 657–666.

Hommel, G., F. Bretz, and W. Maurer (2007). Powerful short-cuts for multiple testing procedures with special reference to gatekeeping strategies. *Statistics in Medicine 26*(22), 4063–4073.

Hung, J. H. and S.-J. Wang (2010). Challenges to multiple testing in clinical trials. *Biometrical Journal 52*(6), 747–756.

Huque, M. F., M. Alosh, and R. Bhore (2011). Addressing multiplicity issues of a composite endpoint and its components in clinical trials. *Journal of Biopharmaceutical Statistics 21*(4), 610–634.

Jennison, C. and B. W. Turnbull (1999). *Group sequential methods with applications to clinical trials*. Boca Raton, FL: Chapman and Hall/CRC.

Julious, S. A. and N. E. McIntyre (2012). Sample sizes for trials involving multiple correlated must-win comparisons. *Pharmaceutical Statistics 11*(2), 177–185.

Klinglmueller, F., M. Posch, and F. König (2014). Adaptive graph-based multiple testing procedures. *Pharmaceutical Statistics 13*(6), 345–356.

Kordzakhia, G. and A. Dmitrienko (2013). Superchain procedures in clinical trials with multiple objectives. *Statistics in Medicine 32*(3), 486–508.

Lan, G. K. and D. L. DeMets (1983). Discrete sequential boundaries for clinical trials. *Biometrika 70*(3), 659–663.

Lawrence, J. (2011). Testing non-inferiority and superiority for two endpoints for several treatments with a control. *Pharmaceutical Statistics 10*(4), 318–324.

Liu, W. (1996). Multiple tests of a non-hierarchical finite family of hypotheses. *Journal of the Royal Statistical Society: Series B (Methodological) 58*(2), 455–461.

Lu, K. (2016). Graphical approaches using a Bonferroni mixture of weighted Simes tests. *Statistics in Medicine 35*(22), 4041–4055.

Marcus, R., E. Peritz, and K. R. Gabriel (1976). On closed testing procedures with special reference to ordered analysis of variance. *Biometrika 63*(3), 655–660.

Maurer, W. and F. Bretz (2013a). Memory and other properties of multiple test procedures generated by entangled graphs. *Statistics in Medicine 32*(10), 1739–1753.

Maurer, W. and F. Bretz (2013b). Multiple testing in group sequential trials using graphical approaches. *Statistics in Biopharmaceutical Research 5*(4), 311–320.

Maurer, W. and F. Bretz (2014). A note on testing families of hypotheses using graphical procedures. *Statistics in Medicine 33*(30), 5340–5346.

Maurer, W., E. Glimm, and F. Bretz (2011). Multiple and repeated testing of primary, coprimary, and secondary hypotheses. *Statistics in Biopharmaceutical Research 3*(2), 336–352.

Maurer, W., L. A. Hothorn, and W. Lehmacher (1995). Multiple comparisons in drug clinical trials and preclinical assays: a-priori ordered hypotheses. *Biometrie in der chemisch-pharmazeutischen Industrie 6*, 3–18.

Maurer, W. and B. Mellein (1988). On new multiple tests based on independent p-values and the assessment of their power. In *Multiple Hypothesenprüfung/Multiple Hypotheses Testing*, pp. 48–66. New York City, NY: Springer.

Millen, B. A. and A. Dmitrienko (2011). Chain procedures: A class of flexible closed testing procedures with clinical trial applications. *Statistics in Biopharmaceutical Research 3*(1), 14–30.

O'Brien, P. C. (1984). Procedures for comparing samples with multiple endpoints. *Biometrics*, 1079–1087.

Rauch, G. and J. Beyersmann (2013). Planning and evaluating clinical trials with composite time-to-first-event endpoints in a competing risk framework. *Statistics in Medicine 32*(21), 3595–3608.

Rohmeyer, K. and F. Klinglmueller (2014). *gMCP: Graph Based Multiple Test Procedures.* R package version 0.8-8. Available at `http://CRAN.R-project.org/package=gMCP`.

Rosenblum, M., T. Qian, Y. Du, H. Qiu, and A. Fisher (2016). Multiple testing procedures for adaptive enrichment designs: combining group sequential and reallocation approaches. *Biostatistics 17*(4), 650–662.

Senn, S. and F. Bretz (2007). Power and sample size when multiple endpoints are considered. *Pharmaceutical Statistics 6*(3), 161–170.

Simes, R. J. (1986). An improved Bonferroni procedure for multiple tests of significance. *Biometrika 73*(3), 751–754.

Sozu, T., T. Sugimoto, and T. Hamasaki (2010). Sample size determination in clinical trials with multiple co-primary binary endpoints. *Statistics in Medicine 29*(21), 2169–2179.

Strassburger, K. and F. Bretz (2008). Compatible simultaneous lower confidence bounds for the Holm procedure and other Bonferroni-based closed tests. *Statistics in Medicine 27*(24), 4914–4927.

Sugitani, T., F. Bretz, and W. Maurer (2016). A simple and flexible graphical approach for adaptive group-sequential clinical trials. *Journal of Biopharmaceutical Statistics 26*(2), 202–216.

Westfall, P. H. and A. Krishen (2001). Optimally weighted, fixed sequence and gatekeeper multiple testing procedures. *Journal of Statistical Planning and Inference 99*, 25–40.

Westfall, P. H., A. Krishen, and S. S. Young (1998). Using prior information to allocate significance levels for multiple endpoints. *Statistics in Medicine 17*(18), 2107–2119.

Westfall, P. H. and S. S. Young (1993). *Resampling-based multiple testing: Examples and methods for p-value adjustment*, Volume 279. Hoboken, NJ: John Wiley & Sons.

Westfall, P. H., S. S. Young, and P. S. Wright (1993). On adjusting p-values for multiplicity. *Biometrics 49*(3), 941–945.

Wiens, B. L. (2003). A fixed sequence Bonferroni procedure for testing multiple endpoints. *Pharmaceutical Statistics 2*, 211–215.

Wiens, B. L. and A. Dmitrienko (2005). The fallback procedure for evaluating a single family of hypotheses. *Journal of Biopharmaceutical Statistics 15*(6), 929–942.

Xi, D. and F. Bretz (2019). Symmetric graphs for equally weighted tests, with application to the hochberg procedure. *Statistics in Medicine 38*(27), 5268–5282.

Xi, D., E. Glimm, and F. Bretz (2016). Multiplicity. In S. L. George, X. Wang, and H. Pang (Eds.), *Cancer Clinical Trials : Current and Controversial Issues in Design and Analysis*, Chapter 3, pp. 69–105. Boca Raton, FL: Chapman and Hall/CRC.

Xi, D., E. Glimm, W. Maurer, and F. Bretz (2017). A unified framework for weighted parametric multiple test procedures. *Biometrical Journal 59*(5), 918–931.

Xi, D. and A. C. Tamhane (2014). A general multistage procedure for k-out-of-n gatekeeping. *Statistics in Medicine 33*(8), 1321–1335.

Xi, D. and A. C. Tamhane (2015). Allocating recycled significance levels in group sequential procedures for multiple endpoints. *Biometrical Journal 57*(1), 90–107.

Xiong, C., K. Yu, F. Gao, Y. Yan, and Z. Zhang (2005). Power and sample size for clinical trials when efficacy is required in multiple endpoints: application to an alzheimer's treatment trial. *Clinical Trials 2*(5), 387–393.

6

Decision Theoretic Considerations of Multiple Comparisons

Arthur Cohen

Rutgers University

Harold Sackrowitz

Rutgers University

CONTENTS

6.1 Introduction

Statisticians have long been faced with settings that have required the simultaneous testing of many different hypotheses. In particular, for the last two decades, scores of multiple testing procedures (MTPs) have emerged to meet the call for new applications in diverse fields such as microarrays, astronomy, genomics, bioweapons use, mutual fund evaluations, proteomics, cytometry, imaging, school evaluations and others. MTPs play an important role in these and many other areas. Among these MTPs, many are stepwise procedures as opposed to procedures (called single step) that test each hypothesis separately. See for example Dudoit, Shaffer and Boldrick (2003), Hochberg and Tamhane (1987) and Lehmann and Romano (2005). The extreme conservativeness of single-step procedures precipitated the important development of many new (mostly stepwise) procedures.

The popularity of criteria on which to evaluate MTPs has varied a bit over time to reflect the popularity of current applications. Almost all criteria are related to the likelihood of Type I and Type II errors actually occurring. The vast majority of MTP research have focused either on the family-wise error rate (FWER) which is the probability of, at least, one Type I error or on the false discovery rate (FDR) which is the expected proportion of rejections that are false. The desire is to control one of these error rates while achieving high

DOI: 10.1201/9780429030888-7

power. This derives from the common practical notion that Type I error is more serious than Type II and that one needs convincing evidence to believe the alternative. This is a logical intuitive consideration, and it is unlikely that any approach would not consider in most applications.

In practice, many situations involve testing hypotheses that are based on comparisons of parameters. The most common of these are pairwise comparisons. That is, each hypothesis testing problem compares two of the parameters. The focus will be mainly on that situation. It will demonstrate the issues considered by a decision theory approach. The same path would be followed even in the more general case of tests involving contrasts.

At the heart of decision theory considerations is the specification of a loss function, $L(\theta, \mathbf{a})$. It specifies the amount that is lost if action $\mathbf{a}$ is taken when, in fact, the true parameter is θ. In multiple testing, both θ and $\mathbf{a}$ would be vectors. The actions $\mathbf{a}$ tell us which hypotheses are rejected. The performance of a decision procedure is then measured by its risk function. That is, the expected value of the loss function when using that procedure. As such, the risk function is a function of θ. One then attempts to find procedures that perform well relative to the risk function. For testing a single hypothesis the classical loss function assigns a loss of 1 to any incorrect decision and a loss of 0 otherwise. This leads to a focus on the standard notions of Type I and Type II errors.

Any loss function that, for each fixed θ, incurs a larger penalty for an incorrect decision than for a correct decision will lead to study of Type I and Type II error probabilities. Decision theory for testing a single individual hypothesis is well established (see, in particular, Lehmann and Romano (2005)). This is much less the case for multiple testing.

Since risk functions depend on θ it is rare that there exists a single procedure that will minimize the risk function for all θ. Thus, once a loss function is specified the notion of admissibility comes into play. A procedure δ is inadmissible if there exists another procedure, δ^*, whose risk function is less than or equal that of δ for every parameter θ and strictly less for some value of θ. When dealing with only a single null and alternative hypothesis, a testing procedure being inadmissible would, essentially, mean that there exists another testing procedure with both smaller probability of Type I and of Type II errors. Thus, for testing a single hypothesis, one might be leery of using an inadmissible procedure. It should be pointed out that admissibility alone does not necessarily make a procedure desirable. Not every admissible procedure is better than every inadmissible procedure. For this reason, at least, one secondary notion of optimality is almost always investigated. For example, unbiasedness and invariance are often used. Uniformly most powerful unbiased (UMPU) tests and uniformly most powerful invariant (UMPI) tests can often be found.

The usual FWER- and FDR-type criteria are in the spirit of size and power considerations of the Neyman - Pearson Lemma (Lehmann and Romano (2005)) for testing a single hypothesis. They are intuitively desirable, but, in the multiple testing setting, they are not really tied to admissibility with respect to some loss function. However, they are so intuitively ingrained in multiple testing that they would almost always be used as secondary properties.

Some researchers have also taken a finite action decision theory problem approach with a variety of loss functions (Genovese and Wasserman (2002)). In these studies, procedures are evaluated and compared by their risk functions. It is important to note that a risk function approach does not always necessitate the need to control a particular type of error rate. Dudoit and Van der Laan (2008) include the study of expected values of functions of numbers of Type I and Type II errors.

Owing to more recent applications in areas such as information theory, machine learning and coding, we notice the use of Hamming loss (see Butucea, C. et.al. (2018) and references therein) in various disciplines. This notion stems from information theory terminology. A Hamming distance in information technology represents the number of points at which two corresponding pieces of data can be different. It is often used in various kinds of error correction or evaluation of contrasting strings or pieces of data. This loss had occasionally been seen in multiple testing literature under different names. When problems are posed in a multiple testing setting, Hamming loss is nothing more than the total number of errors (Types I and II). In Genovese and Wasserman (2002), it was referred to as classification loss. Dudoit and Van der Laan (2008) and Cohen and Sackrowitz (2012) simply refer to the total number of testing errors. Such considerations have thus far played a minor role in the MTP literature.

6.2 Decision Theory Approach

By far, the most popular MTPs are stepwise. Stepwise multiple testing procedures are valuable because they are less conservative than standard single-step procedures, which often rely on Bonferroni critical values. In other words, they are more powerful than their single-step counterparts. In constructing stepwise testing procedures, it is common to begin with tests for the individual hypotheses that are known to have desirable properties. For example, the tests may be UMPU, they may have invariance properties and are almost certain to be admissible. Then, a sequential component is added that tells us which hypothesis to accept or reject at each step and when to stop.

One decision theory approach to the study of general MTPs and stepwise MTPs in particular is as follows. Begin by realizing that all MTPs (including stepwise procedures) induce new tests (likely using all the data) on the individual testing problems. Carrying out any MTP procedure in a multiple hypothesis testing problem is equivalent to applying these induced tests separately to each of the individual hypotheses. That equivalence makes these induced tests worthy of study. If the induced individual tests can be improved in both size and power, then, intuitively, one should feel as though the entire procedure is improved. The sequential nature of stepwise procedures in a multiple testing setting makes study of them extremely difficult, if not impossible, to do in any detail even in a modest number of dimensions. However, it is often the case that some convexity properties with very important implications can be established. Surprisingly, the induced tests frequently do not retain all the desirable properties that made people choose the original tests, used in their MTP construction, in the first place.

6.2.1 Key Property of Induced Tests

Each of the induced tests is used for a particular single hypothesis testing problem. However, these induced tests will typically be based on all the observed data. The necessary and sufficient conditions for admissibility established in Matthes and Truax (1967) can be applied to these induced tests. Those results describe certain convexity requirements on subsections of the acceptance regions.

In these multiple comparison models, these theoretical properties are also very important for practical purposes. They have been called (Cohen and Sackrowitz (2012)) the interval property and have to do with the convexity of acceptance regions. This is a desirable property that the original tests would almost certainly possess but that the stepwise induced tests can easily lose. Informally, the interval property is simply that the resulting test has acceptance sections that are intervals along certain lines.

To clarify, suppose one is constructing a test for a one-sided hypothesis testing problem. In addition to asking for other properties, it is sensible to examine the acceptance and rejection regions. There are often pairs of sample points, $\mathbf{X}$ and $\mathbf{X}^*$ for which there are compelling practical (and sometimes theoretical) reasons for the following to be true. If the point $\mathbf{X}$ is in the rejection region, then the point $\mathbf{X}^*$ should also be in the rejection region. The practical desirability of this property is usually because it is, intuitively, clear that $\mathbf{X}^*$ is a stronger indication of the alternative than is $\mathbf{X}$. In the case of two-sided hypotheses, there are often triples of points, $\mathbf{X}, \mathbf{X}^*$ and $\mathbf{X}^{**}$ (on the same line) such that if both $\mathbf{X}$ and $\mathbf{X}^{**}$ are in the acceptance region then one would also want $\mathbf{X}^*$ to be in the acceptance region if in fact $\mathbf{X}^*$ was not the most indicative of the alternative of the three points.

To further emphasize the compelling nature of an interval property consider an analogous, nonstatistical, situation. Suppose you are teaching a class where final grades are Pass or Fail. It would be totally unacceptable to pass students with grades of 60–70 or 75–100 yet fail those with scores of 71–74. This is in the same spirit of what could happen when having an acceptance region that is not properly convex. Notice that, if you had a small class, it might be unlikely that anyone would score between 71 and 74. That would not justify using such a procedure. Section 6.2.4 presents a similar example using a common MTP.

6.2.2 Models and Notation

A fundamental setting in multiple comparisons is to begin with n populations with the *ith* population characterized by some unknown parameter θ_i. Then there is interest in some number of pairwise comparisons among the θ_i. To fix ideas and demonstrate an approach to incorporating decision theory ideas in such settings, it will suffice to use a simple Normal model.

Let $Z_i, i = 1, \cdots, n$ have a normal distribution with mean θ_i and known variance σ^2. Without loss of generality, we take $\sigma^2 = 1$ for the remainder of section 6.2. The Z_i are independent. To be tested is some collection of hypotheses of the form $H_{ij} : \theta_i = \theta_j$ versus either $K_{ij} : \theta_i \neq \theta_j$ or $K_{ij} : \theta_i > \theta_j$. Thus, there is some set of pairs $Q = \{(ij) : H_{ij} \text{ is to be tested}\}$. This includes a wide variety of common models. For example, in treatments versus control, if we let population 1 be the control, our null hypotheses would be $H_{12}, H_{13}, \cdots, H_{1n}$. In a change point setting, the θ_i would occur sequentially, and we would be looking for a change. This leads to consideration of the hypotheses $H_{12}, H_{23}, \cdots, H_{(n-1)n}$.

If faced with only one of these hypothesis testing problems, one would use a simple, known variance, normal test. That is, in the one-sided case that test would reject if the statistic $T_{ij} = Z_i - Z_j > k_1$ and reject if the statistic $T_{ij} = |Z_i - Z_j| > k_2$ in the two-sided case for some k_1 and k_2. In the one-sided case, this test is uniformly most powerful, and in the two-sided case, there is no UMP test, but, it is UMPU and UMPI. In the face of many such hypotheses, likely choices of MTP would be such as the step-down Holm (1979) procedure or the step-up Benjamini-Hochberg procedure (1995).

Once a problem is specified and a Q is determined, the hypotheses can be relabeled, notation simplified so the general problem is transformed into the following equivalent form. Let M be the number of pairwise differences to be tested and use $X's$ to be the differences in the pairs of $Z's$ corresponding to the pairwise differences to be studied. Then, the random vector $\mathbf{X} = (X_1, \cdots, X_M)$ will have a multivariate normal distribution with mean vector $\nu = (\nu_1, \cdots, \nu_M)$ and some known covariance matrix Σ. Thus, the model consists of (6.1) and (6.2).

$$\mathbf{X} \sim N(\nu, \Sigma) \tag{6.1}$$

and we want to test, for all $i = 1, \cdots, M$,

$$H_i : \nu_i = 0 \quad \text{versus either} \quad K_i : \nu_i \neq 0 \quad \text{or} \quad K_i : \nu_i > 0. \tag{6.2}$$

For example, in the treatments versus control model, although there are n populations, only $n-1$ comparisons are to be made so $M = n-1$. $X_1 = Z_2 - Z_1, X_2 = Z_3 - Z_1, \cdots, X_M = Z_n - Z_1$. $\nu_1 = \theta_2 - \theta_1, \nu_2 = \theta_3 - \theta_1, \cdots, \nu_M = \theta_n - \theta_1$. The covariance matrix Σ has all diagonal elements equal to 2 and all other elements equal to 1. Finally, the test statistics for an individual hypothesis would be $T_i(\mathbf{X}) = X_i$ or $T_i(\mathbf{X}) = |X_i|$

Using this formulation, the generic step-down process and step-up process can be described in a fashion similar to Lehmann and Romano (2005, Sections 9.1 and 9.2). For the stepwise procedures, we suppose that the individual hypothesis H_i has a test based on the test statistic T_i with large values indicating evidence for K_i.

1. **Step-down process.** Fix constants $C_1 < ... < C_M$.

Step 1 Consider $U_{i_1} = \max_{\{1 \le i \le M\}} T_i$. If $U_{i_1} \le C_M$, stop and accept all H_i. If $U_{i_1} > C_M$ reject H_{i_1} and go to step 2.

Step 2 Consider $U_{i_2} = \max_{\{1 \le i \le M, i \ne i_1\}} T_i$. If $U_{i_2} \le C_{M-1}$ stop and accept all $H_i : i \in \{1, \ldots, M\} \backslash \{i_1\}$. If $U_{i_2} > C_{M-1}$ reject H_{i_2} and go to step 3.

Step m Consider $U_{i_m} = \max_{i \in \{1,\ldots,M\} \backslash \{i_1, i_2, \ldots, i_{m-1}\}} T_i$. If $U_{i_m} \le C_{M-m+1}$ stop and accept all $H_i : i \in \{1, \ldots, M\} \backslash \{i_1, \ldots, i_{m-1}\}$. If $U_{i_m} > C_{M-m+1}$ reject H_{i_m} and go to step $(m+1)$.

2. **Step-up process.** Fix constants $C_1 > ... > C_M$.

Step 1 Consider $U_{i_1} = \min_{\{1 \le i \le M\}} T_i$. If $U_{i_1} \ge C_M$, stop and reject all H_i. If $U_{i_1} < C_M$ accept H_{i_1} and go to step 2.

Step 2 Consider $U_{i_2} = \min_{\{1 \le i \le M, i \ne i_1\}} T_i$. If $U_{i_2} \ge C_{M-1}$ stop and reject all $H_i : i \in \{1, \ldots, M\} \backslash \{i_1\}$. If $U_{i_2} < C_{M-1}$ accept H_{i_2} and go to step 3.

Step m Consider $U_{i_m} = \min_{i \in \{1,\ldots,M\} \backslash \{i_1, i_2, \ldots, i_{m-1}\}} T_i$. If $U_{i_m} \ge C_{M-m+1}$ stop and reject all $H_i : i \in \{1, \ldots, M\} \backslash \{i_1, \ldots, i_{m-1}\}$. If $U_{i_m} < C_{M-m+1}$ accept H_{i_m} and go to step $(m+1)$.

6.2.3 Ramifications of Admissibility

When applying a stepwise MTP to a collection of pairwise comparisons problems, the original test statistics being used for different hypotheses are typically not mutually independent. When that happens, the induced tests are usually inadmissible. The only known pairwise comparison setting in which the step-down MTP or step-up MTP yield admissible induced tests is the one-sided treatments versus control model. When an induced test is inadmissible, it means that there exists, at least, one admissible test that is better than it. That is, one with equal or smaller size and greater power. Unfortunately, it is almost impossible to find any of these better tests. For any multiple testing model in which the induced tests are inadmissible there are no known better tests.

The use of Hamming loss has a much stronger connection to the classical $0-1$ loss function for the individual hypothesis testing problems than does either the standard FWER or FDR approaches. With Hamming loss if any one of the induced individual tests is inadmissible, then the entire MTP procedure is inadmissible. With Hamming loss a property like minimaxity might make more sense than FWER or FDR.

The question is then how to proceed. Fortunately, these individual hypothesis testing problems for multiple comparisons have a very relevant decision theory feature. The admissibility criterion of convexity of the acceptance region is also a compelling practical property that one should want their procedure to possess. Thus, if an MTP with admissible induced tests can be found with comparable size and power to an MTP with inadmissible induced tests, it would be more desirable. The families of MTPs described later are all admissible. It is not surprising, however, that they are not uniformly better than the classical stepwise MTPs. This is a common situation in statistical inference. However, after comparing their sizes and powers, they are often seen to be preferable.

6.2.4 Repeated Use Issues

Using some hypothetical numbers in a simple treatments versus control model, it is easy to demonstrate the type of objectionable conclusions that can occur with the repeated use of stepwise procedures that do not have the interval property. For ease of presentation, an example with a control and only two treatments will be used. Suppose that every year quality control assessments are made on companies A and B to see whether their treatments are comparable to a standard government (control) treatment C. They use Treatment A and Treatment B, respectively. If there is sufficient evidence that their treatment is out of compliance that company is fined. As a multiple comparisons testing problem, we have three population means θ_C, θ_A and θ_B and three corresponding observed sample means Z_C, Z_A and Z_B. To be tested are $H_A : \theta_A = \theta_C$ versus $K_A : \theta_A \neq \theta_C$ and $H_B : \theta_B = \theta_C$ versus $K_B : \theta_B \neq \theta_C$. The appropriate statistics for the individual tests are

$$T_A = |Z_A - Z_C| / \sqrt{2} \quad \text{and} \quad T_B = |Z_B - Z_C| / \sqrt{2}. \tag{6.3}$$

Suppose that the same, widely accepted, step-down procedure using critical values $C_1 = 1.64$ and $C_2 = 1.96$ is used each year to make comparisons of the treatments with the control. Possible data and resulting statistics are presented in Table 6.1. Then, with this data, in 2017, only company B is fined as H_B is rejected at stage 1 and H_A is accepted at stage 2. In 2018, both companies are fined as H_B is rejected at stage 1 and H_A is rejected at stage 2. In 2019 neither company is fined as nothing is rejected at stage 1 so the process stops and both H_A and H_B are accepted.

We summarize these results from Company A's point of view. Their focus is on H_A keeping in mind that Z_C, Z_A, Z_B are independent. In 2017 and 2019, this procedure does not find sufficient evidence that Company A and the control are different. That is, Company A and the control would be deemed comparable (accept) and no fine imposed. Remarkably, with a smaller test statistic, in 2018, Company A was deemed to be different from the control and a fine imposed. Any responsible testing agency would have difficulty justifying these actions in the different years.

Despite independence, clearly the performance of Treatment B is impacting the decision about the comparison of Treatment A with the control. This is common with stepwise procedures. Despite its political incorrectness, one may think that this kind of behavior can be attributed to the use of some sort of prior distribution in developing procedures. However, the lack of convexity implies inadmissibility which, in turn, implies that such a procedure cannot be Bayes.

Using the same Table 6.1 data set, it can be seen that the step-up procedure can exhibit the same undesirable phenomenon. Again we take the same two treatments versus a control setting. Recall that, unlike step-down, step-up begins by comparing the smallest statistic with a C_1. Also recall that for step-up $C_1 < C_2$. If that is rejected, the process ends and both are rejected. If it accepts, then the larger test statistic is compared to C_2 to determine acceptance or rejection. Therefore, this time we will take $C_1 = 1.96$ and $C_2 = 1.64$. Now apply to the data in Table 6.1. We again get the same decisions for all three years as we did with step-down.

TABLE 6.1
Yearly Testing Results Using Step-Down or Step-Up

	Treatment Scores			Test Values	
Year	Control	A	B	T_A	T_B
2017	10.00	10.00	13.00	0	2.12
2018	10.00	7.65	12.80	1.66	1.98
2019	10.00	7.60	12.30	1.70	1.63

It should be noted that step-down and step-up procedures can come to the same conclusion for many data sets. However, simply switching the C_i values will result in MTP's with different family-wise error rates.

Since there are only three populations and both test statistics are pairwise differences, it is possible to graphically give a sense of all accept and reject regions. We will do this with the step-down procedure. First note that the problem is invariant under translation. That is, if the same constant is added to each Z, all statistics and actions will remain the same. That implies that the diagram of the accept and reject regions will look the same on every plane of the form $Z_A + Z_B + Z_C = v$ for every value of v. These regions of the sample space on these planes are shown for the Step-down MTP on the left of Figure 6.1. Each region has a pair of letters to indicate the decisions for H_A and H_B, respectively. R indicates reject and A indicates accept. Focusing only on the H_A decision (i.e., the first letter) quickly gives the acceptance (and rejection) region of the induced test for H_A. This is exhibited on the right side of Figure 6.1. In these diagrams, the vertical direction represents Z_B. The horizontal direction represents $Z_A - Z_C$. The issue is the jaggedness of the boundaries caused by the sequential nature of the procedure. Recall that Z_C, Z_A and Z_B are all independent. As indicated on the induced test in Figure 6.1, it can be seen that there are values of Z_B for which the acceptance line is not convex.

It should be pointed out, as in the class grading example, that the regions of non-convexity of the step-down procedure tend to be small. However, their numbers increase considerably as the dimension increases. Thus, the chance of making a politically incorrect claim might be small, unless, the methodology is used repeatedly.

For intuition, Figure 6.2 presents the analogous regions for the procedure called maximum residual down (MRD) that is described in the next section. That MTP has induced tests that have properly convex acceptance regions and are admissible.

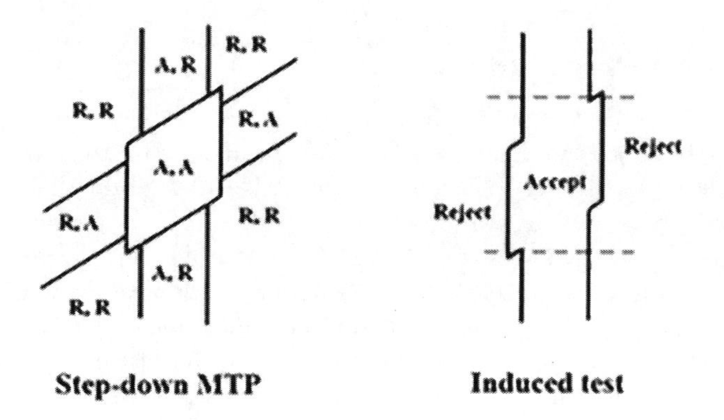

FIGURE 6.1
Regions of action by the Step-down procedure and its induced individual test.

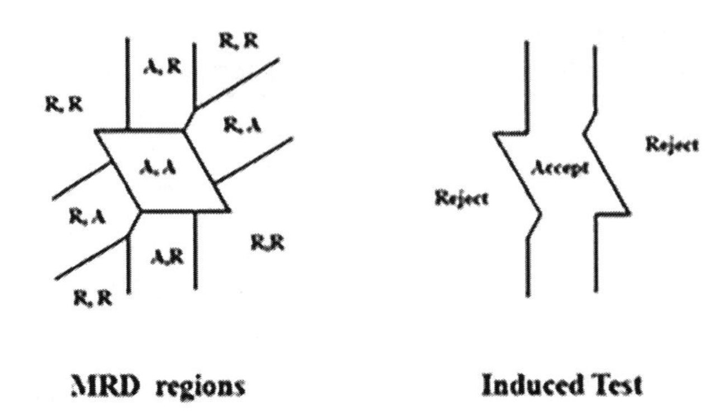

FIGURE 6.2
Regions of action by the MRD procedure and its induced individual test.

6.3 MTPs With the Interval Property

This section presents three different types of MTPs whose induced tests have been shown to have the interval property (and are admissible). The first is a stepwise procedure. The second is a two-stage procedure that allows for a corresponding confidence interval. The last one does all the tests simultaneously and suggests estimates of the parameters. In this setting of pairwise comparisons, the property of Matthes and Truax (1967) needed for admissibility of the induced tests has a simple interpretation. It is simply that the acceptance region for the *ith* test is convex along the direction of $\mathbf{g}_i$ where $\mathbf{g}_i$ is the *ith* column of Σ. The complete specification of these MTPs requires, as do all other MTPs, the specification of a collection of constants. There are many choices of these constants that yield admissible induced tests. In practice these constants would be chosen based on performance in the particular application. A natural way to continue is to find constants that control FWER or FDR and then study the power of those MTPs over relevant portions of the parameter space. Doing all of this would typically require simulation. For example, typically, all three MTPs described below can constructed to control FWER or FDR and outperform commonly used MTPs when the parameter space is sparse (i.e., at most 20% to 25% of the alternatives can be true).

6.3.1 The MRD Procedure

The MRD method, introduced and studied in Cohen, et.al. (2009a), is a stepwise procedure. At each step, it is based on adaptively formed residuals. In that way, MRD utilizes information oftentimes not used by common methods. MRD is in the spirit of the step-down MTP given above and, essentially the same description can be used. Both begin with a set of fixed constants $C_1 < C_2 < \cdots < C_M$. At each step, p say, until the process stops, statistics are computed for each of the remaining variables. If the maximum of these statistics exceeds C_{M-p+1} the hypothesis corresponding to that test statistic is rejected and that variable is removed from consideration. Otherwise, the process stops and accepts all the remaining hypotheses. The only real difference is that the MRD statistics used change from step to step taking into account the correlation among the remaining variables. That is, instead of using the same, more popular, T_i statistics as does step-down, MRD uses residuals obtained from the remaining variables as follows.

Suppose that the process reaches step p and denote the $M-p+1$ remaining variables by $X_{i_1}, X_{i_2}, \ldots, X_{i_{M-p+1}}$. Then, the test statistic corresponding to hypothesis i_j will be

$$X_{i_j} - \sigma'_{i_j} \Sigma_{i_j}^{-1} \mathbf{X}^{(i_j)} \tag{6.4}$$

where σ_{i_j} is the $(M-p) \times 1$ vector of covariances of X_{i_j} with the other remaining X_i, Σ_{i_j} is the $(M-p) \times (M-p)$ covariance matrix for the other remaining X_i and $\mathbf{X}^{(i_j)}$ is the $(M-p) \times 1$ vector made up of the remaining variables other than X_{i_j}.

Note that all the σ_{i_j} and Σ_{i_j} come directly from the original Σ by just removing the corresponding rows and columns as hypotheses are rejected and variables are removed. Using (6.4) is intuitive as it is simply comparing X_{i_j} with its conditional expected value when all nulls are true. When the dimension is large and Σ is somewhat complicated, having to compute matrix inverses many times can be computationally challenging. However, it is often the case that the needed inverses can be easily found, and the entire procedure is easily described. A particularly nice example is the treatments versus control model.

Here the original $M \times M$ covariance matrix Σ is given by

$$\Sigma = \begin{pmatrix} 2 & 1 & 1 & \cdots & 1 & 1 \\ 1 & 2 & 1 & \cdots & 1 & 1 \\ \cdot & \cdot & \cdot & & \cdot & \cdot \\ \cdot & \cdot & \cdot & & \cdot & \cdot \\ \cdot & \cdot & \cdot & & \cdot & \cdot \\ 1 & 1 & 1 & \cdots & 1 & 2 \end{pmatrix} \tag{6.5}$$

Notice that when any collection of corresponding rows and columns are removed all the Σ_{i_j} still have form (6.5) except for dimension. However, the inverse does change with dimension. At step p regardless of which hypotheses had already been rejected, the required $M-p$ dimensional inverse will be

$$\Sigma_{i_j}^{-1} = \frac{1}{M-p+1} \begin{pmatrix} M-p & -1 & -1 & \cdots & -1 & -1 \\ -1 & M-p & -1 & \cdots & -1 & -1 \\ \cdot & \cdot & \cdot & & \cdot & \cdot \\ \cdot & \cdot & \cdot & & \cdot & \cdot \\ \cdot & \cdot & \cdot & & \cdot & \cdot \\ -1 & -1 & -1 & \cdots & -1 & M-p \end{pmatrix} \tag{6.6}$$

Further notice that, at any step p, the $M-p$ dimensional vector $\sigma_{i_j} = (1, 1, \cdots, 1)'$. Thus it follows from (6.6) that the residual in (6.4) is

$$X_{i_j} - \sigma'_{i_j} \Sigma_{i_j}^{-1} = X_{i_j} - \frac{1}{M-p+1} (1, 1, \cdots, 1) \mathbf{X}^{(i_j)} \tag{6.7}$$

A more recognizable version of (6.7) is given when expressed in terms of the Z_i. Recall that the Z_i were the observations that came from the actual treatments and control with Z_1 coming from the control. In this model $X_i = Z_{i+1} - Z_1, i = 1, \cdots, M$. Let $\overline{Z}^{(i_j),p}$ denote the mean of the remaining population Z scores including the control, but, not Z_{i_j+1}. That is,

$$\overline{Z}^{(i_j),p} = (Z_{i_1} + \cdots + Z_{i_{(M-p+1)}} - Z_{i_j} + Z_1)/(M-p) \tag{6.8}$$

Now the test statistic in (6.7) simply becomes

$$Z_{i_j+1} - \overline{Z}^{(i_j),p}. \tag{6.9}$$

TABLE 6.2
Yearly Testing Results Using MRD

	Treatment Scores			Stage 1		Stage 2
Year	**Control**	**A**	**B**	T_A^*	T_B^*	**Test**
2017	10.00	10.00	13.00	1.22	2.45	$T_A = 0.00$
2018	10.00	7.65	12.80	3.06	3.25	$T_A = 1.66$
2019	10.00	7.60	12.30	2.90	2.86	$T_B = 1.63$

A word description is much simpler. At every stage, the standard step-down procedure compares each remaining treatment with the control. At every stage, when using MRD, each remaining treatment is compared, not just to the control, but to the average of the pooled group of the control and all other remaining treatments.

Application to data of Table 6.1: This is the treatments versus control example of Section 6.2.4. Applying MRD to the data of Table 6.1 will indicate the ease in constructing the statistics used by MRD in the treatments versus control model. Of course, it will not suffer from the same undesirable feature of non-convex acceptance region as does step-up and step-down.

In this simple example, there are, at most, two stages. Step-down and step-up both use, at all stages, the standardized statistics given in (6.3) on those samples still under investigation. MRD is different. At stage one it uses the statistics

$$T_A^* = \left| Z_A - \frac{Z_B + Z_C}{2} \right| / \sqrt{3/2} \quad \text{and} \quad T_B^* = \left| Z_B - \frac{Z_A + Z_C}{2} \right| / \sqrt{3/2}. \tag{6.10}$$

and compares them to C_2. If the larger one is greater than C_2 that hypothesis is rejected and the corresponding Z score is removed. Then, at stage two, the statistic in (6.3) based on the remaining Z scores is used and compared to C_1. Table 6.2 summarizes the calculations. In 2017, H_B is rejected at stage 1, and then T_A is used at stage 2 and H_A is accepted. In 2018, again H_B is rejected at stage 1, and then T_A is used at stage 2 and H_A is accepted. In 2019, H_A is rejected at stage 1 and then T_B is used at stage 2 and H_B is accepted.

Since MRD has the interval property it did not (nor could not) exhibit the undesirable feature of not having a convex acceptance region as did step-down and step-up on this data set.

6.3.2 Individualized Two-Stage Process

This approach to constructing MTPs has the bonus that it is accompanied by corresponding interval estimates. It was introduced and studied in Cohen et al. (2013a). This method, in principle, is not a stepwise procedure. It considers all hypotheses individually in no particular order. All hypotheses are treated similarly, but they are each considered, one at a time, with a two-stage process. Suppose we are ready to test H_{i^*}. We do so by subjecting it to the following two-stage process. At stage 1, apply any (M-1 population) MTP to **ONLY** the other hypotheses, i.e., $H_i : \nu_i = 0$ vs $K_i : \nu_i \neq 0,\ i = 1, ..., M, i \neq i^*$, but using **ALL** of the data. At the end of this stage, r_{i^*} = the number of rejections is recorded.

At stage 2, intervals are constructed of the form

$$\widehat{\nu_{i^*}} \pm B(r_{i^*})\sigma_{\widehat{\nu_{i^*}}} \tag{6.11}$$

where $B(r_{i^*})$ is a decreasing function of r_{i^*} , $\widehat{\nu_{i^*}}$ is an estimator of ν_i that can depend on all the data and $\sigma_{\widehat{\nu_{i^*}}}$ is the standard deviation of that estimator. The corresponding test of H_{i^*} is to reject if the interval in (6.11) does not contain 0. Typically this would mimic, except for the use of (a random) r_i^*, what would have been done if only the one population had been observed. That is, (6.11) would correspond to just using X_i as the estimate of ν_i.

This process is repeated for each hypothesis. Although this MTP is being characterized as a two-stage MTP rather than a stepwise procedure, it should be pointed out that one could opt to use a stepwise procedure at the first stage. In any case, the MTP used at stage 1 is referred to as a modified MTP as it makes decisions on all but one of the hypotheses, but, uses the data from all the samples.

It is not difficult to select $B(r_{i^*})$, r_{i^*} and $\widehat{\nu_{i^*}}$ so that the resulting MTP will have admissible induced tests. Recall that by Matthes and Truax (1967) the acceptance region of the induced test for hypothesis i must be convex along the direction $\mathbf{g}_i$ where $\mathbf{g}_i$ is the ith column of σ. By definition, a non-randomized test function $\phi(x)$ is either 0 or 1 depending on whether the null hypothesis is accepted or rejected, respectively, when x is observed. Suppose we desire a test, ϕ_i, of H_i to have the interval property with respect to the direction $\mathbf{g}_i$. This will mean that, as a function of λ, $\phi_i(\mathbf{x}+\lambda\mathbf{g}_i) = 0$ only on an interval. Theorem 4.2 and Lemma 4.3 of Cohen et al. (2013), stated below, are quite useful in finding MTPs having the interval property.

Theorem 4.2 *Suppose for each fixed* $\boldsymbol{x}, r_i(\boldsymbol{x} + \lambda\boldsymbol{g}_i)$ *is constant as a function of* λ *and* $\widehat{\nu_i}(\boldsymbol{x}+\lambda\boldsymbol{g}_i)$ *is first non-increasing and then non-decreasing as a function of* λ*. Then the test and interval estimate will have the interval property with respect to the direction* $\boldsymbol{g}_i$*.*

All such procedures yield admissible induced tests. It should be noted that a natural choice of $\widehat{\nu_1}$ will often depend on only the data involving the parameter ν_1.

Suppose an individualized two-stage test is to have the interval property in the direction $\mathbf{g}$. Lemma 4.3 serves to demonstrate the ease in finding modified MTPs to be used at stage 1 that will guarantee the condition on $r(\cdot)$ in Theorem 4.2.

Lemma 4.3 *If every statistic used during stage 1 modified MTP is based on functions of the form* $\boldsymbol{a}'\boldsymbol{x}$*, i.e., each is a linear combination of the* $\boldsymbol{x}$*'s, where* $\boldsymbol{a}'\boldsymbol{g}_i = 0$ *for each* $\boldsymbol{a}$ *then* $r(\boldsymbol{x}+\lambda\boldsymbol{g}_i)$ *is constant as a function of* λ*.*

Application to data of Table 6.1: This is the treatments versus control example of Section 6.2.4. Since there are only two treatments, the first stage is not really a multiple test as there will be only one hypothesis remaining. However, the idea and spirit of the method is clear. In this setting, both H_A and H_B need to be tested. It does not matter which is done first. Suppose we begin with H_A. Then we need a, particular type of, multiple test (based on all Z_A, Z_B and Z_C) on the remaining hypotheses other than H_A and count the number of those that are rejected. In this simple case only H_B remains. Therefore, the stage 1 MTP is just a single test of H_B. To do this the model must be put in the form in (6.1). Here $X_1 = Z_A - Z_C$ and $X_2 = Z_B - Z_C$ so that $\Sigma = \binom{2\ 1}{1\ 2}$. In general, the more hypotheses that are rejected the smaller the critical value needed to finally reject H_A.

As indicated in the beginning of Section 6.3, the acceptance region for the *ith* test must be convex along the direction of $\mathbf{g_i}$ where $\mathbf{g_i}$ is the *ith* column of Σ. Here, for testing H_A, $\mathbf{g}_1 = \binom{2}{1}$. Following the suggestion of Lemma 4.3 $\mathbf{a} = \binom{1}{-2}$ can be used which leads to the statistic

$$\mathbf{a}'\mathbf{X} = X_A - 2X_B = Z_A - Z_C - 2(Z_B - Z_C) = Z_A + Z_C - 2Z_B.$$

This is equivalent to T_B^* defined in (6.10) and can be used at stage 1. Then, at stage 2, the statistic T_A of (6.3) can be used. Since there is only one hypothesis at stage 1, the number of rejections is either 0 or 1. Thus, in this small model, there would be only a choice of two critical values for use at stage 2. The key idea is that T_A is compared to the smaller critical value if H_B is rejected and to the larger one if H_B is accepted. To test H_B the process begins all over again with H_B and H_A switching roles.

Having an interval estimate for each parameter is a nice feature of this method. Over the years, there has been some interest in obtaining simultaneous interval estimates for the parameters that are considered for multiple testing. Where the number of parameters is large, typical simultaneous interval estimates such as Scheffé, Bonferroni, Tukey pairwise contrasts or Dunnett are excessively wide. See for example Miller (1966) for these latter methods.

A number of authors have commented on the difficulty of the problem of inverting stepwise MTPs, particularly in the case of two-sided hypotheses. These include Lehmann (1986, page 388), Stefansson, Kim, and Hsu (1988) and Benjamini and Stark (1996). Most attempts result in constructions that often lead to non-informative intervals as they contain the entire alternative space. For example, Stefansson et al. (1988) give intervals for a one-sided treatment versus control model but (unless all the parameters are found to be significantly different from zero), the intervals are of the form $(0, \infty)$ for significant parameters.

Attempts at simultaneous confidence intervals for specialized models and some specialized problems have emerged. See for example, Hayter and Hsu (1994), Benjamini and Stark (1996), Benjamini, Hochberg and Stark (1998), Benjamini and Yekutieli (2005) and Guilbaud (2008). Some of those references offer intervals less likely to contain null hypothesis points. An interesting approach with potential appears in Brannath and Schmidt (2014) that builds off a modification of step-down and step-up procedures.

A number of issues may contribute to the difficulties. For one thing, a stepwise procedure typically will not even reach consideration of all hypotheses. Also, the order of consideration need not follow the order of the individual p-values. Confidence intervals for the parameters in the accepted hypotheses are also of value as they indicate how close they came to rejection. Perhaps, most important, is the fact that the individual tests induced by many stepwise procedures do not have convex acceptance regions. This would seem to contradict the possibility of inverting these tests to get confidence intervals.

There is another, surprising, benefit to the individualized two-stage method. Owing to the great flexibility in choices of the modified MTP used at stage 1, the choice of $\widehat{\nu_1}$ and the choice of function $B(\cdot)$, this process generates an extremely large family of MTP's with corresponding interval estimates. Thus every MTP in this family is associated with easily obtained interval estimate. Two MTPs are said to be equivalent if, with probability one, they make the same decisions for all the hypotheses in the collection being tested. The following theorem of Cohen et al. (2013) further indicates the breadth of procedures covered by the new construction.

Theorem 4.1 *The class of MTPs made up of individualized two-stage tests contains MTPs equivalent to the generic step-down and step-up procedures.*

Thus, finding and using the equivalent individualized two-stage tests for step-down and step-up will immediately lead to interval estimates of all parameters. This is done in the proof of Theorem 4.1 in Cohen, et.al. (2013).

6.3.3 Penalized Likelihood Based Procedure

It is well recognized that model selection problems can often be viewed as multiple testing problems and vice versa. Useful contributions to model selection problems have often been made by applying MTP methods. The regression setting is the most common situation for model selection. In those applications, one needs to decide which factors should be included in the model. A good place to see elements of this is Chapter 7 of Dickhaus (2014). However, that is not the issue here.

Recently, a substantively different topic, that of inference following selection has been receiving attention. See, for example, Efron (2014) and Berk et al. (2013). In most of those applications, the ultimate goal is usually the ability to predict. Again the regression setting is the most common situation. In those applications, one needs to decide which factors should be included in the model, how much weight they should be given and how well the final inference task (typically prediction) can be accomplished. Furthermore, in those situations, it is most common to see just a single procedure based on a form of penalized likelihood or least squares expressions to carry out its entire mission. Although not singled out as a separate operation, these penalized methods are, implicitly, doing multiple testing of the potential factors that can enter the model.

In this setting, one rarely sees the usual study of quantities associated with MTPs such as levels of FWER or FDR. Error control or power of this testing component has not been an important consideration. Barber and Candès (2015) construct a variable selection procedure in conjunction with penalized losses that controls FDR. They used knockoff filter methodology. However, those induced tests can also yield non-convex acceptance regions. An interesting decision theory investigation would be to see if a connection can be made between testing error levels and prediction ability.

In Cohen et al. (2019), multiple testing based on penalized loss is studied as a potential, stand alone, MTP that could be used in any multiple testing problem not just regression. Any model of the form (6.1) and (6.2) is considered. The focus there was to modify the, well known, Bayesian Information Criteria (BIC) (Schwartz (1978)) so that it could give some error control and then study its properties. This is somewhat in the spirit of Zheng and Loh (1995) followed by Bunea et al. (2006). Their modification was partially based on p-values, and their focus was on consistency in linear models. In Cohen et al. (2019), it was shown that the induced individual tests of these penalized loss based MTPs do have the interval property (i.e.,convex acceptance regions). Furthermore, in many applications, they have stronger power characteristics than some commonly used procedures. In addition, this approach supplies natural estimates of the parameters. The goal of the BIC method is to maximize the likelihood function by determining which parameters are equal to 0 and to give estimates for the others. This is done subject to some penalty based on the number of non-zero parameters to be included in the model. Suppose there are M factors considered for possible inclusion in the model. That is, M parameters that could possibly be non-zero. In the classical use of BIC the penalty is simply $2\ ln(M)$ times the number of parameters to enter the model. Clearly this is not related to any particular α level and so no particular control can even be expected. The example studied in Cohen et al. (2019) using this penalty function yields operating levels for FWER and FDR at over 0.20.

The MTP called GBIC is a slight generalization of BIC as it allows for a wider class of penalty functions. This makes it possible to achieve a specified error control. Regardless of the particular application, the GBIC is applied as follows. One needs to search through and consider all possible models where each individual model entails the specification of which parameters are zero and which are non-zero. Since there are M parameters, any of which can be zero, there are 2^M such models possible. The GBIC procedure then minimizes

$$-2\ ln\ L\ +\ C(m, M) \tag{6.12}$$

over all possible models where L is the likelihood function, m is the number of non-zero parameters in the particular model being considered and $C(m, M)$ is the penalty for using that model. It is always assumed that

$$C(m_2, M) > C(m_1, M) \quad \text{if} \quad m_2 > m_1. \tag{6.13}$$

Using the simple, variance 1, model in (6.1) the likelihood function is

$$L = Kexp - (1/(2))((\mathbf{X} - \nu)'\Sigma^{-1}(\mathbf{X} - \nu)) \tag{6.14}$$

where $K = (2\pi)^{-M/2}|\Sigma|^{1/2}$ so that

$$-2\ lnL = -2\ ln(K) + (\mathbf{X} - \nu)'\Sigma^{-1}(\mathbf{X} - \nu). \tag{6.15}$$

For each of the 2^M models (i.e., combinations of true and false nulls) minimizing (6.12) begins with minimizing

$$(\mathbf{X} - \nu)'\Sigma^{-1}(\mathbf{X} - \nu) \tag{6.16}$$

over those ν_i not assumed to be zero . Then, upon adding the appropriate amount $C(m, M)$ one can find the minimum value of (6.12).

Note that since the final objective is to obtain some GBIC procedures that can be used as multiple testing methods the penalty function will have to be chosen so that the resulting procedure has some prescribed error control such as FWER or FDR. This can be accomplished by doing simulations for various choices of the constants $C(m, M)$. Of course there are many choices. To find penalties that yield particular levels of control for a fixed value of M some trial and error simulations must be performed. Linear choices are a relatively convenient way to approach the issue for a particular M. That is, to consider penalties of the form $C(m, M) = a\ m + b$ if $m > 0$ and equal 0 if $m = 0$. The power of individual tests is also a consideration and should be checked in the simulations. It should be noted that the choice of the pair (a, b) is not unique. However, the actual shapes of all GBIC acceptance regions are similar, and no one choice can uniformly dominate the others in both size and power.

Application to data of Table 6.1. This is the treatments versus control example of Section 6.2.4. This example can even be done by hand. The maximum likelihood estimates (m.l.e.) can be easily obtained by using the density in terms of the original Z_i and θ_i for $i = A, B, C$. Whenever $\nu_A = 0$ it means that $\theta_A = \theta_C$ and whenever $\nu_B = 0$ it means that $\theta_B = \theta_C$. So the m.l.e. for any θ_i is the pooled mean of all the Zs corresponding to that θ_i and all other means assumed equal to θ_i . For all the other θ_i the m.l.e. is Z_i.

Recall being tested are $H_A : \theta_A = \theta_C$ versus $K_A : \theta_A \neq \theta_C$ and $H_B : \theta_B = \theta_C$ versus $K_B : \theta_B \neq \theta_C$. Here $\nu_1 = \theta_A - \theta_C$ and $\nu_2 = \theta_B - \theta_C$. The above observation about the form of the m.l.e.s is helpful with the data of Table 6.1. Since there are only two treatments, there are only four separate cases in which the minimum needs to be found. Then after the penalty, based on m, is added on the overall minimum among the four cases yields the minimum of (6.12). These are the four cases along with their implication should they turn out to yield the minimal value.

$$\text{Case 1: } (m = 0, \nu_1 = 0, \nu_2 = 0) \quad \Rightarrow \text{Reject neither}$$

$$\text{Case 2: } (m = 1, \nu_1 \neq 0, \nu_2 = 0) \quad \Rightarrow \text{Reject } H_A \text{ only}$$

$$\text{Case 3: } (m = 1, \nu_1 = 0, \nu_2 \neq 0) \quad \Rightarrow \text{Reject } H_B \text{ only}$$

and

$$\text{Case 4: } (m = 0, \nu_1 \neq 0, \nu_2 \neq 0) \quad \Rightarrow \text{Reject both.}$$

Table 6.3 gives the numerical results for GBIC when applied to the data of Table 6.1.

TABLE 6.3
Yearly Testing Results Using GBIC

	Minimum Values				Results
Year	Case 1	Case 2	Case 3	Case 4	
2017	6.0 + 0	4.5 + 1	0.0 + 1	0.0 + 2	Accept H_A only
2018	13.3 + 0	4.9 + 1	3.8 + 1	0.0 + 2	Reject both
2019	11.0 + 0	3.6 + 1	3.9 + 1	0.0 + 2	Reject both

6.4 Discussion

Whenever statisticians face an inference problem, they need to choose or develop an appropriate procedure. To do this, some criteria are needed that reflect desirability in the problem at hand. In a decision theory approach included among these criteria would be a loss function. The theory related to consideration of testing an individual null hypothesis versus an individual alternative hypothesis is very comprehensive even for composite hypotheses. The classical text is Lehmann (1959). The arguments for studying the power function is compelling. Power functions are functions of the parameter so even in this model it is not surprising that there are no nontrivial problems in which there exists a single procedure that is uniformly best for every parameter point. When testing only one hypothesis it has become automatic to focus on size and power. Owing to their intuitive appeal unbiasedness and invariance are often used as additional criteria.

Multiple testing is different. The vast majority of the literature focuses on controlling the FWER or the FDR and achieving good power. This mimics the size and power approach in testing a single hypothesis. They are natural and desirable criteria, but, are not really attached to a loss function. Many MTPs have been developed over the years. However, every MTP, regardless how complicated, whether stepwise or not, will induce individual tests (based on all the data) for the individual testing problems. Using each of these induced tests individually would be equivalent to using the MTP. That is, if it were possible to view the entire multidimensional sample space, one would be able to see an acceptance region for each testing problem. Since they are equivalent, what if the MTP was not specified and we were given only the collection of induced tests? What properties would we demand from them?

Unfortunately, even in modest dimension, these induced tests are very complicated. However, oftentimes, it is possible to establish certain convexity properties of these induced tests and that is enough to determine admissibility. Furthermore, in testing situations, these same convexity properties of acceptance regions are extremely desirable for practical purposes.

The exposition in this chapter presents a decision theory approach to multiple testing that focuses on these induced tests. It is presented in the context of testing of pairwise differences of normal means when the variance is known. This approach extends naturally to more general models. The variances can be unknown, but there would be an independent estimate of the variance. Each individual distribution can be a multivariate normal distribution, and one can look at pairwise differences of vector means (Cohen and Sackrowitz 2016). The distributions can be one-parameter exponential family that includes ordinal data settings (Cohen et al. 2009)). This path can also be followed in some nonparametric settings (Cohen et al. 2013b).

Of course, an MTP simply having admissible induced tests is not sufficient. In the cases discussed above, the evaluation of a multiple testing procedure was viewed from the following perspective. In any particular application one would typically have a sense of desirable criteria as well as those portions of the parameter space that are most relevant. To get a more complete understanding of the behavior of one's procedure, it is recommended that a variety of criteria including error control and risk function properties should be examined. Typically, this would have to be done through simulation.

Bibliography

Barber, R.F. and Candès, E.J. (2015). Controlling the false discovery rate via knockoffs. *Annals of Statistics* **43**, 2055–2085.

Benjamini, Y. and Hochberg, Y. (1995). Controlling the false discovery rate: a practical and powerful approach to multiple testing. *Journal of Royal Statistical Society Series B*, **57**, 289–300.

Benjamini, Y., Hochberg, Y and Stark, P.B. (1998). Confidence intervals with more power to determine the sign. *Journal of the Statistical Association* **93**, 309–317.

Benjamini, Y. and Gavrilov, Y. (2009). A simple forward selection procedure based on false discovery rate control. *Annals of Applied Statistics* **3**, 179–198.

Benjamini, Y. and Stark, P.B. (1996). Nonequivariant Simultaneous Confidence intervals less likely to contain zero. *Journal of the American Statistical Association* **91**, 329–337.

Benjamini, Y. and Yekutieli, Y. (2005). False discovery rate controlling confidence intervals for selected parameters. *Journal of the American Statistical Association* **100**, 71–80.

Berk, R., Brown, L., Buja, A., Zhang, K. and Zhao, L. (2013). Valid post-selection inference. *Annals of Statistics* **41**, 802–837.

Brannath, W. and Schmidt, S. (2014) A new class of powerful and informative simultaneous confidence intervals *Statistics in Medicine* **33**, 3365–3386.

Bunea, F., Wegkamp, M., Auguste, A. (2006) Consistent variable selection in high dimensional regression via multiple testing. *Journal of Statistical Planning and Inference* **136**, 4349–4364.

Butucea, C., Ndaoud, M., Stepanova, N.A. and Tsybakov, A.B.(2018). Variable selection with Hamming loss. *Annals of Statistics* **46**, 1837–1875.

Chen, C., Cohen, A. and Sackrowitz, H.B. (2009b) Multiple testing in ordinal data models. *Electronic Journal of Statistics* **8**, 912–931.

Cohen, A., Sackrowitz, H.B. and Xu, M. (2009a). A new multiple testing method in the dependent case. *Annals of Statistics* **37**, 1518–1544.

Cohen, A., Ma, Y. and Sackrowitz, H. (2014). Individualized two-stage multiple testing procedures with corresponding interval estimates. *Biometrical Journal* **55**, 386–401.

Cohen, A., Ma, Y. and Sackrowitz, H.B. (2013b) Nonparametric multiple testing procedures. *Journal of Statistical Planning and Inference* **14**, 1753–1765.

Cohen, A. and Sackrowitz, H.B. (2012). The interval property in multiple testing of pairwise differences. *Statistical Science* **27**, 294–307.

Cohen, A. and Sackrowitz, H.B. (2016). Convexity issues in multivariate multiple testing of treatments vs. control. *Journal of Multivariate Analysis* **143**, 1–11.

Cohen, A., Kolassa, J. and Sackrowitz, H.B. (2019). Penalized likelihood and multiple testing. *Biometrical Journal* **61**, 62–72.

Dickhaus, T. (2014). *Simultaneous Statistical Inference.* Springer-Verlag, Berlin Heidelberg.

Dudoit, S.,Shaffer, J.P. and Boldrick, J.C. (2003). Multiple hypothesis testing in microarray experiments. *Statistical Science* **18**, 71–103.

Dudoit, S. and van der Laan, M.J. (2008). *Multiple Testing Procedures with Applications to Genomics.* Springer, New York, NY.

Efron, B. (2014). Estimation and accuracy after model selection. *Journal of the American Statistical Association* **109**, 991–1006.

Genovese, C. and Wasserman, L. (2002). Operating characteristics and extensions of the false discovery rate procedure. *J. R. Stat. Soc. Ser. B Stat. Methodol.* **64**, 499–517.

Guilbaud, O. (2008). Simultaneous confidence regions corresponding to Holm's step-down procedure and other closed-testing procedures. *Biometrical Journal* **50**, 678–692.

Hayter, A.J. and Hsu, J.C. (1994). On the relationship between stepwise decision procedures and confidence sets. *Journal of the American Statistical Association* **89,** 128–136.

Hochberg, Y. and Tamhane, A.C. (1987). *Multiple Comparison Procedures.* Wiley, New York, NY.

Holm, S. (1979) A Simple Sequentially Rejective Multiple Test Procedure, *Scandinavian Journal of Statistics* **6**, 65–70.

Lehmann, E.L. (1959). *Testing Statistical Hypotheses*, John Wiley & Sons, New York, NY.

Lehmann, E.L. and Romano, J.P. (2005). *Testing Statistical Hypotheses*, 3rd Ed. Springer Texts in Statistics, Springer Science and Business Media, LLC, New York, NY.

Matthes, T.K.and Truax, D.R. (1967). Tests of composite hypotheses for the multivariate exponential family. *Annals of Mathematical Statistics* **38** , 681–697.

Miller, R.G. (1966). *Simultaneous Statistical Inference.* McGraw-Hill, New York, NY.

Schwarz, G. (1978). Estimating the dimension of a model. *Annals of Statistics* **6** , 461–464.

Stefansson, G., Kim, W., and Hsu, J.C. (1988). On confidence sets in multiple comparisons. In: Statistical Decision Theory and Related Topics IV (Eds.: Gupta, S.S. and Berger, J.O.), Volume 2, pages 89–104. Springer-Verlag, New York, NY.

Zheng, X. and Loh, Y-W.(1995). Consistent Variable Selection in Linear Models. *Journal of the American Statistical Association* **90**, 151–156.

7

Identifying Important Predictors in Large Data Bases - Multiple Testing and Model Selection

Malgorzata Bogdan
University of Wroclaw
Lund University

Florian Frommlet
Medical University Vienna

CONTENTS

This chapter considers a variety of model selection strategies in a high-dimensional setting, where the number of potential predictors p is large compared to the number of available observations n. A typical example of this situation is genome-wide association studies (GWAS). These may include several hundred thousand genetic variants (SNPs), which are used as genetic markers for DNA regions. GWAS are performed to find those variants, which are related to some trait. This could be a binary trait like disease risk or a quantitative trait like height. GWAS are most often analyzed by performing statistical tests for each

DOI: 10.1201/9780429030888-8

individual marker combined with some correction for multiple testing. Often a Bonferroni corrected significance level 5×10^{-8} is recommended. This type of analysis is still prevailing although from a statistical perspective it has many severe drawbacks (Frommlet et al., 2016, 2012b; Renaux et al., 2020).

A major assumption underlying the rationale of performing GWAS is the common disease/common variant assumption. Accordingly, the risk of common diseases should depend on a relatively large number of fairly common genetic variants. In statistical terms, this corresponds to models that include moderate numbers of genetic markers as regressors. It has been shown that under this assumption, testing each marker individually results in a severe loss of power to detect important SNPs Frommlet et al. (2012b). Furthermore, the order of p-values from the individual marker tests may no longer reflect the actual importance of genetic variants and consequently also the chance of false-positive findings is increased. Using model selection approaches to detect important genetic variants can help to overcome these shortcomings. For practical applications with real GWAS data, see for example Dolejsi et al. (2014); Hofer et al. (2017).

For the ease of presentation, the focus will be on the linear model

$$y = X\beta + \epsilon, \tag{7.1}$$

where $y \in \mathbb{R}^n$, $\beta \in \mathbb{R}^p$, $X \in \mathbb{R}^{n \times p}$ and the error terms are independent Gaussian random variables, $\epsilon \sim N_n(0, \sigma^2 I)$. The basic ideas easily extend to more general regression settings, like generalized linear models or generalized linear mixed models. Many high-dimensional model selection strategies make use of penalized likelihood methods, which can be written, for example, in the following form

$$-2\log(\mathcal{L}(\beta)) + \text{Pen}(\beta). \tag{7.2}$$

Here $\mathcal{L}(\beta)$ denotes the likelihood function. In the case of the linear model (7.1) with known σ the first term in (7.2) is up to a constant $\|y - X\beta\|^2/\sigma^2$, which is the residual sum of squares divided by the variance term. There exists a wide range of penalty functions $\text{Pen}(\beta)$ for high-dimensional model selection. This chapter focuses on L_0 penalties and certain weighted L_1 penalties.

In general, model selection might serve two different purposes, identification of the actual data generating model and finding a model which is good for prediction. Depending on the application in mind, the former or the latter goal might be more important, and the most suitable selection strategies might be different. For example, in the context of genetic association studies, one can make use of variable selection methods to identify causal mutations. Correct model identification then corresponds to correctly identifying causal mutations without detecting too many false positives (Frommlet et al., 2016). In statistical terms, one needs a consistent variable selection procedure to achieve this goal.

Section 7.1 will set the stage by considering the simple setting of an orthogonal design matrix X and known error variance σ^2. In that case, estimates of the regression coefficients β_j do not depend on the other components of the vector β and model selection becomes equivalent to multiple testing. Simple results for the classical selection criteria AIC and BIC will illustrate that these are not suitable for model selection when the number of potential regressors p is getting large compared with n. Instead, some L_0 penalties which are modifications of AIC and BIC will be introduced which are designed to control the number of false discoveries. This means that for predictors which are of no relevance, type I error control strategies will be applied which are known from multiple testing. Specifically, selection procedures are introduced which control either the family-wise error rate (FWER) or the false discovery rate (FDR). The simulation results from Section 7.2 then show that these criteria also perform really well when regressors are stochastically independent (but not orthogonal) or strongly correlated.

Using again the framework of orthogonal designs Section 7.3 will first discuss some optimality properties of the introduced penalties in terms of model identification, followed by optimality results in terms of prediction. These theoretical results will indicate that it is often preferable to use methods that control the FDR since these can adapt to the typically unknown level of sparsity. While the L_0 penalties have superb theoretical properties, their practical application leads to a most challenging optimization problem, which is known to be NP-hard. For that reason, there has been a strong interest in L_1-penalties like the LASSO (Tibshirani, 1996), which can be tackled through convex optimization. However, with a fixed penalty weight for all regressors entering the model, LASSO can be compared with a fixed threshold rule in multiple testing. In view of the theoretical results from Section 7.3 and the good performance of the Benjamini Hochberg rule, it would be desirable to have more flexible choices of penalties. Section 7.4 will introduce SLOPE, where a specific choice of weighted L_1-penalties provides another FDR-controlling model selection procedure. Afterwards, Section 7.5 briefly discusses some advanced variable selection procedures controlling FDR, first a Bayesian version of SLOPE and then a procedure that uses the idea of knockoffs. In Section 7.6 the different selection methods are applied to two real data sets. R scripts are available online, which provide the code to perform these analyses.

7.1 Model Selection under an Orthogonal Design

Consider the situation where the columns of the design matrix X are orthogonal and scaled such that $X^T X = nI_p$. Apart from models using wavelets, this will rarely be the case in practice. However, this simple setting allows to see the parallels between multiple testing and model selection. It also provides the basic intuition for the behavior of L_0 penalties in high dimensions. The most important consequence of the orthogonal design is that the estimates of the different components of the coefficient vector β become independent of each other. Denoting the p columns of X by X_j one simply obtains $\hat{\beta}_j = \frac{1}{n} X_j^T y$ and it is fairly easy to see that in case of known σ these estimates are statistically independent and normally distributed, $\hat{\beta}_j \sim N(\beta_j, \sigma^2/n)$. One can test each coefficient using a z-test with the statistic $Z_j := \sqrt{n}\hat{\beta}_j/\sigma, j \in \{1, \ldots, p\}$. Model selection thus reduces to a multiple testing problem.

Alternatively, one can study the properties of model selection based on information criteria. To this end, a particular model is characterized by the index set M corresponding to nonzero coefficients of β. The notation $k_M = \|\beta\|_0$ is used for the corresponding model size. Table 7.1 provides an overview over some important L_0 penalties discussed in more detail below. Historically, the first selection procedures of this type were developed in the 1970s, the Akaike Information Criterion AIC (Akaike, 1974) and the Bayesian Information Criterion (BIC) by Schwarz (1978). AIC uses as penalty $\text{Pen}(\beta) = 2k_M$, whereas BIC has the penalty $\text{Pen}(\beta) = k_M \log n$, which becomes more stringent than the AIC penalty for $n > 7$. A vast literature exists about their statistical properties (see for example Burnham and Anderson (2002)). In particular, AIC has some optimality properties in terms of prediction, and BIC is consistent as long as the number of potential regressors is relatively moderate. However, it will soon become clear that both criteria are not really useful in a high-dimensional setting.

It is well known that performing model selection using AIC in our simple setting is equivalent to performing z-tests. This can be easily seen by considering the fact that under orthogonality it holds that

$$-2\log(\mathcal{L}(\hat{\beta})) = const + \|y - \sum_{j=1}^{p} \hat{\beta}_j X_j\|^2/\sigma^2$$

TABLE 7.1
Different L_0-Penalties and Their Corresponding Properties for High-Dimensional Model Selection. AIC and BIC are classical criteria but not suitable for high dimensions. The four modifications of AIC and BIC, respectively, are the main focus of this presentation. The other mentioned criteria are related, where this list is by no means comprehensive.

Name	Pen$(k_M; n, p)$	Properties
AIC (Akaike, 1974)	$2\ k_M$	Not suitable for $p > n$
BIC (Schwarz, 1978)	$\log n\ k_M$	Not suitable for $p > n$
mBIC (Bogdan et al., 2004)	BIC $+\ 2\log(p/4)\ k_M$	Controls FWER at level $\alpha < n^{-1/2}$
mAIC (Szulc, 2018)	AIC $+\ 2\log(2p)\ k_M$	Controls FWER at level $\alpha < 0.05$
mBIC2 (Żak-Szatkowska and Bogdan, 2011)	mBIC $-2\log k_M!$	Controls FDR at level $\alpha < n^{-1/2}$
mAIC2 (Szulc, 2018)	mAIC $-2\log k_M!$	Controls FDR at level $\alpha < 0.05$
EBIC (Chen and Chen, 2008)	EIC $+2\log\binom{p}{k_M}^{1-\kappa}$	Similar to mBIC2 for $\kappa \approx 1$
RIC (Foster and George, 1994)	$2\log p\ k_M$	Minimal inflation of predictive risk like mAIC with a different constant
Abramovich et al. (2006)	$2k_M \log(p/k_M)$	Minimax optimality similar to mAIC2
Birge and Massart (2001)	$ck_M \log(p/k_M), c > 2$	Bounds on quadratic risk

where the maximum likelihood estimates and the least-squares estimates of the coefficients coincide. It follows that adding regressor X_j to the model reduces the log-likelihood term of AIC by $n\hat{\beta}_j^2/\sigma^2$ and increases the penalty by 2, no matter which other regressors have already entered the model. Therefore, adding X_j decreases AIC if and only if $|Z_j| > \sqrt{2}$, which is equivalent to performing the z-test considered previously. Hence, model selection with AIC under an orthogonal design is equivalent to performing a z-test for each coefficient at the significance level $\alpha = 1 - F_{\chi^2}(2) \approx 0.157$.

Similar considerations hold for BIC, but here the penalty depends on the sample size n. For $n = 8$ selection with BIC corresponds to performing a z-test for each coefficient at a 15% significance level. Owing to the $\log n$ penalty, the significance level α_n decreases with increasing sample size n. More specifically, it holds that $\alpha_n = o(n^{-1/2})$. This is essential for the consistency property of BIC. Mathematically, this follows immediately from the well-known tail bounds of the normal distribution

$$\frac{2\phi(c)}{c}(1 - c^{-2}) \leq P(|Z_j| > c) \leq \frac{2\phi(c)}{c}. \tag{7.3}$$

Model selection with BIC under orthogonality corresponds to the comparison $|Z_j| > \sqrt{\log n}$, which gives according to (7.3) a type I error probability of $\alpha_n \leq \frac{\sqrt{2}}{\sqrt{\pi}}(n \log n)^{-1/2}$.

However, neither AIC nor BIC provides any correction for multiple testing, and it is immediately clear that with growing p, the number of type I errors will increase. In a high-dimensional context, one is typically interested in sparse models, and it follows that under sparsity, both AIC and BIC will massively overfit the data (see for example Broman and Speed (2002); Frommlet and Nuel (2016)). In particular, BIC will be no longer a consistent selection procedure when considering an asymptotic regime where p grows faster than $\sqrt{n}$. This is particularly problematic in applications where one is more interested in correct model identification than in prediction.

The first remedy is provided by the risk inflation criterion (RIC), which was introduced by Foster and George (1994) and has the penalty $2 \log p \; k_M$. In the orthogonal setting this relates to a z-test of the form $|Z_j| > \sqrt{\log p^2}$ with type I error rate controlled at $\alpha_p \leq \frac{\sqrt{2}}{\sqrt{\pi}} p^{-1}(2 \log p)^{-1/2}$. Hence, this penalty is closely related to the Bonferroni rule in multiple testing, where the nominal α level is divided by the number of tests p. Note that RIC is consistent only in the sense that for increasing p the expected number of false detections decreases at the very slow rate $(2 \log p)^{-1/2}$. RIC is not consistent in the classical sense that the probability of identifying the true model will converge to 1 with increasing n. Furthermore, if one is interested in model identification, then RIC has still a rather large rate of false detections as long as p is not exceptionally large. For small $p = 10$, the α_p bound indicates that RIC will control FWER only at approximately 0.35 and for $p = 1{,}000$ FWER is still at approximately 0.2.

7.1.1 Modifications of AIC and BIC

We will now introduce systematically a set of modifications of BIC and AIC, which are suitable for high-dimensional variable selection. The first of these criteria called mBIC was introduced by Bogdan et al. (2004) in the context of QTL mapping. The motivation for this criterion was based on arguments concerning the prior distribution of regressors for Bayesian model selection. According to asymptotic arguments in its classical derivation, BIC neglects the model prior. This is equivalent to giving each possible model M exactly the same prior probability. While such a prior is non-informative for the model, it is highly informative for the model dimension, which can be seen by a simple combinatorial argument. There are only p models of size 1, $\binom{p}{2}$ models of size 2, but there are $\binom{p}{p/2}$ models of size $p/2$.

Consequently, BIC will have a strong bias toward choosing models of intermediate size. Now if p is large and one is interested in sparse models, then BIC will have a tendency to overestimate the model size. This is a Bayesian explanation of the overfitting problem of BIC in high dimensions, which is complementary to the multiple testing perspectives given above.

To overcome this problem, mBIC was derived by using i.i.d priors for the p regressors (Bogdan et al., 2004). This is of course a classical choice in Bayesian variable selection and results in a binomial prior for the model size. The resulting criterion has a penalty of the form

$$\text{mBIC:} \quad \text{Pen}(k_M; n, p) = \log n \; k_M + 2\log(p/E) \; k_M. \tag{7.4}$$

Here the BIC penalty is combined with the penalty from RIC. Clearly, if p is large, the second penalty term will dominate the $\log n$ term. However, the $\log n$ term yields a criterion, which is consistent in the usual sense. Using similar arguments like given above for BIC, it is easy to see that mBIC controls the FWER of false detections under orthogonality at a level $\alpha_n = o(n^{-1/2})$. On the other hand, a similar penalty without the $\log n$ term will control FWER essentially at a constant level. If one is interested in classical consistency, then one should keep the $\log n$ term.

The constant E corresponds to the a-priori expected number of regressors which enter the model. From a frequentist point of view, it can be used as a tuning parameter to calibrate the level α_n. In case of no prior knowledge on the model dimension, a choice of $E = 4$ is recommended, which guarantees that for $n = 150$ the family wise error rate is controlled approximately at a level 0.1 for $p \geq 10$ and the bound drops already to 0.065 for $p = 1,000$. For $n = 500$ FWER is below 0.05 for $p \geq 10$ and below 0.035 for $p = 1,000$.

We have seen that the RIC criterion controls FWER at a constant level with respect to n which is fairly large. FWER decreases with p but at an extremely slow rate. It would take some $p \approx 10^{50}$ to bring down the FWER to 5%. So 'consistency' with respect to p is really fairly theoretical. As an alternative, we introduce the mAIC criterion as a modification of the classical AIC criterion

$$\text{mAIC:} \quad \text{Pen}(k_M; p) = 2k_M + 2\log(p/\text{const}) \; k_M. \tag{7.5}$$

Choosing Euler's number e as the constant this coincides with RIC. To control at the more familiar level $\alpha = 0.05$ for small $p = 10$ (and at 0.035 for $p = 1,000$), one can use const $= 0.5$ which is the recommended choice for our mAIC criterion. As a consequence, the criteria mAIC and mBIC coincide for sample size $n = 473$. Using the constant 1 in (7.5) yields roughly an α level of 0.11 for $p = 10$ (0.07 for $p = 1,000$).

According to the theoretical results from Section 7.3.1, it is desirable to have selection criteria that control the false discovery rate and not the family-wise error rate. This is achieved by the following modifications of BIC and AIC. The first criterion mBIC2 controls FDR at a level, which again depends on the sample size like $\alpha_n \propto (n \log n)^{-1/2}$:

$$\text{mBIC2:} \quad \text{Pen}(k_M; n, p) = \log n \; k_M + 2\log(p/E) \; k_M - 2\log k_M!. \tag{7.6}$$

In accordance with the definition of mBIC, we recommend the choice of $E = 4$, though FDR levels are then slightly higher than the FWER levels for mBIC (see the simulation results below for details). The additional penalty term $-2\log k_M!$ relaxes the penalty of mBIC and is closely related to the Benjamini Hochberg (BH) procedure, hence the control of the false discovery rate. In fact $-2\log k_M!$ is a first-order approximation of the penalty

$$\text{Pen}_{BH} := \sum_j q_N^2(\alpha j/2p), \tag{7.7}$$

where q_N denotes the quantile of the normal distribution. This penalty was introduced by Abramovich et al. (2006) in their seminal paper on minimax optimality of FDR-controlling model selection rules. Details of the derivation of mBIC2 and its theoretical properties are provided in Frommlet et al. (2011), which also considers a second-order approximation mBIC1 of Pen_{BH}. However, for all practical purposes, mBIC2 performs just as well and is much easier to compute. The extended Bayesian Information Criterion (EBIC) from Chen and Chen (2008) provides another family of BIC modifications suitable for high-dimensional variable selection. It depends on a parameter κ which varies between 0 and 1. EBIC with $\kappa = 0$ coincides with the original BIC, whereas for κ being large EBIC behaves very similar to mBIC2. However, there is not such an immediate interpretation of the parameter κ in terms of controlling the FDR level.

Augmenting the mAIC criterion with $-2 \log k_M!$ yields criteria which control FDR roughly at a fixed level:

$$\text{mAIC2:} \quad \text{Pen}(k_M; p) = 2k_M + 2\log(p/\text{const})\ k_M - 2\log k_M!. \tag{7.8}$$

To control FDR at a level close to $\alpha = 0.05$, we recommend once again to use const $= 0.5$. The penalty of mAIC2 is extremely similar to the penalty $2k_M \log(p/k_M)$ suggested by Abramovich et al. (2006) as an approximation of Pen_{BH}. The difference (up to a constant) between $k_M \log k_M$ and $\log k_M!$ is due to Sterling's approximation and for small values of k_M mAIC2 is actually closer to Pen_{BH} than $2k_M \log(p/k_M)$. Similar penalties of the form $ck_M \log(p/k_M)$ with $c > 2$ have been studied by Birge and Massart (2001).

Figure 7.1 illustrates the functional form of different L_0 penalties. The two penalties *mBIC* and *mAIC* which control the FWER are much more severe than the other penalties. For $n = 200$ mAIC (mAIC2) penalizes stronger than mBIC (mBIC2), for $n = 1,000$, the opposite is true. The FDR-controlling criteria mAIC2 and mBIC2 have been designed for sparse model selection and should only be used for $k < p/4$. For larger p, they can actually penalize less than BIC. It is also important that $k \ll n$. Otherwise, one might run into the problem of getting saturated models where the log-likelihood is converging to infinity. In high-dimensional applications (with $p > n$) L_0, penalties will typically have their global minimum at $k = n$. What one is looking for in practice is a local minimum with $k \ll n$.

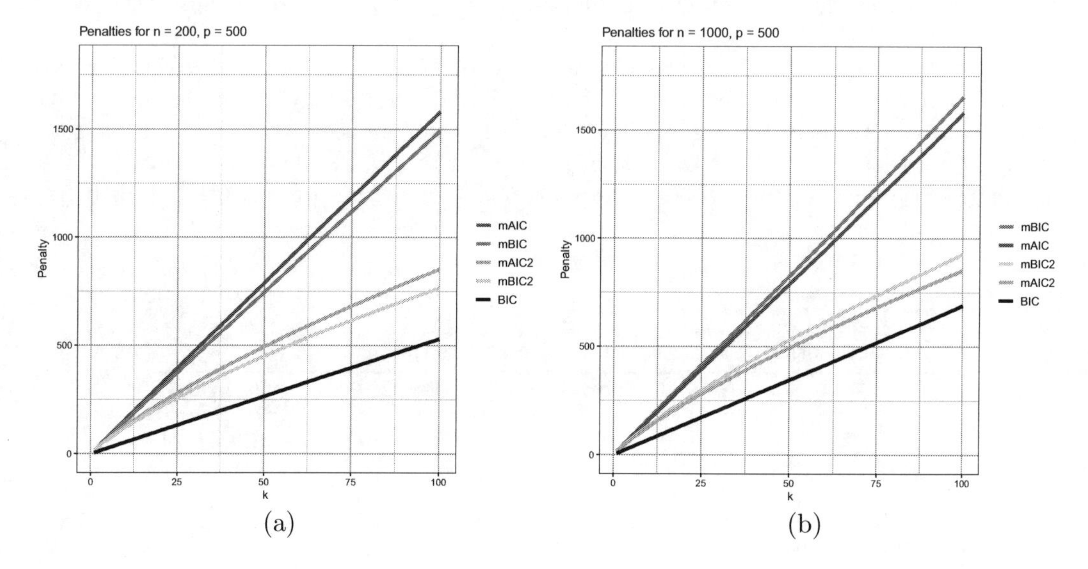

FIGURE 7.1
Different L_0-Penalties for $n = 200$ and $n = 1,000$ as a function of k.

7.2 Simulation Study for L_0-Penalties

The following simulations illustrate the properties of the four L_0 penalties (7.4)–(7.8) introduced above in comparison with Schwarz BIC. Data are simulated according to the linear model (7.1). The first part is concerned with independent regressors and illustrates that in that case type I error control of the different criteria is very similar to the orthogonal setting. The second part studies a specific scenario with correlated regressors and illustrates that even in such a setting our L_0 penalties perform quite well as long as correlations between regressors are not getting excessively large. Analysis was performed with the R package big step, which is available at CRAN (Szulc, 2018).

7.2.1 Independent Regressors

In the scenarios considered here, both the columns of X and the error term ϵ are i.i.d. standard normal. Scenario 0 is concerned with the type I error rate (number of false discoveries) under the assumption that there are no regressors associated with the dependent variable (that is $k = 0$). The other three scenarios consider sparse data generating models, where the total number of regressors p behaves differently with growing sample size n. In the first scenario, p remains constant with growing n; in the second scenario, p is proportional to $\sqrt{n}$ and in the final scenario p equals n. In Scenario 1, the number of regressors in the model k remains also constant, whereas in the other two scenarios k is mildly growing. Table 7.2 provides more details.

All coefficients from β were set to 0.4 for those regressors, which enter the data generating model. In Section 7.3.1, we will pay more attention to the effect sizes, which can actually be detected with different model selection criteria. Here our main focus is rather on the type I error rates where the simulations are supposed to illustrate the control rates claimed above.

To estimate FWER and FDR for each scenario, 1,000 simulation runs were performed. Regressors selected by some criterion are counted as true positives (TP) if they are part of the data generating model, otherwise they are counted as false positives (FP). FWER is then estimated as the average number of simulation runs with at least one FP detection. FDR is defined as the average over simulation runs of the proportion of false discoveries $\#FP/\max(1, \#FP + \#TP)$.

The panel (a) of Figure 7.2 shows the dependence of FWER on the sample size. Clearly BIC has a much larger type I error rate than the other four criteria. One can see that mAIC nicely controls FWER at the nominal level 0.05, while under the global null mAIC2 has an FWER (and thus FDR) closer to 0.08. mBIC has larger FWER than mAIC for $n < 500$ and smaller type I error for $n > 500$. The same relationship hold for mAIC2 and mBIC2.

TABLE 7.2
Characteristics of the Four Simulation Scenarios

	Scen 0		**Scen 1**		**Scen 2**		**Scen 3**	
n	p	k	p	k	p	k	p	k
49	49	0	49	5	49	5	49	5
100	49	0	49	5	70	7	100	7
225	49	0	49	5	105	10	225	10
529	49	0	49	5	161	13	529	15
1024	49	0	49	5	224	16	1024	20
2048	49	0	49	5				

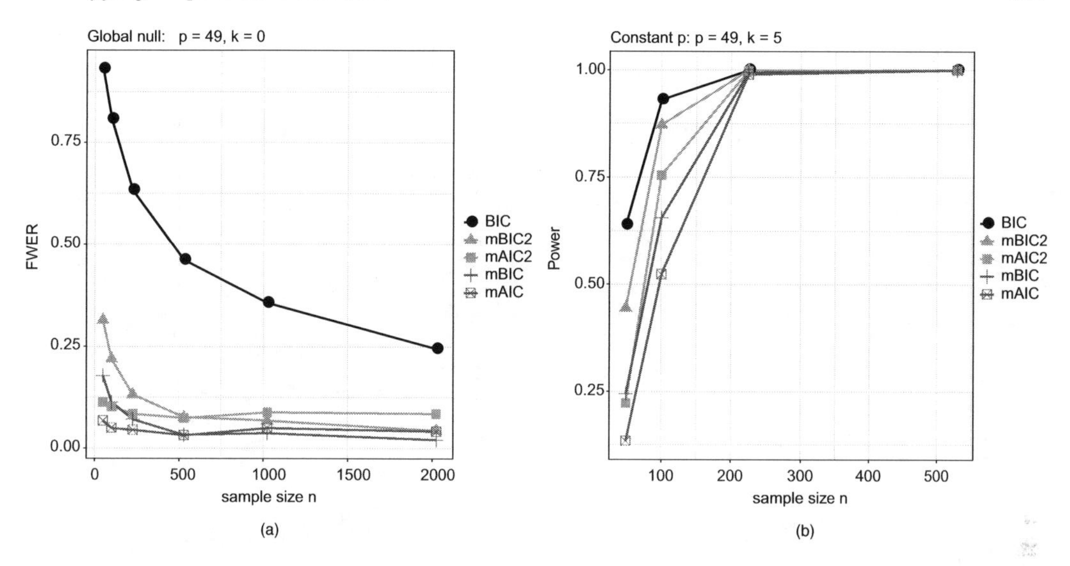

FIGURE 7.2
(a): Family wise error rate for different L_0-penalties depending on sample size n under the global null model of Scenario 0. (b): Power depending on n for Scenario 1 with constant $p = 49$ and constant $k = 5$.

The panel (b) of Figure 7.2 provides the power for Scenario 1, which is defined here as the percentage of correctly detected regressors from the data-generating model. The corresponding plots for Scenarios 2 and 3 look fairly similar and are not presented. BIC has the largest power followed by mBIC2, but already for $n = 500$ all criteria achieve a power of 1. In terms of consistency, it is therefore for these scenarios of primary importance to look at the type 1 error rates depicted in Figure 7.3.

The three left panels of Figure 7.3 show FWER depending on the sample size. For constant p, it is known that BIC is consistent and consequently FWER keeps on decreasing with increasing n. However, even for $n = 2,000$ the average number of FP detections is still at 0.27. In comparison, FWER of mBIC and mAIC are really small already for quite moderate sample size with values, which are in accordance with the results for the orthogonal design. Note that the FWER of mAIC remains at about 0.04 even for large n and is actually not expected to get smaller for arbitrary large n. Just like AIC, mAIC is not consistent, whereas mBIC is.

Looking at the two lower panels of Figure 7.3 shows that when p is growing with n BIC is no longer consistent at all. For $p \propto \sqrt{n}$, the average number of false detections still decreases very slowly but for $p = n$ it actually keeps on growing with n. Therefore, even for $p = n$, BIC is already completely unsuitable as a selection criterion if one is interested in model identification and for $p > n$ things are only getting worse. In contrast, both mAIC and mBIC are doing a very good job in controlling FWER for all our simulation scenarios, and they work similarly well in the case of $p > n$. The FWER estimated from simulations for independent regressors are remarkably close to the theoretical values from the orthogonal design.

The three right panels of Figure 7.3 provide the FDR. One can observe in all three scenarios that at least for larger n mAIC2 nicely controls FDR at the level 0.05. Similarly, the FDR of mBIC2 drops with growing n even below 0.05. mAIC and mBIC tend to have extremely small FDR and in view of the discussion of Section 7.3.1 are therefore potentially too conservative for many high-dimensional applications.

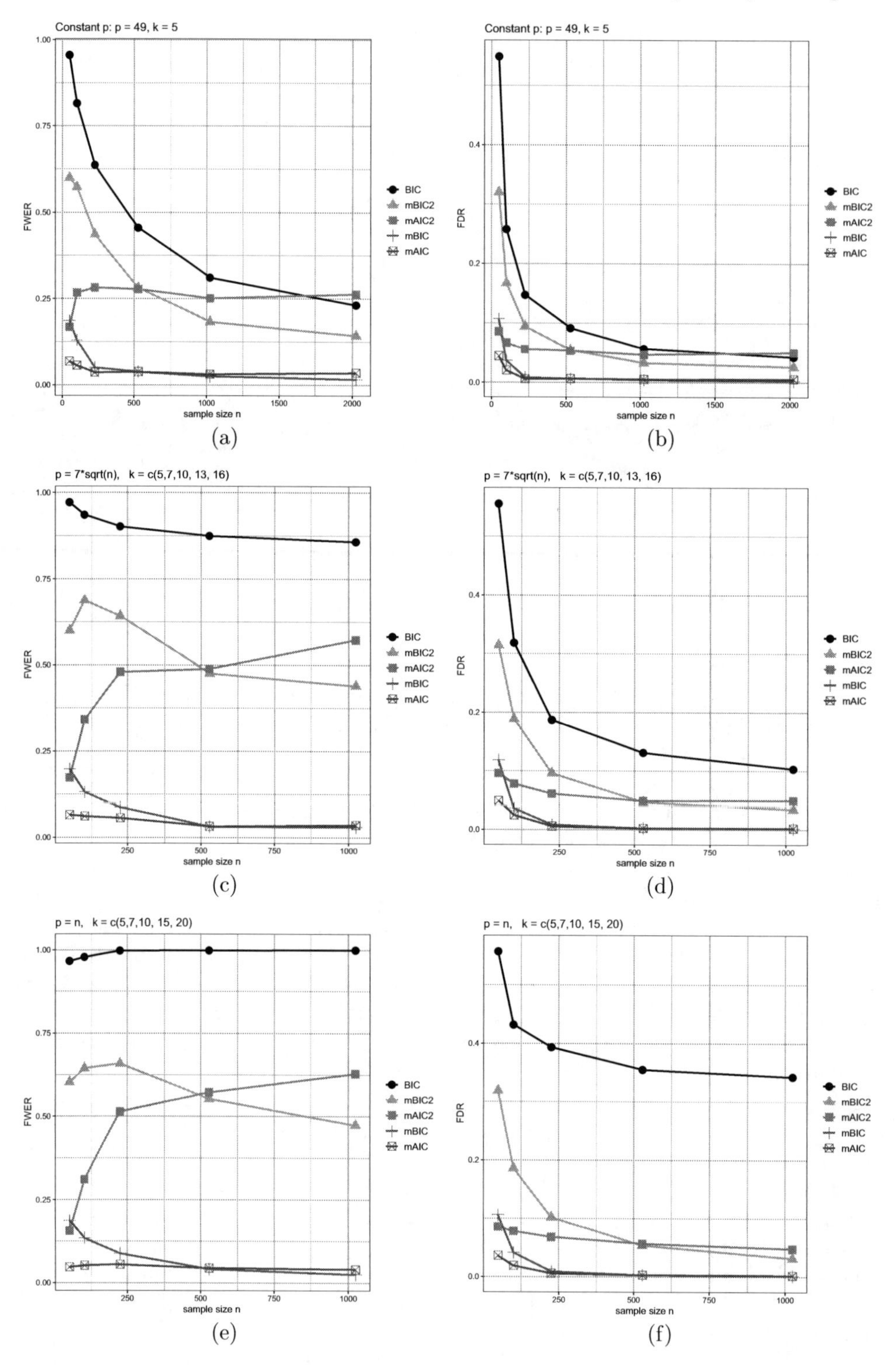

FIGURE 7.3
Family wise error rate and false discovery rate depending on n for three different simulation scenarios. (a and b): Scenario 1 with constant $p = 49$ and constant $k = 5$; (c and d): Scenario 2 with $p \propto \sqrt{n}$ and k growing mildly with n; (e and f): Scenario 3 with $p = n$ and k growing mildly with n.

7.2.2 Correlated Regressors

The previous simulations were performed for statistically independent regressors. For highly correlated predictor variables, it usually becomes difficult to distinguish between correctly identified predictors from an underlying data generating model and variables which are only highly correlated. Nevertheless, mBIC and mBIC2 have been repeatedly applied successfully in the context of genetic association studies (Dolejsi et al., 2014; Frommlet et al., 012a, 2012b; Szulc et al., 2017). The following simulation scenario taken from Frommlet (2012) gives an impression to which extent model selection based on our modifications of AIC and BIC still performs well in the case of correlated regressors.

Consider 256 potential regressor variables with a specific block correlation structure sketched in Figure 7.4. The first four blocks include each 32 variables, then come four blocks with 16 variables, four blocks with 8 variables and four blocks with 4 variables, respectively. Within each block, one has compound symmetry with variance 1 and correlation ρ, where depending on the simulation run ρ is ranging from 0 to 0.6. Otherwise, the blocks of variables are independent from each other, and there are some additional 16 variables which are also independent. This block structure is inspired by correlation patterns one might find in genome-wide association studies, though apparently it is a simplified setting.

Regressors from the data generating model are referred to as 'causal' in this example. For each block size, there is one block with three, one with two and one with one causal variable, respectively. In addition, there are four causal variables among the 16 independent variables. In summary, our data generating model, thus, has $k^* = 4*3+4*2+4*1+4 = 28$ causal variables. Effect sizes are randomly drawn from a normal distribution with mean 0 and variance 0.5. For these scenarios, 2,000 data sets are simulated to assess the performance of different selection criteria.

Figure 7.5 presents the results of the simulation study for correlated regressors. Apart from Power, FDR and FWER, it also shows the average number of misclassifications obtained with each selection criterion. For $\rho = 0$ the potential regressor variables are all independent and the type I error rates are similar to those from Scenario 3 of the previous

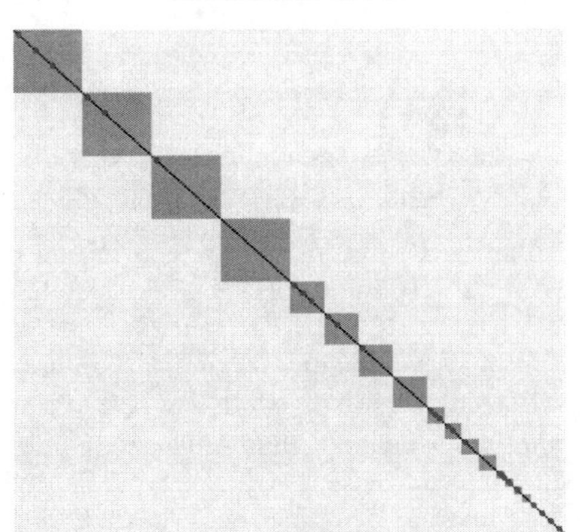

FIGURE 7.4
Block correlation structure of the matrix X. Four different block sizes and four blocks per size, where within each block all variables have pairwise correlation ρ. Simulations are run with ρ ranging from 0 to 0.6. Blue dots indicate the causal variables from the data-generating model.

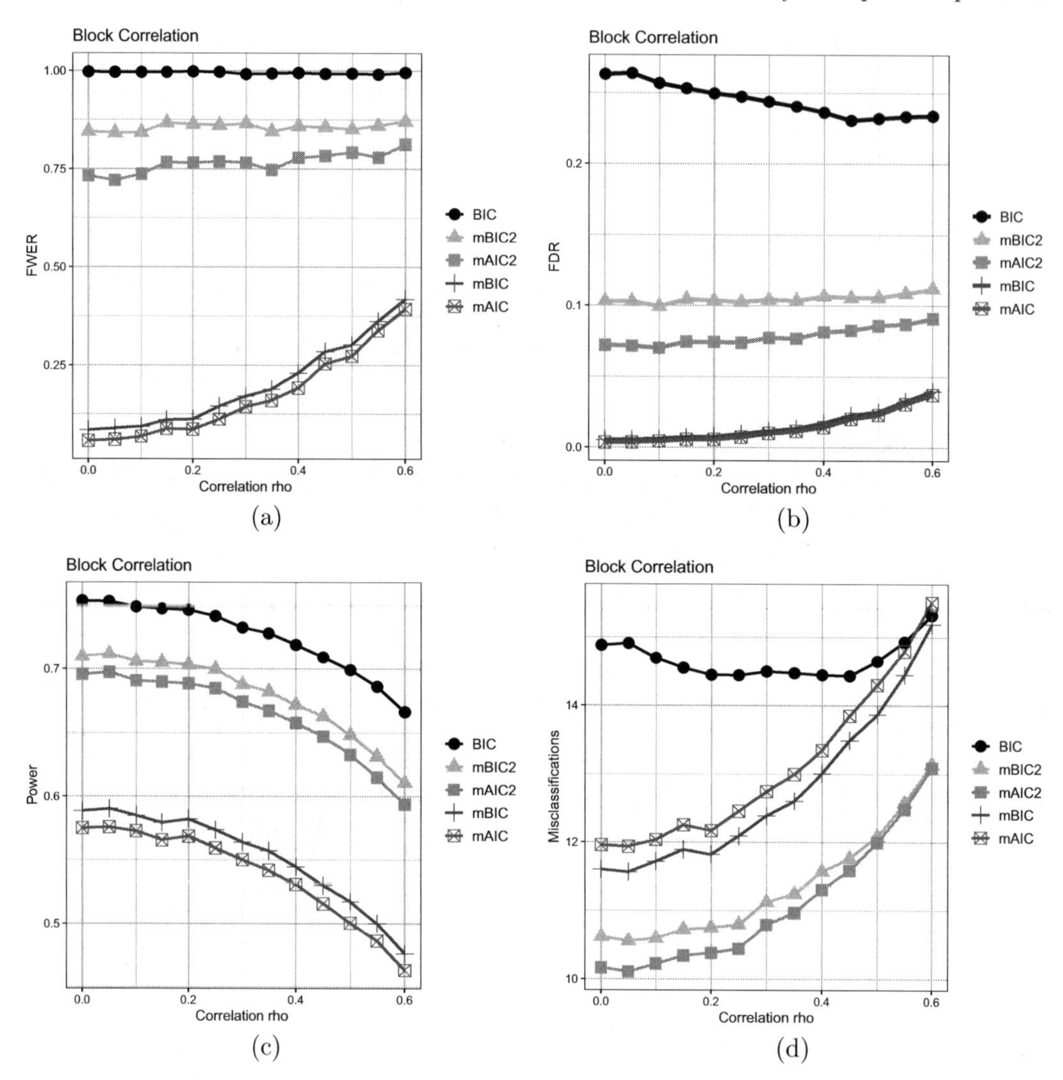

FIGURE 7.5
Family-wise error rate, false discovery rate, power and number of misclassifications, depending on within block correlation ρ for different L_0 penalties.

simulation study. In particular mAIC and mBIC are controlling FWER at levels 0.06 and 0.08, respectively. Furthermore, mAIC2 and mBIC2 are controlling FDR at levels 0.07 and 0.1, whereas BIC has FWER close to 1 and FDR at 0.26.

BIC has with 75% the largest power but due to a large number of false positives also the largest number of misclassifications. Note that this scenario is not particularly sparse, and hence BIC is performing not that bad. Increasing the number of potential regressors p while keeping the same number of causal variables would provide scenarios where BIC would perform much worse compared with the other criteria. Note that mAIC2 and mBIC2 have the lowest number of misclassifications which corresponds to the optimality properties discussed in Section 7.3.1. Controlling FDR gives the best trade-off between controlling the number of false positives and having sufficient power to detect true positives.

Increasing the correlation within blocks has the following effects. For all criteria, the power decreases. This is natural because for larger ρ, it becomes more likely that a causal variable is substituted by a highly correlated variable in the model. Eventually for very large ρ correlated variables will become more or less indistinguishable and the concept of true and false positives breaks down for variables within the same block. For the very same reason, mAIC and mBIC suffer from an increasing number of false-positive detections with increasing ρ and are no longer controlling FWER. Now interestingly, the average number of false positives detected by mAIC2 and mBIC2 remains fairly stable with increasing ρ while for BIC the number of false positives is actually decreasing. FDR of mAIC2 and mBIC2 is increasing only slowly with increasing ρ, and they keep on having the lowest misclassification rate.

7.3 Optimality Results

In the last simulation scenario, the FDR-controlling criteria mAIC2 and mBIC2 performed best in terms of the number of misclassifications. The following section briefly recaps theoretical results from multiple testing, which corroborate this observation. Subsequently, some results from the literature are presented, which show that mAIC2 is asymptotically optimal with respect to prediction.

7.3.1 Optimality for Model Identification

Bogdan et al. (2011) introduced the concept of Asymptotic Bayes Optimality under Sparsity (ABOS) for multiple testing in the context of normal mixture distributions. The optimality results obtained there were later extended to more general multiple testing settings (Frommlet and Bogdan, 2013; Neuvial and Roquain, 2012) and in the case of orthogonal designs can be directly transferred to the model selection criteria which were introduced above (Frommlet et al., 2011). Here we will only sketch the ABOS results for multiple testing to provide an intuition about the advantages of using FDR-controlling model selection criteria in a high-dimensional setting.

Consider a set of normally distributed populations $T_j|\mu_j \sim N(\mu_j, \sigma^2)$, $j = 1, \ldots, p$. A two groups normal mixture model

$$T_j \sim (1-\eta)N(0, \sigma^2) + \eta N(0, \sigma^2 + \tau^2) \tag{7.9}$$

can be used for testing the hypotheses

$$H_{0j} : \mu_j = 0 \quad \text{against} \quad H_{Aj} : \mu_j \sim N(0, \tau^2). \tag{7.10}$$

This is similar to the classical two-sided test, but like in the simulation scenario of Section 7.2.2 the effect size under the alternative is a random variable. The parameter η gives the proportion of true alternative hypotheses in the population. It is assumed to be small and will be referred to as sparsity parameter.

The concept of ABOS is based on a decision theoretical framework. For each individual test, a type I error imposes a loss of δ_0 and a type II error a loss of δ_A. The total loss is then defined as the additive loss over all individual tests. This is a common choice in classification tasks. For a specific multiple testing procedure let t_{1j} and t_{2j} be the probabilities of type I and II errors for the j-th hypothesis, respectively. The corresponding Bayes risk is then defined as the expected total loss,

$$R = p\left((1-\eta)t_{1j}\delta_0 + \eta t_{2j}\delta_A\right). \tag{7.11}$$

In the case of $\delta_0 = \delta_A = 1$, this is simply the expected number of misclassifications.

Assuming that the p tests based on model (7.9) are independent and that σ is known it is easy to explicitly obtain the risk R_{opt} of the Bayes classifier which minimizes the Bayes risk (7.11). In Bogdan et al. (2011), certain asymptotic regimes were studied and a multiple testing procedure in this setting was classified as ABOS if the ratio between its Bayes risk and the risk of the Bayes classifier converges to one, i.e. $R/R_{opt} \to 1$.

Without including all the technicalities of Bogdan et al. (2011) the main result can be illustrated by considering the asymptotic setting where the number of tests p converges to infinity and the sparsity parameter η is decreasing with p according to $\eta \propto p^{-\beta}$ for some $\beta \in (0, 1]$. Here β describes the asymptotic level of sparsity. Given some other technical conditions, it turns out that Bonferroni correction is ABOS only when $\eta \propto 1/p$. This is the assumption of "extreme" sparsity, under which the expected number of true signals does not increase with p. Instead, Benjamini-Hochberg correction is ABOS for any $\beta \in (0, 1]$. When the number of true signals is very small, then the results of BH are not substantially different from those obtained using the Bonferroni correction. However, BH can adapt to the unknown sparsity level and is ABOS for a wide range of scenarios under which $\eta \to 0$.

For a loss with $\delta_0 = \delta_A = 1$, it is also necessary that with increasing p the FDR level α of BH converges to 0 at a certain rate to obtain ABOS. In Frommlet and Bogdan (2013) model (7.9) was extended to consider the situation of tests based on random samples of size n and conditions are given under which BH with FDR level $\alpha \propto n^{-1/2}$ is ABOS. In terms of model selection criteria, this rate of decrease directly corresponds to mBIC2 as we have seen in Section 7.1.

7.3.2 Optimality for Prediction

Abramovich et al. (2006) analyzed the properties of BH for estimating a vector of random variables with expected values μ. Specifically, they consider the hard-thresholding estimator

$$\tilde{\mu}_j = \begin{cases} X_j & \text{if BH rejects } H_{0j} : \mu_j = 0 \\ 0 & \text{otherwise} \end{cases} \tag{7.12}$$

The optimality of BH is analyzed with respect to the estimation risk over "sparse" balls in the space $\mu \in R^p$. In Abramovich et al. (2006) different notions of sparsity are considered and highly technical results are proven. In essence, it is shown that the hard-thresholding rule based on BH is asymptotically minimax over a wide range of sparsity levels or, in other words, it adapts to the unknown sparsity and optimally selects those components of μ for which the signal strength substantially exceeds the variance of the estimation error.

In Wu and Zhou (2013), these results are extended to the class of estimators of the form:

$$\hat{\mu} = \underset{\mu}{\text{argmin}} \left\{ ||Y - \mu||^2 + \sigma^2 Pen\left(||\mu||_0\right) \right\}, \tag{7.13}$$

where $Pen\left(||\mu||_0\right)$ is the penalty for the number of nonzero elements of μ. Specifically, in Wu and Zhou (2013) it is shown that $\hat{\mu}$ is asymptotically minimax if the penalty grows like $2k \log(p/k)$. It is easy to check that mAIC2 is asymptotically equivalent to such a penalty. Note also that under orthogonality with $X'X = I$ a multiple regression model can be represented as

$$\hat{\beta} = X'Y = \beta + \tilde{\epsilon},$$

where $\tilde{\epsilon} = X'\epsilon \sim N(0, \sigma^2 I)$. Thus, the results of Wu and Zhou (2013) prove also asymptotic optimality of mAIC2 for minimizing the estimation and prediction error in a multiple regression model when σ is known and the design matrix is orthogonal and normalized such that $X'X = I$.

7.4 Model Selection with the Sorted L-One Norm

The last section has shown that there is a lot of theoretical underpinning for FDR-controlling model selection criteria like mBIC2 or mAIC2. The main difficulty in applications with high-dimensional data is the computational complexity. Identifying the model that minimizes any of these criteria is an NP-hard problem. In the context of genetic data, which exhibit rather low range spatial correlations, very good results were obtained by certain modifications of a simple stepwise search (Dolejsi et al., 2014; Frommlet et al., 2012b; Szulc et al., 2017). Available software includes the C++ program *MOSGWA* by Dolejsi et al. (2014) which is particularly designed to analyze GWAS data and the more generally applicable R package *bigstep* by Szulc (2018). Another interesting possibility is to use some adaptive Ridge regression to obtain good models according to the modified information criteria (Frommlet and Nuel, 2016). However, all these optimization algorithms give no guarantee that the optimal model has been identified.

In contrast, convex optimization problems have a unique solution which can be solved efficiently by a number of different algorithms. There is a strong interest in considering penalized likelihood methods of the form

$$\hat{\beta} = \text{argmin}_\beta\{-\log \mathcal{L}(\text{Y}|\text{X}, \beta) + \|\beta\|\}, \tag{7.14}$$

where $\|\cdot\|$ is some norm, because for classical generalized linear models (GLM) $\hat{\beta}$ then becomes a convex function of the parameter vector β. Note that in contrast to (7.2) we are no longer multiplying the log-likelihood term with a factor 2. This is quite common when working with L_1 penalties.

The most popular model selection procedure of this type is the Least Absolute Shrinkage and Selection Operator (LASSO) (Tibshirani, 1996), which uses the standard L_1 norm multiplied by a tuning parameter λ:

$$\hat{\beta}^L = \text{argmin}_\beta \left\{\frac{1}{2}\|\text{Y} - \text{X}\beta\|_2^2 + \lambda\|\beta\|_1\right\}, \tag{7.15}$$

with $\|\beta\|_1 = \sum_{j=1}^p |\beta_j|$. It is easy to check that in case of $X'X = I$ it holds that

$$\hat{\beta}_j^L = 0 \quad \text{if and only if} \quad |X_j'Y| \leq \lambda,$$

and furthermore one has $X_j'Y \sim N(\beta_j, \sigma^2)$. Hence, a Bonferroni-like tuning parameter $\lambda = \lambda_{Bon} = \sigma\sqrt{2\log p}(1 + o_p)$ is needed to control FWER. This provides the intuition why most of the theoretical results on consistency and optimality of LASSO require that λ is proportional to $\sqrt{\log p}$.

Thus, similarly to mBIC or mAIC, under orthogonality LASSO can be interpreted as a fixed threshold multiple testing procedure. The theoretical results for multiple testing under sparsity show that procedures based on decaying sequences of thresholds (like Benjamini-Hochberg) perform better than fixed threshold rules (like Bonferroni). Furthermore, we have seen that for high-dimensional model selection, mAIC and mBIC are outperformed by the nonlinear penalties mAIC2 and mBIC2. Therefore, it is quite natural to consider replacing the single tuning parameter λ from LASSO with a decaying sequence of tuning parameters.

This idea was used by Bogdan et al. (2015, 2013) to propose the SLOPE (Sorted L-One Penalized Estimation) procedure. For any nonzero and non-increasing sequence $\lambda_1 \geq \cdots \geq \lambda_p \geq 0$ the SLOPE estimator is given by

$$\hat{\beta}^{SL} = \text{argmin}_\beta \left\{\frac{1}{2}\|\text{Y} - \text{X}\beta\|_2^2 + \text{J}_\lambda(\beta)\right\}, \tag{7.16}$$

where $J_\lambda(\beta) = \sum_{j=1}^{p} \lambda_j |\beta|_{(j)}$, and $|\beta|_{(1)} \geq \ldots \geq |\beta|_{(p)}$ is the vector of sorted absolute values of elements of β. It is easy to check that the function $J_\lambda(b) = \sum_{j=1}^{p} \lambda_j \left|b\right|_{(j)}$ is a norm (see Bogdan et al. (2015, 2013)) and hence (7.16) can be solved with convex optimization tools.

Figure 7.6 illustrates different shapes of the unit balls corresponding to different versions of the Sorted L-One Norm. Since the solutions of SLOPE tend to occur on the vertices of respective balls, Figure 7.6 demonstrates large flexibility of SLOPE with respect to dimensionality reduction. For $\lambda_1 = \cdots = \lambda_p$ SLOPE coincides with LASSO and reduces dimensionality by shrinking the coefficients to zero. In contrast for $\lambda_1 > \lambda_2 = \cdots = \lambda_p = 0$ the reduction of dimensionality is performed by shrinking coefficients towards each other (since the vertices of the l_∞ ball correspond to vectors b such that at least two coefficients are equal to each other). When the sequence of thresholding parameters is monotonically decreasing SLOPE reduces the dimensionality both ways: it shrinks them towards zero and towards each other. Thus, it returns sparse and stable estimators.

We are particularly interested in decaying parameter sequences which result in FDR-controlling selection procedures. This is once again achieved by translating the thresholds from the BH multiple testing procedure into penalty parameters. The BH penalty (7.7) from Abramovich et al. (2006) corresponds to the SLOPE parameter sequence

$$\lambda_j^{BH}(c, q) = c\, \Phi^{-1}\left(1 - \frac{jq}{2p}\right), \quad j \in \{1, \ldots, p\}, \quad q \in (0, 1), \tag{7.17}$$

where c is some tuning parameter to be discussed later and q corresponds formally to the FDR level in BH. Note however that the nominal FDR level of SLOPE will depend both on c and q which is the reason why we change here notation and do not use α in (7.17). Also note that for the limit $q = 0$ the parameter sequence $\lambda_j^{BH}(c, 0)$ is constant and the procedure turns into LASSO with tuning parameter $\lambda = c\, \Phi^{-1}(1)$.

A second-order approximation, which was also used by Abramovich et al. (2006), yields a similar sequence of tuning parameters of the form

$$\lambda_j \propto \sqrt{2 \log(p/j)}, \quad j \in \{1, \ldots, p\}. \tag{7.18}$$

The following two sections present different properties of SLOPE using these BH parameter sequences. We will first focus on prediction and then on model identification. In particular, we will see that different choices of the tuning parameter c in (7.17) are necessary to achieve these different goals.

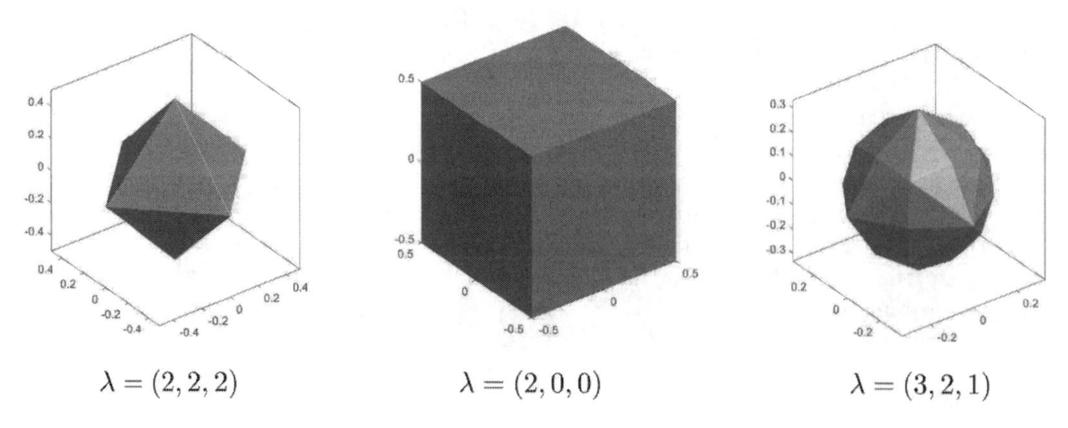

FIGURE 7.6
Shapes of different SLOPE balls

7.4.1 Prediction Properties of SLOPE

To evaluate the estimation and prediction properties of SLOPE, we consider two different mean squared errors. With the notation $\hat{\mu} = X\hat{\beta}$ and $\mu = E(Y) = X\beta$, the mean squared prediction error is defined as

$$MSE(\hat{\mu}) = E||X\hat{\beta} - X\beta||_2^2, \tag{7.19}$$

whereas for the coefficients we consider

$$MSE(\hat{\beta}) = E||\hat{\beta} - \beta||_2^2, \tag{7.20}$$

For the convergence of $MSE(\hat{\beta})$, some theoretical results for high-dimensional linear and logistic regression are available, which were proven for SLOPE with the sequence (7.18), see e.g. Abramovich and Grinshtein (2017); Bellec et al. (2018); Su and Candés (2016). Specifically, under some assumptions on the sparsity of β and the structure of the design matrix X, SLOPE achieves an asymptotic minimax rate $k \log\left(\frac{p}{k}\right)$, where $k = ||\beta||_0$ is the number of nonzero coefficients of β. Since the optimal rate of convergence of the LASSO estimator with a fixed tuning parameter λ is only $k \log p$ one would expect that SLOPE can outperform LASSO in terms of the estimation rate for larger values of k. According to the following simulation study, similar theoretical results should also hold for the mean squared prediction error.

Simulation: In the following simulation study, the estimation and prediction properties of SLOPE and LASSO are compared. The sample size and number of potential predictors is $n = p = 1,000$. The rows of the design matrix are simulated as independent random vectors from a multivariate normal distribution $N(0, \frac{1}{n}\Sigma)$. In the first scenario, predictors are *independent*, that means $\Sigma = I$. In the second scenario *correlated*, predictors are simulated using a compound symmetry matrix with $\Sigma_{i,i} = 1$ and $\Sigma_{i,j} = 0.5$ for $i \neq j$. Values of Y are generated according to the linear model (7.1) with

$$\begin{cases} \beta_1 = \cdots = \beta_k = \sqrt{2\log\left(\frac{1000}{k}\right)} \\ \beta_{k+1} = \cdots = \beta_{1000} = 0, \end{cases}$$

with $\sigma = 1$. For the model size two different values $k \in \{20, 100\}$ are considered. Estimation of $MSE(\hat{\mu})$ and $MSE(\hat{\beta})$ is based on 100 independent replicates of the whole experiment.

Figure 7.7 presents hypsometric maps of the prediction error of SLOPE with parameter sequence (7.17) using a range of values for c and q. The white triangle marks the combination of (c, q) values for which $MSE(\hat{\mu})$ is optimal. For $\Sigma = I$ and $k = 20$, the optimal q is very close to zero. This illustrates that LASSO has good prediction properties when the regressors are independent and β is very sparse. However, when $k = 100$ the optimal value of q is close to 0.4 and the prediction error of the optimal version of SLOPE is substantially smaller than the prediction error of the optimal version of LASSO. The advantage of SLOPE over LASSO is even more pronounced when regressors are correlated. Here, SLOPE offers a much lower $MSE(\hat{\mu})$ even when $k = 20$. For $k = 100$, the difference between these two methods becomes very large.

Figure 7.8 compares confidence intervals of $MSE(\hat{\beta})$ for LASSO with optimal λ and SLOPE with optimal choice of (c, q). Results for $k = 20$ and independent regressors are not shown because in that case LASSO is more or less identical with SLOPE. For all other scenarios, SLOPE is performing substantially better than LASSO.

These simulations illustrate the great potential of SLOPE, particularly when the signal is relatively strong and the number of true regressors is moderately large or when predictors

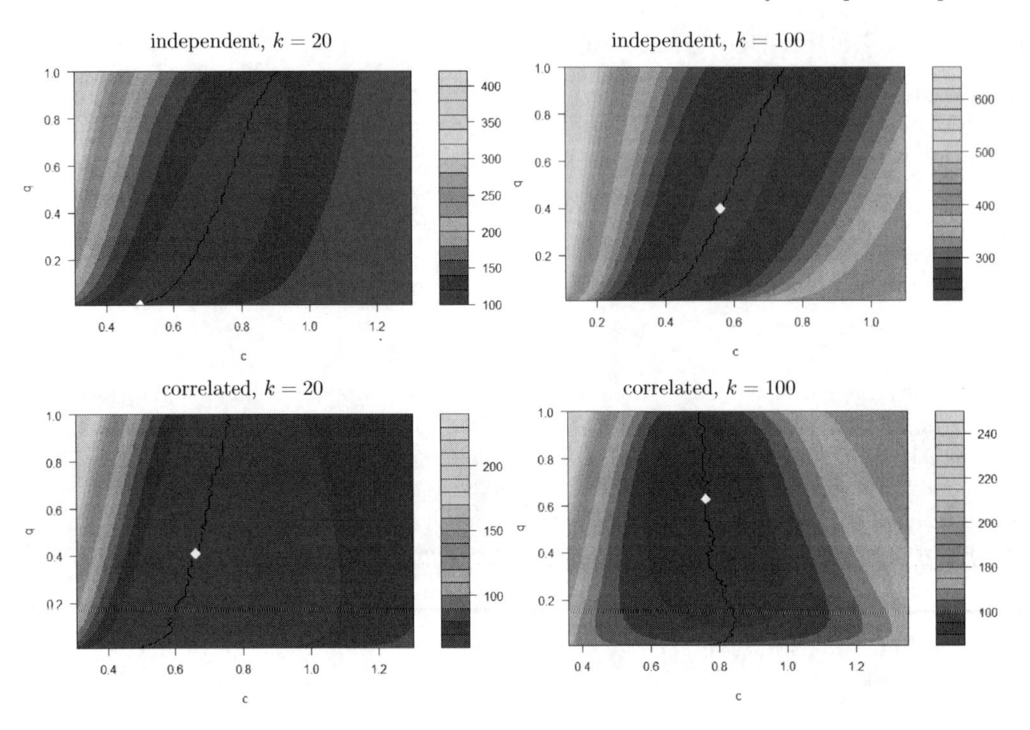

FIGURE 7.7
Hypsometric maps of $MSE(\hat{\mu})$ defined in (7.19) for SLOPE with the sequence of tuning parameters (7.17) and different values of parameters c and q. The white triangle marks the positions with the minimal $MSE(\hat{\mu})$. For $q = 0$ one obtains $MSE(\hat{\mu})$ from LASSO. Here $n = p = 1,000$ and k denotes the number of nonzero elements in β.

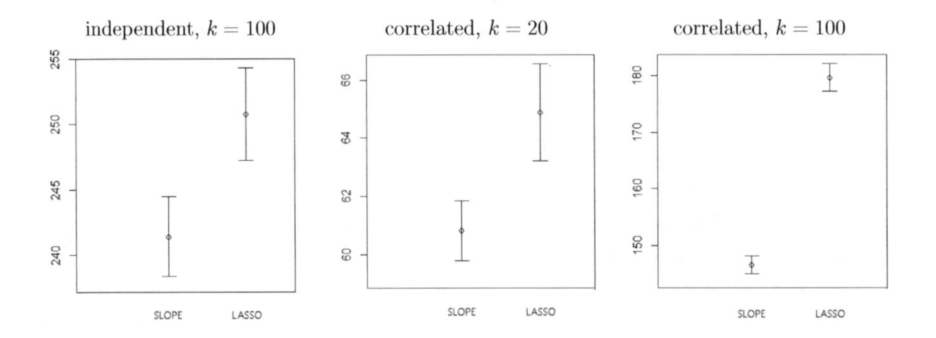

FIGURE 7.8
95% confidence intervals for MSE for optimal versions of LASSO and SLOPE based on 100 independent experiment replicates.

are strongly correlated. For the practical application of SLOPE with real data, there remains to be explored to which extent the optimal values of c and q can be obtained via cross-validation, which can provide almost unbiased estimators of prediction error (see e.g., Burman (1989); Fushiki (2011); Yanagihara et al. (2006)).

The good performance of SLOPE under correlated designs confirms the conjectures of Bondell and Reich (2008), who developed Octagonal Selection and Clustering Algorithm for

Regression (OSCAR), a predecessor of SLOPE. They proposed to use a linearly decaying sequence of tuning parameters to improve the properties of LASSO. According to results from extensive simulation studies (not reported here) SLOPE with the sequence λ^{BH} (7.17) usually performs substantially better than OSCAR. Here also the work of Zeng and Figueiredo (2014) should be mentioned, who independently developed a version of SLOPE (called OWL) as an extension of OSCAR to deal with correlated designs.

7.4.2 Model Identification Properties of SLOPE

Like in the case of other selection procedures, it is to be expected that cross-validation for SLOPE with $\lambda^{BH}(c, q)$ will result in values of c and q which will give too large models, that is the corresponding selection procedure will include too many false positive regressors. It has been shown that under an orthogonal design with known error variance, the choice of $c = \sigma$ provides FDR control at the level q (Bogdan et al., 2015, 2013). Unfortunately, this is no longer true if the inner products between columns of the design matrix are different from zero, which will almost always be the case in practical applications.

LASSO suffers from a very similar problem, and to better understand what is going on here, it is worth looking at the general solution of the LASSO procedure:

$$\hat{\beta}_j^L = \eta_\lambda(\beta_j + X_j'\epsilon + v_j), \tag{7.21}$$

where

$$\eta_\lambda(t) = sign(t)(|t| - \lambda)_+$$

and

$$v_j = \left\langle X_j, \sum_{l \neq j} X_l(\beta_l - \hat{\beta}_l^L) \right\rangle.$$

Only for an orthogonal design matrix X it holds that $v_j = 0$, otherwise this term contributes to the variance of the estimator of β. The magnitude of this additional noise depends on the model sparsity, the bias of large regression coefficients and the inner products between columns of the design matrix. As long as the correlations and the number of nonzero elements in β are small enough, this additional variance can be controlled by increasing the value of the tuning parameter λ. However, increasing λ leads to increased bias and the whole process gets out of control when the number of nonzero elements in β exceeds some limiting value.

This phenomenon is captured, e.g., by Theorem 2 of Wainwright (2009), which says that no matter how large the nonzero regression coefficients of the data-generating model are, the probability that LASSO can identify the true model is smaller than 0.5 unless a stringent *irrepresentability* condition is satisfied (van de Geer and Bühlmann, 2009; Wainwright, 2009; Zhao and Yu, 2006). This condition in principle sets a limit on the sparsity of β, which depends on the correlations between columns in X. A thorough discussion of this condition, with examples of *irrepresentability* curves for different design matrices, can be found in Tardivel and Bogdan (2018). These issues were also thoroughly analyzed in Bogdan et al. (2015) and Su et al. (2017) for design matrices with i.i.d. standard normal columns. Specifically, Bogdan et al. (2015) used the theory of Approximate Message Passing Algorithms (Bayati and Montanari, 2012) to provide sparsity limits needed for FDR control when LASSO is used with an arbitrary but fixed λ. Su et al. (2017) discuss the trade-off between FDR and Power provided by LASSO when λ is chosen adaptively based on the data.

The theory describing the limitations of SLOPE for model identification still needs to be fully developed, but its behavior under orthogonality and the cited results for LASSO suggest that SLOPE can efficiently control FDR if β is sparse enough and the regressors are roughly independent. There are some theoretical results available which point in that direction. Kos (2019) considered SLOPE with a sequence of tuning parameters $(1+\delta_n)\lambda^{BH}$, where δ_n is slowly converging to zero and showed that it asymptotically controls FDR at the level q when the design matrix is random with uncorrelated predictors and p is fixed while n diverges to infinity. Kos (2019) and Kos and Bogdan (2020) prove also that if the columns of the design matrix are i.i.d random variables from a Gaussian distribution then SLOPE with the sequence of tuning parameters

$$(1+\delta)\lambda^{BH},\ \delta > 0, \tag{7.22}$$

has FDR converging to zero and power converging to one if the number of true nonzero regression coefficients k satisfies

$$\frac{k^2 \log p}{n} \to 0, \tag{7.23}$$

and the magnitude of these nonzero coefficients is large enough.

To improve FDR-controlling properties of SLOPE, Bogdan et al. (2015) used equation (7.21) to derive a heuristic adjustment of the λ^{BH} sequence, which is well justified for design matrices with i.i.d normally distributed columns:

$$\lambda_i^{ad}(q) = \begin{cases} \sigma\Phi^{-1}(1-q/2p) & if \quad i=1 \\ \min\left(\lambda_{i-1}, \sigma\Phi^{-1}(1-qi/2p)\sqrt{1+\frac{\sum_{j<i}\lambda_j^2}{n-i-2}}\right) & if \quad i>1. \end{cases} \tag{7.24}$$

Simulation: We want to compare the performance of LASSO and SLOPE for model identification in different simulation scenarios. For SLOPE, we use four different tuning parameter sequences: $\lambda^{BH}(\sigma,q)$ according to (7.17), then $(1+\delta)\lambda^{BH}(\sigma,q)$ according to (7.22) with $\delta = 0.05$ and $\delta = 0.1$, respectively, and finally $\lambda^{ad}(q)$ according to (7.24). As tuning parameter for LASSO, the Bonferroni threshold $\lambda = \Phi^{-1}\left(1-\frac{q}{2p}\right)$ was chosen.

Simulations were performed as before using a linear model with design matrix X having independent Gaussian columns. The signal magnitude to generate Y was $\beta_1 = \cdots = \beta_k = 0.9\sqrt{2\log p}$. To study the asymptotic behavior of different procedures, simulations were performed for sample size $n \in \{100, 200, 500, 1000, 2000\}$. The number of potential regressors was set accordingly to $p = 0.05n^{1.5}$. Three scenarios of different sparsity levels were studied by setting the model size to $k = round(n^{\alpha})$, with $\alpha \in \{0.3, 0.4, 0.5\}$.

Figure 7.9 presents FDR and power as a function of n for $q = 0.2$. For $\alpha = 0.3$ and $\alpha = 0.4$ assumption (7.23) is satisfied for the two SLOPE versions with $\delta > 0$, and it is noticeable that in those cases FDR seems to converge and is close to or below the nominal level $q = 0.2$ for the whole range of considered values of n. For $\alpha = 0.5$, the assumption is violated though FDR is still a decreasing function of n, but the rate of this decrease is rather slow, and it is difficult to predict if it would converges to 0 with increasing n.

Apparently larger values of δ lead to more conservative versions of SLOPE. For $\delta = 0$, one has the original λ^{BH} sequence and FDR is above the nominal level. For $\alpha < 0.5$ FDR is slowly decreasing with n, but it remains to be checked if it actually converges to $q = 0.2$. When $\delta = 0$ and $\alpha = 0.5$ FDR seems to stabilize substantially above the nominal value of 0.5, which suggests that condition (7.23) is indeed necessary for the asymptotic FDR control with the original λ^{BH} sequence.

Figure 7.9 shows that LASSO with Bonferroni tuning parameter is substantially more conservative than all the different versions of SLOPE. In all three scenarios, its FDR converges to zero, which in the case of such moderate signals leads to a substantial decrease in power compared with SLOPE. Interestingly, SLOPE with the heuristic choice of tuning

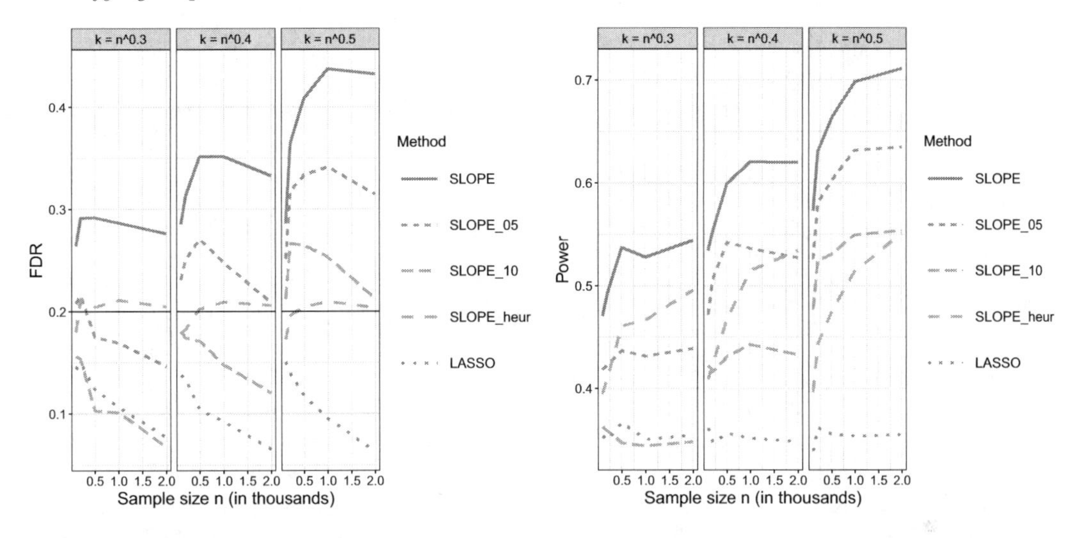

FIGURE 7.9
Comparison of LASSO and four different SLOPE versions. In the figure labels, SLOPE refers to the tuning sequence $\lambda^{BH}(\sigma, q)$, SLOPE_05 and SLOPE_10 to the adjusted sequence (7.22) with $\delta = 0.05$ and $\delta = 0.1$, respectively and SLOPE_heur to the heuristic sequence (7.24). FDR and Power are given as a function of n for $q = 0.2$ and for three different sparsity levels $k = round(n^{\alpha})$. Estimates were obtained by averaging the false or true positive rates over 500 independent simulation replicates.

parameters provides a stable FDR control for all considered scenarios. This suggests that the upper bound on k provided in assumption (7.23) could be relaxed when working with this heuristic sequence. In practical applications, SLOPE with the heuristic sequence (7.24) has been shown to control FDR when the design matrix contains genotypes of independent or strongly correlated SNPs and the number k of nonzero elements in the sequence β is sufficiently small (Bogdan et al., 2015, 2013).

7.4.3 Extensions of SLOPE

Clustered regressors: In Brzyski et al. (2017), SLOPE was combined with an algorithm for clustering of correlated SNPs to control FDR for spatially correlated GWAS data. After a preliminary selection of representatives of groups of correlated genetic markers, SLOPE is used to identify significant representatives. It is proven that the suggested method of identifying representatives does not impair FDR control. The method has been implemented in the publicly available package *geneSLOPE*, and the simulations presented in Brzyski et al. (2017) show good properties of *geneSLOPE* in terms of FDR control and power to identify relevant genes.

It is possible to go one step further and use SLOPE to select groups of predictors (Brzyski et al., 2018). Let $I = \{I_1, \ldots, I_m\}$ be a partition of the set $\{1, \ldots, p\}$ and rewrite the multiple regression model (7.1) in the form

$$y = \sum_{i=1}^{m} X_{I_i} \beta_{I_i} + \epsilon, \tag{7.25}$$

where X_{I_i} is a sub-matrix of X consisting of columns with indices from the set I_i, and β_{I_i} consists of elements of β with indices from I_i. Then the norm $\|X_{I_i}\beta_{I_i}\|_2$ will serve as a measure for the influence of the i-th group on the response variable. We will say that the i-th group has an impact on Y if and only if $\|X_{I_i}\beta_{I_i}\|_2 > 0$. Thus, the task of identifying significant groups of variables boils down to identifying the support of the vector

$$[[\beta]]_I := \left(\|X_{I_1}\beta_{I_1}\|_2, \dots, \|X_{I_m}\beta_{I_m}\|_2\right)^{\mathsf{T}}.$$

For any non-negative and non-increasing sequence $\lambda_1, \dots, \lambda_m$ and any positive sequence of weights $w_1, \dots, w_m$ the group SLOPE (gSLOPE) estimator is defined as

$$\beta^{\text{gS}} := argmin_b \left\{\frac{1}{2}\|y - Xb\|_2^2 + \sigma J_\lambda\left(W[[b]]_I\right)\right\}, \tag{7.26}$$

where W is a diagonal matrix with elements $W_{i,i} := w_i$.

FDR control can then be obtained by selecting a sequence λ compatible with chi-square distribution quantiles (Brzyski et al., 2018). A conservative selection of this sequence allows FDR control when the columns of the design matrix belonging to different groups are mutually orthogonal. In that case, asymptotic optimality of estimation of $[[\beta]]_I$ has been proved. In addition, a heuristic adaptation of the sequence λ has been proposed which allows for FDR control when variables in different groups are mutually independent.

gSLOPE has been successfully applied to the problem of gene localization, where groups consist of two variables corresponding to the additive effect and the dominance effect of a given gene. The proposed method works particularly well for identifying so-called rare recessive variants, where the dominance effect is of particular importance.

Outlier detection: One approach to outlier detection and robust estimation of regression coefficients is the mean-shift model (Candès and Randall, 2008; Gannaz, 2006; McCann and Welsch, 2007; She and Owen, 2011):

$$Y = X\beta + I\mu + \varepsilon, \tag{7.27}$$

where I is the $n \times n$ identity matrix, $\mu = (\mu_1, \dots, \mu_n) \in R^n$ and $\mu_i \neq 0$ means that observation i is an outlier. In Virouleau et al. (2017) an extension of SLOPE is used to estimate $\beta \in R^p$ and $\mu = (\mu_1, \dots, \mu_n)$ according to

$$(\hat{\beta}, \hat{\mu}) = argmin_{\beta \in R^p, \mu \in R^n}\left\{\|y - X\beta - \mu\|_2^2 + 2\rho_1 J_{\tilde{\lambda}}(\beta) + 2\rho_2 J_\lambda(\mu)\right\}, \tag{7.28}$$

where ρ_1 and ρ_2 are two positive constants. It is shown that under a proper adaptation of the Restricted Eigenvalue condition (Bickel et al., 2009; van de Geer, 2008; Ye and Zhang, 2010) SLOPE with a sequence of tuning parameters proportional to λ^{BH} or to the sequence with the elements $\lambda_i = \sigma\sqrt{\log\left(\frac{2n}{i}\right)}$ satisfies

$$\|\hat{\beta} - \beta^*\|_2^2 + \|\hat{\mu} - \mu^*\|_2^2 \quad \leq \quad C_1\left(k\log\left(\frac{p}{k}\right) + s\log\left(\frac{n}{s}\right)\right), \tag{7.29}$$

where s is the number of outliers. Thus, when $p > n$ and the number of outliers s is smaller than the number of nonzero coefficients of β, the version (7.28) of SLOPE for the mean-shift model allows to obtain a minimax rate $k\log\left(\frac{p}{k}\right)$ to estimate β. Moreover, it was shown that SLOPE with the sequence (7.22) asymptotically controls FDR with respect to outlier detection if

$$\frac{\left(s\log(n/s) \vee k\log(p/k)\right)^2}{n} \to 0.$$

Here FDR control can be obtained even when the columns of the design matrix X are strongly correlated. This is because that in the mean-shift model (7.27) the vector μ is accompanied by the identity matrix I.

7.5 Advanced Methods for Model Identification and Prediction

Convex optimization methods, like LASSO or SLOPE, are shrinking estimates of regression coefficients toward zero. Model size is reduced when the values of tuning parameters λ are getting large enough that coefficients are shrunk to zero. However, large values of λ result in shrinkage of all parameters and consequently in a large bias of the estimators of the most important regression coefficients. It follows that estimation and prediction properties can be rather poor as we have seen in the real data example.

Therefore, it is practically impossible to tune LASSO or SLOPE in such a way that one obtains both good prediction and selection properties at the same time. One practical solution suggested by Bogdan et al. (2015) and Brzyski et al. (2018) consists of applying a two-stage procedure: (a) Use LASSO or SLOPE to detect significant predictors; (b) apply standard least-squares methods for the selected predictors to estimate coefficients. This two-stage procedure allows to correct for bias of LASSO and SLOPE estimates under a variety of scenarios, but it still does not solve the problem of deterioration of the model selection properties of these methods, which occurs when the number of nonzero coefficients in the true vector of regression coefficients is sufficiently large. A brief look at the term v_j from equation (7.21) is enough to see that this additional noise will typically increase with λ, the correlation between columns of the design matrix X and the sparsity of the true vector of regression coefficients. When the noise variable v_j becomes too large then LASSO or SLOPE are not capable of recovering the true order of the magnitude of regression coefficients and will therefore not classify true and false discoveries correctly. This, in turn, results in a deterioration of the model selection properties.

Two different solutions have been developed to solve this problems for LASSO. In the adaptive or reweighted LASSO (Candès et al., 2008; Zou, 2006), the values of the tuning parameters become different for different variables and depend on some initial estimators of the vector of regression coefficients. Large expected values of regression coefficients are assigned smaller weights, which results in debiasing the large regression coefficients and decreasing the value of the noise variable v_j. This allows the adaptive LASSO to recover the true model for a much wider range of realistic scenarios than the regular LASSO. The second solution relies on applying LASSO with a relatively small value of the tuning parameter λ, such that the ordering of false and true discoveries is optimal. False discoveries are then eliminated by using an appropriate threshold or some model selection criterion. Specifically, in Pokarowski and Mielniczuk (2015) the consistency of such a two-step procedure with thresholding based on the Generalized Information Criterion (GIC, Konishi and Kitagawa (1996)) is proved and very good properties of the Extended BIC or Modified BIC are shown via simulations. We have also seen in our real data example that this is a viable strategy. Good model selection properties of adaptive and thresholded LASSO are reported in Tardivel and Bogdan (2018) and Rejchel and Bogdan (2020). In case when one aims at FDR control then thresholding can also be performed by using the knockoff filter (Barber and Candès, 2015; Candès et al., 2018), which provably controls FDR.

The rest of this chapter is devoted to have a look at these advanced methods. First, we will briefly describe the Adaptive Bayesian Slope, which uses the Bayesian framework for the selection of weights in reweighted SLOPE. Then we will introduce the knockoff methodology. A final short simulation study will compare these different methods of convex optimization with respect to model identification and estimation properties.

7.5.1 Adaptive Bayesian SLOPE

To address the described problems with model identification and estimation properties of SLOPE, a new synergistic procedure called adaptive Bayesian SLOPE (ABSLOPE) was proposed (Jiang et al., 2019). ABSLOPE effectively combines SLOPE with the Spike-and-Slab LASSO method of Ročková and George (2018), which provides a Bayesian version for LASSO reweighting. ABSLOPE performs simultaneous variable selection and parameter estimation based on data which can contain missing values. As with the Spike-and-Slab LASSO, the regression coefficients are regarded as arising from a hierarchical model consisting of two groups: (a) the spike for the nonactive variables or negligibly small signals and (b) the slab for large signals. In contrast to the Spike-and-Slab LASSO, the "ABSLOPE" spike prior is fixed and relies on the sequence λ^{BH} of the SLOPE tuning parameters in order to control FDR. Other prior parameters like the signal sparsity or the "average" magnitude of the large signals are treated as latent variables and are iteratively updated in the spirit of a Stochastic Approximation EM algorithm (SAEM). The algorithm can handle missing data under the Missing at Random (MAR) assumption and estimates the variance of the error term. The prior is designed in such a way that the Maximization step of the algorithm is performed by invoking a reweighted SLOPE, with weights dependent on the current conditional probability that a given variable is a large signal. According to simulation results in Jiang et al. (2019), ABSLOPE allows to control FDR under a much wider range of scenarios than SLOPE. One also obtains good model identification and estimation properties at the same time.

7.5.2 Model Selection with Knockoffs

Barber and Candès (2015) and Candès et al. (2018) proposed the so-called, *knockoff* methodology to control the number of false discoveries. The method can be used with almost any measure of importance for the explanatory variables, like e.g. the marginal correlation with the response variable or the estimate of regression coefficients provided by any regularization method. The main idea is to create a matrix of *fake* explanatory variables in such a way that its correlation structure corresponds to the correlation structure of X. More specifically, in case of a random design matrix X with independent rows, swapping any set of columns X with the same columns in $\tilde{X}$ should not affect the distribution of the extended matrix and the fake variables should be conditionally independent of Y, given X. The knockoff matrix $\tilde{X}$ is attached to X and the method to evaluate the importance of explanatory variables is run on the extended design matrix.

Knockoffs are then used to define a measure of importance W_j, $j \in \{1, \ldots, p\}$ in such a way that the signs of $\{W_j : \beta_j = 0\}$ are i.i.d. coin flips. Here, it is important to note that this property is usually not satisfied for statistics calculated based on the matrix X only, since in this case the sign of β_j will depend on the correlations between X_j and true predictors. To construct our importance measure, suppose that a statistic

$$T = (U, \widetilde{U}) = (U_1, ..., U_p, \widetilde{U}_1, ..., \widetilde{U}_p) \tag{7.30}$$

is computed from $(Y, X, \tilde{X})$, where T has the natural property that swapping the j and $j + p$ columns in $\mathbb{X}$ results in swapping the corresponding components of T. Suppose that the feature importance statistics are formed as

$$W_j = f(T_j, T_{j+p}), \quad j = 1, ..., p \tag{7.31}$$

where f is an anti-symmetric function; for example, we can take $W_j = T_j - T_{j+p}$. Then it is easy to see that the statistic $W = (W_1, ..., W_p)$ has a *flip-sign* property, namely, swapping the j and $j + p$ columns in $\mathbb{X} = (X, \tilde{X})$ has the effect of changing the sign of W_j. Then, according to the results of Barber and Candès (2015) and Candès et al. (2018) the knockoff filter defined as

$$\widehat{\mathcal{S}} = \{j : W_j \geq \hat{t}\}, \tag{7.32}$$

where

$$\hat{t} = \min\left\{t > 0 : \frac{1 + \#\{j : W_j \leq -t\}}{\#\{j : W_j \geq t\}} \leq q\right\}, \tag{7.33}$$

controls FDR at the level q.

Knockoff thresholding allows for FDR control with SLOPE or LASSO for any choice of tuning parameters. Here the tuning parameter should be selected to provide a proper ranking of explanatory variables rather than to execute their selection.

7.5.3 Simulation Study

In this section, we provide results of a simulation study comparing different convex optimization methods for selection of important variables and estimation of parameters in the multiple regression model

$$Y = X\beta + \epsilon.$$

In all our simulations $n = p = 500$, ϵ is a random vector from the standard multivariate normal distribution $N(0, I)$ and the number k of nonzero regression coefficients in the vector β takes values from the set $k \in \{10, 20, 40, 60, 80, 100\}$. We consider weak signals with

$$\beta_1 = \cdots = \beta_k = 1.3\sqrt{2\log p} \tag{7.34}$$

and strong signals with

$$\beta_1 = \cdots = \beta_k = 2\sqrt{2\log p}. \tag{7.35}$$

In all our simulations, the rows of the design matrix are independent random vectors from the multivariate normal distribution $N\left(0, \frac{1}{n}\Sigma\right)$. We consider two scenarios, one with *independent* regressors, where the correlation matrix $\Sigma = I$, the other one with *correlated* regressors, where Σ is the compound symmetry matrix with $\Sigma_{i,j} = 0.5$ for $i \neq j$.

We compare five different estimation and model selection methods:

- **mBIC2**: mBIC2 with an advanced stepwise search procedure implemented in the *bigstep* package,
- **SLOPE**: SLOPE with the vector of tuning parameters (7.24) with $q = 0.2$ for the *independent* scenario and the regular λ^{BH} sequence (7.17) with $c = 1$ and $q = 0.2$ for the *correlated* scenario,
- **SLOBE**: a simplified version of ABSLOPE described in Section 3.4 of Jiang et al. (2019) with $q = 0.1$,
- **Lcv**: LASSO with λ selected by 10-fold cross-validation aimed at minimizing the prediction error,
- **knLcv**: model free knockoffs (Candès et al., 2018) based on estimates of LASSO applied to the augmented design matrix $[X, \tilde{X}]$ and with the tuning parameter λ selected by cross-validation. For the independent case the iid rows of the knockoff matrix are generated from $N(0, I_{p\times p})$ distribution. For the correlated case, the knockoff matrix is generated using the equicorrelated construction from Section 3.4.2 of Candès et al. (2018) with the parameter s equal to the minimal eigenvalue of Σ. According to our simulations, this choice of s allows to achieve a higher power in the considered example than the choice suggested in Candès et al. (2018).

The performance of the different methods to identify models correctly is assessed like in the previous simulation study by the FDR and Power. These are estimated based on 200 simulation runs. In addition, we consider here the relative estimation and prediction errors, defined as

$$MSE = \frac{MSE(\hat{\beta})}{||\beta||^2}$$

and

$$MSP = \frac{MSE(\hat{\mu})}{||X\beta||^2}.$$

Results:

As illustrated in Figures 7.10 and 7.11 mBIC2 controls FDR below 0.1 for the independent regressors. It outperforms all other procedures when the regressors are independent and the signal is weak and sparse. However, it suffers from a loss of power and diminished estimation and prediction properties when the number of causal regressors increases. In the case of the correlated design FDR of mBIC2 increases, particularly when the signals are weak. Here, in most cases, mBIC2 has a slightly larger FDR, smaller power and worse estimation and prediction properties than SLOBE. We believe that this is a drawback of the search procedure, which has problems with identifying the optimal model when predictors are strongly correlated.

SLOPE based on the heuristic sequence of tuning parameters (7.24) keeps FDR close to the nominal level of $q = 0.2$ when the regressors are independent. It also has a pretty good power but suffers from relatively large estimation and prediction errors. This is due to excessive shrinkage by a large λ sequence needed for FDR control. In the case of correlated regressors, SLOPE does no longer control FDR. It performs similarly to cross-validated LASSO but has a larger FDR and worse estimation and prediction properties. This is due to the specific sequence of tuning parameters, which is not ideal for the purpose of prediction. Interestingly, in the scenario with correlated predictors and weak signals, both cross-validated LASSO and SLOPE have better estimation and prediction properties than the methods which control FDR.

SLOBE, the adaptive version of SLOPE, exhibits very good properties under all considered scenarios. It nicely keeps FDR close to the nominal level 0.1, even when predictors are correlated. When the signals are strong, SLOBE has superior predictive and estimation properties. For weak signals, SLOBE outperforms mBIC2 when the predictors are correlated and when they are independent and the number of causal markers increases. However, it needs to be mentioned that the comparison between mBIC2 and SLOBE may depend on the actual values of n and p. In the simulations reported in Frommlet et al. (2020) mBIC2 outperforms SLOBE when the predictors are independent and $n = 1,000$ and $p = 2,000$.

As expected, cross-validated LASSO has a very large FDR, which allows to obtain a high power and good estimation and prediction properties when signals are weak. When signals are strong, then cross-validated LASSO has worse estimation and prediction properties than SLOBE, which nicely controls FDR. Due to the results reported in Section 7.4.1 we expect that the cross-validated version of SLOPE will outperform LASSO with respect to prediction properties. This, however, requires the development of an efficient SLOPE algorithm, which is currently under development (see e.g. Bao et al. (2020); Larsson et al. (2020)).

As expected, knockoffs always control FDR. They provide large power and good estimation properties when regressors are independent and k is sufficiently large. In cases when k is very small or regressors are strongly correlated, knockoffs have less power than other model selection methods. We believe that the lack of power for small k is not necessarily an inherent property of the knockoff methodology and could be eliminated by applying some modifications. On the other hand, the lack of power for correlated regressors

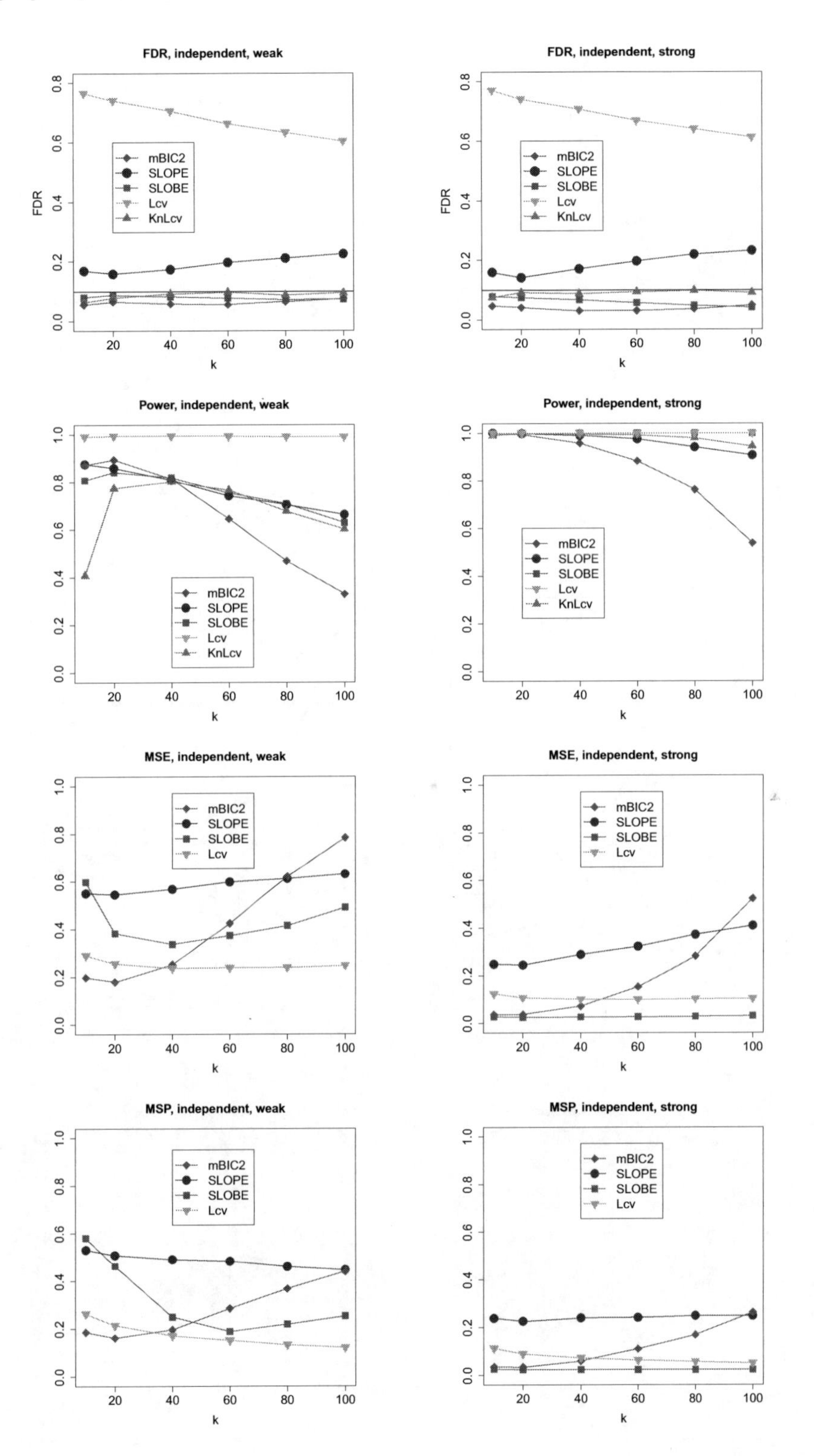

FIGURE 7.10
Results for *independent* regressors and weak (7.34) and strong (7.35) signals.

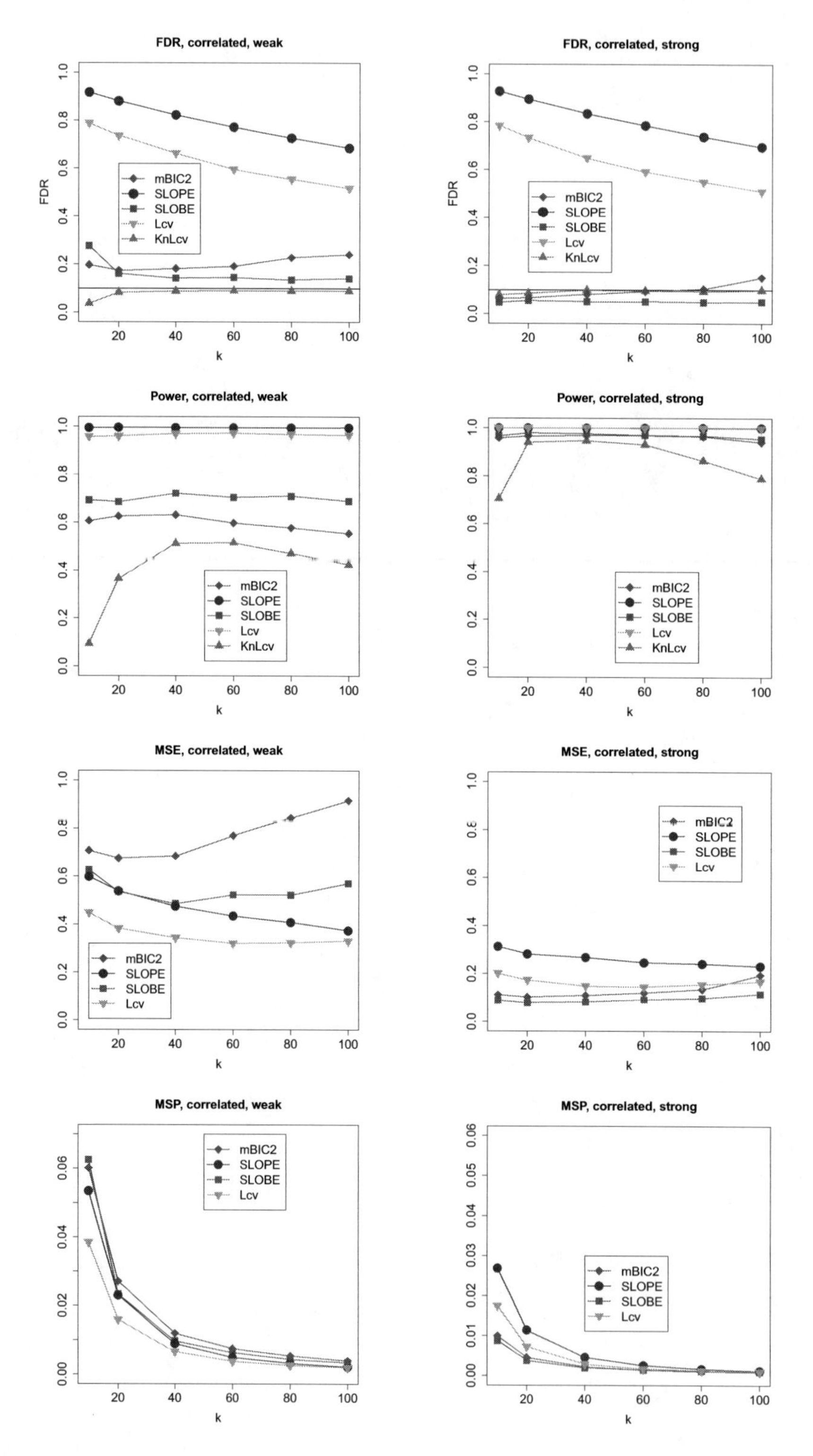

FIGURE 7.11
Results for *correlated* regressors and weak (7.34) and strong (7.35) signals.

seems to result from an increased variance of LASSO estimates after the design matrix is augmented with knockoff variables (i.e. p changes from 500 to 1,000). In the case of some specific designs, like equicorrelated predictors, this could be solved, for example, by using the counting knockoffs of Weinstein et al. (2017) (see also Weinstein et al. (2020)), which use less than p knockoff copies. In other cases, one could resort to the conditional randomization tests of Candès et al. (2018), which, however, are much more computationally intensive.

7.6 Real Data Examples

In this section, the different model selection methodologies are applied to analyze two real data sets, one with a continuous and one with a binary response. The *R* codes which illustrate how to use the different *R* packages to perform the analysis are available as Supplementary Material.

7.6.1 Continuous Response

The first data set is concerned with gene expression levels in lymphoblastoid cell lines of 210 unrelated HapMap individuals (The International HapMap Consortium, 2005) from four populations (60 Utah residents with ancestry from northern and western Europe, 45 Han Chinese in Beijing, 45 Japanese in Tokyo, 60 Yoruba in Ibadan, Nigeria) first presented by Stranger et al. (2007). This data set is available at *ftp://ftp.sanger.ac.uk/pub/genevar/* and was previously studied e.g. in Bradic et al. (2011), Fan et al. (2014) and Rejchel and Bogdan (2020). The goal is to identify genes by whose expression levels can be used to predict the expression level of the gene CCT8, which appears within the Down syndrome critical region on human chromosome 21. Such analyses are performed to identify genes that regulate the expression of CCT8.

The original data set contains expression levels measured for 47,293 probes. Following Wang et al. (2012) and Rejchel and Bogdan (2020) we preprocess the data by removing probes for which the maximum expression level among the 210 individuals is smaller than the 25-th percentile of all measured expression levels and for which the range of expression levels among the 210 individuals is smaller than 2. After this preprocessing, we are left with $p = 3220$ probes, which will be used to predict the expression level of CCT8.

We begin our analysis by identifying interesting explanatory variables using an advanced stepwise search strategy to minimize the mBIC2 criterion. Similarly, as in the simulation study, we first eliminate variables with a marginal p-value larger than 0.15, then perform a liberal forward selection with BIC followed by backward elimination using mBIC2, and finally a stepwise selection on the whole data set using mBIC2 as selection criterion. These computations are performed with the R package *bigstep* Szulc (2018).

This search strategy identified five important variables, which are listed in Table 7.3. Looking at the p-values when testing coefficients of the corresponding multiple regression model, one observes that all five variables are significant after Benjamini Hochberg correction at an FDR level $\alpha = 0.05$, but V1354 would not be significant after Bonferroni correction. The multiple R^2 of this model is equal to 0.581 and the value of the mBIC2 criterion equals 392.8249.

Table 7.3 illustrates that variables V1004 and V682 are strongly correlated with the expression of CCT8 and represent a large cluster of at least 215 probes with strongly correlated expression levels. Probe V1370 has just one strongly correlated probe, while the other two selected probes are not strongly correlated to any other probes in the data set.

TABLE 7.3
Properties of the Five Variables Selected with mBIC2 for the Sanger Data

Name	**p val**	**R(Y)**	**R(V1004)**	**R(V682)**	**R(V1370)**	**R(V1354)**	$\#(\lvert R \rvert > 0.65)$
V1004	6.84e-07	0.59					214
V682	2.52e-08	0.58	0.64				160
V1370	7.94e-09	0.51	0.35	0.28			1
V1354	5.02e-05	0.43	0.26	0.26	0.36		0
V206	8.79e-06	0.25	0.49	0.48	0.37	0.26	0

Probe V206 is interesting since it is not that strongly correlated with CCT8 and would not be selected based on the marginal p-value. However, it is relatively strongly correlated with V1004 and V682 and is significant in the multiple regression model built by mBIC2.

When applying SLOBE for the analysis of our data set, we obtain an empty model. This result is in agreement with the simulation results from Frommlet et al. (2020), which show that SLOBE might have rather low power when $p \gg n$. The reason for this is quite well understood and has to do with how regularization techniques like LASSO operate. Compared to the stepwise regression with mBIC2, which needs to estimate regression coefficients only in very small models, SLOPE, LASSO and SLOBE need to estimate all p coefficients. This leads to excessive variance when $p \gg n$, which results in problems with identifying the optimal model. Therefore, regularization techniques are usually applied only after the number of variables has been substantially reduced by some screening procedure, like, e.g., Sure Independence Screening (Fan and Lv, 2008). For our data set this technique was used previously by Wang et al. (2012) and Rejchel and Bogdan (2020), who preselected 300 predictors based on their marginal correlations with CCT8. In our analysis, we additionally include V206, which was selected by mBIC2 but has a relatively small marginal correlation with CCT8.

Applying the described *bigstep* mBIC2 procedure on the reduced data set yields nine selected variables, which include all five variables selected by mBIC2 on the full data set. This model is likely to be too large since now the penalty in mBIC2 is adjusted only to $p = 301$ and does not consider that the explanatory variables were preselected using marginal correlations with Y. After performing backward elimination with mBIC2 adjusted to the number of variables in the full data set $p = 3,220$ we obtain exactly the same model as the one selected by *bigstep* on the full data set.

To perform SLOBE on the reduced data set, we use the estimator provided by the cross-validated LASSO as starting point and identify seven interesting variables: V2524, V980, V1370, V3173, V315, V1354 and V206. Only three of these variables coincide with variables selected by mBIC2. Fitting a multiple regression model with these variables gives fairly large p-values for the variables V2524 ($p = 0.003071$) and V3173 ($p = 0.000155$). This is because of their strong correlation ($R = 0.74$). Backward elimination with mBIC2 removes V2524 and consequently the p-value for V3173 drops to 3.69e-14. This suggests that the large group of variables strongly correlated with V3173 contains some important predictors.

As shown in Table 7.4, all p-values in the resulting model with six variables are small enough to be rejected by the Benjamini-Hochberg procedure adjusted to the number of variables $p = 3,220$ in the original data set, whereas V206 and V315 would not be rejected by the Bonferroni procedure (V315 is pretty much at the decision boundary). The multiple R^2 for this model is equal to 0,6116 and the value of mBIC2 is equal to 392.0696, which is smaller than the value of mBIC2 for the model selected by the extended stepwise procedure from *bigstep*. Hence, this model would also be preferred according to mBIC2 but could not be identified with the initial search strategy.

Comparing the models from Tables 7.3 and 7.4, we observe that they have three variables in common: V206, V1354 and V1370. Probe V3173, which is the strongest predictor in the SLOBE model, replaced V1004 and V682 selected by the stepwise procedure. The two remaining variables in the SLOBE model V980 and V315 are also strongly correlated with V1004 (their marginal correlations with V1004 exceed 0.5) and still somewhat correlated with V682. Hence, the SLOBE model substituted two strongly correlated variables, which were also strongly correlated with V206 by three less correlated variables which are also slightly less correlated with V206 The six variables in the reduced SLOBE model have a maximal pairwise correlation that does not exceed 0.4, whereas in the model selected by the stepwise procedure three pairwise correlations exceed 0.45.

TABLE 7.4
Properties of the Six Variables Selected with SLOBE for the *Sanger* Data Set

Name	**p val**	**R(Y)**	**R(V980)**	**R(V1370)**	**R(3173)**	**R(V315)**	**R(V1354)**	$\#(\lvert R\rvert > 0.65)$
V980	6.80e-06	0.51						0
V1370	4.25e-09	0.51	0.34					1
V3173	3.69e-14	0.47	0.30	0.10				270
V315	1.56e-05	0.44	0.37	0.31	0.14			1
V1354	4.95e-07	0.43	0.19	0.36	0.10	0.19		0
V206	4.53e-05	0.25	0.34	0.37	0.35	0.30	0.26	0

TABLE 7.5
Correlation between mBIC2 and SLOBE Variables for the *Sanger* Data Set

Name	R(V980)	R(V1370)	R(3173)	R(V315)	R(V1354)	R(V206)
V1004	0.51	0.36	0.64	0.59	0.26	0.49
V682	0.43	0.28	0.69	0.31	0.26	0.48
V1370	0.34	1.00	0.10	0.31	0.36	0.37
V1354	0.19	0.36	0.10	0.19	1.00	0.26
V206	0.34	0.37	0.35	0.30	0.26	1.00

The above comparison illustrates the difficulties that emerge when one wants to identify the "best" model among a set of predictors which are strongly correlated. Both mBIC2 and SLOBE tend to select only small subsets of correlated groups and depending on the context it might or might not be desirable to have only a few representatives for clusters of correlated predictors. One solution to this problem is provided by SLOPE, which tends to select larger subsets of correlated variables.

To verify the performance of SLOPE on the reduced data set, we used SLOPE combined with cross-validation as implemented in the R package *SLOPE*. SLOPE identified 50 variables that contained all variables selected by mBIC2 and SLOBE apart from variable V1004, which was previously selected by the stepwise procedure although it had only weak marginal correlation with CCT8. SLOPE selected 21 variables from the cluster of correlated variables related to V1004, V682 and V3173, which reflects the importance of this cluster. Compared to the models selected by mBIC2 and SLOBE, cross-validated SLOPE included additional 23 variables whose marginal correlation with CCT8 varied between 0.57 and 0.43. Based on the results of simulations, we would expect that many of these additional findings are false positives. Cross-validated LASSO with the *glmnet* package identified 40 variables, including 18 variables which represent the large cluster of correlated probes. Furthermore, LASSO included 17 variables where marginal correlation with CCT8 was ranging between 0.43 and 0.57. Again, we would expect that most of these additional detections are false positives.

The scatter plot from Figure 7.12 compares regression coefficients estimated by SLOPE and LASSO. The eight variables which have coefficients with the largest absolute value coincide for both methods. These include the six variables V1370, V206, V1354, V3173, V980, and V315 which were previously identified by mBIC2 or SLOBE and two other variables which are strongly correlated with V3173 (R>0.71). The only large negative coefficient corresponds to V206, the probe which is only weakly correlated with CTT8 but which is an important regressor in all models built by the advanced methods used in this section. The scatter plot shows also that the shrinkage to zero is stronger for SLOPE coefficients than for LASSO coefficients. This is due to the inclusion of a larger number of correlated variables by SLOPE. We can also see that nonzero SLOPE estimates have a tendency to cluster around horizontal lines because SLOPE shrinks similar coefficients towards each other. The simulation results from Section 7.4.1 indicate that these shrinkage properties potentially improve the prediction properties of SLOPE, but this assertion remains to be verified based on more extensive comparisons on other real data sets.

7.6.2 Binary Response Variable

As an example of a data set with a binary response we consider the ARCENE data set, which is available at the UCI Machine Learning Repository Repository (2008). The data set was used for the NIPS 2003 challenge on feature selection (Guyon et al., 2006). The website of the workshop provides details about the challenge and further description of the data sets (Guyon, 2003).

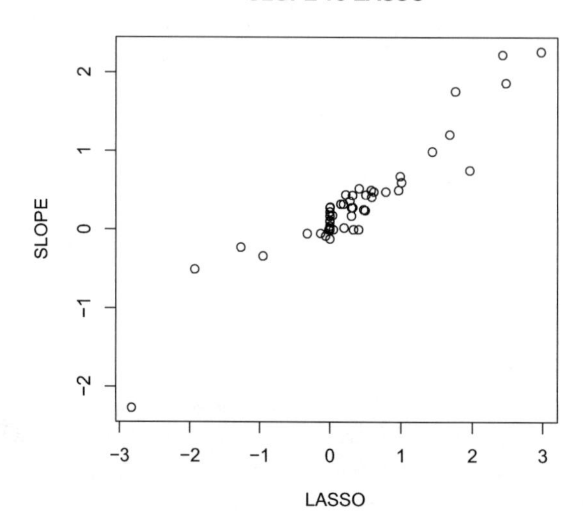

FIGURE 7.12
Slope estimates vs LASSO estimates for the Sanger data.

The ARCENE data set consists of mass-spectrometry data for cancer patients and for controls. For the NIPS 2003 challenge, there was a training set and a validation set, each with 44 cancer patients and 56 controls. In addition, there was also a test set of 310 cases and 390 controls. This is however not available at the UCI repository. There are 10,000 metric explicatory variables, most of them measuring the abundance of proteins of a given mass but some of them were created randomly. The task of the challenge was to predict cancer cases using the mass-spectrometry features. Here we want to focus on the question which features are related to cancer. In other words, we want to illustrate the behavior or our methods in terms of model identification and not so much in terms of prediction.

In a preprocessing step, 39 variables are removed which have only zero measurements and 139 variables are removed which yield quasi-complete separation. For the remaining 9,822 features, we perform logistic regression analysis considering the following methods for model selection.

- **Multiple testing**: Testing individual features with simple logistic regression and applying Bonferroni correction
- **L_0 penalties**: Model selection with the original BIC and with its modifications discussed in Section 7.1.1 using the R package *bigstep*
- **LASSO**: Considering the LASSO search path and models obtained with cross validation using the R package *glmnet*
- **SLOPE**: Considering the SLOPE search path and models obtained with cross validation using the R package *SLOPE*

After Bonferroni correction, 478 features are significant at a nominal significance level $\alpha = 0.05$. The analysis of the ARCENE data set is challenging due to the large amount of correlation between explicatory variables, particularly among those which are significantly associated with cancer status. Pairwise correlations among the 478 significant features are larger than 0.5 in 84.4%, and larger than 0.8 in 22% of all cases.

This has an interesting consequence when performing model selection with the four modifications of BIC we have introduced before. These L_0 penalties tend to select a small number of representatives of clusters of highly correlated features. For the sample size $n = 200$ mAIC is the most conservative choice and the *stepwise* function of bigstep selectes only two features. For the other three criteria (mAIC2, mBIC, mBIC2) stepwise selects three features. However, a plain *stepwise* search is not guaranteed to find the best model because there is the danger of getting stuck in local optima. For the criterion mBIC2, we performed some very simple strategies to escape the potential local minimum from the stepwise procedure. Adding two forward steps with the milder criterion BIC and then swiching back to stepwise with mBIC2 gave an improved model with five features. Two other strategies we tested did not further improve this model.

Table 7.6 provides some information about the five selected variables. Interestingly only the first and the third variable have very small marginal p-values, whereas particularly the second and the fifth feature are marginally not that strongly associated with cancer status. The strongest correlation is between the first and the third variable ($R = 0.49$). Particularly, the fifth variable is not correlated to any of the others. The first three variables have a large number of strongly correlated features. Specifically the first variable represents more than 200 other variables with an absolute correlation larger than 0.8. Only the fifth variable has no strong correlation with any other features. Knowing about the data generation for the challenge indicates that it might be a random variable which was added to the set of features. Among the 478 features which were significant after Bonferroni 95% have a correlation larger than 0.5 with at least one of the first three features. This illustrates how well the first three features selected by the stepwise procedure represent the marginally significant variables. The two features which were additionally selected with the multiple forward steps are not strongly correlated to the marginally significant features, having a maximum correlation of 0.31 and 0.13, respectively.

To present the results of LASSO, we initially focus on the beginning of the LASSO search path. The first variables to enter are V3365 and V5005, followed by V7748, followed by V5005 and V7748, followed by V9215 and V9585. In this initial phase no variables are removed along the search path. Here are the correlations between the 5 variables selected by mBIC2 and the first seven variables along the LASSO search path (in the order they have entered).

	V1936	V3365	V3629	V4973	V7748	V9215	V9585
V3365	−0.48	1.00	−0.43	0.52	0.50	−0.05	0.52
V729	−0.03	−0.29	−0.04	−0.15	−0.20	0.37	−0.15
V698	−0.32	0.49	−0.33	0.70	1.00	0.15	0.70
V6584	0.04	−0.19	−0.02	−0.09	−0.02	0.89	−0.09
V3161	0.08	−0.05	0.03	−0.08	−0.07	0.11	−0.08

What we see is quite typical for the LASSO search path. The first three variables to enter are strongly correlated with V3365, the next two and the last are strongly correlated with V698. LASSO selects more correlated variables than the L_0 penalties do. Due to shrinkage, each variable in the model tends to explain less variance than it would do in an ordinary regression model. This allows to include several strongly correlated predictors which typically stabilizes and enhances the prediction properties and to some extent prevents loss of important predictors due to their replacement by some strongly correlated regressors. On the other hand, there is the danger that false positives may be included in the model.

TABLE 7.6
Properties of the Five Variables Selected with mBIC2

Name	**p val**	**R(V3365)**	**R(V729)**	**R(V698)**	**R(V6584)**	$\#(\lvert R\rvert > 0.5)$	$\#(\lvert R\rvert > 0.8)$
V3365	3.54 e-08					1,389	207
V729	0.0043	−0.29				785	23
V698	6.12 e-08	0.49	−0.20			1,638	34
V6584	0.00085	−0.19	0.34	−0.01		160	9
V3161	0.0062	−0.05	0.11	−0.08	0.14	0	0

A rather tricky question is which model to choose along the LASSO search path. In terms of prediction, the most common answer is to make use of cross validation. For model identification however this approach has some serious drawbacks. First of all, the model one obtains depends on the random selection of subsamples for cross validation and the effect on the selected model can be dramatic. In our example, the model sizes obtained in different cross-validation runs vary between 40 and 80. In the R script, we look in more detail at a representative model of size 50. This model includes three of the five features found with mBIC2 (V3365, V729 and V3161) and there are two more features (V7748 and V9275), which are strongly correlated with the remaining two features from the mBIC2 model ($|R| > 0.95$). Furthermore, there are quite a number of additional features correlated with features from the mBIC2 models but also many which are uncorrelated. It is more than likely that this model includes many false positives. A viable strategy is to perform some model selection among the features obtained with LASSO cross-validation based on our modifications of BIC. Using mBIC2 this approach results in a model with the five features mentioned above. This model is almost identical to the previous mBIC2 model but has a slightly lower criterion.

Finally, we want to discuss the results obtained by SLOPE. The first observation is that the SLOPE search path has a very different initial behavior from the LASSO search path. The LASSO search path starts with very small models which are incrementally increased and only relatively rarely features from the search path are removed when decreasing the penalty. In contrast, Figure 7.13) illustrates that the SLOPE search path immediately starts with a rather large number of features which are then thinned out before the model size increases again. The large number of nonzero coefficients at the beginning of the SLOPE path results from clustering of similar regression coefficients. This behavior is more pronounced when there exist large gaps between consecutive elements of the SLOPE sequence. The SLOPE path is obtained by multiplying the basic SLOPE sequence (BH with $q = 0.2$) with a constant c. Large values of c lead to larger gaps between the elements of λ. Thus, at the

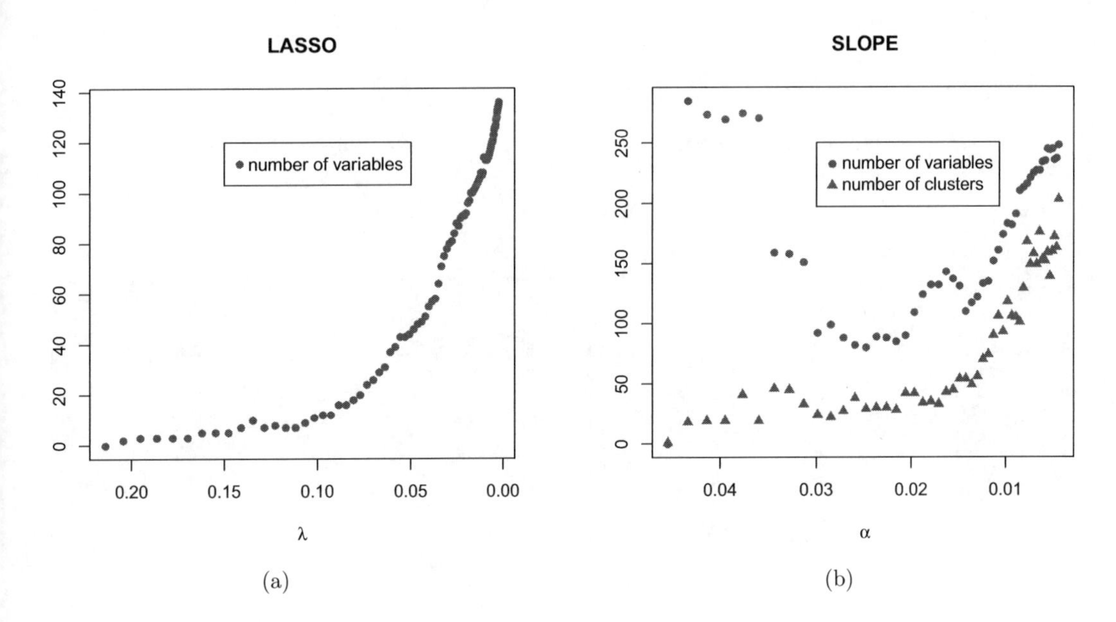

FIGURE 7.13
Number of variables along the LASSO path and numbers of variables and clusters along the SLOPE path. In case of LASSO clustering does not occur.

beginning of the path SLOPE has a tendency to replace single predictors with clusters of predictors. The corresponding regression coefficient all have almost the same (very small) value. Thus, the initial 285 features are in fact grouped in only 18 clusters with unique values of regression coefficients. The clusters tend to become smaller and their number tends to increase along the SLOPE path and at the end of the path, the number of clusters becomes similar to the number of selected variables.

The left panel of Figure 7.14 presents a scatter plot of the cross-validated LASSO estimates vs the SLOPE estimates, where we considered that model on the SLOPE path with the smallest number of nonzero regression coefficients. There is only a fairly small correlation between SLOPE and LASSO estimates which is due to the rather large variance of estimators for both methods when $p > n$. The right panel of Figure 7.14 provides the scatter plot for a reduced data set, where 300 variables with the smallest marginal p-value were preselected. These were augmented with three variables selected by mBIC2 although they had relatively large marginal p-values. Here the estimates of SLOPE were obtained using cross-validation as implemented in the *SLOPE* package. After reduction of the number of features, the correlation between SLOPE and LASSO estimates increases. The five most important predictors obtained with both methods coincide, among them the three variables selected by mBIC2 with large marginal p-values (V729, V6584 and V3161) and two other variables (V7734 and V2804) from the cluster strongly correlated with V3665. The resulting model including these five top variables selected by LASSO and SLOPE has only slightly larger residual deviance than the model identified by mBIC2. Comparing SLOPE to LASSO we note that SLOPE selects 40 variables from the cluster related to V3665 while LASSO includes only seven variables from this cluster. Like for the first data example with a continuous response, SLOPE estimates are again smaller and more shrinked toward each other than LASSO estimates. It remains to be validated if this goes along with an improvement of the prediction/classification accuracy.

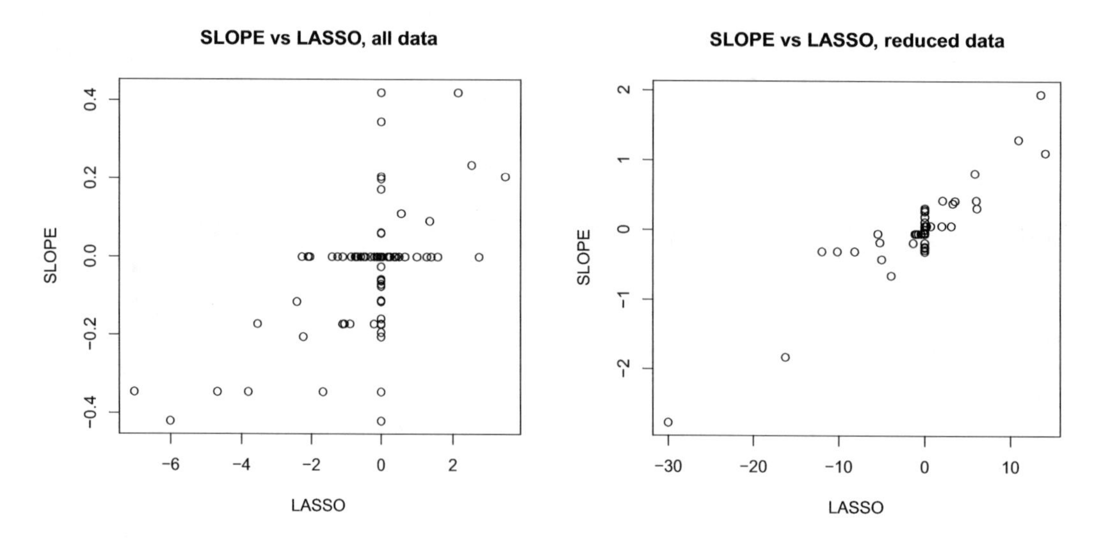

FIGURE 7.14
Slope estimates vs LASSO estimates for the Arcene data.

7.7 Summary

The theoretical results reported in this chapter as well as the results from simulations and real data analysis illustrate that there does not exist a uniformly optimal model selection method for high-dimensional regression problems. The choice of the methodology depends on the study purpose, on the actual values of n and p and on the amount of correlations between explanatory variables.

When n is larger or comparable to p, then the convex optimization methods like LASSO, SLOPE or their extensions like SLOBE perform very well. Here SLOBE or knockoffs based on cross-validated LASSO (Lcv) can be used for preventing false discoveries, with Lcv knockoffs providing exact FDR control but having lower power than SLOBE when signals are very sparse or when the explanatory variables are strongly correlated. Our simulations illustrate that SLOBE also keeps FDR close to the nominal level under a wide range of scenarios and yields very good estimation and prediction properties when the signal is strong. Cross-validated versions of LASSO and SLOPE include many false positives but yield very good prediction properties when the signal is weak. As shown in the real data analysis this goal is accomplished by selecting many representatives from a set of correlated predictors. Our simulation study from Section 7.4.1 suggests that the prediction properties of SLOPE are better than those of LASSO for a wide range of scenarios, particularly when there is a larger number of weak and correlated predictors. Here SLOPE has a tendency to include more of the correlated predictors than LASSO, which is related to its ability to cluster regression coefficients of variables which have a similar influence on the loss function (Kremer et al., 2019; Schneider and Tardivel, 2020). This property brings the potential for the identification of low-dimensional models, where some regression coefficients are equal to each other, which remains an interesting topic for further exploration. Concerning model selection properties of LASSO, it is important to be aware of the fact that selecting variables according to the order in which they appear on the LASSO path is definitely not an optimal solution (e.g., see Weinstein et al. (2020)). Concerning model selection properties it is much better to order variables according to the magnitudes of their estimates from the cross-validated LASSO and then threshold them using some version of the knockoff methodology or using some model selection criteria, like mBIC2.

The major drawback of the abovementioned regularization techniques is that for $p \gg n$ they lose their superior estimation and model selection properties because the respective estimators have too large variance due to a large number of estimated parameters. Model selection based on modifications of the Bayesian Information Criterion, like mBIC2, is based on the least-squares fitting of the compared sub-models and does not suffer much from increased p when the signal is sufficiently sparse to guarantee a low variance of least squares estimators. Our simulations reported here as well as in Dolejsi et al. (2014); Frommlet et al. (2012b, 2020); Szulc et al. (2017) show that selection based on mBIC2 allows to control FDR under a wide range of scenarios. Furthermore, already relatively simple extensions of the stepwise selection strategy implemented in *bigstep* (Szulc, 2018) can discover representatives of important clusters of variables even when $p \gg n$ and when predictors are correlated. Since mBIC2 has a tendency to include only very few correlated variables, the stepwise selection strategy has the potential to include predictors which have small marginal correlations with the response and which tend to be missed by simple screening strategies like Sure Independence Screening (SIS, Fan and Lv (2008)). This happened in both our real data examples, where the final optimal models selected by SLOBE or LASSO used variables identified by mBIC2 but missed by SIS.

The main limitation of performing model selection with mBIC2 for $p \gg n$ is that if p is getting too large then the optimization problem is getting extremely challenging and heuristic search methods based on stepwise selection strategies might no longer be efficient enough. However, there is currently a lot of interesting research going on to overcome these limitations (e.g. see Bertsimas et al. (2016)). Finally, our results illustrate that the analysis of high-dimensional data might always remain a complex task. The best results are often obtained by the combination of different methods. Ideally, such an analysis will lead to coherent and consistent descriptions of the underlying mechanisms but one has to be aware of the limitations of what one can expect from modeling of phenomena in a high dimensional setting.

7.8 Acknowledgments

We would like to thank the editor and the reviewers for suggestions that helped to improve the presentation of this chapter. We would also like to thank Dominik Nowakowski for performing simulations for Section 7.4.1 and to Wojciech Rejchel for suggesting and preprocessing the data set used in Section 7.6.2. M.Bogdan acknowledges the support of the Polish National Center of Science with grant Nr 2016/23/B/ST1/00454.

7.9 Glossary

FWER: Family-wise Error Rate

FDR: False Discovery Rate

GLM: Generalized Linear Models

MSE: Mean Squared Error

LASSO: Least Absolute Shrinkage and Selection Operator

SLOPE: Sorted L-One Penalized Estimation

Bibliography

Abramovich, F., Y. Benjamini, D. Donoho, and I. Johnstone (2006). Adapting to unknown sparsity by controlling the false discovery rate. *Ann. Statist. 34*(2), 584–653.

Abramovich, F. and V. Grinshtein (2017, 06). High-dimensional classification by sparse logistic regression. *IEEE Transactions on Information Theory PP*.

Akaike, H. (1974). A new look at the statistical model identification. *IEEE Transactions on Automatic Control 19*(6), 716–723.

Bao, R., B. Gu, and H. Huang (2020, July). Fast OSCAR and OWL regression via safe screening rules. In *Proceedings of the 37th International Conference on Machine Learning*, Volume 119, pp. 11. PMLR, Vienna, Austria.

Barber, R. F. and E. J. Candès (2015). Controlling the false discovery rate via knockoffs. *The Annals of Statistics 43*(5), 2055–2085.

Bayati, M. and A. Montanari (2012). The Lasso risk for Gaussian matrices. *IEEE Trans. Inform. Theory 58*(4), 1997–2017.

Bellec, P., G. Lecué, and A. Tsybakov (2018). Slope meets lasso: Improved oracle bounds and optimality. *Annals of Statistics 46*(6B), 3603–3642.

Bertsimas, D., A. King, and R. Mazumder (2016). Best subset selection via a modern optimization lens. *Ann. Stat. 2*(44), 813–852.

Bickel, P., Y. Ritov, and A. B. Tsybakov (2009). Simultaneous analysis of Lasso and Dantzig selector. *Ann. Statist. 37*, 1705–1732.

Birge, L. and P. Massart (2001). Gaussian model selection. *Journal of the European Mathematical Society 3*(3), 208–268.

Bogdan, M., A. Chakrabarti, F. Frommlet, and J. Ghosh (2011). Asymptotic Bayes optimality under sparsity of some multiple testing procedures. *Annals of Statistics 39*, 1551–1579.

Bogdan, M., J. Ghosh, and R. Doerge (2004). Modifying the schwarz bayesian information criterion to locate multiple interacting quantitative trait loci. *Genetics 167*, 989–999.

Bogdan, M., E. van den Berg, C. Sabatti, W. Su, and E. J. Candès (2015). Slope – adaptive variable selection via convex optimization. *Annals of Applied Statistics 9*(3), 1103–1140.

Bogdan, M., E. van den Berg, W. Su, and E. J. Candès (2013). Statistical estimation and testing via the ordered ℓ_1 norm. *arXiv:1310.1969*.

Bondell, H. and B. Reich (2008, March). Simultaneous regression shrinkage, variable selection, and supervised clustering of predictors with OSCAR. *Biometrics 64*(1), 115–123.

Bradic, J., J. Fan, and W. Wang (2011). Penalized composite quasilikelihood for ultrahigh-dimensional variable selection. *J.R. Statist. Soc. Ser. B 73*, 325–349.

Broman, K. and T. Speed (2002). A model selection approach for the identification of quantitative trait loci in experimental crosses. *J. Royal Stat. Soc.: Series B (Statist. Meth.) 64*(4), 641–656.

Brzyski, D., A. Gossmann, W. Su, and M. Bogdan (2018). Group slope - adaptive selection of groups of predictors. *Journal of the American Statistical Association*. DOI: 10.1080/01621459.2017.1411269.

Brzyski, D., C. Peterson, P. Sobczyk, E. Candès, M. Bogdan, and C. Sabatti (2017). Controlling the rate of gwas false discoveries. *Genetics 205*, 61–75.

Burman, P. (1989). A comparative study of ordinary cross-validation, v-fold cross-validation and the repeated learning-testing methods. *Biometrika 76*(3), 503–514.

Burnham, K. and D. Anderson (2002). *Model Selection and Multimodel Inference*. Second Edition, Springer, New York, NY.

Candès, E., Y. Fan, J. L., and L. J. (2018). Panning for gold: model-free knockoffs for high-dimensional controlled variable selection. *Journal of the Royal Statistical Society 80*(3), 551–577.

Candès, E. and P. A. Randall (2008). Highly robust error correction byconvex programming. *IEEE Transactions on Information Theory 54*(7), 2829–2840.

Candès, E. J., M. B. Wakin, and S. P. Boyd (2008). Enhancing sparsity by reweighted l1 minimization. *Journal of Fourier analysis and applications 14*(5-6), 877–905.

Chen, J. and Z. Chen (2008). Extended Bayesian Information criteria for model selection with large model spaces. *Biometrika 95*(3), 759–771.

Dolejsi, E., B. Bodenstorfer, and F. Frommlet (2014). Analyzing genome-wide association studies with an fdr controlling modification of the bayesian information criterion. *PLoS ONE 7*(9), e103322. DOI: 10.1371/journal.pone.0103322.

Fan, J., Y. Fan, and E. Barut (2014). Adaptive robust variable selection. *Ann. Statist. 42*, 324–351.

Fan, J. and J. Lv (2008). Sure independence screening for ultrahighdimensional feature space. *J. R. Statist. Soc. B 70*, 849–911.

Foster, D. and E. George (1994). The risk inflation criterion for multiple regression. *Ann. Stat. 22*(4), 1947–1975.

Frommlet, F. (2012). Modifications of bic for data mining under sparsity. In *Operations Research Proceedings 2011*, pp. 243–248.

Frommlet, F. and M. Bogdan (2013). Some optimality properties of FDR controlling rules under sparsity. *Electronic Journal of Statistics 7*, 1328–1368.

Frommlet, F., M. Bogdan, and D. Ramsey (2016). *Phenotypes and Genotypes: Search for Influential Genes.* Springer Series in Computational Biology, Springer - Verlag, London.

Frommlet, F., A. Chakrabarti, M. Murawska, and M. Bogdan (2011). Asymptotic bayes optimality under sparsity for generally distributed effect sizes under the alternative. *Technical Report.* arXiv:1005.4753.

Frommlet, F., I. Ljubic, H. Arnardottir, and M. Bogdan (2012a). Qtl mapping using a memetic algorithm with modifications of bic as fitness function. *Statistical Applications in Genetics and Molecular Biology 11*(4), Art.2.

Frommlet, F. and G. Nuel (2016). An adaptive ridge procedure for l0 regularization. *PLoS ONE 11*(2), e0148620. DOI: 10.1371/journal.pone.0148620.

Frommlet, F., F. Ruhaltinger, P. Twaróg, and M. Bogdan (2012b). Modified versions of bayesian information criterion for genome-wide association studies. *Computational Statistics and Data Analysis 56*(5), 1038–1051.

Frommlet, F., P. Szulc, F. König, and M. Bogdan (2020). Selecting predictive biomarkers from genomic data. Technical report. Technical Report in preparation, University of Wroclaw.

Fushiki, T. (2011). Estimation of prediction error by using k-fold cross-validation. *Statistics and Computing 21*, 137–146.

Gannaz, I. (2006). Robust estimation and wavelet thresholding in partial linear models. *Technical report, University Joseph Fourier, Grenoble, France.*

Guyon, I. (2003). Design of experiments for the nips 2003 variable selection benchmark. http://clopinet.com/isabelle/Projects/NIPS2003/Slides/NIPS2003-Datasets.pdf.

Guyon, I., S. Gunn, H. A.B., and D. G. (2006). Design and analysis of the nips2003 challenge. In I. Guyon, M. Nikravesh, S. Gunn, and L. A. Zadeh (Eds.), *Feature extraction: foundations and applications*, Volume 207 of *Studies in Fuzziness and Soft Computing.* Springer-Verlag, Berlin, Heidelberg.

Hofer, P., M. Hagmann, S. Brezina, E. Dolejsi, K. Mach, G. Leeb, A. Baierl, S. Buch, H. Sutterlüty-Fall, J. Karner-Hanusch, et al. (2017). Bayesian and frequentist analysis of an austrian genome-wide association study of colorectal cancer and advanced adenomas. *Oncotarget 8*(58), 98623.

Jiang, W., M. Bogdan, J. Josse, B. Miasojedow, V. Ročková, and T. Group (2019). Adaptive Bayesian SLOPE – High-dimensional Model Selection with Missing Values. arXiv:1909.06631. To appear in Journal of Computational and Graphical Statistics.

Konishi, S. and G. Kitagawa (1996). Generalised information criteria in model selection. *Biometrika 83*, 875–890.

Kos, M. (2019). Identification of Statistically Important Predictors in High-Dimensional Data. Theoretical Properties and Practical Applications. PhD thesis, Institute of Mathematics, University of Wroclaw, Poland.

Kos, M. and M. Bogdan (2020). On the asymptotic properties of SLOPE. *arXiv:1708.03950v1, Sankhya A, 82*(2), 499–532.

Kremer, P., D. Brzyski, M. Bogdan, and S. Paterlini (2019, June). Sparse index clones via the sorted L1-norm. To appear in Quantitative Finance.

Larsson, J., M. Bogdan, and J. Wallin (2020). The strong screening rule for SLOPE. Advances in Neural Information Processing Systems. NeurIPS 2020. *33*, 14592–14603

McCann, L. and R. E. Welsch (2007). Robust variable selection using least angle regression and elemental set sampling. *Computational Statistics and Data Analysis 52*(1), 249–257.

Neuvial, P. and E. Roquain (2012). On false discovery rate thresholding for classification under sparsity. *Annals of Statistics 40*, 2572–2600.

Pokarowski, P. and J. Mielniczuk (2015). Combined l_1 and greedy l_0 penalized least squares for linear model selection. *Journal of Machine Learning Research 16*, 961–992.

Rejchel, W. and M. Bogdan (2020). Rank-based Lasso–efficient methods for high-dimensional robust model selection. *Journal of Machine Learning Research, 21*(244), 1–47.

Renaux, C., L. Buzdugan, M. Kalisch, and P. Bühlmann (2020). Hierarchical inference for genome-wide association studies: a view on methodology with software. *Computational Statistics 35*(1), 1–40.

Repository, U. M. L. (2008). Arcene data set. https://archive.ics.uci.edu/ml/datasets/Arcene.

Ročková, V. and E. I. George (2018). The Spike-and-Slab LASSO. *Journal of the American Statistical Association 113*(521), 431–444.

Schneider, U. and P. Tardivel (2020, April). The geometry of uniqueness and model selection of penalized estimators including SLOPE, LASSO, and basis pursuit. *arXiv:2004.09106 [math, stat]*.

Schwarz, G. (1978). Estimating the dimension of a model. *The Annals of Statistics 6*(2), 461–464.

She, Y. and A. Owen (2011). Outlier detection using nonconvex penalized regression. *Journal of the American Statistical Association 106*(494), 626–639.

Stranger, B., S. Forrest, M. Dunning, C. Ingle, C. Beazley, N. Thorne, R. Redon, C. Bird, A. de Grassi, C. Lee, C. Tyler-Smith, N. Carter, S. Scherer, S. Tavar, P. Deloukas, M. Hurles, and E. Dermitzakis (2007). Relative impact of Nucleotide and Copy Number Variation on Gene Expression Phenotypes. *Science 315*, 848–853.

Su, W., M. Bogdan, and E. Candès (2017). False discoveries occur early on the lasso path. *Annals of Statistics 45*(5), 2133–2150.

Su, W. and E. Candés (2016, 06). Slope is adaptive to unknown sparsity and asymptotically minimax. *Ann. Statist. 44*(3), 1038–1068.

Szulc, P. (2018). R-package bigstep: Stepwise selection for large data sets. https://cran.r-project.org/web/packages/bigstep/index.html.

Szulc, P., M. Bogdan, F. Frommlet, and H. Tang (2017). Joint genotype- and ancestry-based genome-wide association studies in admixed populations. *Genetic Epidemiology 41*(6), 555–566.

Tardivel, P. and M. Bogdan (2018). On the sign recovery by LASSO, thresholded LASSO and thresholded Basis Pursuit Denoising. Technical report. arXiv:1812.0573.

The International HapMap Consortium (2005). A haplotype map of the human genome. *Nature 437*, 1299–1320.

Tibshirani, R. (1996). Regression shrinkage and selection via the lasso. *Journal of the Royal Statistical Society, Series B 58*, 267–288.

van de Geer, S. (2008). High-dimensional generalized linear models and the lasso. *Annals of Statistics 36*, 614–645.

van de Geer, S. and P. Bühlmann (2009). On the conditions used to prove oracle results for the lasso. *Electron. J. Statist. 3*, 1360–1392.

Virouleau, A., A. Guilloux, S. Gaiffas, and M. Bogdan (2017). High-dimensional robust regression and outliers detection with slope. arXiv:1712.02640.

Wainwright, M. J. (2009). Sharp thresholds for high-dimensional and noisy sparsity recovery using constrained quadratic programming (lasso). *IEEE transactions on information theory 55*(5), 2183–2202.

Wang, L., Y. Wu, and R. Li (2012). Quantile regression for analyzing heterogeneity in ultra-high dimension. *Journal of the American Statistical Association 107*, 214–222.

Weinstein, A., R. Barber, and E. J. Candès (2017). A power and prediction analysis for knockoffs with lasso statistics. arXiv:1712.06465.

Weinstein, A., W. J. Su, M. Bogdan, R. Barber, and E. J. Candès (2020). A power analysis for knockoffs with the lasso coefficient-difference statistic. arXiv:2007.15346.

Wu, Z. and H. H. Zhou (2013). Model selection and sharp asymptotic minimaxity. *Probab. Theory Relat. Fields 156*, 165–191.

Yanagihara, H., T. Tonda, and C. Matsumoto (2006). Bias correction of cross-validation criterion based on kullback-leibler information under a general condition. *J. Multivar. Anal. 97*, 1965–1975.

Ye, F. and C. H. Zhang (2010). Rate minimaxity of the lasso and Dantzig selector for the l_q loss in l_r balls. *J. Mach. Learn. Res. 11*, 3519–3540.

Żak-Szatkowska, M. and M. Bogdan (2011). Modified versions of bayesian information criterion for sparse generalized linear models. *Computational Statistics and Data Analysis 55*, 2908–2924.

Zeng, X. and M. Figueiredo (2014). Decreasing weighted sorted l_1 regularization. *IEEE Signal Processing Letters 21*(10), 1240–1244.

Zhao, P. and B. Yu (2006). On model selection consistency of lasso. *Journal of Machine Learning Research 7*, 2541–2563.

Zou, H. (2006). The adaptive lasso and its oracle properties. *Journal of the American statistical association 101*(476), 1418–1429.

8

Prevalence Estimation

Jonathan D. Rosenblatt

Ben-Gurion University of the Negev

CONTENTS

8.1 Summary

The proportion of non-null coordinates in a multivariate parameter is known as the *signal's prevalence*, denoted $1 - \pi_0$. Prevalence estimation is of direct interest for itself but also as a plugin parameter for adaptive multiple testing procedures and ranking algorithms. For more on the importance of π_0 as a plugin parameter, see Chapter 2 in this book. In this contribution, we review the vast literature on π_0 estimation.

We try to categorize many π_0 estimators along their design principles and statistical guarantees. Recurring design principles include:

DOI: 10.1201/9780429030888-9

1. Likelihood estimators in a parametric two-group mixture model.
2. Estimators in a semi-parametric two-group mixture model.
3. Estimators based on the number of rejections of some statistical test.
4. Interval estimators.

By the end of this text, it will be clear that these principles are not granular enough, so they are revisited in our discussion (8.11).

Estimators differ not only in their motivation but also in their statistical guarantees. These include:

1. Consistency, rates, efficiency, etc. for point estimators.
2. Coverage probabilities and tightness for interval estimators.
3. Multiple-testing guarantees such as FDR, FWER, etc. for plugin parameter in adaptive procedures.

These properties are studied in the literature in various problem setups: assuming a prior-distribution of effects or without it; finite sample versus asymptotically; under independence versus various types of dependence.

We will not recommend a one-size-fits-all estimator of π_0. The main contribution of this text is in its comprehensive bibliography, with pointers to the practitioner regarding the choice of an adequate estimator. A full detail of the results in the field would require a book-length treatment.

8.2 Motivation

When studying multiple measurements simultaneously, it is possible that they are sampled from a nonhomogeneous population. This is the case in genetics, where some genes are associated to a disease and some are not. This is also the case in neuroimaging, where function does not always follow anatomy, meaning that not all brains will show response at a given location. The idea of nonhomogeneous populations is not restricted to these modern applications. The toxicology literature deals with affected and nonaffected populations since the 1970s (e.g. Good, 1979). The epidemiology literature deals with healthy and sick populations (e.g. Rogan and Gladen, 1978). The astrophysics literature deals with pulsed signals in a series of photon arrival times (e.g. Swanepoel et al., 1996). In all the above applications, a quantity of interest is the *prevalence* of effects: what proportion of the population has a non-null effect?

Prevalence estimation is also of interest as an intermediary step when solving a multiple testing problem. Indeed, it is easier to discover genes that are associated to a disease if one knows how many to look for. Also, if we know that the proportion of nonassociated genes is Smaller than the permissible FDR level, a valid strategy is declaring all genes associated. This observation is the heart of *adaptive* multiple testing procedures where one tunes some testing algorithm using an estimate of the signal's prevalence. The seminal works of Hochberg and Benjamini (1990) and Benjamini and Hochberg (2000) are early examples of such algorithms, which have spurred great interest in prevalence estimation. The first is for adaptive FWER control, and the second is for adaptive FDR control.

Another problem that may benefit from knowledge of the effects' prevalence is the *ordering problem.* It turns out that it is easier to order effects when we know how many are non-null; at least when doing it in a Bayesian framework. Indeed, various ordering algorithms use the posterior probability of the null as the ordering score, itself requiring an estimate of the signal's prevalence (e.g. Lönnstedt and Speed, 2002; Thomas et al., 1985).

8.3 Problem Setup

Consider a parameter $\theta = \theta_1, \dots, \theta_p$. The mean vector of a multivariate distribution is the canonical example. One may now ask several questions on this unknown θ. Ordered along their (increasing) statistical difficulty, these questions include:

1. Is θ different than some θ_0? We call this *signal detection*, or a *global test*, also known as an *omnibus test*, and possibly other names.

2. How many entries in θ differ from θ_0? We call this *prevalence estimation*, or π_0-*estimation.* It is also known as *sparsity estimation*, *signal counting*, and possibly other names.

3. Which entries in θ differ from θ_0? We call this *multiple testing* or *selection*, but it may also be known as *signal identification*, and possibly other names.

4. What is the magnitude of θ in entries that differ from θ_0? We call this a *selective estimation*, but it is also known as *post-hoc estimation*, *circular-inference*, *double-dipping*, and possibly other names.

Definition 8.1 (π_0 for Simple Null Hypotheses) *Denote* $p_0 := p - \|\theta - \theta_0\|_0$, *the number of null coordinates. Then*

$$\pi_0 := p_0/p, \tag{8.1}$$

where $\|x\|_0$ *is the* l_0 *vector norm. The signal's prevalence is the proportion of coordinates that depart from their null values, and equals* $1 - \pi_0$.

Remark: In this text, we treat "prevalence estimation" and "π_0 estimation" as the same problem because one completely defines the other.

Without loss of generality, we will typically assume that $\theta_0 = 0$. By *signal detection* we mean that we test $\pi_0 = 1$ against $\pi_0 < 1$. By *prevalence estimation* we mean that we wish to estimate π_0.

In general, one may have access to n_j observations that carry information on each θ_j. We denote x_{ij} the i'th observation on θ_j. It is often assumed that all n_j are equal, leading to a rectangular $n \times p$ array of x_{ij}, denoted X. Alternatively, we may also assume that we only have access to variable-wise summaries. That is, each n_j measurements have been collapsed to $y_j = y_j(x_{1j}, \dots, x_{n_j j})$. In the case θ is a location parameter, we can denote $y_j := \theta_j + \varepsilon_j$.

One may also assume a distribution over effects, θ_j, or over summaries, y_j. A common approach is to assume a two-group mixture: a first component for the distribution of y_j's for which θ_j is null, and a second component for the distribution of y_j, for non-null θ_j's. Formally:

$$y_j \sim F = \pi_0 F_0 + (1 - \pi_0) F_1. \tag{8.2}$$

Assuming distributions F, F_0 and F_1 admit density functions, we can write

$$f(y) = \pi_0 f_0(y) + (1 - \pi_0) f_1(y). \tag{8.3}$$

Here are some common names for these various setups.

- **Multivariate Setup**: When a rectangular $n \times p$ array, X is available.
- **Multiple Testing Setup**: When only variable-wise summaries, y_j, are available on each θ_j. These may be Z-statistics, t-statistics, p-values, etc. For this reason, it is also known as the *massively-univariate* setup.
- **Normal Means Setup**: When y_j is Gaussian and θ_j its location parameter. I.e. $y_j := \theta_j + \varepsilon_j$, with $\varepsilon \sim \mathcal{N}(0, \sigma_j^2)$. Typically, σ_j are assumed equal for all j.
- **Two-Group Setup**: When adding a prior on the distribution of y_j, as in Equation 8.3. Also known as the *two-group prior*, the *empirical Bayes* setup, the *random-effect* model, or the *non-conditional* model.

We cover various estimators of π_0 from the literature in (approximately) chronological order. Discussion of statistical properties of the various algorithms if deferred to Section 8.10. The estimation algorithms we cover differ in their motivation and assumed inputs. We thus distinguish between:

1. Estimators in the parametric two-group mixture model.
2. Estimators in the semi-parametric two-group mixture model.
3. Interval estimators.
4. Estimators in the normal-means setup.
5. Estimators in the multivariate setup .

8.4 Parametric Two-Group Mixture

By *parametric* estimation of π_0 it is usually understood that one adopts the two-group mixture in Equation 8.3, with parametric assumptions on f_0 and f_1. Assumptions on f_0 are quite straightforward. When y_j's are p-values, then f_0 is typically assumed uniform over $[0, 1]$. When y_j's are in Z-scale, then f_0 is assumed $\mathcal{N}(0, 1)$. This practice implies a point-null and knowledge of the null distribution of the test statistic. Assuming a parametric form of f_1 is less evident. Choices are often a matter of computational convenience and statistical flexibility. With parametric f_1 and f_0, estimation will be typically performed through maximum likelihood.

When neuroimaging was still at its infancy, and genome studies were far from popular, Thomas et al. (1985) were concerned with large-scale inference in the search for new occupational carcinogens from large databases. In a contribution that is ahead of its time, Thomas et al. (1985) proposed an empirical Bayes approach to large-scale inference and estimate π_0 in Z-scale assuming f_0 and f_1 are Gaussian. In particular, they propose sorting candidate discoveries along the posterior probability of the null (see their Appendix 2); a quantity that years later will become known as the *local fdr*.

Roughly at the same time as Thomas et al. (1985), ray astronomers derived similar ideas. The arrival times of cosmic rays were modeled as a two-group mixture: periodic rays from pulsars (f_1) and aperiodic rays from background radiation (f_0). A parametric approach to this problem would assume that f_1 has a Cardioid or a von-Mises density (Loots, 1995, Sec.4.9). A pulsar's strength, $1-\pi_0$, would then be estimated using maximum likelihood. This approach can be found in De Jager et al. (1989, Sec.4) and references therein. Interestingly, many of the ideas that were developed by the statistical community for π_0 estimation, were developed by astronomers for ray astronomy. This link is explored further in our semi-parametric Section (8.5).

Returning to the motivating application of genetics, Pounds and Morris (2003) proposed estimating π_0 in p-value scale, where f_1 is a Beta distribution. Tamhane and Shi (2009) later found that this model is hard to estimate, and semi-parametric estimators are better— both for estimating π_0, and for plugging in adaptive procedures. Markitsis and Lai (2010) propose a variation of Pounds and Morris (2003), by censoring very small p-values.

Are Beta distributions justified by theory? This question was investigated by Liao et al. (2004), which derived f_1 assuming *proportional hazard* alternatives, also known as *Lehman alternatives*. Yu and Zelterman (2017) later derive the parametric form of f_1 assuming effects are random shifts of Chi-square variables.

Diaconis and Ylvisaker (1985) show that any distribution on $[0,1]$ may be approximated as a mixture of Betas. Parker and Rothenberg (1988), Allison et al. (2002) and Gadbury et al. (2004) apply this idea to multiple testing and model f_1 as a *mixture* of betas. This parametric approach can be viewed as a nonparametric approximation of f_1.

Operating in Z-scale Pan et al. (2002), Wu et al. (2006), McLachlan et al. (2006) and Lee and Bjørnstad (2013) proposed a finite Gaussian scale-location mixture. Formally: $f_1(y) := \sum_{h=1}^{q} \pi_h \phi_h(y)$, where $\phi_h(y)$ is the Gaussian PDF, with location μ_h and scale σ_h. The motivation of such a model is that the Gaussian scale-location class is flexible enough to approximate a large class of distributions. It can thus be considered as nonparametric.

Pawitan et al. (2005) propose estimating π_0 in t-scale. They define an effect to be the non-centrality of a t-statistic and approximate the effect distribution as a location mixture of t-distributions.

Estimation of π_0 may also be done in the original effects scale, instead of Z-scale, or p-values. This is typically, but not necessarily, encountered when viewing the two group-mixture as a sampling distribution, and not a subjective prior. Lee et al. (2000) propose a mixture of Gaussians to model observable gene expression levels. Newton et al. (2001) propose a mixture of Gamma. Kendziorski et al. (2003) propose both a mixture of Gamma and log-normals. Lee et al. (2002) and Newton et al. (2004) propose a three-component parametric mixture: a component for the null, for overexpression, and for under-expression. We note that this type of assumption on effects, invalidates many of the assumptions favored by theoreticians such as the monotonicity of f_0/f_1. We also note that Lee et al. (2002) is an early and excellent exposition of "Bayesian inference" versus "frequentist inference".

In Lönnstedt and Speed (2002), the authors propose a two-group mixture in effect-scale. Their mixture distribution is clearly a sampling distribution, which may deal with different effects for different genes. Their model allows to estimate π_0, for ordering genes. Lönnstedt and Speed (2002) also includes an interesting comparison of their mixture, with the semi-parametric two-group mixture of Efron et al. (2000).

8.5 Semi-Parametric Two-Group Mixture

By *semi-parametric* estimation of π_0, it is usually understood that one adopts the two-group mixture in Equation 8.3, and the unknowns are split into a *Euclidean part*, and a *functional part*. Typically f_1 is the functional part, π_0 is the Euclidean part, and f_0 fully known. Some authors, however, do not commit to knowledge of f_0. This setup is known as the *empirical null* (Section 8.9.2). The semi-parametric setup is attractive, since one rarely knows a-priori the distribution of effects, f_1.

Unlike the parametric models, in the semi-parametric models, maximum likelihood estimation is impossible. The authors thus seek to adopt some weaker-than-parametric assumptions on the functional part and to harness these assumptions to design and analyze their estimators. Assumptions may include symmetry, monotonicity, etc.

Benjamini and Hochberg (2000) introduce *adaptive* FDR control, and thus aim at estimating π_0 for the purpose of plugging in the BH procedure. Motivated by the graphical "algorithm" of Schweder and Spjøtvoll (1982), they compare various estimators of π_0 based on ordered p-values. Their preferred estimator is the *least slope line* (LSL). Denoting $S_j := (1 - p_{(j)})/(p + 1 - j)$, the LSL is defined as:

$$\hat{\pi}_0 := \min_j \left\{ \left(\frac{1}{S_j} + 1 \right) /p, 1 \right\}, \tag{8.4}$$

where $p_{(j)}$ is the j'th smallest p-value. Why this enigmatic construction? The rationale, appearing already in Hochberg and Benjamini (1990) or Rogan and Gladen (1978), goes as follows: Let $R(\lambda)$, be the number of rejections of an algorithm with variable-wise false positive rate λ. Rejections comprise of an unknown number of true positives, and roughly $p_0\lambda$ false positives. The number of true nulls, p_0 is roughly equal to the number of nonrejected, plus false positives: $p - R(\lambda) + p_0\lambda$. We may now equate the two equations for p_0 and solve for p_0 to get

$$p_0(\lambda) = \frac{p - R(\lambda)}{1 - \lambda}. \tag{8.5}$$

One may note that if $R(\lambda)$ is simply "reject p-values below λ", then dividing by p yields the so-called *Storey's estimator*, which we present later. To arrive from Equation 8.5 to 8.4 one allows λ to takes values in $\{p_{(1)}, \ldots, p_{(p)}\}$, add the conservativeness constant $+1$ in the numerator, and take a minimum for a data-driven choice of λ. Other equivalent ways to derive Equation 8.4 are discussed in Benjamini and Hochberg (2000, p.70).

A similar idea, up to adaptations in Equation 8.5, will reappear in Benjamini et al. (2006). In there, the rejection rule $R(\lambda)$ is replaced with the BH procedure. BH is thus applied twice: First is for estimating π_0, and second is for adaptive selection with FDR guarantees. For this reason, the whole procedure is termed "two-stage".

Most of the plugin procedures aim at FDR control. Knowledge of π_0 is useful, however, also for FWER control. This idea is explored in Finner and Gontscharuk (2009), which add some mild conservatism to Equation 8.5, that ensures FWER control when used in a plug-in Bonferroni correction.

History fans may enjoy discovering that the first to consider the estimator in Equation 8.5 are probably epidemiologists. Indeed, Equation 8.5 already appears in Rogan and Gladen (1978, p.73), for estimating an effect's prevalence based on the results of some screening test[1]. Even the bias, variance, and the Gaussian limit of this estimator, are already stated in Rogan and Gladen (1978). Equation 8.5 is thus an instance of the "Rogan-Gladen estimator", when assuming the sensitivity of the screening test is 1.

An estimator like Equation 8.5 was also well known in the ray-astrophysics community. Equation 8.5 is used in Swanepoel et al. (1996) as a baseline estimator, to be compared to their own maximal-spacing estimator. Swanepoel et al. (1996), based on Hester Loots

[1] I thank Dr. Carsten Allefeld for bringing the Rogan-Gladen estimator to my attention.

PhD thesis (Loots, 1995), is indeed an impressive contribution. The authors prove their own estimator's consistency, convergence rate, limiting Gaussianity, and show its superiority over other estimators in a massive simulation study. In particular, they show the superiority of the maximal-spacing estimator over various "indirect" estimators of π_0, which first attempt to estimate the density f. They also introduce ideas like bootstrapping for balancing the bias and variance; ideas that will reappear in Storey (2002), for instance. For an up-to-date introduction to the ray-astronomy mixture problem, see Schutte and Swanepoel (2016).

Another idea that appears in the astrophysics community alongside its appearance in the statistical community is the link between lowest density regions of f and the spacings of order statistics (Swanepoel, 1999). The intuition is that in low density regions, the spacings between order statistics will be the largest. This link appears in the statistical literature in many guises: some authors estimate π_0 by looking at the $\min \hat{f}(y)$; some authors look for bending points in the ECDF of p-values; some authors look at large spacings between ordered p-values. These ideas are roughly equivalent, a fact that is exploited in Swanepoel (1999).

Returning to the statistical literature, Mosig et al. (2001) propose an iterative algorithm based on the observation that large p-values are probably from the null. They iteratively remove large p-values until they find π_0 which best explains the number of observed p-values near 1. Nettleton and Hwang (2003) and Nettleton et al. (2006) later show the convergence of this algorithm, propose a non-iterative shortcut, and study statistical properties via simulation. Like Mosig et al. (2001), Scheid and Spang (2004) also remove p-values until they resemble a sample from the uniform distribution, with respect to a penalized Kolmogorov-Smirnov criterion.

Tusher et al. (2001) use a permutation scheme to discover significant genes. The signal's prevalence is determined using the proportion of scores in an interval, divided by its expectation under the null; this is essentially what Equation 8.5 does in p-value scale. Scheid and Spang (2003) use this same estimator when operating on Wilcoxon RankSum statistics. Being distribution-free, the expectation of the RankSum statistic under the null is known and does not require permutations.

Efron et al. (2001) is a seminal contribution to the topic, which introduced various ideas. It advocates the analysis in Z-scale as a principled scale. It also coins the term *Local fdr* for the posterior probability of the null.

Definition 8.2 (Local fdr)

$$fdr(y) := P(null|y) = \frac{\pi_0 f_0(y)}{f(y)}. \tag{8.6}$$

Efron et al. (2001) also point at the links between the *fdr* statistic, and FDR control. With respect to π_0 estimation, Efron et al. (2001), introduce a smoothing-spline Poisson-regression to estimate $fdr(y)$ from a histogram of Z-statistics and take $\hat{\pi}_0 := \max_y\{\hat{fdr}(y)\}$. The rationale being that if f_0 and f_1 disagree on their support, then $\max_y\{f_0(y)/f(y)\} = 1$ so that $\max_y\{fdr(y)\} = \pi_0$. In their Remark F, they also propose the estimator of π_0 in Equation 8.5, which will later be known as *Storey's estimator*, and is discussed in detail in our Section 8.5.2.

Broët et al. (2002) operate in the original data scale, and model gene expressions as an infinite Gaussian mixture. Dalmasso et al. (2004) operate in p-value scale. They propose a class of moment estimators shown to be conservative. Pounds and Cheng (2004) operate in p-value scale. Their method, known as *SPLOSH*, consists of LOESS smoothing of a histogram estimator of f. Given an estimate of f, π_0, is recovered using $\hat{\pi}_0 := \min_{p_i}\{\hat{f}(p_i)\}$.

Cruz-Medina and Hettmansperger (2004) operate in Z-scale and propose an estimator that builds on the assumed symmetry of the mixing components. Their estimator can be seen as a hybrid between a quantile matching estimator and a minimal distance histogram estimator.

Langaas et al. (2005) operate in p-value scale. They propose two estimators. The first, assuming F is increasing, amounts to a Grenander estimator: a piece-wise fixed, and monotonically decreasing function, that best fits the p-values' ECDF. Then $\hat{f}$ is recovered, and π_0 is estimated by evaluating $\hat{f}$ at the largest p-value. Langaas et al. (2005) show this to be equivalent to Equation 8.5, with a data-dependent λ that equals the largest p-value. A second estimator of π_0, based on the Grenander estimator of f, is $\hat{f}$ evaluated at the top of the largest level set of $\hat{f}$. A third estimator in Langaas et al. (2005) is an adaptation for convex f.

When viewing the two-group mixture as a sampling distribution, and not a subjective prior, one may analyze the convergence rates of estimators. Asymptotics will be with respect to p, not n (for asymptotics in n see, for instance, Chi et al. (2007)). Also, being semi-parametric, the usual $p^{-1/2}$ rates are not guaranteed, even for the Euclidean part.

When adopting the two group-mixture, and assuming nothing on f, identifiability issues arise. For instance, if $f_1 = f_0$, then π_0 may take any value in $[0, 1]$. The analysis of π_0 estimators in the two-group prior framework, thus requires some assumption on f. Bordes et al. (2006) show that identifiability may be resolved by assuming the symmetry of f_1 and exploit this symmetry for designing two consistent estimators of π_0. These estimators are shown to converge at a rate arbitrarily close to $p^{-1/4}$. Maiboroda and Sugakova (2010) propose a generalized estimating equations (GEE) estimator for the Euclidean part of this model and show its error to have a Gaussian limit at $p^{-1/2}$ rate. Bordes and Vandekerkhove (2010) complete these results by showing that a kernel-smoothed version of the minimum-contrast estimator in Bordes et al. (2006), converges to a Gaussian process limit (jointly for the Euclidean and functional parts).

Patra and Sen (2016) relax the symmetry assumption on f_1. They take a conservative approach to resolve identifiability, and define π_0 to be the largest of its permissible values. For estimation, they use the fact that F_1 is nondecreasing, and borrow ideas from isotonic regression to propose several estimators of F_1 and π_0. The first estimator of π_0 can be arbitrarily close to a parametric rate of $p^{-1/2}$, but depends on an oracle tuning parameter. A second estimator is a feasible heuristic with very favorable simulated properties. Both their estimators rely on the observation that when estimating F_1 by subtracting the known component, F_0, from the empirical $\hat{F}$, the returned function is not a proper CDF. They thus propose various ways of shape-restricting F_1 to be a valid CDF.

Cheng (2006) operate in p-value scale. Their estimator exploits the fact that bendpoint in the ECDF of F depends on the true π_0. They use this observation to select a data driven λ to be plugged in Equation 8.5. They further propose an adaptive procedure that uses their π_0 estimator and prove it to guarantee weak-FWER control under positive-orthant dependence.

Ruppert et al. (2007) operate in t-scale. They estimate f_1 with a B-spline. For computational efficiency, Ruppert et al. (2007) first bin the t-statistics and apply a penalized-least squares estimator to fit the spline to a histogram of t-statistics. Qu et al. (2012) do not bin the t-statistics. They generalize Ruppert et al. (2007) by allowing different degrees of freedom to each t-statistic and replace the least-squares estimation, with maximum likelihood. A similar idea is promoted by Ji and Wong (2005), which allow each Gaussian to be sampled with a different variance, and propose a shrinkage estimator for the denominator of the t-statistics.

Jin (2008) operates in Z-scale. He argues that a *purity condition*, required by various authors to ensure identifiability, is rarely met in practice. He thus designs two estimators of π_0, in the frequency domain. He proves various favorable properties of his estimator, in the normal-means setup, but also in the two-group mixture of Gaussians setup: His

Theorem 8 shows that if the effects, θ_j in the normal-means problem satisfy some regularity conditions, then the spectral estimator proposed converges to π_0 in the two-group mixture. A decade later, Chen (2019) generalized this approach in order to deal with discrete test statistics. In doing so, Chen (2019) provides a prevalence estimator for all two-group mixture distributions, not only Gaussian location mixtures, provided that their Lebesgue-Stieltjes integral equations may be solved.

Celisse and Robin (2010) propose a cross-validated histogram estimator in p-value scale. The estimator relies on the relative frequency of p-values in an interval assumed to contain only true nulls, which is selected using cross-validation. They prove its consistency when independently sampling from the two-group mixture. A version of the *purity condition* is assumed. They prove asymptotic FDR control when using their estimator in a plug-in BH procedure. Their simulation finds that their estimator is also valid under moderate dependence and provides FDR control when plugged in the BH procedure.

Tong et al. (2013) start with a simulation study, from which they conclude the quality of many estimators under dependence leaves room to desire (their Table 1). They thus propose various histogram estimators, which estimate π_0 based on the frequencies of p-values for various λ. The choice of λ is resolved, in the spirit of Storey et al. (2004), Nettleton et al. (2006), and others, by minimizing a resampled estimate of $MSE[\pi_0, \hat{\pi}_0(\lambda)]$. They perform a simulation study, using the empirical covariance matrix from gene expression data. Oyeniran and Chen (2016) also propose a histogram estimator in p-value scale. They bin p-values and then maximize the likelihood of their uniform-multinomial mixture.

Cheng et al. (2014) are concerned with the bias of $\hat{\pi}_0(\lambda)$ in our Equation 8.7 and propose a debiasing correction. Their correction depends on the tail distribution under the alternative, for which they use the noncentral t distribution. For robustness to dependence Cheng et al. (2014) adopt the "λ averaging" approach in Jiang and Doerge (2008), and average their estimator over a sequence of λ's. Hwang et al. (2014) propose an estimator that deals with dependence by combining step-up and step-down estimators. Nguyen and Matias (2014) is an extraordinary theoretical analysis of the two group model (8.2) in p-value scale. They show that π_0 may be estimated with a parametric rate, only if the support f_1 differs from the support of f_0, which justifies the need of various authors for the purity condition.

Qiao et al. (2017) propose two new estimators of π_0: a maximal-spacing estimator, and a moment-matching estimator. Shen et al. (2018) assure identifiability by assuming the coefficient of variation of f_1 is strictly increasing. They also design an majorization-minimization algorithm to estimate π_0 and f_1.

8.5.1 Fully Bayesian Estimators

By "fully Bayesian" we mean that a prior is set on the parameters of the two-group mixture. Contributions include Tang et al. (2007), which operate in p-value scale. They model f_1 as an infinite beta mixture. They propose a Dirichlet prior on mixing weights. Do et al. (2005) proposes a similar approach in Z-scale. Such priors allow using the posterior distribution of $\hat{\pi}_0$ for inference.

Ince et al. (2020) propose a (subjective) Bayesian prevalence estimator which puts a Beta prior on the rejection probability of some statistical test.

8.5.2 The Rogan-Gladen AKA Storey's Estimator

Version of the estimator known as *Storey's estimator* already appears in Rogan and Gladen (1978), Hochberg and Benjamini (1990), and others. It was a series of publications, starting with Storey (2002), that cemented its place in statistical literature, thus gaining its unofficial name. This estimator has motivated so much research, that we feel it deserves its own section.

The estimator is tailored to estimating π_0 in p-value scale. Its underlying rationale is the following: Because f_1 represents p-values from the alternative, it is stochastically smaller than f_0. The largest p-values should thus originate from f_0 and not f_1. Formally, given p p-values $p_1, \ldots, p_p$, and some $\lambda \in (0,1)$, then

$$\hat{\pi}_0(\lambda) := \frac{\#\{p_j > \lambda\}}{(1-\lambda)p}. \tag{8.7}$$

This estimator can also be viewed as a kernel estimator of f_1, evaluated at 1 (Neuvial, 2013). This class of estimators has attracted a lot of attention with the following major questions: How to choose λ? Are they consistent? Under which asymptotics? At what rate? Do they control the FDR in adaptive procedures?

Swanepoel et al. (1996), Storey (2002), and others, propose to choose λ so that it minimizes a Bootstrapped MSE of the estimated proportion of false discoveries. Storey and Tibshirani (2003) propose setting $\lambda = 1/2$, or spline smoothing of $\hat{\pi}(\lambda)$, and evaluating it at $\lambda = 1$. Motivated by possible dependence between p-values, Blanchard and Roquain (2009) propose setting $\lambda = \alpha$, where α is the desired FDR control in an adaptive procedure. Also motivated by dependence, Neumann et al. (2020) set $\lambda = 1/2$, but apply the estimator to Bootstrap samples, drawn under an independence assumption. They prove that under a wide range of dependence structures, this strategy allows consistent prevalence estimation.

Clearly, the number of observations in the estimator decreases as λ increases. Balancing between the bias in small λ, and the variance in large λ was studied quite thoroughly, starting with Swanepoel et al. (1996). Jiang and Doerge (2008), for instance, propose to compute $\hat{\pi}_0(\lambda)$, over various values of λ. Wang et al. (2010) compute $\hat{\pi}_0(\lambda)$ in ranges of λ. In which case Equation 8.7 should be written $\hat{\pi}_0([\lambda_1, \lambda_2])$, for $1 > \lambda_2 > \lambda_1 > 0$. Each such range provides an estimate of π_0, of which Wang et al. (2010) propose a way to aggregate.

The expectation of Equation 8.7 with respect to f equals:

$$\mathbb{E}[\hat{\pi}_0(\lambda)] = \pi_0 + (1-\pi_0)\frac{1-F_1(\lambda)}{1-\lambda}. \tag{8.8}$$

Genovese and Wasserman (2004) showed that $\hat{\pi}_0(\lambda)$ has a limiting Gaussian distribution about its expectation:

$$\sqrt{p}(\hat{\pi}_0 - \mathbb{E}[\hat{\pi}_0(\lambda)]) \rightsquigarrow \mathcal{N}\left(0, \frac{F(\lambda)(1-F(\lambda))}{(1-\lambda)^2}\right).$$

This estimator thus has a parametric $p^{-1/2}$ rate, but it is inconsistent. Indeed, consistency of this semi-parametric estimator occurs at slower than $p^{-1/2}$ rates (Neuvial, 2013).

If λ in Equation 8.7 is not fixed, but rather data-driven, then the expectation in Equation 8.8 is not correct. Black (2004), among others, showed that the bootstrapped estimator in Storey (2002) is actually downward biased, i.e., overestimates the signal's prevalence.

For identifiability to be satisfied, an infinitesimally small interval near 1 needs to contain only null p-values. This assumption appears in various references, starting with Swanepoel (1999), and was formalized by Genovese and Wasserman (2004), proving that $f_1(1) = 0$ is sufficient for identifiability. They called it the *purity condition*, also referred to as the *strong zero assumption* in Turnbull (2007). This purity condition is not satisfied by many popular models (Neuvial, 2010).

The intuition that the smallest number of null p-value occur near 1 motivated Storey and Tibshirani (2003) to design an estimator in which $\hat{\pi}_0(\lambda)$ is evaluated for various λ, and extrapolated to λ=1 using a cubic spline (their Remark B). Alas, the vanishing of λ, and the purity condition, are not enough to guarantee the consistency of $\hat{\pi}_0$ (Neuvial, 2013, Proposition 14).

8.6 Interval Estimators

The main application that has been driving research in prevalence estimation is genetics. In particular, how many genes are associated with a given phenotype? For this purpose, it is of great interest to provide not only point estimates of π_0 but also confidence bounds. Particularly statements of the type "π_0 is no larger than a", which means that a proportion of at least $1 - a$ genes are associated with the phenotype. The interval-estimation problem may also benefit the point-estimation problem: one may use the lower bound of a one-sided interval as a point estimate. In particular, the 0.5 one-sided interval guarantees a median unbiased point estimate.

Genovese and Wasserman (2004) construct a lower confidence bound on π_0 with:

$$\hat{\pi}_0 := \sup_{t \in (0,1)} \left\{ \frac{\hat{F}(t) - t - \varepsilon_{p,\alpha}}{1 - t} \right\}. \tag{8.9}$$

If $\varepsilon_{p,\alpha}$ is set to be $\sqrt{1/(2p)\log(2/\alpha)}$, then by the Dvoretzky-Kiefer-Wolfowitz (DKW) inequality, $\hat{\pi}_0$ in Equation 8.9 is an upper confidence bound for π_0 with $1 - \alpha$ coverage under independence of p-values. The idea in Equation 8.9 was generalized by Meinshausen and Bühlmann (2005) and Meinshausen and Rice (2006), who replace the DKW inequality, and construct $\varepsilon_{p,\alpha}$ by means of *bounding functions.*

Meinshausen and Bühlmann (2005) use a label permutation scheme to construct valid bounding functions under arbitrary dependence. For this purpose, Meinshausen and Bühlmann (2005) assume a nonstandard setup where a vector of summary measurements is available for each subpopulation. For instance, a set of gene-wise Z-statistics for healthy, and a set of Z-statistics for sick. This is not the usual multiple-testing setup discussed so far, but it is quite realistic and paves the way for new algorithms.

Meinshausen and Rice (2006) study a more standard setup, where a summary statistic is not available for each class so that label permutation is not possible. Using the idea of bounding functions, they propose a whole class of finite-sample confidence bounds for π_0; the DKW bounds of Genovese and Wasserman (2004) are recovered as a particular instance. In their remarkable analysis, Meinshausen and Rice (2006) answer various statistical matters regarding feasibility and optimality of π_0 estimators. In particular, they study the interplay between signal's strength and prevalence, for consistent estimation of π_0 (Meinshausen and Rice, 2006, Thm.3).

Partial conjunction tests may also be used to bound π_0. A partial-conjunction is a hypothesis that at least r of the p hypotheses are non-null. Heller et al. (2007) propose testing this hypothesis by aggregating the largest p-values. The rationale is that if the "no signal in the $p - r$ largest p-values" hypothesis is rejected, then clearly there is more than r variables carrying signal. Heller et al. (2007) propose estimating π_0 by testing a sequence of values of r. The error guarantees in Heller et al. (2007) are thus inherited from the multiple testing procedure in which they are embedded. This partial conjunction estimator was later generalized in Wang and Owen (2019).

Goeman and Solari (2011) operate in p-value scale and provide lower bounds for π_0, with finite sample guarantees. In their remarkable contribution, they provide an estimate of π_0 not only for the initial set of p variables, but rather, for any subset of these. It is actually a "meta-algorithm": any signal detection algorithm with a provable false-positive rate may be plugged in a *closed testing procedure*, to construct confidence bounds for π_0. This means that the estimator may also deal with any dependence: simply by plugging in a signal detector that is valid under dependence. Goeman et al. (2019), for instance, plug in the signal detector due to Simes, for this purpose. Because Simes's test is valid under a wide range of dependence, this means that the coverage probabilities of the lower bound of π_0 is valid under this dependence. Rosenblatt et al. (2018) have applied this estimator in the context of neuroimaging to estimate the prevalence of brain activation. Ebrahimpoor et al. (2020) used it to analyze feature sets in genomics.

8.7 The Normal Means Setup

In the normal-means setup, one assumes that the observables, y_j's, are generated through

$$y_j := \theta_j + \varepsilon_j, \tag{8.10}$$

where $(\theta_j)_{j=1}^p$ is an unknown $p-$vector of real parameters, and the sampling distribution of ε_j is centered Gaussian with fixed variance, σ^2. This variance is sometimes assumed known, and sometimes assumed unknown. Selection, i.e., the identification of the non-null θ_j's, is typically done by means of thresholding y_j's. The normal means setup is used to study the statistical performance of some estimator $\hat{\theta}$, compared to the true θ.

The normal means problem with l_0 loss, is a selection problem. Indeed, the loss $\|\hat{\theta} - \theta_0\|_0$ is a count of false positives and negatives. Given a selection algorithm, the link to prevalence estimation has been laid in Equation 8.5, and consists of discounting the number of selected by the false-positive rate. A second link to prevalence estimation is done by means of sequential testing. If one constructs a test for $\|\theta\|_0 \geq k$, then k can be increased until the test is no longer rejected. The statistical guarantees on $\hat{\pi}_0$ are inherited from the guarantees of the sequential test. This is the strategy employed by Carpentier and Verzelen (2019).

8.8 The Multivariate Setup

In the previous sections, the estimation of π_0 was implicitly decomposed into: (1) Compute variable-wise statistics, y_j. (2) Estimate π_0 as a univariate problem. If, however, a $n \times p$ data matrix X is available[2], then one may attempt estimating π_0 in a bona fide multivariate fashion. For instance, by maximizing a likelihood function. Much like multiple testing in the multivariate setup, discussed in Chapter 3 of this book, this approach has the promise to deal with dependencies, which cannot be estimated if x_{ij}s have been collapsed to $y_j(x_{1j}, \ldots, x_{nj})$.

When assuming a two-group prior, the likelihood will no longer enjoy the "cleanness" of Gaussian random-effects theory encountered in mixed models. This may explain why this multivariate likelihood approach is rarely encountered in the literature. Lai (2007), for instance, do not write the complete likelihood but do assume access to X. They design an algorithm for estimating π_0 by splitting the raw data, X, into two subsets. They then solve the moment equations of the p-values of the two-data sets simultaneously. When looking at

[2]It is not crucial that the data be rectangular. Allowing n_j to vary with j does not change much.

their estimator of π_0 it seems that they assume a within-variable mixture random effect, and not a between-variable mixture. Put differently, they assume subjects are nonhomogeneous, not (only) genes.

Lu and Perkins (2007) estimate π_0 based on raw observations (X), in p-value scale. They observe, like others, that for FDR control of a plug-in procedure, some conservative bias in $\hat{\pi}_0$ is required. They introduce conservatism, and robustness to correlation, by resampling the data, and returning an upper quantile of the resampling distribution of $\hat{\pi}_0$. Within each re-sample, they use $\hat{\pi}_0(\lambda)$ in Equation 8.7.

Friguet and Causeur (2011) assume a factor-analysis type covariance, derive the bias and variance of the estimator in Equation 8.7 as a function of the dependence structure, and use this relation to debias the prevalence estimator.

8.8.1 To Whiten or not to Whiten?

There is a gap between theory and application with respect to dependence. Most proofs of statistical guarantees of prevalence estimators assume independence between y_j's, or between columns of X. This is despite the fact that many applications, and certainly in genetics and neuroimaging, the sampling distribution does not admit an independence assumption.

In the multivariate setup, when one has access to p-variate samples x_{ij}, there is hope of estimating correlations and whitening the data. Say x_{ij} is Gaussian, and $\Sigma_{jj'} := Cov[x_{ij}, x_{ij'}]$ is estimable, one may try to deal with dependence by whitening. Denote A the Cholesky decomposition of the precision matrix, i.e. $AA' = \Sigma^{-1}$. Then $(Ax)_j$ is uncorrelated to $(Ax)_{j'}$. Having whitened the data, one may apply various estimation algorithms above. The downside of this whole process is that the transformation Ax will change the signal's prevalence. Formally, $\|\theta - \theta_0\|_0 \neq \|A\theta - A\theta_0\|_0$.

We understand that the definition of π_0 is not invariant to a change of basis, and we need to deal with dependence in the original coordinate system, or whiten the null and non-null coordinates separately. Put differently: unlike signal detection, for prevalence estimation we cannot whiten the data. This may justify the route taken by Friguet and Causeur (2011): instead of whitening the data, they compute the prevalence in the original basis, and then debias it.

A second way to go about is to use algorithms with valid statistical guarantees under dependence. Benjamini et al. (2006) and Goeman et al. (2019), for instance, use the number of hypotheses rejected to estimate π_0. Because their tests are valid under various sorts of dependence, then so are their guarantees on π_0. Another avenue for valid guarantees under dependence is discussed in Dickhaus et al. (2012) and references therein. It consists of estimating the "effective number of tests" and then using a plug-in procedure that is adapted to this number.

8.9 Some Related Matters

8.9.1 Compound Decision Theory

Compound Decision Theory (CDT), pioneered by Robbins (1951), deals with the analysis of multiple decisions, where the decision maker incurs an additive loss. This means that if considering some $\|\hat{\theta} - \theta\|_q$ loss, and $q = 0$ in particular, then CDT collapses to estimation and selection. FDR, on the other hand, is not additive, thus does not fall within CDT. For a comprehensive treatment of CDT, including implications to hypothesis testing, see Marits and Lwin (1989).

8.9.2 Empirical Nulls

Almost all authors mentioned previously, assume f_0 is fully known. In a large corpus of work, Efron advocates abandoning full specification of f_0, in favor of some milder, non-parametric assumptions, such as symmetry about 0. This is known as an *empirical null.*

> *Use of the theoretical null is mandatory in classic one-at-a-time testing, where theory provides the only information available for null behavior. However, things change in large-scale simultaneous testing situations: serious defects in the theoretical null may become obvious, while empirical Bayes methods can provide more realistic null distributions. (Efron, 2008)*

Hohmann and Holzmann (2013), for instance, allow f_0 to have an unknown location parameter. Lee et al. (2000), Xiang et al. (2014), Ma and Yao (2015), and Ma et al. (2011) define f_0 parametrically, up to some general unknown parameters. In Rosenblatt et al. (2014) the null distribution is assumed to be symmetric about 0, and approximated using a scale-mixture of centered Gaussians.

Cai et al. (2007) operate in Z-scale. They assume f_0 is Gaussian with unknown scale and location. They also assume f_1 is an infinite Gaussian scale-location mixture. They propose to estimate π_0 in the frequency domain. In Cai and Jin (2010), the same authors show that their estimator achieves minmax-optimal rates in a class of mixture densities.

Strimmer (2008) designs a procedure that does not a-priori assume the scale of the statistic. Put differently, he designs a single procedure that is valid in p-scale, Z-scale, or t-scale. Such a procedure naturally implies the null is theoretically unknown, thus empirical. His prevalence estimator, like Equation 8.7, is based on the proportion of test statistics in a region. Because Strimmer (2008) is mostly motivated by adaptive testing, his choice of λ is motivated by maximizing power (false non-discovery rate, to be precise).

8.9.3 Composite Nulls, Discrete Statistics, and Super-Uniform Nulls

The uniform distribution of p-values under the null will be invalidated if the test statistic has a discrete distribution or if the null is composite. Pounds and Cheng (2006) were among the first to focus on this setup, and their prevalence estimator is a simple average of transformed p-values. The transformation depends on the setup: one-sided and/or discrete statistics.

Chi (2010) deals with composite nulls in the two-group mixture setup. He assumes the null is a finite mixture of known components. In this setup, a p-value is taken to be the most conservative between the various mixing components of the null. Given these p-values, prevalence estimation, and plugin FDR control follows as usual.

Another way to deal with discrete test statistics is by randomized tests, that exhaust the permissible type I error rate. This approach is applied in Dickhaus et al. (2012) for contingency tables, and in Dickhaus (2013) and Hoang and Dickhaus (2020) for composite nulls.

A different avenue is taken by Chen (2019), who generalizes the frequency domain estimator of Jin (2008) to deal with mixtures of discrete test statistics. He proposes to estimate prevalence by solving the Lebesgue-Stieltjes integral equations of the mixture. He proves uniform consistency, convergence rates, and limiting variances in a wide class of mixtures. Chen's introduction is a highly recommended read for anyone interested in prevalence estimation.

8.10 Evaluating the Performance of an Estimator

At this point, it is clear that many estimators of π_0 are around. However, an algorithm is just an algorithm. It is only truly useful when its statistical properties are understood. A comprehensive comparison of various properties of various estimators is impossible. Not only because of the amount of estimators, and the amount of possible statistical properties, but mostly because of the many problem setups considered by different authors. We thus content ourselves with a catalog of the variety of the properties studied, and the problem setups out there. We do so after some musings on the two-group mixture setup, which plays an important role.

8.10.1 What Is This Two-Group Mixture?

The *two-group* mixture has attracted much interest but also some controversy. The following two sections present these matters, which ultimately deal with "what does the mixture distribution represent?"

8.10.1.1 Priors Are on Parameters

First, note that the distribution, f, is not on parameters, θ_j, but rather on statistics, y_j. Many Bayesians would thus not call this a "prior" distribution. If θ_j is a location parameter, this matter may be easily remedied. For instance, by assuming a "spike and slab" prior on θ_j. Formally:

$$h_j(\theta) = \pi_0 h_0(\theta) + (1 - \pi_0) h_1(\theta), \tag{8.11}$$

where $h_0(\theta) = \delta_0(\theta)$ is the Dirac delta at 0. If θ_j is a location parameter then $y_j = \theta_j + \varepsilon_j$. This means that f_0 and f_1 in Equation 8.3 are simply $h_0(\theta)$ and $h_1(\theta)$ convolved with the density of ε_j. Furthermore, if ε_j is Gaussian, we recover the normal means problem. This, for instance, is the construction used in the empirical Bayes approach to the normal means problem of Johnstone and Silverman (2004), or Cai et al. (2007). For more on spike-and-slab priors, see the references in van der Vaart (2019).

8.10.1.2 Interpretation

The prior in Equation 8.3 has gained various interpretations and is still a matter of debate:

> *"Empirical Bayes" is in the paper's title, said in Section 3 to be a "bipolar" methodology that draws on frequency and Bayes, but otherwise with a meaning left for us to infer from the paper's example datasets.— (Morris, 2008)*

The interpretation of the two-group prior has implications on the statistical guarantees of an algorithm. Is this mixture a sampling distribution? An empirical prior? A subjective prior? We elaborate this matter in Section 8.10.3, but certainly not resolve it.

8.10.2 Statistical Properties in the Literature

When judging $\hat{\pi}_0$ as a point estimator, properties of interest include bias, variance, consistency, convergence rate, efficiency, etc. When judging $\hat{\pi}_0$ as a confidence bound, properties of interest include coverage probabilities, tightness, etc. When judging $\hat{\pi}_0$ as a plug-in parameter in a selection algorithm, properties of interest are inherited from the selection guarantees: FWER, FDR, optimality, etc. When judging $\hat{\pi}_0$ as a plug-in parameter in a ranking algorithm, properties of interest may include optimality w.r.t. a ranking loss[3].

Various desirable properties are not always compatible. Enough evidence has been accumulated to confirm that some conservativeness in $\hat{\pi}_0$ is desired when used as a plugin parameter (Benjamini and Hochberg, 2000; Black, 2004; Hsueh et al., 2003; Lu and Perkins, 2007; Storey, 2002). A rigorous treatment of this matter can be found in Neuvial (2013). This means that a good plugin parameter may be a bad point estimate and vice versa.

8.10.3 Problem Setups in the Literature

The study of the statistical properties of $\hat{\pi}_0$ requires an assumed problem setup. A first assumption to look for is independence: are the provided guarantees valid under dependence or independence. When dealing with convergence rates, dependence is rarely a concern. Only strong long range dependencies invalidate convergence rates. Short-range dependencies are typically absorbed in constants, i.e. affect only variances but not rates.

The two-group mixture assumption is a more delicate matter. This is because different authors use the two-group mixture for different purposes: as a sampling distribution, as a subjective prior, and as a frequentist empirical prior.

Assuming a two-group mixture as a sampling distribution is a popular setup. It is necessary when studying the properties of a point estimator. The two-group mixture, however, may not be the sampling distribution of choice when studying a selection algorithm. Indeed, the early works on adaptive FWER and FDR, such as Hochberg and Benjamini (1990) and Benjamini and Hochberg (2000), assume nothing on the distribution of effects (f_1). Later work on selection algorithm such as Abramovich et al. (2006); Donoho et al. (2006) and Johnstone and Silverman (2004), assume effects are non-random, but *sparse* (in some way). More recent work, that tries to reconcile the random two-group mixture setup, and the non-random setup, typically recur to asymptotics. Indeed, asymptotically in p, the randomness in the proportion of null hypotheses converges to its limit, π_0.

From the previous discussion, we see that an important property of the problem setup is *sparsity*. The two-group mixture with fixed π_0 implies that the number of non-nulls increases indefinitely with p. This, in contrast with other problem setups, and various applications, where the number of non-nulls, $p - p_0$, does not increase indefinitely as p grows.

Another property of the two-group mixture is the assumed distribution under the alternative, f_1. Early works on adaptive selection procedures assume nothing on f_1. Assuming f_0 is enough to provide FWER, FDR, and coverage probabilities. It is not enough to study the properties of a point estimator or to claim the optimality of a selection procedure. Soon enough, parametric and weaker-than-parametric assumptions on f_1 began appearing: Gaussianity, concavity, symmetry, etc. While clearly essential for any type of optimality to be declared, we observe that the applied literature has an affinity to modeling f_1 as a mixture. Such a mixture assumption is incompatible with most of the assumptions found in the theoretical literature.

The view of the two-group mixture as an *Empirical-Bayes* prior does not alleviate the above concerns. The term "Empirical Bayes" itself has subsumed many interpretations throughout history: Robbins (1956)'s empirical Bayes prior is subjective. Efron and Morris (1973)'s empirical Bayes prior is empirical, i.e., a sampling distribution. Efron et al. (2000)'s empirical Bayes prior seems to be a regularization device, à-la James-Stein, which caused some confusion when introduced (see Section 8.10.1.2).

[3]The problem of loss functions for ranking is not discussed at all in our text. See Wakefield (2009) for further details.

Part of the difficulty in viewing the two-group prior as a sampling distribution is due to the *exchangeability* assumption. Exchangeability means that a multivariate distribution is invariant to a coordinate permutation. For sampling distributions, it means all measurements have the same distribution. For subjective beliefs, it means that we have equal uncertainty on all measurements.

In his comment to Efron (2008) (Section 8.10.1.2), it seems that Morris (2008) interprets the two-group mixture as a sampling-distribution. This is why he is concerned with the sampling variance of the different statistics; if not equal *exchangeability* is invalid. Similar concerns are also stated in Greenshtein and Ritov (2019). Thomas et al. (1985) on the other hand, seem to advocate exchangeability, from a subjective-prior point of view:

> *One might reasonably wonder how one could justify assuming that two associations, concerning different exposures and/or different diseases, share a common distribution. The answer lies in the concept of "exchangeability", which essentially states that in the absence of prior knowledge about either association, one is as likely to be true as another.*

A similar argument is found in Marits and Lwin (1989):

> *It has been argued that in many simultaneous decision problems, the parameters could be exchangeable in that the prior opinion of any particular parameter is the same as that for any other member of the sequence.*

Efron himself contradicts exchangeability with the idea of "hot prospect genes", which allows π_0 to vary over genes (Efron, 2008, Eq.(2.15)). This is not only an epistemological concern, but one that has implications on the type of statistical guarantees provided by an estimator.

Reconciling various interpretations of the two-group prior, and the implied guarantees of a selection procedure is well beyond the scope of this work. For the interested reader, we point at some references in the Bibliographic Notes (8.12). Also see Neuvial et al. (2008) and Sun and Cai (2007), which prove their results both in the two-group random setup, and in a non-random setup.

8.11 Discussion

At this point, it is quite clear that literature offers a tremendous variety of π_0 estimators, with varying statistical guarantees. Here is a discussion, in the form of Q&A, of some matters of interest the practitioner.

Q: Are there some patterns in this jungle of estimators?

In the problem setup (Section 8.3), we proposed the classification into "multivariate", "multiple-testing", "normal-means", and "two-group". At this stage, it is clear that in order to understand what a method does, a much finer classification is in order. Other attempts at the ambitious task of classifying estimators include Chen (2019, Table 1), and Strimmer (2008, Table 2). Our scope, however, is broader, so we propose our own classification. We distinguish between the classification of the estimators and their statistical guarantees. We demonstrate our classification on a selection of statistics in Table 8.1. We do not intend for this table to be comprehensive; it is merely intended to demonstrate the variability in the literature.

TABLE 8.1
A Demonstration of Our Classification Criteria, Applied on Selected Contributions

	Estimator					Statistical Guarantees			
Reference	**Data**	**Scale**	**Null**	**Effect**	**Principle**	**Type**	**Effects**	**Null**	**Dependence**
Rogan and Gladen (1978)	–	–	–	–	Rejections	Point	Fixed	Exact	Independence
Swanepoel et al. (1996)	Summary	p-scale	Assumed	Non-para.	Spacing	Point	Random	Exact	Independence
Benjamini and Hochberg (2000)	Summary	p-scale	Assumed	Non-para.	Rejections	Plug-in	Fixed	Exact	Independence
Efron and Tibshirani (2002)	Summary	Z-scale	Empirical	Non-para.	Spline	Plug-in	Unclear	–	Independence
Storey (2002)	Summary	p-scale	Assumed	Non-para.	Rejections	Point & Plug-in	Random	Exact	Independence
Pounds and Morris (2003)	Summary	p-scale	Assumed	Para.	ML	Point	Random	Exact	Independence
Genovese and Wasserman (2004)	Summary	p-scale	Assumed	Non-para.	Various	Point & Interval & Plug-in	Random	Exact	Independence
Tang et al. (2007)	Summary	p-scale	Assumed	Non-para.	Bayesian	Point & Plug-in	Random	Exact	Independence
Strimmer (2008)	Summary	All	Empirical	Non-para.	Deconv.	Point & Plug-in	Random	Exact	Independence
Goeman and Solari (2011)	Summary	p-scale	Assumed	Non-para.	Closed Testing	Interval	Fixed	Exact	Depends on Test
Lu and Perkins (2007)	X	Raw	Assumed	Non-para.	Rejections	Plug-in	Random	Exact	Empirical
Dickhaus (2013)	Summary	p-scale	Assumed	Non-para.	Rejections	Point & Plug-in	Fixed	Composite	Factor-Analysis
Patra and Sen (2016)	Summary	All	Assumed	Non-para.	Deconv.	Point	Random	Exact	Independence

Classifying estimators:

- **Data**: Do we observe X? The marginal summaries y? Or some other quantity (e.g. group-wise margins).
- **Scale**: Do we operate on p-values? Z-scores? t-statistics? Or effect-scale?
- **Null distribution**: Is the null distribution, f_0, assumed or empirical? We include composite nulls in the "assumed" class, since it is certainly not empirical.
- **Effect distribution**: Is the effect's distribution, f_1, assumed parametric or non-parametric?
- **Estimation Principle**: What is the estimation principle? Maximum Likelihood (ML)? Using the spacings between statistics? Using the minimal density? Using the rejections of some test? Deconvolving ECDFs? Something else?

Classifying statistical guarantees:

- **Type**: Are the statistical guarantees as a point estimator ("point")? As an interval estimator? ("interval") As a plug-in adaptive procedure? We do not distinguish between simulated guarantees to analytical guarantees.
- **Effect distribution**: Are guarantees provided with respect any effect configuration, a.k.a. the *conditional* model, or only on average, a.k.a. the *two-group prior*, *empirical Bayes*, *random-effects*, or the *non-conditional* model. Lacking an agreed upon terminology, we borrow from the mixed-models literature and denote the former with "fixed" and the latter with "random".
- **Exact null**: Are guarantees valid for exact nulls or also for composite nulls?
- **Dependence**: What is the assumed dependence between coordinates? Independence? Exchangeability? Factor-analysis and spiked-covariance? Other?

Q: Why such lengthy discussions of the two-group mixture?

The reader may wonder why we give so much emphasis to the interpretation of the two-group mixture. This is because the nature of statistical guarantees rests on the role assumed by this mixture, and we find this matter did not receive enough attention in the literature.

Q: Why identifiability is only a concern in the two-group mixture?

Why the identifiability of π_0 is a concern in the two-group prior (Equation 8.3), and not in the multiple-testing/normal-means/compound-decision setup (Equation 8.1)? The answer is that identifiability comes "for free" in the definition of Equation 8.1: θ_j governs the distribution of y_j, so if θ differs from θ_0 in some j, it is implied that the null distribution of y_j differs from non-null y_j.

Q: Should I collapse X variable-wise?

If one has access to n independent samples from a p-variate distribution, and not only their marginal summaries (y_j), is there anything to be gained by analyzing X directly? The answer is affirmative. By analyzing raw measurements, X, there is hope of modeling dependencies in the likelihood function. This may come at some computational cost, but nowadays, this is a lesser concern. Then again, if n is not much larger than p, then only very simple correlation structures may be estimated, and there may be not much statistical gain.

The n samples may not independent. This was indeed the case in the canonical motivating example of microarrays, where each gene j, was sampled using various probes and, plates (Efron et al., 2001). In this case, the full likelihood can deal with dependence between j's and i's, but dependence needs to be modeled between rows and not only between columns. It may thus be easier, as argued by Efron et al. (2001), to collapse over rows at the possible cost of some power, than to try to model this dependence.

Q: Is my estimator Bayesian or Frequentist?

Some of the proposed estimators place a prior on π_0, and infer using the posterior $\pi_0|y_1,\dots,y_p$. As such, they may be termed "Bayesian". On the other hand, if one studies an algorithm with respect to its properties over repeated experiments, one is clearly frequentist. Such is the case if we care about consistency, bias, and FDR control of plug-in procedures.

Q: At what scale should I estimate π_0?

Different authors operate at different scales. Z-scale, and p-values, have the advantage that they are a principled scale. They are also defined with respect to some null, θ_0, which is not necessarily the case when operating in observation scale. An important aspect of these scales is that f_0 is known, which is crucial for many algorithms.

Some authors argue that Z-scale is to be preferred. Sun and Cai (2007) argue that operating in Z-scale allows a richer class of selection rules. Neuvial (2013) argues that Z-scale is better equipped to deal with the directionality of hypotheses.

A second consideration is the validity of f_0. If one adopts the criticism of Efron (2008) on theoretical nulls, then f_0 may not be trusted, and Z-scale or p-scale lose much of their appeal. One possible remedy, in the lines of Efron and Tibshirani (2002) or Strimmer (2008), is to abandon theoretical nulls, in favor of empirical nulls and the symmetry principle. Another possible remedy is justifying f_0 by working with permutation p-values, or distribution-free test statistics such as ranks.Other considerations include:

- Validity of f_1: It may be possible that effects are easier to state in one scale over another.
- Correlations: It may be possible that correlation is easier to model in one scale over another. For instance, correlations in Z-scale are more popular than copulas in p-scale.

Q: Why Symmetry?

Some authors are not willing to assume a particular parametric form of the null distribution, f_0, but are willing to assume that f_0 is symmetric about 0 (e.g. Efron et al., 2000; Neuvial et al., 2008; Rosenblatt et al., 2014). This view can be justified by the belief that noise tends to balance itself (in some scale), while effects do not. In Z-scale, this may manifest in symmetry about 0. In p-value scale, this may manifest in symmetry about 1/2.

Q: Which π_0 estimator should I use?

The purpose of this whole text is to advise the practitioner regarding the preferred prevalence estimator. It is also clear by now, that there many options out there, and we apologize before the reader for not providing simple do's and don'ts. We do hope, however, that the references and considerations discussed in this review will aid in choosing an adequate estimator.

8.12 Bibliographic Notes

Hsueh et al. (2003) compare five estimators of π_0: They find that for the purpose of adaptive FDR control with the BH algorithm, the LSL estimator in Equation 8.4, is preferred. For a large simulated comparison of various π_0 estimators at the time, see Broberg (2005). For a simulation comparison of coverage probabilities of confidence intervals for two-group mixtures see Xiang et al. (2006). For an excellent review of semi-parametric estimators mixture models, see Xiang et al. (2018).

The origin of the *compound decision* problems is undoubtedly Robbins (1951). Five years later he presented the empirical Bayes view of this same problem (Robbins, 1956). Robbins was probably aware, from the start, of the link between the two problems. Quoting Robbins' student, James Hannan:

> *I believe that Robbins published his compound decision problem in the Second Berkeley Symposium. The only inversion was that Robbins had both the empirical Bayes formulation and the set compound formulation in mind in 1950. He could only give one talk and thought that the compound decision result would create a bigger splash. — Gilliland et al. (2010)*

For more on the link between compound decisions and multiple testing see Sun and Cai (2007), and also the exceptional treatment of Marits and Lwin (1989). Marits and Lwin (1989) summarize the link between compound decisions such as hypothesis testing, and empirical Bayes in the following:

> *The compound decision rules are found to be formally identical to empirical Bayes rules, the main difference being in the criteria on which their performance is judged.*

For a more concise and recent discussion of this link, see Greenshtein and Ritov (2019).

Bibliography

Abramovich, F., Y. Benjamini, D. L. Donoho, and I. M. Johnstone (2006). Adapting to unknown sparsity by controlling the false discovery rate. *The Annals of Statistics 34*(2), 584–653.

Allison, D. B., G. L. Gadbury, M. Heo, J. R. Fernández, C.-K. Lee, T. A. Prolla, and R. Weindruch (2002). A mixture model approach for the analysis of microarray gene expression data. *Computational Statistics & Data Analysis 39*(1), 1–20.

Benjamini, Y. and Y. Hochberg (2000). On the adaptive control of the false discovery rate in multiple testing with independent statistics. *Journal of Educational and Behavioral Statistics 25*(1), 60–83.

Benjamini, Y., A. M. Krieger, and D. Yekutieli (2006). Adaptive linear step-up procedures that control the false discovery rate. *Biometrika 93*(3), 491–507.

Black, M. A. (2004). A note on the adaptive control of false discovery rates. *Journal of the Royal Statistical Society: Series B (Statistical Methodology) 66*(2), 297–304.

Blanchard, G. and É. Roquain (2009). Adaptive false discovery rate control under independence and dependence. *Journal of Machine Learning Research 10*(Dec), 2837–2871.

Bordes, L., C. Delmas, and P. Vandekerkhove (2006). Semiparametric estimation of a two-component mixture model where one component is known. *Scandinavian Journal of Statistics 33*(4), 733–752.

Bordes, L. and P. Vandekerkhove (2010). Semiparametric two-component mixture model with a known component: an asymptotically normal estimator. *Mathematical Methods of Statistics 19*(1), 22–41.

Broberg, P. (2005). A comparative review of estimates of the proportion unchanged genes and the false discovery rate. *BMC Bioinformatics 6*(1), 199.

Broët, P., S. Richardson, and F. Radvanyi (2002). Bayesian hierarchical model for identifying changes in gene expression from microarray experiments. *Journal of Computational Biology 9*(4), 671–683.

Cai, T. T. and J. Jin (2010). Optimal rates of convergence for estimating the null density and proportion of nonnull effects in large-scale multiple testing. *The Annals of Statistics 38*(1), 100–145.

Cai, T. T., J. Jin, and M. G. Low (2007). Estimation and confidence sets for sparse normal mixtures. *The Annals of Statistics 35*(6), 2421–2449.

Carpentier, A. and N. Verzelen (2019). Adaptive estimation of the sparsity in the gaussian vector model. *The Annals of Statistics 47*(1), 93–126.

Celisse, A. and S. Robin (2010). A cross-validation based estimation of the proportion of true null hypotheses. *Journal of Statistical Planning and Inference 140*(11), 3132–3147.

Chen, X. (2019). Uniformly consistently estimating the proportion of false null hypotheses via lebesgue–stieltjes integral equations. *Journal of Multivariate Analysis 173*, 724–744.

Cheng, C. (2006). An adaptive significance threshold criterion for massive multiple hypotheses testing. In *Optimality*, pp. 51–76. Institute of Mathematical Statistics.

Cheng, Y., D. Gao, and T. Tong (2014). Bias and variance reduction in estimating the proportion of true-null hypotheses. *Biostatistics 16*(1), 189–204.

Chi, Z. (2010). Multiple hypothesis testing on composite nulls using constrained p-values. *Electronic Journal of Statistics 4*, 271–299.

Chi, Z. et al. (2007). Sample size and positive false discovery rate control for multiple testing. *Electronic Journal of Statistics 1*, 77–118.

Cruz-Medina, I. R. and T. P. Hettmansperger (2004). Nonparametric estimation in semi-parametric univariate mixture models. *Journal of Statistical Computation and Simulation 74*(7), 513–524.

Dalmasso, C., P. Broët, and T. Moreau (2004). A simple procedure for estimating the false discovery rate. *Bioinformatics 21*(5), 660–668.

De Jager, O. C., C. B. Raubenheimer, and J. W. Swanepoel (1989). The rayleigh statistic in the case of weak signals—applications and pitfalls. In *Data Analysis in Astronomy III*, pp. 21–30. Springer.

Diaconis, P. and D. Ylvisaker (1985). Quantifying prior opinion in "bayesian statistics 2. In *Proceedings of the second Valencia international meeting", September*, Volume 6, pp. 1983.

Dickhaus, T. (2013). Randomized p-values for multiple testing of composite null hypotheses. *Journal of Statistical Planning and Inference 143*(11), 1968–1979.

Dickhaus, T., K. Straßburger, D. Schunk, C. Morcillo-Suarez, T. Illig, and A. Navarro (2012). How to analyze many contingency tables simultaneously in genetic association studies. *Statistical Applications in Genetics and Molecular Biology 11*(4).

Do, K.-A., P. Müller, and F. Tang (2005). A bayesian mixture model for differential gene expression. *Journal of the Royal Statistical Society: Series C (Applied Statistics) 54*(3), 627–644.

Donoho, D., J. Jin, et al. (2006). Asymptotic minimaxity of false discovery rate thresholding for sparse exponential data. *The Annals of Statistics 34*(6), 2980–3018.

Ebrahimpoor, M., P. Spitali, K. Hettne, R. Tsonaka, and J. Goeman (2020). Simultaneous enrichment analysis of all possible gene-sets: unifying self-contained and competitive methods. *Briefings in Bioinformatics 21*(4), 1302–1312.

Efron, B. (2008). Microarrays, empirical bayes and the two-groups model. *Statistical Science 23*(1), 1–22.

Efron, B. and C. Morris (1973). Stein's estimation rule and its competitors—an empirical bayes approach. *Journal of the American Statistical Association 68*(341), 117–130.

Efron, B. and R. Tibshirani (2002). Empirical bayes methods and false discovery rates for microarrays. *Genetic Epidemiology 23*(1), 70–86.

Efron, B., R. Tibshirani, V. Goss, G. Chu, and S. G. Chu (2000). Microarrays and their use in a comparative experiment.

Efron, B., R. Tibshirani, J. D. Storey, and V. Tusher (2001). Empirical bayes analysis of a microarray experiment. *Journal of the American Statistical Association 96*(456), 1151–1160.

Finner, H. and V. Gontscharuk (2009). Controlling the familywise error rate with plug-in estimator for the proportion of true null hypotheses. *Journal of the Royal Statistical Society: Series B (Statistical Methodology) 71*(5), 1031–1048.

Friguet, C. and D. Causeur (2011). Estimation of the proportion of true null hypotheses in high-dimensional data under dependence. *Computational statistics & data analysis 55*(9), 2665–2676.

Gadbury, G. L., G. P. Page, J. Edwards, T. Kayo, T. A. Prolla, R. Weindruch, P. A. Permana, J. D. Mountz, and D. B. Allison (2004). Power and sample size estimation in high dimensional biology. *Statistical Methods in Medical Research 13*(4), 325–338.

Genovese, C. and L. Wasserman (2004). A stochastic process approach to false discovery control. *The Annals of Statistics 32*(3), 1035–1061.

Gilliland, D., R. Ramamoorthi, and J. Hannan (2010). A conversation with james hannan. *Statistical Science*, 126–144.

Goeman, J. J., R. J. Meijer, T. J. Krebs, and A. Solari (2019). Simultaneous control of all false discovery proportions in large-scale multiple hypothesis testing. *Biometrika 106*(4), 841–856.

Goeman, J. J. and A. Solari (2011). Multiple testing for exploratory research. *Statistical Science 26*(4), 584–597.

Good, P. I. (1979). Detection of a Treatment Effect When Not All Experimental Subjects Will Respond to Treatment. *Biometrics 35*(2), 483–489. 00071.

Greenshtein, E. and Y. Ritov (2019). Comment: Empirical bayes, compound decisions and exchangeability. *Statistical Science 34*(2), 224–228.

Heller, R., Y. Golland, R. Malach, and Y. Benjamini (2007). Conjunction group analysis: an alternative to mixed/random effect analysis. *Neuroimage 37*(4), 1178–1185.

Hoang, A.-T. and T. Dickhaus (2020). On the usage of randomized p-values in the schweder-spjotvoll estimator. *arXiv preprint arXiv:2004.08256*.

Hochberg, Y. and Y. Benjamini (1990). More powerful procedures for multiple significance testing. *Statistics in Medicine 9*(7), 811–818.

Hohmann, D. and H. Holzmann (2013). Semiparametric location mixtures with distinct components. *Statistics 47*(2), 348–362.

Hsueh, H.-m., J. J. Chen, and R. L. Kodell (2003). Comparison of methods for estimating the number of true null hypotheses in multiplicity testing. *Journal of Biopharmaceutical Statistics 13*(4), 675–689.

Hwang, Y.-T., H.-C. Kuo, C.-C. Wang, and M. F. Lee (2014). Estimating the number of true null hypotheses in multiple hypothesis testing. *Statistics and Computing 24*(3), 399–416.

Ince, R. A., J. W. Kay, and P. G. Schyns (2020). Bayesian inference of population prevalence. *bioRxiv*.

Ji, H. and W. H. Wong (2005). Tilemap: create chromosomal map of tiling array hybridizations. *Bioinformatics 21*(18), 3629–3636.

Jiang, H. and R. Doerge (2008). Estimating the proportion of true null hypotheses for multiple comparisons. *Cancer Informatics 6*, 117693510800600001.

Jin, J. (2008). Proportion of non-zero normal means: universal oracle equivalences and uniformly consistent estimators. *Journal of the Royal Statistical Society: Series B (Statistical Methodology) 70*(3), 461–493.

Johnstone, I. M. and B. W. Silverman (2004). Needles and straw in haystacks: Empirical bayes estimates of possibly sparse sequences. *The Annals of Statistics 32*(4), 1594–1649.

Kendziorski, C., M. A. Newton, H. Lan, and M. N. Gould (2003). On parametric empirical bayes methods for comparing multiple groups using replicated gene expression profiles. *Statistics in Medicine 22*(24), 3899–3914.

Lai, Y. (2007). A moment-based method for estimating the proportion of true null hypotheses and its application to microarray gene expression data. *Biostatistics 8*(4), 744–755.

Langaas, M., B. H. Lindqvist, and E. Ferkingstad (2005). Estimating the proportion of true null hypotheses, with application to dna microarray data. *Journal of the Royal Statistical Society: Series B (Statistical Methodology) 67*(4), 555–572.

Lee, M.-L. T., F. C. Kuo, G. Whitmore, and J. Sklar (2000). Importance of replication in microarray gene expression studies: statistical methods and evidence from repetitive cdna hybridizations. *Proceedings of the National Academy of Sciences 97*(18), 9834–9839.

Lee, M.-L. T., W. Lu, G. Whitmore, and D. Beier (2002). Models for microarray gene expression data. *Journal of Biopharmaceutical Statistics 12*(1), 1–19.

Lee, Y. and J. F. Bjørnstad (2013). Extended likelihood approach to large-scale multiple testing. *Journal of the Royal Statistical Society: Series B (Statistical Methodology) 75*(3), 553–575.

Liao, J. J., Y. Lin, Z. E. Selvanayagam, and W. J. Shih (2004). A mixture model for estimating the local false discovery rate in dna microarray analysis. *Bioinformatics 20*(16), 2694–2701.

Lönnstedt, I. and T. Speed (2002). Replicated microarray data. *Statistica Sinica*, 31–46.

Loots, H. (1995). *Nonparametric estimation of the antimode and the minimum of a density function.* Ph. D. thesis, Potchefstroom University.

Lu, X. and D. L. Perkins (2007). Re-sampling strategy to improve the estimation of number of null hypotheses in fdr control under strong correlation structures. *BMC Bioinformatics 8*(1), 157.

Ma, J., S. Gudlaugsdottir, and G. Wood (2011). Generalized em estimation for semi-parametric mixture distributions with discretized non-parametric component. *Statistics and Computing 21*(4), 601–612.

Ma, Y. and W. Yao (2015). Flexible estimation of a semiparametric two-component mixture model with one parametric component. *Electronic Journal of Statistics 9*(1), 444–474.

Maiboroda, R. and O. Sugakova (2010). Generalized estimating equations for symmetric distributions observed with admixture. *Communications in Statistics—Theory and Methods 40*(1), 96–116.

Marits, J. S. and T. Lwin (1989). *Empirical Bayes Methods* (2nd ed.). Chapman and Hall, London, UK..

Markitsis, A. and Y. Lai (2010). A censored beta mixture model for the estimation of the proportion of non-differentially expressed genes. *Bioinformatics 26*(5), 640–646.

McLachlan, G. J., R. Bean, and L. B.-T. Jones (2006). A simple implementation of a normal mixture approach to differential gene expression in multiclass microarrays. *Bioinformatics 22*(13), 1608–1615.

Meinshausen, N. and P. Bühlmann (2005). Lower bounds for the number of false null hypotheses for multiple testing of associations under general dependence structures. *Biometrika 92*(4), 893–907.

Meinshausen, N. and J. Rice (2006). Estimating the proportion of false null hypotheses among a large number of independently tested hypotheses. *The Annals of Statistics 34*(1), 373–393.

Morris, C. N. (2008). Comment: Microarrays, empirical bayes and the two-groups model. *Statistical Science*, 34–40.

Mosig, M. O., E. Lipkin, G. Khutoreskaya, E. Tchourzyna, M. Soller, and A. Friedmann (2001). A whole genome scan for quantitative trait loci affecting milk protein percentage in israeli-holstein cattle, by means of selective milk dna pooling in a daughter design, using an adjusted false discovery rate criterion. *Genetics 157*(4), 1683–1698.

Nettleton, D. and J. Hwang (2003). Estimating the number of false null hypotheses when conducting many tests. *Preprint Series 9.*

Nettleton, D., J. G. Hwang, R. A. Caldo, and R. P. Wise (2006). Estimating the number of true null hypotheses from a histogram of p values. *Journal of Agricultural, Biological, and Environmental Statistics 11*(3), 337.

Neumann, A., T. Bodnar, and T. Dickhaus (2020). Estimating the proportion of true null hypotheses under dependency: A marginal bootstrap approach. *Journal of Statistical Planning and Inference*.

Neuvial, P. (2010). Intrinsic bounds and false discovery rate control in multiple testing problems. Technical report, Citeseer.

Neuvial, P. (2013). Asymptotic results on adaptive false discovery rate controlling procedures based on kernel estimators. *The Journal of Machine Learning Research 14*(1), 1423–1459.

Neuvial, P. et al. (2008). Asymptotic properties of false discovery rate controlling procedures under independence. *Electronic Journal of statistics 2*, 1065–1110.

Newton, M. A., C. M. Kendziorski, C. S. Richmond, F. R. Blattner, and K.-W. Tsui (2001). On differential variability of expression ratios: improving statistical inference about gene expression changes from microarray data. *Journal of Computational Biology 8*(1), 37–52.

Newton, M. A., A. Noueiry, D. Sarkar, and P. Ahlquist (2004). Detecting differential gene expression with a semiparametric hierarchical mixture method. *Biostatistics 5*(2), 155–176.

Nguyen, V. H. and C. Matias (2014). On efficient estimators of the proportion of true null hypotheses in a multiple testing setup. *Scandinavian Journal of Statistics 41*(4), 1167–1194.

Oyeniran, O. and H. Chen (2016). Estimating the proportion of true null hypotheses in multiple testing problems. *Journal of Probability and Statistics 2016.*

Pan, W., J. Lin, and C. T. Le (2002). Model-based cluster analysis of microarray gene-expression data. *Genome Biology 3*(2), research0009–1.

Parker, R. A. and R. Rothenberg (1988). Identifying important results from multiple statistical tests. *Statistics in Medicine 7*(10), 1031–1043.

Patra, R. K. and B. Sen (2016). Estimation of a two-component mixture model with applications to multiple testing. *Journal of the Royal Statistical Society: Series B (Statistical Methodology) 78*(4), 869–893.

Pawitan, Y., K. R. K. Murthy, S. Michiels, and A. Ploner (2005). Bias in the estimation of false discovery rate in microarray studies. *Bioinformatics 21*(20), 3865–3872.

Pounds, S. and C. Cheng (2004). Improving false discovery rate estimation. *Bioinformatics 20*(11), 1737–1745.

Pounds, S. and C. Cheng (2006). Robust estimation of the false discovery rate. *Bioinformatics 22*(16), 1979–1987.

Pounds, S. and S. W. Morris (2003). Estimating the occurrence of false positives and false negatives in microarray studies by approximating and partitioning the empirical distribution of p-values. *Bioinformatics 19*(10), 1236–1242.

Qiao, Y., W. Yu, and W. Xu (2017). Two new estimators for the proportion of true null hypotheses in multiple test. *Journal of Statistical Computation and Simulation 87*(4), 712–723.

Qu, L., D. Nettleton, and J. C. Dekkers (2012). Improved estimation of the noncentrality parameter distribution from a large number of t-statistics, with applications to false discovery rate estimation in microarray data analysis. *Biometrics 68*(4), 1178–1187.

Robbins, H. (1951). Asymptotically subminimax solutions of compound statistical decision problems. In *Proceedings of the second Berkeley symposium on mathematical statistics and probability.* The Regents of the University of California.

Robbins, H. (1956). An empirical bayes approach to statistics. *Herbert Robbins Selected Papers*, 41–47.

Rogan, W. J. and B. Gladen (1978). Estimating prevalence from the results of a screening test. *American Journal of Epidemiology 107*(1), 71–76.

Rosenblatt, J. D., L. Finos, W. D. Weeda, A. Solari, and J. J. Goeman (2018). All-resolutions inference for brain imaging. *NeuroImage 181*, 786–796.

Rosenblatt, J. D., M. Vink, and Y. Benjamini (2014). Revisiting multi-subject random effects in fmri: Advocating prevalence estimation. *NeuroImage 84*, 113–121.

Ruppert, D., D. Nettleton, and J. G. Hwang (2007). Exploring the information in p-values for the analysis and planning of multiple-test experiments. *Biometrics 63*(2), 483–495.

Scheid, S. and R. Spang (2003). A false discovery rate approach to separate the score distributions of induced and non-induced genes. In *3rd International Workshop on Distributed Statistical Computing (DSC 2003)*.

Scheid, S. and R. Spang (2004). A stochastic downhill search algorithm for estimating the local false discovery rate. *IEEE/ACM Transactions on Computational Biology and Bioinformatics (TCBB) 1*(3), 98–108.

Schutte, W. D. and J. W. Swanepoel (2016). Sopie: an r package for the non-parametric estimation of the off-pulse interval of a pulsar light curve. *Monthly Notices of the Royal Astronomical Society 461*(1), 627–640.

Schweder, T. and E. Spjøtvoll (1982). Plots of p-values to evaluate many tests simultaneously. *Biometrika 69*(3), 493–502.

Shen, Z., M. Levine, and Z. Shang (2018). An mm algorithm for estimation of a two component semiparametric density mixture with a known component. *Electronic Journal of Statistics 12*(1), 1181–1209.

Storey, J. D. (2002). A direct approach to false discovery rates. *Journal of the Royal Statistical Society: Series B (Statistical Methodology) 64*(3), 479–498.

Storey, J. D., J. E. Taylor, and D. Siegmund (2004). Strong control, conservative point estimation and simultaneous conservative consistency of false discovery rates: a unified approach. *Journal of the Royal Statistical Society: Series B (Statistical Methodology) 66*(1), 187–205.

Storey, J. D. and R. Tibshirani (2003). Statistical significance for genomewide studies. *Proceedings of the National Academy of Sciences 100*(16), 9440–9445.

Strimmer, K. (2008). A unified approach to false discovery rate estimation. *BMC Bioinformatics 9*(1), 303.

Sun, W. and T. T. Cai (2007). Oracle and adaptive compound decision rules for false discovery rate control. *Journal of the American Statistical Association 102*(479), 901–912.

Swanepoel, J. W. (1999). The limiting behavior of a modified maximal symmetric $2s$-spacing with applications. *The Annals of Statistics 27*(1), 24–35.

Swanepoel, J. W., C. De Beer, and H. Loots (1996). Estimation of the strength of a periodic signal from photon arrival times. *The Astrophysical Journal 467*, 261.

Tamhane, A. C. and J. Shi (2009). Parametric mixture models for estimating the proportion of true null hypotheses and adaptive control of fdr. *Lecture Notes-Monograph Series*, 304–325.

Tang, Y., S. Ghosal, and A. Roy (2007). Nonparametric bayesian estimation of positive false discovery rates. *Biometrics 63*(4), 1126–1134.

Thomas, D., J. Siemiatycki, R. Dewar, J. Robins, M. Goldberg, and B. Armstrong (1985). The problem of multiple inference in studies designed to generate hypotheses. *American Journal of Epidemiology 122*(6), 1080–1095.

Tong, T., Z. Feng, J. S. Hilton, and H. Zhao (2013). Estimating the proportion of true null hypotheses using the pattern of observed p-values. *Journal of Applied Statistics 40*(9), 1949–1964.

Turnbull, B. B. (2007). Optimal estimation of false discovery rates. In *Tech rep.*

Tusher, V. G., R. Tibshirani, and G. Chu (2001). Significance analysis of microarrays applied to the ionizing radiation response. *Proceedings of the National Academy of Sciences 98*(9), 5116–5121.

van der Vaart, A. (2019). Comment: Bayes, oracle bayes and empirical bayes. *Statistical Science 34*(2), 214–218.

Wakefield, J. (2009). Bayes factors for genome-wide association studies: comparison with p-values. *Genetic Epidemiology: The Official Publication of the International Genetic Epidemiology Society 33*(1), 79–86.

Wang, H.-Q., L. K. Tuominen, and C.-J. Tsai (2010). Slim: a sliding linear model for estimating the proportion of true null hypotheses in datasets with dependence structures. *Bioinformatics 27*(2), 225–231.

Wang, J. and A. B. Owen (2019). Admissibility in partial conjunction testing. *Journal of the American Statistical Association 114*(525), 158–168.

Wu, B., Z. Guan, and H. Zhao (2006). Parametric and nonparametric fdr estimation revisited. *Biometrics 62*(3), 735–744.

Xiang, Q., J. Edwards, and G. L. Gadbury (2006). Interval estimation in a finite mixture model: Modeling p-values in multiple testing applications. *Computational Statistics & Data Analysis 51*(2), 570–586.

Xiang, S., W. Yao, and J. Wu (2014). Minimum profile hellinger distance estimation for a semiparametric mixture model. *Canadian Journal of Statistics 42*(2), 246–267.

Xiang, S., W. Yao, and G. Yang (2018). An overview of semiparametric extensions of finite mixture models. *arXiv preprint arXiv:1811.05575*.

Yu, C. and D. Zelterman (2017). A parametric model to estimate the proportion from true null using a distribution for p-values. *Computational Statistics & Data Analysis 114*, 105–118.

9

On Agnostic Post-Hoc Approaches to False Positive Control

Gilles Blanchard

Université Paris-Saclay, CNRS, Inria

Pierre Neuvial

Université de Toulouse and CNRS

Etienne Roquain

Sorbonne Université

CONTENTS

Classical approaches to multiple testing grant control over the amount of false positives for a specific method prescribing the set of rejected hypotheses. On the other hand, in practice, many users tend to deviate from a strictly prescribed multiple testing method and follow ad-hoc rejection rules, tune some parameters by hand, compare several methods and pick from their results the one that suits them best, etc. This will invalidate standard statistical guarantees because of the selection effect. To compensate for any form of such

DOI: 10.1201/9780429030888-10

"data snooping", an approach which has garnered significant interest recently is to derive "user-agnostic", or post-hoc, bounds on the false positives valid uniformly over all possible rejection sets; this allows arbitrary data snooping from the user. In this chapter, we start from a common approach to post-hoc bounds considering the p-value level sets for any candidate rejection set and explain how to calibrate the bound under different assumption concerning the distribution of p-values. We then build toward a general approach to this problem using a family of candidate rejection subsets (call this a reference family) together with associated bounds on the number of false positives they contain, the latter holding uniformly over the family. It is then possible to interpolate from this reference family to find a bound valid for any candidate rejection subset. This general program encompasses, in particular, the p-value level sets considered initially in the chapter; we illustrate its interest in a different context where the reference subsets are fixed and spatially structured. These methods are then applied to a genomic example (differential gene expression) and a neuromaging example (functional Magnetic Resonance Imaging). Code vignettes to reproduce these examples using the R R Core Team (2020) package sansSouci Neuvial et al. (2020) are provided as Supplementary Materials[1]. In this chapter, all references are gathered in Section 9.11.

9.1 A Motivating Example

Differential gene expression studies in cancerology aim at identifying genes whose activity differs significantly between two (or more) cancer populations, based on a sample of measurements from individuals from these populations. The activity of a gene is usually quantified by its level of expression in the cell. We consider here a microarray data set[2] consisting of expression measurements for more than 12,000 genes for biological samples from $n = 79$ individuals with B-cell acute lymphoblastic leukemia (ALL). A subset of cardinal $n_1 = 37$ of these individuals harbor a specific mutation called BCR/ABL, while the remaining $n_2 = 42$ do not. One of the goals of this study is to identify, from this sample, those genes for which there is a difference in the mean expression level between the mutated and non-mutated populations. This question can be addressed, after relevant data preprocessing, by performing a statistical test of equality in means for each gene. A classical approach is then to derive a list of "differentially expressed" genes (DEG) as those passing a FDR correction by the Benjamini-Hochberg (BH) procedure at a user-defined level. This is illustrated by Figure 9.1 for the Leukemia data set, where 163 genes are called "differentially expressed" at FDR level $q = 0.05$.

9.2 Setting and Basic Assumptions

Let us observe a random variable X with distribution P belonging to some model $\mathcal{P}$. Consider m null hypotheses $H_{0,i} \subset \mathcal{P}$, $i \in \mathbb{N}_m = \{1, \dots, m\}$, for P. We denote $\mathcal{H}_0(P) = \{i \in \mathbb{N}_m : P \text{ satisfies } H_{0,i}\}$ the set of true null hypotheses and $\mathcal{H}_1(P) = \mathbb{N}_m \backslash \mathcal{H}_0(P)$ its

[1]The corresponding source files are available from the package web site: https://pneuvial.github.io/sanssouci/.

[2]Taken from Chiaretti et al., *Clinical Cancer Research*, 11(20):7209–7219, 2005.

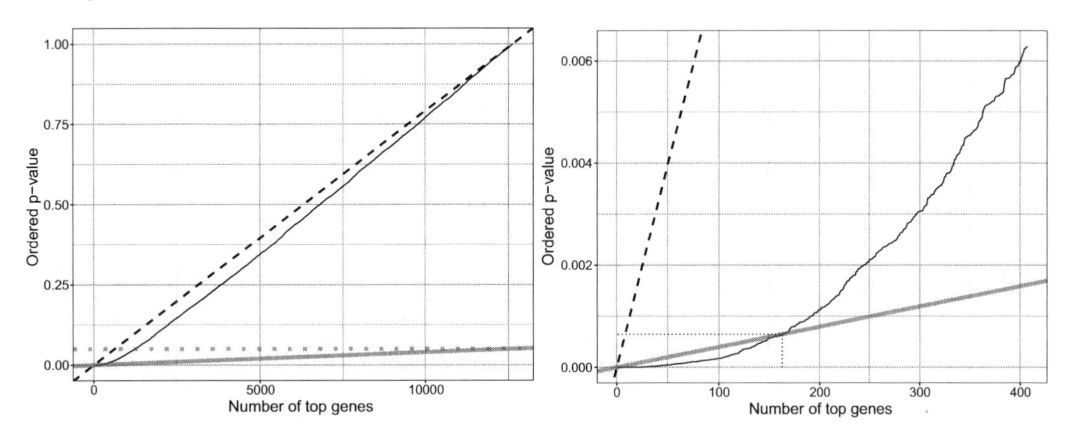

FIGURE 9.1
Left: Sorted p-values for the Leukemia data set (thin solid line). Right: Zoom on the smallest 400 p-values. Dashed line: $y = x/m$; bold solid line: $y = qx/m$ (for $q = 0.05$), whose intersection with the p-value curve determines the rejections of the BH procedure at level q. Dotted line: $y = q$. Here, 163 genes are declared as differentially expressed.

complement. We assume that a p-value $p_i(X)$ is available for each null hypothesis $H_{0,i}$, for each $i \in \mathbb{N}_m$.

We introduce the following assumptions on the p-values and the distribution P, which will be useful in the sequel:

$$\forall i \in \mathcal{H}_0(P), \ \forall t \in [0,1], \ \mathbb{P}(p_i(X) \le t) \le t; \qquad \text{(Superunif)}$$

$$\begin{array}{c}\{p_i(X)\}_{i\in\mathcal{H}_0(P)} \text{ is a family of independent variables,} \\ \text{which is independent of } \{p_i(X)\}_{i\in\mathcal{H}_1(P)}.\end{array} \qquad \text{(Indep)}$$

9.3 From Confidence Bounds ...

Consider some fixed deterministic $S \subset \mathbb{N}_m$. A $(1-\alpha)$-confidence bound $V = V(X)$ for $|S \cap \mathcal{H}_0(P)|$, the number of false positives in S, is such that

$$\forall P \in \mathcal{P}, \qquad \mathbb{P}_{X\sim P}\Big(|S \cap \mathcal{H}_0(P)| \le V\Big) \ge 1-\alpha.$$

The first example is given by the k_0-Bonferroni bound $V(X) = \sum_{i\in S} \mathbf{1}\{p_i(X) \ge \alpha k_0/|S|\} + k_0 - 1$, for some fixed $k_0 \in \mathbb{N}_m$ such that $k_0 \le |S|$ (otherwise the bound is trivial). The coverage probability is ensured under (Superunif) by the Markov inequality:

$$\begin{aligned}
&\mathbb{P}(|S \cap \mathcal{H}_0(P)| \ge V+1) \\
&\qquad \le \mathbb{P}\Bigg(|S \cap \mathcal{H}_0(P)| \ge \sum_{i\in S\cap\mathcal{H}_0(P)} \mathbf{1}\{p_i(X) \ge \alpha k_0/|S|\} + k_0\Bigg) \\
&\qquad = \mathbb{P}\Bigg(\sum_{i\in S\cap\mathcal{H}_0(P)} \mathbf{1}\{p_i(X) < \alpha k_0/|S|\} \ge k_0\Bigg) \\
&\qquad \le \frac{|S \cap \mathcal{H}_0(P)|\alpha k_0/|S|}{k_0} \le \alpha.
\end{aligned}$$

However, in practice, S is often chosen by the user and possibly depends on the same data set, then denoted $\widehat{S}$ to emphasize this dependence; it typically corresponds to items of potential strong interest. The most archetypal example is when $\widehat{S}$ consists of the s_0 smallest p-values $p_{(1:m)}, \ldots, p_{(s_0:m)}$, for some fixed value of $s_0 \in \mathbb{N}_m$. In that case, it is easy to check that the above bound does not have the correct coverage: for instance, when the p-values are i.i.d. $U(0,1)$ and $\mathcal{H}_0(P) = \mathbb{N}_m$, we have for $k_0 \leq s_0$ (i.e., when the bound is informative),

$$\begin{aligned}\mathbb{P}(|\widehat{S} \cap \mathcal{H}_0(P)| \leq V) &= \mathbb{P}\Big(k_0 - 1 + \textstyle\sum_{i \in \hat{S}} \mathbf{1}\left\{p_i(X) \geq \alpha k_0/s_0\right\} \geq s_0\Big) \\ &= \mathbb{P}\Big(\textstyle\sum_{i \in \hat{S}} \mathbf{1}\left\{p_i(X) < \alpha k_0/s_0\right\} \leq k_0 - 1\Big) \\ &= \mathbb{P}\left(p_{(k_0:m)}(X) \geq \alpha k_0/s_0\right) \\ &= \mathbb{P}(\beta(k_0, m - k_0 + 1) \geq \alpha k_0/s_0),\end{aligned}$$

where $\beta(k_0, m-k_0+1)$ denotes the usual beta distribution with parameters k_0 and $m-k_0+1$. For instance, taking $s_0 = 10$, $k_0 = 5$, $\alpha = 0.05$ and $m = 500$, the latter is approximately equal to 0.005, while the intended target is $1 - \alpha = 0.95$.

This phenomenon is often referred to as the selection effect: after some data driven selection, the probabilities change and thus the usual statistical inferences are not valid.

9.4 ... to Post-Hoc Bounds

To circumvent the selection effect, one way is to aim for a function $V(X, \cdot) : S \subset \mathbb{N}_m \mapsto V(X, S) \in \mathbb{N}$ (denoted by $V(S)$ for short) satisfying

$$\forall P \in \mathcal{P}, \qquad \mathbb{P}_{X \sim P}\Big(\forall S \subset \mathbb{N}_m,\ |S \cap \mathcal{H}_0(P)| \leq V(S)\Big) \geq 1 - \alpha, \tag{9.1}$$

that is, a $(1 - \alpha)$ confidence bound that is valid *uniformly* over all subsets $S \subset \mathbb{N}_m$. As a result, for any particular algorithm $\widehat{S}$, inequality (9.1) entails $\mathbb{P}(|\hat{S} \cap \mathcal{H}_0(P)| \leq V(\hat{S})) \geq 1-\alpha$, and thus does not suffer from the selection effect. Such a bound will be referred to as a $(1-\alpha)$-*post-hoc confidence bound* throughout this chapter, "post hoc" meaning that the set S can be chosen after having seen the data, and possibly using the data several times.

As a first example, the k_0-*Bonferroni post-hoc bound* is

$$V^{k_0 \mathrm{Bonf}}(S) = |S| \wedge \left(\sum_{i \in S} \mathbf{1}\left\{p_i(X) \geq \alpha k_0/m\right\} + k_0 - 1\right). \tag{9.2}$$

Following the same reasoning as above, it has a coverage at least $1 - \alpha$ under (Superunif):

$$\begin{aligned}&\mathbb{P}(\exists S \subset \mathbb{N}_m : |S \cap \mathcal{H}_0(P)| \geq V^{k_0 \mathrm{Bonf}}(S) + 1) \\ &\qquad \leq \mathbb{P}\left(\exists S \subset \mathbb{N}_m : \sum_{i \in S \cap \mathcal{H}_0(P)} \mathbf{1}\left\{p_i(X) < \alpha k_0/m\right\} \geq k_0\right) \\ &\qquad = \mathbb{P}\left(\sum_{i \in \mathcal{H}_0(P)} \mathbf{1}\left\{p_i(X) < \alpha k_0/m\right\} \geq k_0\right) \\ &\qquad \leq \frac{|\mathcal{H}_0(P)| \alpha k_0/m}{k_0} \leq \alpha.\end{aligned}$$

Remark 9.1 *Compared to the k_0-Bonferroni confidence bound of Section 9.3, α has been replaced by $\alpha|S|/m$, so that the post-hoc bound is much more conservative than a (standard, non uniform, S fixed) confidence bound when $|S|/m$ gets small, which is well expected. This scaling factor is the price paid here to make the inference post hoc. We will see in Sections 9.5 and 9.8 that it can be diminished when considering bounds of a different nature.*

Coming back to the motivating example of Section 9.1, if we choose $k_0 = 100$, the k_0-Bonferroni post-hoc bound (9.2) ensures that with probability at least 90%, the number of false positives among the 163 genes selected by the BH procedure at level $q = 0.05$ is upper bounded by 99.

Example 9.1 *For $k_0 = 1$, when the p-values are i.i.d. $U(0,1)$ and $\mathcal{H}_0(P) = \mathbb{N}_m$, the coverage probability of the k_0-Bonferroni post-hoc bound is equal to $(1-\alpha/m)^m = e^{m\log(1-\alpha/m)}$, which is very close to $1-\alpha$ when α is small.*

The Bonferroni post-hoc bound, while it is valid under no assumption on the dependence structure of the p-value family, may be conservative, in the sense that $V(S)$ will be large for many subsets S. For instance, one has $V^{k_0\text{Bonf}}(S) = |S|$ (trivial bound) for all the sets S such that $S \subset \{i \in \mathbb{N}_m : p_i(X) > \alpha k_0/m\}$.

The Bonferroni bound can be further improved under some dependence restriction, with the *Simes post-hoc bound:*

$$V^{\text{Sim}}(S) = \min_{1\leq k\leq |S|}\left\{\sum_{i\in S}\mathbf{1}\{p_i(X)\geq \alpha k/m\} + k - 1\right\} = \min_{1\leq k\leq |S|}\{V^{k\text{Bonf}}(S)\}. \tag{9.3}$$

Its coverage can be computed as follows (using arguments similar as above):

$$\begin{aligned}
&\mathbb{P}(\exists S \subset \mathbb{N}_m : |S\cap \mathcal{H}_0(P)| \geq V^{\text{Sim}}(S)+1)\\
&\leq \mathbb{P}\left(\exists S \subset \mathbb{N}_m\,,\, \exists k \in \{1,\dots,m\} : \sum_{i\in S\cap\mathcal{H}_0(P)} \mathbf{1}\{p_i(X) < \alpha k/m\} \geq k\right)\\
&= \mathbb{P}\left(\exists k \in \{1,\dots,|\mathcal{H}_0(P)|\} : p_{(k:\mathcal{H}_0(P))} < \alpha k/m\right). \qquad (9.4)
\end{aligned}$$

Under (Superunif) and (Indep), this is lower than or equal to $\alpha|\mathcal{H}_0(P)|/m \leq \alpha$ by using the *Simes inequality*. More generally, the Simes post-hoc bound is valid in any setting where the Simes inequality holds. This is the case under a specific positive dependence assumption called Positive Regression Dependency on a Subset of hypotheses (PRDS), which is also the assumption under which the Benjamini-Hochberg (BH) procedure has been shown to control the false discovery rate (FDR).

While it uses more stringent assumptions, $V^{\text{Sim}}(S)$ can be much less conservative than $V^{k_0\text{Bonf}}$. For instance, if $S = \{i \in \mathbb{N}_m : 5\alpha/m \leq p_i(X) < 10\alpha/m\}$, we have $V^{5\text{Bonf}}(S) = |S|$ and $V^{\text{Sim}}(S) \leq |S| \wedge 9$, which can lead to a substantial improvement. Coming back to the motivating example of Section 9.1, the Simes post-hoc bound (9.3) ensures that with probability at least 90%, the number of false positives among the 163 genes selected by the BH procedure at level $q = 0.05$ is upper bounded by 78.

From Example 9.2 below, the Simes bound has a nice graphical interpretation: $|S| - V^{\text{Sim}}(S)$ can be interpreted as the smallest integer u for which the shifted line $v \mapsto \alpha(v-u)/m$ is strictly below the ordered p-value curve, see Figure 9.2.

Example 9.2 *Starting from (9.3) and writing $|S| - V^{\text{Sim}}(S) \leq u$ for some u, one can show that for all $S \subset \mathbb{N}_m$, $|S| - V^{\text{Sim}}(S)$ is equal to*

$$\min\{u \in \{0,\dots,|S|\} : \forall v \in \{u+1,\dots,|S|\} : p_{(v:S)} \geq \alpha(v-u)/m\}, \tag{9.5}$$

where $p_{(1:S)},\dots,p_{(|S|:S)}$ denote the ordered p-values of $\{p_i(X), i \in S\}$.

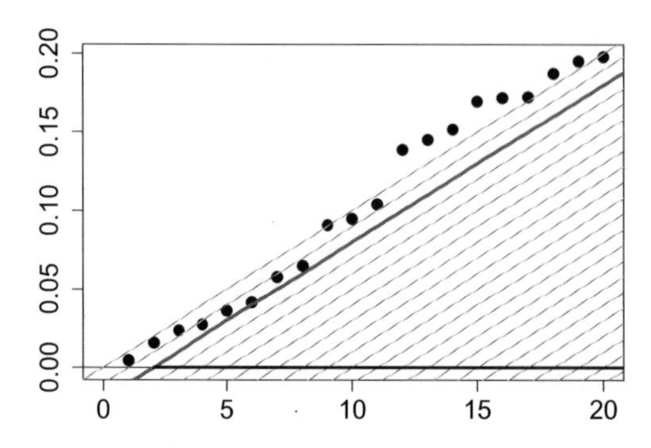

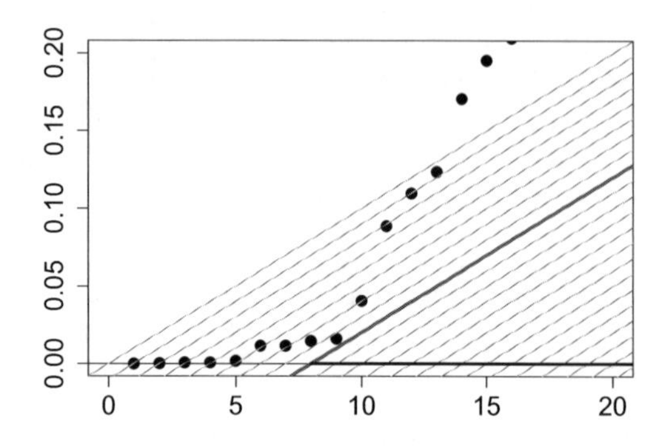

FIGURE 9.2
Illustration of the Simes post-hoc bound (9.3) according to the expression (9.5), for two subsets of $\mathbb{N}_m$ (left display/right display), both of cardinal 20 and for $m = 50$. The level is $\alpha = 0.5$ (taken large only for illustration purposes). Dots: sorted p-values in the respective subsets. Lines: thresholds $k \in \{u+1, \ldots, |S|\} \mapsto \alpha(k-u)/m$ (in bold for $u = |S| - V(S)$). The post-hoc bound $V^{\text{Sim}}(S)$ corresponds the length of the bold line on the X-axis.

Example 9.3 *In Figure 9.2, we have* $V^{\text{Sim}}(S) = 18$ *(resp.* $V^{\text{Sim}}(S) = 12$*) in the left (resp. right) situation. Instead,* $V^{k_0\text{Bonf}}$ *for* $k_0 = 7$ *is equal to* 18 *(resp.* 16*).*

The Simes post-hoc bound (9.3) has, however, some limitations: first, the coverage is only valid when the Simes inequality holds. This imposes restrictive conditions on the model used, which are rarely met or provable in practice. As noted above, the same caveat applies to the BH procedure.

Second, even in that case, the bound does not incorporate the dependence structure, which may yield conservativeness (see Example 9.4 below). Finally, this bound intrinsically compares the ordered p-values to the threshold $k \mapsto \alpha k/m$ (possibly shifted). We can legitimately ask whether taking a different threshold (called template below) does not provide a better bound.

Example 9.4 *Consider the case* $\mathcal{H}_0(P) = \mathbb{N}_m$*, for which* m *is even, and denote* $\overline{\Phi}$ *the upper-tail distribution function of a standard* $\mathcal{N}(0,1)$ *variable. Consider the one-sided testing situation where* $p_i = \overline{\Phi}(X_1)$, $1 \leq i \leq m/2$ *and* $p_i = \overline{\Phi}(X_2)$, $m/2+1 \leq i \leq m$*, for a 2-dimensional Gaussian vector* (X_1, X_2) *that is centered, with covariance matrix having* 1 *as diagonal elements and* $\rho \in [-1,1]$ *as off-diagonal elements. In this case, one can show that the coverage probability of the Simes post-hoc bound is equal to*

$$\alpha/2 + \int_{\alpha/2}^{\alpha} \overline{\Phi}\left(\frac{\overline{\Phi}^{-1}(\alpha) - \rho\overline{\Phi}^{-1}(w)}{(1-\rho^2)^{1/2}}\right) dw + \int_{\alpha}^{\infty} \overline{\Phi}\left(\frac{\overline{\Phi}^{-1}(\alpha/2) - \rho\overline{\Phi}^{-1}(w)}{(1-\rho^2)^{1/2}}\right) dw \tag{9.6}$$

The above quantity is displayed in Figure 9.3 for $\alpha = 0.2$*, as a function of* ρ*.*

9.5 Threshold-Based Post-Hoc Bounds

This section presents the λ-calibration method, which allows to derive more accurate threshold-based post-hoc bounds under mild assumptions. This is of major interest from

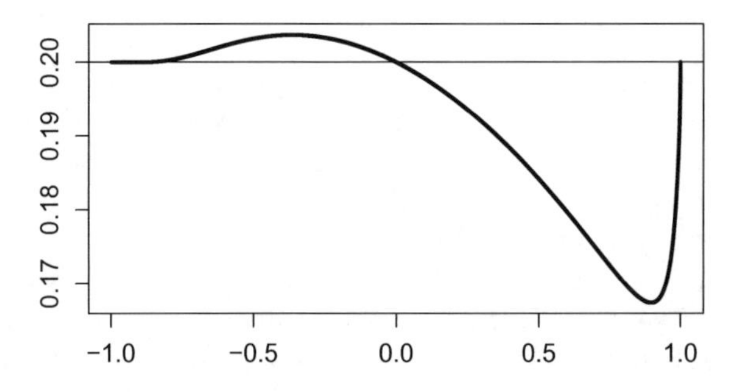

FIGURE 9.3
Coverage of the Simes post-hoc bound (9.6) in the setting of Example 9.4 as a function of ρ and for $\alpha = 0.2$.

a practical perspective since these assumptions are met in the two-sample multiple testing setting, which is often encountered in applications.

Let us consider bounds of the form

$$V^\lambda(S) = \min_{1\leq k\leq |S|} \left\{ \sum_{i\in S} \mathbf{1}\left\{p_i(X) \geq t_k(\lambda)\right\} + k - 1 \right\}, \quad \lambda \in [0,1], \tag{9.7}$$

where $t_k(\lambda)$, $\lambda \in [0,1]$, $1 \leq k \leq m$, is a family of functions, called a template. A template can be seen as a spectrum of curves, parametrized by λ. We focus here on the two following examples:

- Linear template: $t_k(\lambda) = \lambda k/m$, $t_k^{-1}(y) = ym/k$;
- Beta template: $t_k(\lambda) =\lambda$-quantile of $\beta(k, m-k+1)$, $t_k^{-1}(y) = \mathbb{P}(\beta(k, m-k+1) \leq y)$.

An illustration for the above templates is provided in Figure 9.4.

For a fixed template, the idea is now to choose one of these curves, that is, one value of the parameter $\lambda = \lambda(\alpha)$, so that the overall coverage is larger than $1-\alpha$. Following exactly the same reasoning as the one leading to (9.4), we obtain

$$\begin{aligned} \mathbb{P}(\exists S \subset \mathbb{N}_m \,:\, & |S \cap \mathcal{H}_0(P)| \geq V^\lambda(S)+1) \\ &\leq \mathbb{P}\left(\exists k \in \{1,\ldots,|\mathcal{H}_0(P)|\} \,:\, p_{(k:\mathcal{H}_0(P))} < t_k(\lambda)\right) \qquad (9.8) \\ &= \mathbb{P}\left(\min_{k\in\{1,\ldots,|\mathcal{H}_0(P)|\}} \left\{t_k^{-1}(p_{(k:\mathcal{H}_0(P))})\right\} < \lambda\right), \qquad (9.9) \end{aligned}$$

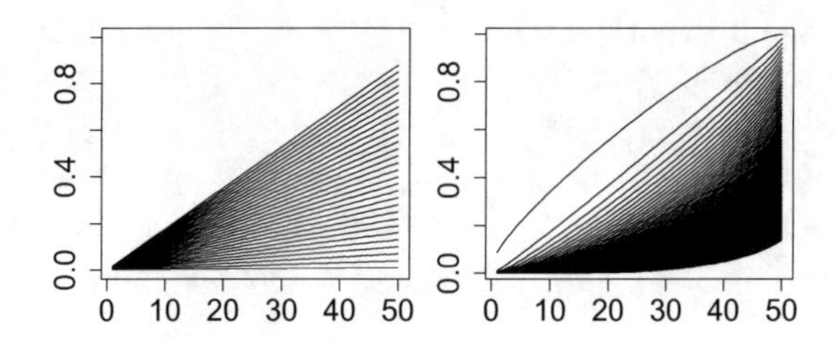

FIGURE 9.4
Curves $k \mapsto t_k(\lambda)$ for a wide range of λ values. Left: Linear template. Right: Beta template.

by letting $t_k^{-1}(y) = \max\{x \in [0,1] : t_k(x) \le y\}$ the generalized inverse of t_k (in general, this is valid provided that for all $k \in \{1,\ldots,m\}$, $t_k(0) = 0$ and $t_k(\cdot)$ is non-decreasing and left-continuous on $[0,1]$, as in the case of the two above examples). What remains to be done is thus to calibrate $\lambda = \lambda(\alpha, X)$ such that the quantity (9.9) is below α.

Several approaches can be used for this. It is possible that for the model under consideration, the joint distribution of $(p_i(X))_{i\in\mathcal{H}_0(P)}$ is equal to the restriction of some known, fixed distribution on $[0,1]^{\mathbb{N}_m}$ to the coordinates of $\mathcal{H}_0(P)$ (this is a version of the so-called subset-pivotality condition). It is met under condition (Indep), but it is also possible that the dependence structure of the p-values is known (e.g., in genome-wide association studies, the structure and strength of linkage disequilibrium can be tabulated from previous studies and give rise to a precise dependence model). In such a situation, the calibration of $\lambda = \lambda(\alpha, X)$ can be obtained either by exact computation, numerical approximation or Monte-Carlo approximation under the full null.

Another situation of interest, on which we focus for the remainder of this section, is when the null corresponds to an invariant distribution with respect to a certain group of data transformations, which is the setting for (generalized) permutation tests, allowing for the use of an exact randomization technique. More precisely, assume the existence of a finite transformation group $\mathcal{G}$ acting onto the observation set $\mathcal{X}$. By denoting $p_{\mathcal{H}_0}(x)$ the null p-value vector $(p_i(x))_{i\in\mathcal{H}_0(P)}$ for $x \in \mathcal{X}$, we assume that the joint distribution of the transformed null p-values is invariant under the action of any $g \in \mathcal{G}$, that is,

$$\forall P \in \mathcal{P},\ \forall g \in \mathcal{G},\ (p_{\mathcal{H}_0}(g'.X))_{g'\in\mathcal{G}} \sim (p_{\mathcal{H}_0}(g'.g.X))_{g'\in\mathcal{G}}, \qquad \text{(Rand)}$$

where $g.X$ denotes X that has been transformed by g.

Let us consider a (random) $B-$tuple $(g_1, g_2, \ldots, g_B)$ of $\mathcal{G}$ (for some $B \ge 2$), where g_1 is the identity element of $\mathcal{G}$ and $g_2, \ldots, g_B$ have been drawn (independently of the other variables) as i.i.d. variables, each being uniformly distributed on $\mathcal{G}$. Now, let for all $x \in \mathcal{X}$, $\Psi(x) = \min_{1\le k\le m}\left\{t_k^{-1}\left(p_{(k:m)}(x)\right)\right\}$ and consider $\lambda(\alpha, X) = \Psi_{(\lfloor \alpha B\rfloor+1)}$ where $\Psi_{(1)} \le \Psi_{(2)} \le \cdots \le \Psi_{(B)}$ denote the ordered sample $(\Psi(g_j.X), 1 \le j \le B)$. The following result holds.

Theorem 9.1 *Under* (Rand), *for any deterministic template, the bound* $V^{\lambda(\alpha,X)}$ *is a post-hoc bound of coverage* $1-\alpha$. *This level is to be understood as a joint probability with respect to the data and the draw of the group elements* $(g_i)_{2\le i\le B}$.

As a case in point, let us consider a two-sample framework where

$$X = (X^{(1)}, \ldots, X^{(n_1)}, X^{(n_1+1)}, \ldots, X^{(n_1+n_2)}) \in (\mathbb{R}^m)^n$$

is composed of $n = n_1 + n_2$ independent m-dimensional real random vectors with $X^{(j)}$, $1 \le j \le n_1$, i.i.d. $\mathcal{N}(\theta^{(1)}, \Sigma)$ (case) and $X^{(j)}$, $n_1+1 \le j \le n$, i.i.d. $\mathcal{N}(\theta^{(2)}, \Sigma)$ (control). Then we aim at testing the null hypotheses $H_{0,i}$: "$\theta_i^{(1)} = \theta_i^{(2)}$", simultaneously for $1 \le i \le m$, without knowing the covariance matrix Σ. Consider any family of p-values $(p_i(X))_{1\le i\le m}$ such that $p_i(X)$ only depends on the i-th coordinate $(X_i^{(j)})_{1\le j\le n}$ of the observations (e.g., based on difference of the coordinate means of the two groups). Note that $p_{\mathcal{H}_0}(X)$ is thus a measurable function of $(X_i^{(j)})_{i\in\mathcal{H}_0, 1\le j\le n}$. Now, the group $\mathcal{G}$ of permutations of $\{1,\ldots,n\}$ is naturally acting on $\mathcal{X} = (\mathbb{R}^m)^n$ via the permutation of the individuals: for all $\sigma \in \mathcal{G}$,

$$\sigma.X = (X^{(\sigma(1))}, \ldots, X^{(\sigma(n_1))}, X^{(\sigma(n_1+1))}, \ldots, X^{(\sigma(n))}).$$

Since the variables $(X_i^{(1)})_{i\in\mathcal{H}_0}, \ldots, (X_i^{(n)})_{i\in\mathcal{H}_0}$ are i.i.d., it is clear that (Rand) holds in this case.

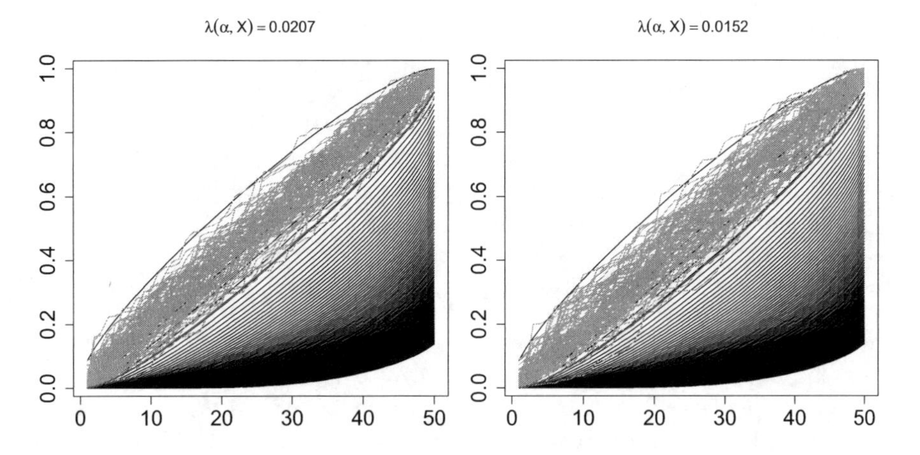

FIGURE 9.5
Illustration of the $\lambda = \lambda(\alpha, X)$ calibration method on one realization of the data X. Gray curves: ordered p-values (after permutation) $k \mapsto p_{(k:m)}(g_j.X)$ for $1 \leq j \leq B = 1000$. Smooth curves: Beta template $k \mapsto t_k(\lambda)$ for some range of λ values, with $k \mapsto t_k(\lambda(\alpha, X))$ in bold. Left: Full null, Right: half of true nulls (see text). (Parameters $m = 50$, $\alpha = 0.2$, $n_1 = 50$, $n_2 = 50$, $\delta = 3$.)

The practical interest of Theorem 9.1 is illustrated in Section 9.9 for the differential gene expression study introduced in Section 9.1, and in Section 9.10 for functional magnetic resonance imaging (fMRI) data. These numerical results demonstrate that substantial gains in power may be obtained by λ-calibration: in both cases, the lower bounds on the number of true positives are two to three times higher than with the classical Simes bounds.

An illustration of the above λ-calibration method is provided in Figure 9.5 in the case where $\Sigma = I_m$,

$$p_i(X) = 2\left(1 - \Phi\left(s_{n_1,n_2}^{-1}\left|n_2^{-1}\sum_{j=n_1+1}^{n_1+n_2} X_i^{(j)} - n_1^{-1}\sum_{j=1}^{n_1} X_i^{(j)}\right|\right)\right),$$

for $s_{n_1,n_2} = (n_1^{-1} + n_2^{-1})^{1/2}$ and using a Beta template. In the left panel (full null), we have $\theta^{(1)} = \theta^{(2)} = 0$, so that $\mathcal{H}_0(P) = \mathbb{N}_m$. In the right panel (half of true nulls), we have $\theta_i^{(1)} = \theta_i^{(2)} = 0$ for $1 \leq i \leq m/2$ and $\theta_i^{(1)} = 0$, $\theta_i^{(2)} = \delta/s_{n_1,n_2}$ for $m/2+1 \leq i \leq m$, for some $\delta > 0$, so that $\mathcal{H}_0(P) = \{1, \ldots, m/2\}$. Following expression (9.8), $k \mapsto t_k(\lambda(\alpha, X))$ is the highest Beta curve such that at most $B\alpha$ sorted p-value curves (in gray) have a point situated below it. This also shows that the above λ-calibration is slightly more severe when part of the data follows the alternative distribution. This is a commonly observed phenomenon: although the permutation approach is valid even when part of the null hypotheses are false, their inclusion in the permutation procedure tends to yield test statistics that exhibit more variation under permutation, thus inducing more conservativeness in the calibration.

9.6 Reference Families

We cast the previous bounds in a more general setting, where $(1-\alpha)$–post-hoc bounds are explicitly based on a *reference family* with some *joint error rate* (JER in short) controlling property. This general point of view offers more flexibility and allows us to consider post-hoc bounds of a different nature, as for instance those incorporating a spatial structure, see Section 9.8.

In general, a reference family is defined by a collection $\mathfrak{R} = \big((R_1(X), \zeta_1(X)), \ldots, (R_K(X), \zeta_K(X))\big)$, where the R_k's are data-dependent subsets of $\mathbb{N}_m$ and the ζ_k's are data dependent integer numbers (we will often omit the dependence in X to ease notation). The reference family $\mathfrak{R}$ is said to control the JER at level α if

$$\forall P \in \mathcal{P}, \qquad \mathbb{P}_{X\sim P}(\forall k \in \mathbb{N}_K \, : \, |R_k(X) \cap \mathcal{H}_0| \leq \zeta_k(X)) \geq 1 - \alpha. \tag{9.10}$$

Markedly, (9.10) is similar to (9.1), but restricted to some subsets R_k, $k \in \mathbb{N}_K$. The rationale behind this approach is that, while the choice of S is let completely free in (9.1) (to accommodate any choice of the practitioner), the choice of the R_k's and ζ_k's in (9.10) is done by the statistician and is part of the procedure. Once we obtain a reference family $\mathfrak{R}$ satisfying (9.10), we obtain a post-hoc bound by interpolation:

$$V^*_{\mathfrak{R}}(S) = \max\{|S \cap A| \, : \, A \subset \mathbb{N}_m, \forall k \in \mathbb{N}_K, |R_k \cap A| \leq \zeta_k\}, \quad S \subset \mathbb{N}_m\,. \tag{9.11}$$

We call $V^*_{\mathfrak{R}}$ the *optimal post-hoc bound* (built upon the reference family $\mathfrak{R}$). Computing the bound $V^*_{\mathfrak{R}}(S)$ can be time-consuming, it has NP-hard complexity in a general configuration. We can introduce the following computable relaxations: for $S \subset \mathbb{N}_m$,

$$\overline{V}_{\mathfrak{R}}(S) = \min_{k \in \mathbb{N}_K} (|S \setminus R_k| + \zeta_k) \wedge |S|; \tag{9.12}$$

$$\widetilde{V}_{\mathfrak{R}}(S) = \left(\sum_{k \in \mathbb{N}_K} |S \cap R_k| \wedge \zeta_k + \left| S \setminus \bigcup_{k \in \mathbb{N}_K} R_k \right| \right) \wedge |S|. \tag{9.13}$$

One can easily check that $V^*_{\mathfrak{R}}(S) \leq \overline{V}_{\mathfrak{R}}(S)$ and $V^*_{\mathfrak{R}}(S) \leq \widetilde{V}_{\mathfrak{R}}(S)$ for all $S \subset \mathbb{N}_m$. Moreover, provided that (9.10) holds, $V^*_{\mathfrak{R}}$, $\overline{V}_{\mathfrak{R}}$ and $\widetilde{V}_{\mathfrak{R}}$ are all valid $(1-\alpha)$–post-hoc bounds. The details are left to the reader.

In addition, the following result shows that the relaxed versions coincide with the optimal bound if the reference sets have some special structure:

Lemma 9.1

- *In the nested case, that is, $R_k \subset R_{k+1}$, for $1 \leq k \leq K-1$, we have $\overline{V}_{\mathfrak{R}} = V^*_{\mathfrak{R}}$;*
- *In the disjoint case, that is, $R_k \cap R_{k'} = \emptyset$ for $1 \leq k \neq k' \leq K$, we have $\widetilde{V}_{\mathfrak{R}} = V^*_{\mathfrak{R}}$.*

We can briefly revisit the post-hoc bounds of the previous sections in this general framework. The k_0-Bonferroni post-hoc bound (9.2) derives from the one-element reference family $(R = \{i \in \mathbb{N}_m : p_i(X) < \alpha k_0/m\}, \zeta = k_0 - 1)$. The Simes post-hoc bound (9.3) derives from the reference family comprising the latter reference sets for all $k_0 \in \mathbb{N}_m$. More generally, the threshold-based post-hoc bounds V^λ of the form (9.7) are equal to the optimal bound $V^*_{\mathfrak{R}}$ with $R_k = \{i \in \mathbb{N}_m \, : \, p_i(X) < t_k(\lambda)\}$ and $\zeta_k = k-1$, $k \in \mathbb{N}_m$(indeed, these reference sets are nested so that $V^*_{\mathfrak{R}} = \overline{V}_{\mathfrak{R}}$).

How to choose a suitable reference family in general? A general rule of thumb is to choose the reference sets R_k of the same qualitative form as the sets S for which the bound is expected to be accurate. For instance, the Simes post-hoc bound will be more accurate for sets S with the smallest p-values. In Section 9.8, we will choose reference sets R_k with a spatial structure, which will produce a post-hoc bound more tailored for spatially structured subsets S.

9.7 Case of a Fixed Single Reference Set

It is useful to focus first on the case of a single fixed (non-random) reference set R_1, with (random) ζ_1 satisfying (9.10), that is,

$$\mathbb{P}(|\mathcal{H}_0(P) \cap R_1| \leq \zeta_1(X)) \geq 1 - \alpha.$$

(In contrast with the k_0-Bonferroni bound (9.2) where ζ was fixed and R variable, here R_1 is fixed and ζ_1 is variable.) In other words, $\zeta_1(X)$ is a $(1-\alpha)$–confidence bound of $|\mathcal{H}_0(P) \cap R_1|$. Several examples of such $\zeta_1(X)$ can be built, under various assumptions.

Example 9.5 *For $R_1 \subset \mathbb{N}_m$ fixed, the following bounds are $(1-\alpha)$–confidence bounds for $|\mathcal{H}_0(P) \cap R_1|$:*

- *under* (Superunif), *for some fixed $t \in (0, \alpha)$,*

$$\zeta_1(X) = |R_1| \wedge \left\lfloor \sum_{i \in R_1} \mathbf{1}\{p_i(X) > t\} / (1 - t/\alpha) \right\rfloor, \tag{9.14}$$

  where $\lfloor x \rfloor$ denotes the largest integer smaller than or equal to x (this is a simple application of the Markov inequality).

- *under* (Superunif) *and* (Indep),

$$\zeta_1(X) = |R_1| \wedge \min_{t \in [0,1)} \left\lfloor \frac{C}{2(1-t)} + \left(\frac{C^2}{4(1-t)^2} + \frac{\sum_{i \in R_1} \mathbf{1}\{p_i(X) > t\}}{1-t} \right)^{1/2} \right\rfloor^2, \tag{9.15}$$

  where $C = \sqrt{\frac{1}{2} \log\left(\frac{1}{\alpha}\right)}$ (this can deduced by using the DKW inequality, that is, for any integer $n \geq 1$, for $U_1, \ldots, U_n$ i.i.d. $U(0,1)$, we have $\sup_{t \in [0,1]} \left\{ n^{-1} \sum_{i=1}^n \mathbf{1}\{U_i > t\} -(1-t) \right\} \geq -\sqrt{\log(1/\lambda)/(2n)}$ with probability at least $1 - \lambda$).

In addition to the two above bounds (9.14) and (9.15), we can elaborate another bound in the generalized permutation testing framework (Rand), as described in Section 9.5. Applying the result of that section, the following bound is also valid:

$$\zeta_1(X) = \min_{1 \leq k \leq |R_1|} \left\{ \sum_{i \in R_1} \mathbf{1}\{p_i(X) \geq t_k(\lambda(\alpha, X))\} + k - 1 \right\}, \tag{9.16}$$

where $t_k(\lambda)$ denotes the λ-quantile of a $\beta(k, |R_1| - k + 1)$ distribution and $\lambda(\alpha, X) = \Psi_{(\lfloor \alpha B \rfloor + 1)}$, where $\Psi_{(1)} \leq \Psi_{(2)} \leq \cdots \leq \Psi_{(B)}$ denote the ordered sample $(\Psi(g_j.X), 1 \leq j \leq B)$ for which $\Psi(x) = \min_{1 \leq k \leq |R_1|} \left\{ t_k^{-1}\left(p_{(k:|R_1|)}(x)\right) \right\}$ (see the λ-calibration method of Section 9.5).

Once a proper choice of $\zeta_1(X)$ has been done, the optimal post-hoc bound can be computed as follows: for any $S \subset \mathbb{N}_m$, $V^*_{\mathfrak{R}}(S) = \overline{V}_{\mathfrak{R}}(S) = \widetilde{V}_{\mathfrak{R}}(S) = |S \cap R_1^c| + \zeta_1(X) \wedge |S \cap R_1|$. When S is large and does not contain very small p-values, this bound can be sharper than the Simes bound. For instance, let us consider the single reference family $R_1 = \mathbb{N}_m$ and $\zeta_1(X)$ as in (9.15) (choosing $t = 1/2$). For S such that $S \subset \{i \in \mathbb{N}_m : p_i(X) > \alpha |S|/m\}$, we have $V^{\mathrm{Sim}}(S) = |S|$ and $V^*_{\mathfrak{R}}(S) = |S| \wedge \zeta_1(X) \leq |S| \wedge 2\left(\log\left(\frac{1}{\alpha}\right) + 2\sum_{i \in \mathbb{N}_m} \mathbf{1}\{p_i(X) > 1/2\}\right)$. The latter can be smaller than $|S|$ when many p-values are below $1/2$.

Finally, while the case of a single reference set can be considered as an elementary example, the bounds developed in this section will be useful in the next section, for which several fixed reference sets R_k are considered, and thus several (random) ζ_k should be designed.

9.8 Case of Spatially Structured Reference Sets

We consider here the case where the null hypotheses $H_{0,i}$, $1 \leq i \leq m$, have a spatial structure, and we are interested in obtaining accurate bounds on $|S \cap \mathcal{H}_0(P)|$ for subsets S of the form $S = \{i \in \mathbb{N}_m : i_0 \leq i \leq j_0\}$, for some $1 \leq i_0 < j_0 \leq m$.

In that case, it is natural to choose R_k formed of contiguous indices. To be concrete, consider reference sets consisting of disjoint intervals of the same size : assume $m = Ks$ for some integers $K > 0$ and $s > 0$ and let

$$R_k = \{(k-1)s+1, \ldots, ks\},\ k \in \mathbb{N}_K. \tag{9.17}$$

When each of these regions is considered in isolation, Section 9.7 suggested several approaches (in the appropriate settings (Superunif), (Indep) or (Rand)) of a specific form $\zeta_k(X) = f(R_k, \alpha, X)$, to underline the dependence of $\zeta_k(X)$ in R_k and α. By using a simple union bound, it is then straightforward to show that the JER control (9.10) is satisfied for

$$\zeta_k(X) = f(R_k, \alpha/K, X),\ k \in \mathbb{N}_K. \tag{9.18}$$

When the reference regions R_k are disjoint as in the example (9.17) above, we can use the proxy $\widetilde{V}_{\mathfrak{R}}(S)$ (see (9.13)) which is known to coincide with the optimal bound $V^*_{\mathfrak{R}}(S)$. This gives rise to a post-hoc bound that accounts for the spatial structure of the data.

Example 9.6 *In the case where $\zeta_1(X) = f(R_1, \alpha, X)$ is given by (9.14) ($t = \alpha^2$, $K < 1/\alpha$), we obtain*

$$\zeta_k(X) = |R_k| \wedge \left\lfloor \sum_{i \in R_k} \mathbf{1}\left\{p_i(X) > \alpha^2\right\}/(1-\alpha K)\right\rfloor.$$

Note that this bound quickly increases as the size of the family K increases. By contrast, when $\zeta_1(X) = f(R_1, \alpha, X)$ is given by (9.15), one can derive

$$\zeta_k(X) = |R_k| \wedge \min_{t \in [0,1)} \left\lfloor \frac{C^2}{2(1-t)} + \left(\frac{C^2}{4(1-t)^2} + \frac{\sum_{i \in R_k} \mathbf{1}\{p_i(X) > t\}}{1-t}\right)^{1/2}\right\rfloor^2,$$

for $C = \sqrt{\frac{1}{2}\log\left(\frac{K}{\alpha}\right)}$. The size of the family K appears here in a logarithmic term, which makes this bound less sensitive to the parameter K.

When considering the reference regions defined by segments (9.17), we have to prescribe a scale (s here, the size of the segments). It is possible to extend this to a multiscale approach, choosing overlapping reference intervals R_k at different resolutions arranged in a tree structure, where parent sets are formed by taking union of (disjoint) children sets taken at a finer resolution. Furthermore, the proxy (9.13) has to be replaced by a more elaborate one, minimizing over all possible multi-scale partitions made of such reference regions. This can still be computed efficiently by exploiting the the tree structure. Doing so, the post-hoc bound will be more scale adaptive to sets S with possibly various sizes. The price to pay lies in the cardinality K of the family, which gets larger. However, this does not necessarily make the corresponding bound much larger, as Example 9.6 shows when using the bound (9.15), since the level α only enters it logarithmically.

9.9 Application to Differential Gene Expression Studies

In this section, we illustrate how the post-hoc inference framework introduced in the preceding sections can be applied to the case of differential gene expression introduced in Section 9.1 to build confidence envelopes for the proportion of false positives (Section 9.9.1), and to obtain bounds on data-driven sets of hypotheses (Section 9.9.2), and on sets of hypotheses defined by an a priori structure (Section 9.9.3). These numerical results were obtained using the R package `sansSouci`, version 0.9.0. A Rmarkdown vignette [3] to reproduce results and plots from this section is provided as Supplementary Material.

9.9.1 Confidence Envelopes

In the absence of specific prior information on relevant subsets of hypotheses to consider, it is natural to focus on subsets consisting of the most significant hypotheses. Specifically, we define the k−th p-value level set S_k as the set of the k most significant hypotheses, corresponding to the p-values $(p_{(1:m)}, p_{(2:m)}, \ldots, p_{(k:m)})$, and consider post-hoc bounds associated to S_k for $k \in \mathbb{N}_m$. Figure 9.6 provides *post-hoc confidence envelopes* for the ALL data set, for $\alpha = 0.1$. While $(1-\alpha)$-lower confidence bounds on the proportion of false positives $\left\{\left(k, \overline{V}(S_k)/\left|S_k\right|\right) : k \in \mathbb{N}_m\right\}$ are displayed in the left panel, $(1-\alpha)$-upper confidence bounds on the number of true positives of the form $\left\{\left(k, |S_k| - \overline{V}(S_k)\right) : k \in \mathbb{N}_m\right\}$ are shown in the right panel.

The confidence envelopes are built from the Simes bound (9.3) (long-dashed dark gray curve), and from two bounds obtained from Theorem 9.1 by λ-calibration using $B = 1,000$ permutation of the sample labels, based on the two templates introduced in Section 9.5: the dashed black curve corresponds to the linear template with $K = m$, and the solid light gray curve to the Beta template with $K = 50$. Note that Assumption (Rand) holds because we are in the two-sample framework described after Theorem 9.1.

The vertical line in Figure 9.6 corresponds to the 163 genes selected by the BH procedure at level 5%. The Simes bound ensures that the FDP of this subset is not larger than 0.48. As noted above concerning the BH procedure, we have a priori no guarantee that this bound is valid because such multiple two-sample testing situations have not been shown to satisfy the PRDS assumption under which the Simes inequality is valid[4]. In contrast, the λ-calibrated bounds built by permutation are by construction valid here. Moreover, both are much sharper than the Simes bound while the λ-calibrated bound using the linear template is twice smaller, ensuring FDP< 0.23, and even smaller for the Beta template with $K = 50$. The bound obtained by λ-calibration of the linear template is uniformly sharper that the original Simes bound (9.3), which corresponds to $\lambda = \alpha$. This illustrates the adaptivity to dependence achieved by λ-calibration. The bound obtained from the Beta template is less sharp for p-value level sets S_k of cardinal less than $k = 120$, and then sharper. This is consistent with the shape of the threshold functions displayed in Figure 9.4.

9.9.2 Data-Driven Sets

A common practice in the biomedical literature is to only retain, among the genes called significant after multiple testing correction, those whose "fold change" exceeds a prescribed level. The fold change is the ratio between the mean expression levels of the two groups.

[3]See https://github.com/rstudio/rmarkdown.

[4]In this particular case, λ-calibration with the linear template yields $\lambda(\alpha) > \alpha$, which a posteriori implies that the Simes inequality was indeed valid.

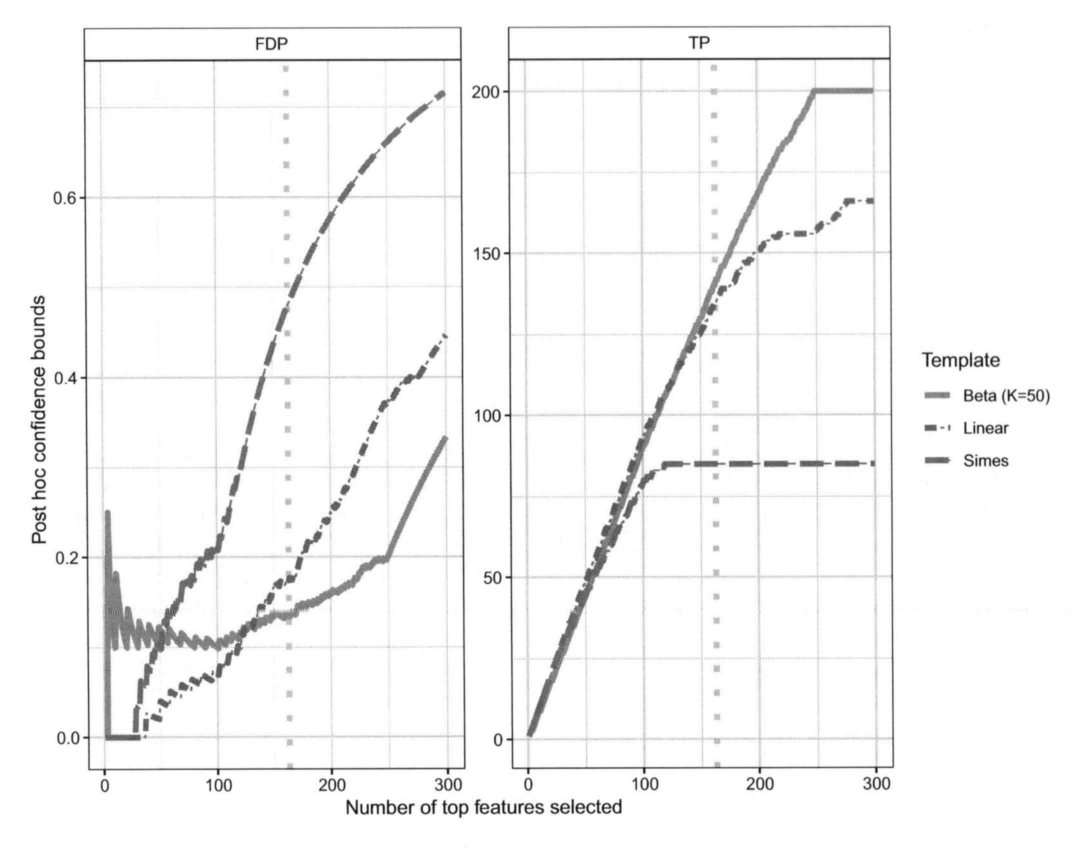

FIGURE 9.6
Confidence bounds on the proportion of false positives (left) and on the number of true positives (right) for the Leukemia data set. Reference families: Simes reference family (long-dashed dark gray curve), linear template after λ-calibration (dashed black curve), and Beta template after λ-calibration (solid light gray curve).

With the notation of Section 9.5, the fold-change of gene i is given by $\Delta_i = \overline{X}_i^{(2)}/\overline{X}_i^{(1)}$, where $\overline{X}_i^{(1)} = n_1^{-1}\sum_{j=1}^{n_1} X_i^{(j)}$ and $\overline{X}_i^{(2)} = n_2^{-1}\sum_{j=1}^{n_2} X_i^{(j)}$.

This is illustrated by Figure 9.7, where each gene is represented as a point in the $(\log(\text{fold change}), -\log(p))$ plan. This representation is called a "volcano plot" in the biomedical literature. Among the 163 genes selected by the BH procedure at level 0.05, 151 have an absolute log fold change larger than 0.3. As FDR is not preserved by selection, FDR controlling procedures provide no statistical guarantee on such data-driven lists of hypotheses.

In contrast, the post-hoc bounds proposed in this chapter are valid for such data-driven sets. The two shaded boxes in Figure 9.7 correspond to the data-driven subsets $S^{\text{BH}} \cap S^-$ and $S^{\text{BH}} \cap S^+$, where S^{BH} is the set of 163 genes selected by the BH procedure at level 0.05, $S^- = \{i \in \mathbb{N}_m, \log(\Delta_i) < -0.3\}$ and $S^+ = \{i \in \mathbb{N}_m, \log(\Delta_i) > +0.3\}$. The post-hoc bounds on the number of true positives in $S^{\text{BH}} \cap S^+$, $S^{\text{BH}} \cap S^-$ and $S^{\text{BH}} \cap (S^+ \cup S^-)$ obtained by the Simes bound and by the λ-calibrated linear and Beta templates are given in Table 9.1. Both λ-calibrated bounds are more informative than the Simes bound, in the sense that they provide a higher bound on the number of true confidence. Moreover, they have proven $(1-\alpha)$-coverage, whereas the coverage of the Simes bound is a priori unknown for multiple two-sample tests. None of the two λ-calibrated bounds dominates the other one, which is in

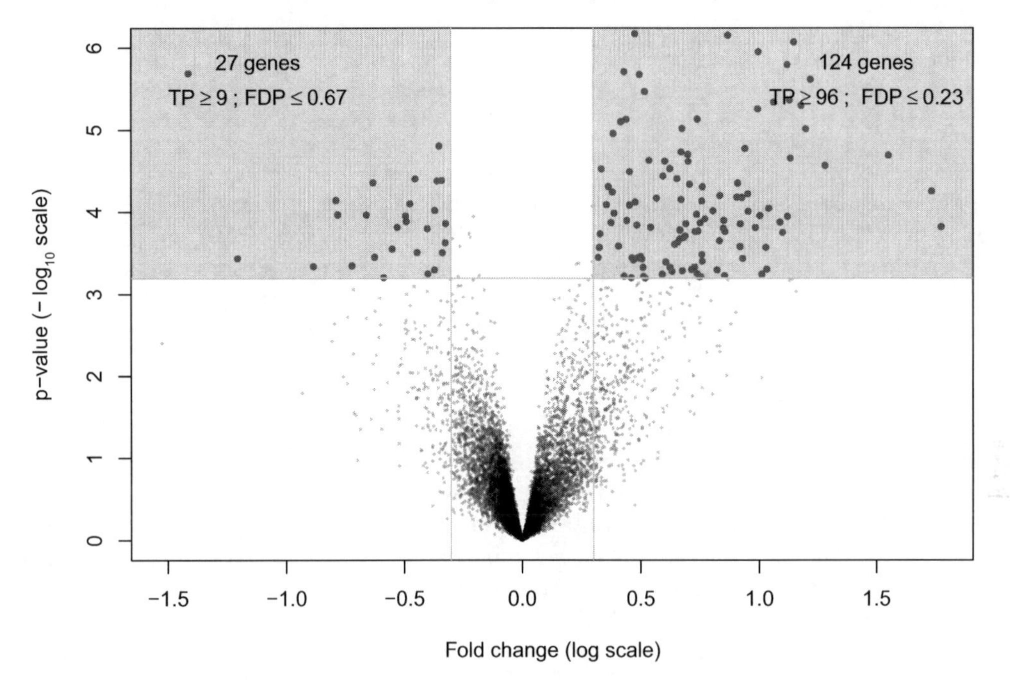

FIGURE 9.7
Post-hoc inference for volcano plots. Each dot corresponds to a gene in the (fold change, p-value space) on a logarithmic scale. Bold dots corresponds to 151 genes that (i) are selected by the BH procedure at level $q = 0.05$, and (ii) have a absolute log fold change larger than 0.3.

TABLE 9.1
Post-Hoc Bounds on the Number of True Positives in $S^{\mathrm{BH}} \cap S^+, S^{\mathrm{BH}} \cap S^-$ and $S^{\mathrm{BH}} \cap (S^+ \cup S^-)$ Obtained by the Post-Hoc Bounds Displayed in Figure 9.6

S	$\lvert S \rvert$	Simes	Linear	Beta(K=50)
$S^{\mathrm{BH}} \cap S^-$	124	62	96	103
$S^{\mathrm{BH}} \cap S^+$	27	1	9	7
$S^{\mathrm{BH}} \cap (S^+ \cup S^-)$	151	79	123	130

line with the fact that the linear template is well-adapted to situations with smaller p-value level sets than the Beta template.

Finally, we also note that the bound obtained for $S^+ \cup S^-$ is systematically larger than the sum of the two individual bounds, which, again, is in accordance with the theory.

9.9.3 Structured Reference Sets

In this section, we give an example of application of the bounds mentioned in Section 9.8. Our biological motivation is the fact that gene expression activity can be clustered along the genome.

The m individual hypotheses are naturally partitioned into 23 subsets, each corresponding to a given chromosome. Within each chromosome, we consider sets of $s = 10$ successive genes as in (9.17). Hence, we focus on a reference family with the following elements

$$R_{c,k} = \{(k-1)s+1, \ldots, \min(ks, m_c)\}, \quad k \in \mathbb{N}_{K_c}, \quad c \in \{1, \ldots, 23\},$$

where, in chromosome c, m_c denotes the number of genes, $K_c = \lceil m_c/s \rceil$ the number of corresponding regions. In addition, for each (c,k) we use $\zeta_{c,k}(X) = f(R_{c,k}, \alpha_c/K_c, X)$ coming from the union bound (9.18) in combination with the device (9.15) and $\alpha_c = \alpha m_c/m$. This choice accounts for a union bound over all the chromosomes. As shown in Example 9.5, $\zeta_{c,k}(X)$ is a valid upper confidence bound for $|\mathcal{H}_0(P) \cap R_{c,k}|$ under (Superunif) and (Indep). In this genomic example, (Indep) may not hold, so we have no formal guarantee that this bound is valid. Therefore, the results obtained below are merely illustrative of the approach and may not have biological relevance.

We report the results for chromosome $c = 19$, which contains $m_c = 626$ genes. In this particular case, we obtain trivial bounds $\zeta_{c,k}(X) = |R_{c,k}|$ for all $k \in \mathbb{N}_{K_c}$. Therefore, the proxy $\tilde{V}^*_{\mathfrak{R}}$ defined in (9.13) for disjoint sets does not identify any signal for this chromosome. However, non-trivial bounds can be obtained via the multiscale approach briefly mentioned in Section 9.8. The idea is to enrich the reference family by recursive binary aggregation of the neighboring $R_{c,k}$. The total number of elements in this family is less than $2K_c$. In our example, it turns out that (9.15) yields six true discoveries in the interval $R_{17:24}$ and one true discovery in the interval $R_{53:54}$, where we have denoted

$$R_{u:v} = \bigcup_{u \leq k \leq v} R_{c,k}.$$

This is illustrated by Figure 9.8 where the individual p-values are displayed (on the $-\log_{10}$ scale) as a function of their order on chromosome 19. The sets $R_{17:24}$ and $R_{53:54}$ are highlighted, with the corresponding number of true discoveries marked in each region.

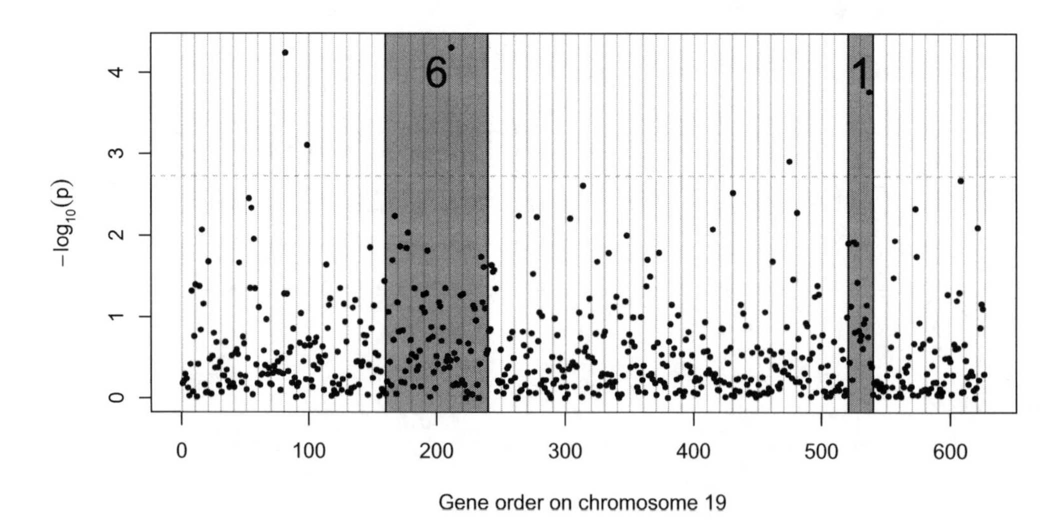

FIGURE 9.8
Evidence of locally-structured signal on chromosome 19 detected by the bound (9.15). The numbers correspond to the lower bound on the false positives in each of the highlighted regions.

We obtain a nontrivial bound not because of the large effect of any individual gene, but because of the presence of sufficiently many moderate effects. In particular, in the rightmost highlighted region in Figure 9.8, the distribution of $-\log_{10}(p)$ is shifted away from 0 when compared to the rest of chromosome 19. In comparison, we obtain trivial bounds $\overline{V}_{\mathfrak{R}}(R_{53:54}) = |R_{53:54}| = 2s$ and $\overline{V}_{\mathfrak{R}}(R_{17:24}) = |R_{17:24}| = 8s$ from (9.12) both for the linear or the Beta template. These numerical results illustrate the interest of the bounds introduced in Section 9.8 in situations where one expects the signal to be spatially structured.

9.10 Application to fMRI Studies

We focus on the problem of detecting brain regions whose activity is significantly different between two motor tasks performed by subjects: left versus right click. The fMRI data have been extracted from the Localizer data set[5]. A Rmarkdown vignette to reproduce results and plots from this section is provided as Supplementary Material.

9.10.1 Confidence Envelopes

As in Section 9.9, we begin by constructing confidence envelopes for top-k feature lists. Figure 9.9 provides *post-hoc confidence envelopes* for the Localizer data set, for $\alpha = 0.1$. While $(1-\alpha)$-lower confidence bounds on the proportion of false positives $\left\{\left(k, \overline{V}(S_k)/\,|S_k|\right) : k \in \mathbb{N}_m\right\}$ are displayed in the left panel, $(1-\alpha)$-upper confidence bounds on the number of true positives of the form $\left\{\left(k, |S_k| - \overline{V}(S_k)\right) : k \in \mathbb{N}_m\right\}$ are shown in the right panel.

The confidence envelopes are built from the Simes bound (9.3) (long-dashed dark gray curve), and from two bounds obtained from Theorem 9.1 by λ-calibration using $B = 1,000$ permutation of the sample labels, based on the two templates introduced in Section 9.5: the dashed black curve corresponds to the linear template with $K = m$, and the solid light gray curve to the Beta template with $K = 500$. Assumption (Rand) holds because we are in the two-sample framework described after Theorem 9.1. The Simes bound is also called All Resolution Inference (ARI) in that context, see Supplementary Material 2 for more details and references.

The results are qualitatively similar as for the genomic example given in Section 9.9. Both permutation-based post-hoc bounds are much sharper than the Simes bound, illustrating the adaptivity to dependence achieved by λ-calibration.

9.10.2 Post-Hoc Bounds on Brain Atlas Areas

The goal in this section is to calculate post-hoc bound on user-defined brain regions. One definition of such regions is given by the Harvard-Oxford brain atlas[6]. We have calculated the post-hoc bounds associated to each of these 48 areas: by definition, these are confidence bounds valid *simultaneously* for all areas. In this particular example, evidence of signal is obtained for three of these atlases, as summarized in Table 9.2.

[5] Orfanos, D. P. et al. *Neuroimage*, 181:786–796 (2017).
[6] https://fsl.fmrib.ox.ac.uk/fsl/fslwiki/Atlases.

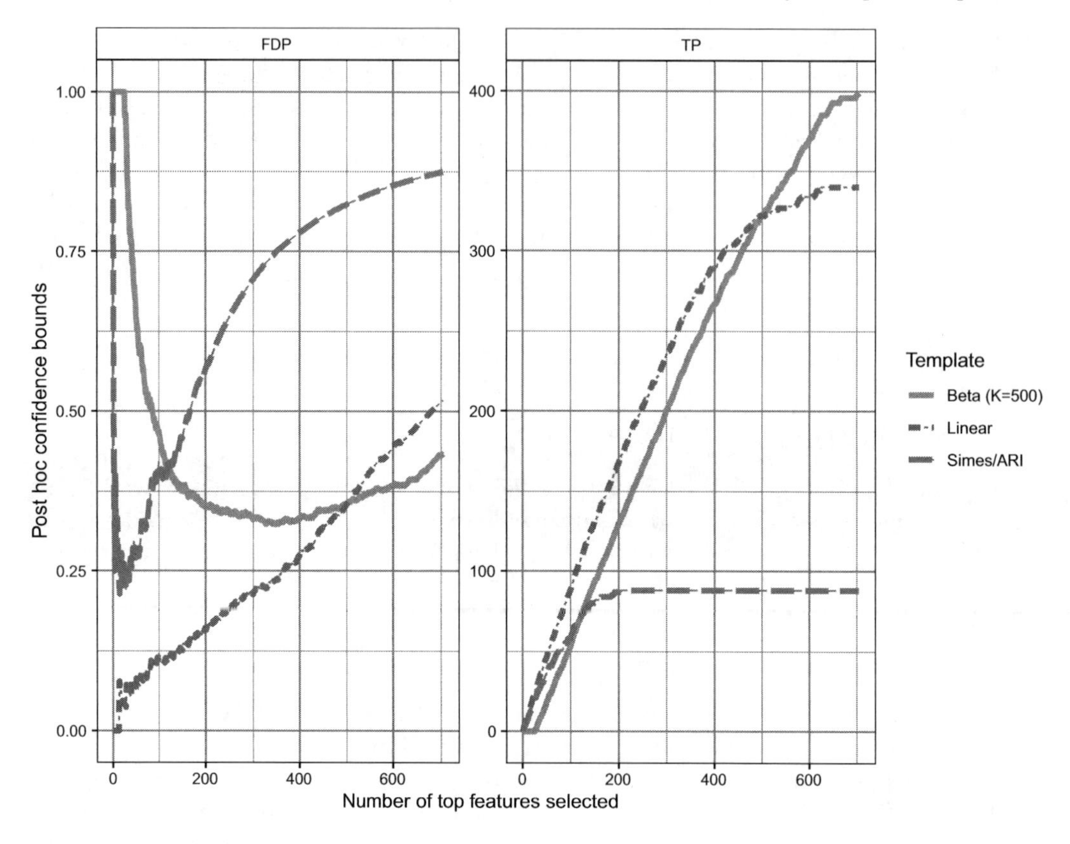

FIGURE 9.9
Confidence bounds on the proportion of false positives (left) and on the number of true positives (right) for the fMRI data set. Reference families: Simes reference family (long-dashed dark gray curve), linear template after λ-calibration (dashed black curve), and Beta template after λ-calibration (solid light gray curve).

TABLE 9.2
Lower Bounds on the Number of True Positives in Three Brain Regions. For the same target risk level ($\alpha = 0.1$), both permutation-based bounds are much less conservative than the classical Simes/ARI bound, showing that permutation-based approaches are able to adapt to unknown dependency.

	Area	Size	Simes/ARI	Linear	Beta (K=500)
7	Precentral Gyrus	6640	79	254	281
17	Postcentral Gyrus	5030	69	202	196
19	Supramarg. Gyrus (ant. div.)	2017	0	3	0

9.11 Bibliographical Notes

The material exposed in this chapter is mainly a digested account of the article Blanchard et al. (2020). The seminal work Genovese and Wasserman (2006) introduced the idea of false positive bounds for arbitrary rejection sets. It started from the idea of building a confidence set on the set of null hypotheses $\mathcal{H}_0(P)$, and introduced the concepts of *augmentation procedure* and *inversion procedure*. The latter consists of building a confidence set based on the inversion of tests for $\mathcal{H}_0(P) = \mathcal{A}$ for all $\mathcal{A} \subset \mathbb{N}_m$. The former starts from a set R with controlled k-family-wise error rate, and the proposed associated post-hoc bound is (9.10) (for the one-element reference family $(R, \zeta = k - 1)$). The name *augmentation* refers to a similar idea found in Dudoit and van der Laan (2008). The relaxation (9.10) can in this sense be called "generalized augmentation procedure". A post-hoc bound for an arbitrary rejection set based on a closed test principle was proposed in Goeman and Solari (2011). It can also be seen as a reformulation of the inversion procedure of Genovese and Wasserman (2006). Post-hoc bounds over a large class of reference families extracted from classical FDR control procedures combined with martingale techniques were recently proposed in Katsevich and Ramdas (2018). The principle of the graphical representation used in Figure 9.2 to visualize the Simes inequality-based bound originates from J. Goeman.

The use of generalized permutation procedures in a multiple testing framework has been explored in several landmark works Dudoit and van der Laan (2008); Hemerik and Goeman (2018); Hemerik et al. (pear); Meinshausen (2006); Romano and Wolf (2005); Westfall and Young (1993). The subset-pivotality condition has been defined in Westfall and Young (1993). Assumption (Rand) has been introduced in Hemerik and Goeman (2017) and is a weaker version of the randomization hypothesis of Romano and Wolf (2005). The phenomenon of conservativeness in the permutation-based calibration mentioned at the end of Section 9.5, when not all the null hypotheses are true, can be in part alleviated by using a step-down principle (see Romano and Wolf (2005) for a seminal work on this topic and Blanchard et al. (2020) for more details on this approach in the specific setting considered here). The choice of the size K of the reference family, which can be crucial in practice, is also discussed in Blanchard et al. (2020).

Multiple testing for spatially structured hypotheses is in itself a very active and broad area of research. It has been specifically considered in conjunction with post-hoc bounds in Meijer et al. (2015). The use of the reference family approach for post-hoc bounds in combination with spatially structured hypotheses has been studied in Durand et al. (2020), where the notion of tree- (or forest-)structured reference regions is introduced, along with an efficient algorithm to compute the optimal bound $V^*_{\mathfrak{R}}$ in this setting.

The Simes inequality Simes (1986) is a particularly nice and elegant theoretical device with manifold applications in multiple testing which is still a very active research area, see, e.g., Block et al. (2013); Bodnar and Dickhaus (2017); Finner et al. (2017). The DKW inequality with optimal constant was proved in Massart (1990). The Benjamini-Hochberg (BH) procedure has been introduced in Benjamini and Hochberg (1995), where it is also proved to control the false discovery rate (FDR). A huge literature on FDR control has followed this seminal paper.

The data used for the genomics application in Section 9.9 are taken from Chiaretti et al. (2005). The fact that the signal is clustered along the genome is motivated by previous studies showing possible links between gene expression and DNA copy number changes or other regulation mechanisms Reyal et al. (2005); Stransky et al. (2006). The data used for the neuroimaging application in Section 9.10 were obtained from the Brainomics/Localizer database Orfanos et al. (2017) and the Harvard-Oxford brain atlas[7], using the Python package `nilearn` Abraham et al. (2014). The ARI method mentioned in that section Rosenblatt et al. (2018) corresponds to the Simes post-hoc bound of Goeman and Solari (2011).

[7]https://fsl.fmrib.ox.ac.uk/fsl/fslwiki/Atlases.

Acknowledgements

We would like to thank referees for their helpful comments. PN would like to thank Alexandre Blain who contributed to the application to fMRI studies, and Bertrand Thirion for constructive feedback on this application. This work has been supported by ANR-16-CE40-0019 (SansSouci), ANR-17-CE40-0001 (BASICS) and by the GDR ISIS through the "projets exploratoires" program (project TASTY). GB acknowledges the support from the german DFG under the Collaborative Research Center SFB-1294 "Data Assimilation". GB and PN acknowledge the support from the Franco-German University through the binational Doktorandenkolleg CDFA 01-18.

Supplementary Materials

Vignette "Permutation-based post-hoc inference for differential gene expression studies":

This Rmarkdown vignette demonstrates how the R package `sansSouci` may be used to obtain post-hoc confidence bounds on false positives in the case of differential gene expression analysis. In particular, it contains the R code to reproduce Figures 9.1, 9.6 and 9.7, and Table 9.1.

Vignette: "Permutation-based post-hoc inference for fMRI studies"

This Rmarkdown vignette demonstrates how the R package `sansSouci` may be used to obtain post-hoc confidence bounds on false positives in the case of functional Magnetic Resonance Imaging (fMRI) studies. In particular, in contains R code to reproduce Figure 9.9 and Table 9.2.

Bibliography

A. Abraham, F. Pedregosa, M. Eickenberg, P. Gervais, A. Mueller, J. Kossaifi, A. Gramfort, B. Thirion, and G. Varoquaux (2014). Machine learning for neuroimaging with scikit-learn. *Frontiers in Neuroinformatics 8*, 14.

Y. Benjamini and Y. Hochberg (1995). Controlling the false discovery rate: a practical and powerful approach to multiple testing. *Journal of the Royal Statistical Society: Series B 57*(1), 289–300.

G. Blanchard, P. Neuvial, and E. Roquain (2020). Post hoc confidence bounds on false positives using reference families. *Annals of Statistics 48*, 1281–1303.

H. W. Block, T. H. Savits, J. Wang, and S. K. Sarkar (2013). The multivariate-t distribution and the Simes inequality. *Statistics & Probability Letters 83*(1), 227–232.

T. Bodnar and T. Dickhaus (2017). On the Simes inequality in elliptical models. *Annals of the Institute of Statistical Mathematics 69*(1), 215–230.

S. Chiaretti, X. Li, R. Gentleman, A. Vitale, K. S. Wang, F. Mandelli, R. Foa, and J. Ritz (2005). Gene expression profiles of b-lineage adult acute lymphocytic leukemia reveal genetic patterns that identify lineage derivation and distinct mechanisms of transformation. *Clinical Cancer Research 11*(20), 7209–7219.

S. Dudoit and M. J. van der Laan (2008). *Multiple Testing Procedures with Applications to Genomics.* Springer Series in Statistics. New York: Springer.

G. Durand, G. Blanchard, P. Neuvial, and E. Roquain (2020). Post hoc false positive control for spatially structured hypotheses. *Scandinavian Journal of Statistics 47*(4), 1114–1148.

H. Finner, M. Roters, and K. Strassburger (2017). On the Simes test under dependence. *Statistical Papers 58*(3), 775–789.

C. R. Genovese and L. Wasserman (2006). Exceedance control of the false discovery proportion. *Journal of the American Statistical Association 101*(476), 1408–1417.

J. J. Goeman and A. Solari (2011). Multiple testing for exploratory research. *Statistiscal Science 26*(4), 584–597.

J. Hemerik and J. Goeman (2018). Exact testing with random permutations. *TEST 27*(4), 811–825.

J. Hemerik and J. J. Goeman (2017). False discovery proportion estimation by permutations: confidence for significance analysis of microarrays. *Journal of the Royal Statistical Society: Series B (Statistical Methodology).*

J. Hemerik, A. Solari, and J. J. Goeman (to appear). Permutation-based simultaneous confidence bounds for the false discovery proportion. *Biometrika.*

E. Katsevich and A. Ramdas (2018). Simultaneous high-probability bounds on the false discovery proportion in structured, regression, and online settings. arXiv preprint 1803.06790.

P. Massart (1990). The tight constant in the Dvoretzky-Kiefer-Wolfowitz inequality. *Annals of Probability 18*(3), 1269–1283.

R. J. Meijer, T. J. Krebs, and J. J. Goeman (2015). A region-based multiple testing method for hypotheses ordered in space or time. *Statistical Applications in Genetics and Molecular Biology 14*(1), 1–19.

N. Meinshausen (2006). False discovery control for multiple tests of association under general dependence. *Scandinavian Journal of Statistics 33*(2), 227–237.

P. Neuvial, B. Sadacca, G. Blanchard, G. Durand, and E. Roquain (2020). sanssouci: Post hoc multiple testing inference. R package version 0.9.0.

D. P. Orfanos, V. Michel, Y. Schwartz, P. Pinel, A. Moreno, D. Le Bihan, and V. Frouin (2017). The brainomics/localizer database. *Neuroimage 144*, 309–314.

R Core Team (2020). *R: A Language and Environment for Statistical Computing.* Vienna, Austria: R Foundation for Statistical Computing.

F. Reyal, N. Stransky, I. Bernard-Pierrot, A. Vincent-Salomon, Y. de Rycke, P. Elvin, A. Cassidy, A. Graham, C. Spraggon, Y. Désille, A. Fourquet, C. Nos, P. Pouillart, H. Magdelénat, D. Stoppa-Lyonnet, J. Couturier, B. Sigal-Zafrani, B. Asselain, X. Sastre-Garau, O. Delattre, J. P. Thiery, and F. Radvanyi (2005, February). Visualizing chromosomes as transcriptome correlation maps: evidence of chromosomal domains containing co-expressed genes - a study of 130 invasive ductal breast carcinomas. *Cancer Research 65*(4), 1376–1383.

J. P. Romano and M. Wolf (2005). Exact and approximate stepdown methods for multiple hypothesis testing. *Journal of the American Statistical Association 100*(469), 94–108.

J. D. Rosenblatt, L. Finos, W. D. Weeda, A. Solari, and J. J. Goeman (2018). All-resolutions inference for brain imaging. *Neuroimage 181*, 786–796.

R. J. Simes (1986). An improved Bonferroni procedure for multiple tests of significance. *Biometrika 73*(3), 751–754.

N. Stransky, C. Vallot, F. Reyal, I. Bernard-Pierrot, S. G. D. de Medina, R. Segraves, Y. de Rycke, P. Elvin, A. Cassidy, C. Spraggon, A. Graham, J. Southgate, B. Asselain, Y. Allory, C. C. Abbou, D. G. Albertson, J.-P. Thiery, D. K. Chopin, D. Pinkel, and F. Radvanyi (2006). Regional copy number-independent deregulation of transcription in cancer. *Nature Genetics 38*, 1386–1396.

P. H. Westfall and S. S. Young (1993). *Resampling-Based Multiple Testing.* New York, NY: Wiley.

Part II

Applications in Medicine

10

Group Sequential and Adaptive Designs

Ekkehard Glimm

Novartis Pharma AG

Lisa V. Hampson

Novartis Pharma AG

CONTENTS

10.1 Introduction

Group sequential (GS) designs are characterized by repeated analyses of data that accrue in a clinical trial. In general, the number and timing of interim analyses (IAs) are prespecified. At the times of IAs, the data are inspected and the trial is stopped if a beneficial treatment effect is established ("stop for efficacy") or if the experimental treatment is deemed insufficiently beneficial or if it is harmful ("stop for futility"). Otherwise, the trial continues. The treatment effect is judged by statistical tests with control of the family-wise

DOI: 10.1201/9780429030888-12

error rate (FWER) of false rejection under the null hypothesis of no treatment effect. GS designs are special cases of adaptive designs. In GS designs, the only permitted decisions at interim are to stop the study for efficacy or futility. In adaptive designs, additional design modifications are permitted. In practice, sample size reestimation, dropping of treatment arms, and restriction of subsequent randomization to subpopulations of patients are the most common design modifications. To guarantee FWER control, the most frequently used techniques are combination test approaches such as p-value combination approaches and conditional error function approaches. In recent years, trial designs are increasingly using combinations of these ideas with other multiplicity adjustment methods such as the closed test principle. For example, some recent trials investigated several treatment arms with several IAs and the possibility to drop treatment arms.

10.2 Example Application

To illustrate the statistical approaches presented in the following sections, we will use a randomized, double-blind clinical trial for a drug to treat Alzheimer's disease (AD). The trial was originally designed with three treatment arms: (a) a high-dose arm (H) of a new experimental treatment, (b) a low dose arm (L) of the new treatment, and (c) the standard-of-care (C). Two primary endpoints are the most relevant indicators of treatment efficacy: (a) change from baseline in the score of Alzheimer's Preclinical Composite Cognitive test (APCC) after 5 years of treatment and (b) time-to-progression to fully developed AD. The trial was planned with $2:1:2$ randomization to arms $H:L:C$. An IA was planned after $120+60+120=300$ patients had completed 5 years of study. At this point in time, it was planned to unblind and investigate the data and to consider the following adaptations:

- Stop the study for futility if results in both H and L do not indicate a treatment benefit.
- Continue the study with $2:1:2$ randomization until $800+400+800=2,000$ patients are available for a final analysis.
- If H is unsafe, drop it and continue the trial only with L and C. In this case, the randomization ratio in the second stage will be adapted in such a way that an identical number of patients (a target of 800 each, subject to possible further sample size modification) will be available for the final analysis in groups L and C.
- The stage 2 sample size may be modified from the original plan.

While the trial protocol was approved by the American and European health authorities, it was ultimately not conducted. However, we will use its general setup to discuss various aspects of GS and adaptive design methodology in the following. We will not always stick with the design of the real trial as it was planned. More details on the planned design can be found in Glimm et al. (2018).

10.3 Group Sequential Testing

10.3.1 Introduction

Suppose we are planning a two-arm randomized controlled trial (RCT) comparing an experimental therapy versus control, where we represent the effect of the new therapy on the primary endpoint by θ. Depending on the distribution of the primary endpoint, θ could, for example, be a log hazard ratio, a log odds ratio, or the difference in expected outcomes. A clinical trial following a fixed sample design must recruit and follow-up all patients for their primary endpoint before the null hypothesis, $H_0 : \theta \leq 0$, can be tested and success (or futility) declared for the experimental therapy. In contrast, a group sequential (GS) test monitors the trial data as they accumulate, re-testing the null hypothesis periodically until either a maximum sample size is reached or a definitive conclusion about H_0 can be drawn. Early stopping can be triggered either for efficacy or for futility. Stopping boundaries are calibrated to avoid inflation of the overall type I error rate in the presence of repeated significance testing while the maximum sample size (or more generally information level) is chosen to meet a target power. A GS test will require a larger maximum sample size than the fixed sample design with the same type I and type II error rates. In simple cases, this can be explained by the fact that by the Neymann-Pearson Lemma (Cox and Hinkley (1974), Chapter 4), the fixed sample design is the uniformly most powerful test of H_0.

GS testing is appealing because it reduces the expected sample size and time needed to reach a conclusion about H_0 (Food and Drug Administration, 2019). This means that, on average, fewer trial participants will be randomized to an inferior treatment and ensures patients outside the trial have speedier access to an effective medicine. However, investigators should always ensure the maximum sample size is feasible since this must be recruited in the absence of early stopping. For example, see the PREMILOC trial, designed using a triangular test (Whitehead, 1997) which was stopped early due to financial and resource limitations (Baud et al., 2016). The statistical benefits of GS testing increase with the number of IAs (Eales and Jennison, 1992). However, practically important gains can be still made with as few as one or two IAs, when the logistical burden associated with cleaning data and convening an Independent Data Monitoring Committee (IDMC) for each analysis is manageable.

The first GS procedures arose in the field of industrial sampling. Dodge and Romig (1929) developed a two-stage "double sampling plan" to test whether the proportion of defectives in a batch is less than a prespecified threshold. Attempts were made at proposing fully sequential monitoring procedures for clinical trials (see the overviews by Ghosh and Sen 1991 and Jennison and Turnbull 1999), but these were quickly supplanted by GS tests after the landmark papers of Pocock (1977) and O'Brien and Fleming (1979). Since then, GS tests have been explored extensively in the statistical literature and implemented in many clinical studies (Rosenblatt, 2017; Todd, 2007). Several monographs on GS tests have been written, including Whitehead (1997); Jennison and Turnbull (1999); Proschan, Lan and Wittes (2006); and more recently Wassmer and Brannath (2016). Furthermore, several tutorials have also appeared in high-impact medical journals, either exclusively on the topic of GS tests (Schulz and Grimes, 2005a) or mentioning GS tests in the broader context of adaptive designs (Bhatt and Mehta, 2016). GS designs are recognized by health authorities (European Medicines Agency, 2007; Food and Drug Administration, 2019), and their implementation is supported in standard statistical software using, for example, the SEQDESIGN and SEQTEST procedures in SAS (2016) or the R packages `rpact` (Wassmer and Pahlke, 2019) and `gsDesign` (Anderson, 2016). Specialist software for adaptive designs such as EAST (Cytel, 2016) and ADDPLAN (ICON, 2020) also exists.

A wide variety of frequentist and Bayesian two-sided and one-sided GS tests have been proposed. We begin by focusing on frequentist K-stage one-sided tests of $H_0 : \theta \leq 0$ with type I error rate α at $\theta = 0$ and power $1 - \beta$ at $\theta = \delta$. In recent years, there has been some controversy over whether clinical trials should be permitted to stop early for benefit on the primary endpoint (Schulz and Grimes, 2005b). This is due to concerns that effect estimates derived from trials stopped early for efficacy are subject to an upward selection bias and that there may be insufficient data to draw inferences on secondary hypotheses or adverse events. However, Korn et al. (2009) found that these biases may not be large enough to be practically important, while Goodman (2007) noted that there is also a selection bias inherent in the estimates arising from positive fixed sample tests. This topic is discussed in more detail in Section 10.3.6.

GS tests can be designed with binding or nonbinding futility boundaries. If a test's futility boundaries are binding, the test's stopping rule and maximum sample size are set in tandem to control the type I error rate at $\theta = 0$ and the type II error rate at $\theta = \delta$ at their nominal values. This results in less stringent efficacy boundaries because under H_0, some test statistic sample paths that would otherwise lead to incorrect rejection of H_0 instead cross a futility boundary first and stop without rejection of the null. However, when the GS test is implemented, the binding futility boundaries must be strictly adhered to in order to prevent type I error rate inflation. GS tests with nonbinding futility boundaries are designed with efficacy boundaries that control the type I error rate assuming the trial will not stop early for futility. Then the DMC has the flexibility to deviate from the futility stopping rule without the risk of inflating the type I error rate. Health authorities tend to discourage the use of binding futility boundaries.

For $k = 1, \ldots, K$, define $\hat{\theta}_k$ as the maximum likelihood estimate (MLE) of θ obtained from fitting a generalized linear model to the data at analysis k. Furthermore, let $\mathcal{I}_k = var(\hat{\theta}_k)^{-1}$ denote the Fisher information for θ at analysis k and let $Z_k = \sqrt{\mathcal{I}_k}\hat{\theta}_k$ represent the corresponding standardized test statistic. Then, given $\mathcal{I}_1, \mathcal{I}_2, \ldots$, the statistics $Z_1, Z_2, \ldots, Z_K$ follow asymptotically a multivariate normal joint distribution (Jennison and Turnbull, 1997; Scharfstein et al., 1997) with:

$$\begin{aligned}&\text{(i) } Z_k \sim N(\theta\sqrt{\mathcal{I}_k}, 1), \text{ for } k = 1, 2, \ldots \\ &\text{(ii) } Cov(Z_{k_1}, Z_{k_2}) = \sqrt{\mathcal{I}_{k_1}/\mathcal{I}_{k_2}}, \text{ for } k_1 \leq k_2.\end{aligned} \tag{10.1}$$

This joint distribution also holds asymptotically if test statistics are based on a log-rank test or partial likelihood estimates obtained from fitting a Cox proportional hazards model. This joint distribution is often referred to as the "canonical joint distribution", since it applies in a wide variety of testing scenarios. This result implies that it is the sequence of information levels that drives a test's operating characteristics and ensures that the same GS designs can be applied in many different contexts.

Several extensions have been made to standard GS tests. While early GS designs planned for a fixed sequence of group sizes, Lan and DeMets "error spending" designs are flexible enough to accommodate unpredictable group sizes while maintaining exact control of the type I error rate (Lan and DeMets, 1983). In a setting where k stages of the clinical trial have been performed and yielded information levels $\{\mathcal{I}_1, \ldots, \mathcal{I}_k\}$, future recruitment rates and analysis scheduling can be adapted in response to these recruitment patterns so long as $\mathcal{I}_{k+1}$ remains conditionally independent of $\{\hat{\theta}_1, \ldots, \hat{\theta}_k\}$ given $\{\mathcal{I}_1, \ldots, \mathcal{I}_k\}$ (Jennison and Turnbull (1999), p.146). Mehta and Tsiatis (2007) extend error-spending tests to cases with unknown "nuisance" parameters, such as the standard deviation of a continuous endpoint or the placebo response rate for a binary endpoint. At each analysis k of the information monitoring procedure, one calculates updated estimates of all unknown nuisance parameters and uses these to estimate $\mathcal{I}_k$. Critical values defining the error spending test at stage k are then found based on the estimated information sequence $\hat{\mathcal{I}}_1, \ldots, \hat{\mathcal{I}}_k$, and the target sample size can be refined to ensure the target information level will be reached in the absence of early stopping. GS tests monitoring several endpoints (Kosorok et al., 2004) or which test a pair of efficacy and safety endpoints (Jennison and Turnbull, 1993) are also available. Repeated confidence intervals (RCIs) are a sequence of intervals that have a simultaneous coverage probability of $1 - \alpha$ (Jennison and Turnbull, 1989, 1999). We can construct RCIs by inverting a parent GS test with type I error rate α, such as a one-sided error spending test; see Section 10.3.7 for more details. Using RCIs to monitor a clinical trial (rather than the parent GS test itself) ensures that, on termination, we obtain a confidence interval for θ with a coverage rate which is at least as large as the nominal level, regardless of whether the test is stopped in accordance with the stopping rule or not. This implies the type I error rate of the trial monitored using the RCIs will be at most α.

10.3.2 Bayesian GS Tests

Much of the theory underpinning GS tests has been developed in the frequentist paradigm, with stopping rules calibrated to control overall type I and type II error rates. In a highly powered trial, statistical significance can be achieved with a point estimate of θ some way below the target effect the trial was powered for. However, while significance is essential, evidence of a clinically relevant treatment effect (δ) may also be needed for a clinical trial to be deemed successful. Although we could require that $H_{10} : \theta \leq 0$ and $H_{20} : \theta \leq \delta$ be rejected at significance levels $\alpha = 0.025$ and $\alpha = 0.5$, respectively, these success criteria can also be conveniently captured by a Bayesian GS design basing stopping decisions on the posterior distribution for θ. For example, Gsponer et al. (2014) propose stopping rules of the form:

Stop at IA k if

$$Pr\{\theta > 0 \mid Z_k\} > p_k \qquad \text{and} \qquad Pr\{\theta > \delta \mid Z_k\} > 0.5 \tag{10.2}$$

where if data are normally distributed and a noninformative prior is placed on θ, the second criterion corresponds to requiring that $\hat{\theta}_k > \delta$. When using these decision criteria in practice, one should check whether one criterion always dominates the other and if this is the case, drop the redundant condition. One of the advantages of adopting a Bayesian design is that we can place informative priors on parameters for which relevant information is available. This information could be comparative data from a previous RCT, for example an estimate of the treatment effect in adults when planning a pediatric trial. In this case, to introduce some degree of skepticism (or shrinkage), we could base inferences on a mixture prior for θ placing prior weight ω on the distribution based on the historical effect estimate and weight $(1 - \omega)$ on a weakly informative distribution centered at the null. Alternatively, if we only have information on the control arm and θ is a difference in means, say $\theta = \mu_E - \mu_C$, we can place a weakly informative prior on μ_E and a robust meta-analytic-predictive (MAP) prior (Schmidli et al., 2014) on μ_C to incorporate the historical controls. To reflect prior scepticism that the treatment effect will be as large as δ say, other authors have considered, for example, basing inferences on a prior for θ with mean 0 and a variance which incorporates a 'handicap' to reflect the strength of prior scepticism; this handicap can be calibrated to ensure control of the frequentist type I error rate at a certain level under a given monitoring scheme (Grossman et al., 1994).

Numerical integration, or more generally Monte Carlo simulation, can be used to verify the frequentist operating characteristics of the trial using decision rule (10.2), for a given sequence of thresholds $p_1, \ldots, p_K$ (Gsponer et al., 2014). The challenge of using simulation to verify adequate type I error rate (or FWER) control is that it may be impossible to guarantee that the scenarios considered include the case maximizing the error rate. To reduce the computational burden of simulations, one can use a normal approximation to the prior for θ and assume either sufficient statistics for θ follow the canonical asymptotic joint distribution shown in (10.1), or assume aggregate summaries of responses on each arm are normally distributed (see the R package gsbDesign (Gerber and Gsponer, 2019)). When investigators perform the trial itself, these approximations can be discarded and decisions are based on exact inferences.

Stochastic curtailment procedures basing early stopping decisions on Bayesian predictive power are also examples of Bayesian GS designs. These procedures recommend early stopping for efficacy (futility) when the Bayesian predictive probability of meeting the prespecified success criteria at the final analysis given the interim data is sufficiently high (low). More generally, we can base stopping decisions on the predictive probability of meeting any prespecified efficacy criteria, such as (10.2), given the interim data. In contrast to rules based on conditional power, which assume a particular value for the unknown treatment effect, rules on predictive power assume that the unknown treatment effect is from a distribution. One challenge when designing a futility-stopping rule based on predictive power is how to set the threshold(s) triggering early termination. Several authors (Gallo et al., 2014; Lachin, 2005) have pointed out that a futility stopping rule expressed on one scale can be translated onto several other scales. For example, considering a trial designed to test $H_0 : \theta \leq 0$ once with power 0.9 at $\theta = 0.4$ and type I error rate $\alpha = 0.025$, with an IA scheduled at information fraction $\mathcal{I}_1/\mathcal{I}_{max} = 0.5$:

$$\text{Stop if Bayesian predictive power} \leq \gamma$$

is equivalent to

$$\begin{aligned} &\text{Stop if } Pr\{\text{Reject } H_0 \text{ at final analysis } \mid \theta = \delta\} \\ &\qquad \leq \Phi\left(\frac{\Phi^{-1}(\gamma) + \delta\sqrt{\mathcal{I}_{max}}}{\sqrt{2}} - \frac{\Phi^{-1}(1-\alpha)}{\sqrt{(2\mathcal{I}_{max})}}\right) \end{aligned}$$

assuming we evaluate predictive power in the normal case under an improper prior for θ with density equal to a fixed constant across the real line. In light of this, futility stopping rules should be calibrated to ensure acceptable operating characteristics and expressed on whichever scale best facilitates communication with clinical collaborators (Gallo et al., 2014).

Öhrn (2011) seeks standard GS tests with nonbinding futility rules which are optimal in the sense of minimizing the expected trial sample size subject to control type I and type II error rates at specified values. Öhrn finds each optimal frequentist design as the solution to a carefully constructed Bayes decision problem. Several other authors have adopted a similar approach to find optimal frequentist GS tests for a variety of testing scenarios such as: one-sided (Barber and Jennison, 2002) and two-sided (Eales, 1995) GS tests; "inner-wedge" GS tests allowing early stopping to declare superiority, inferiority or noninferiority (Öhrn and Jennison, 2010); adaptive one-sided GS tests which permit an adaptive choice of group size (Jennison and Turnbull, 2006a); and GS tests for delayed responses (Hampson and Jennison, 2013). Optimal designs that minimize the expected sample size on termination averaged across a prior distribution for the treatment effect have been found, as have designs that minimize the expected sample size under a fixed treatment effect or a range of effects. To illustrate how this approach works, consider the relatively straightforward case of finding an optimal one-sided GS design, where the optimal frequentist test is found as the solution to a Bayes decision problem defined by: (a) a prior for θ; (b) a sampling cost function representing the cost of recruiting each patient which depends on θ; and (c) a decision loss function assigning a loss to each possible hypothesis decision (stop the trial without rejecting H_0; reject H_0), which is also dependent on θ. For a particular combination of prior, sampling cost function and decision loss function, the corresponding Bayes test selects at each IA the action (stop and reject H_0; continue; stop without rejecting H_0) associated with the lowest expected additional loss, where we take expectations with respect to the future unobserved data and θ in light of the current observations. We then search to find a combination of prior, sampling cost and decision loss functions such that the Bayes GS test attains the smallest average expected sample size in the class of GS tests controlling the frequentist type I and type II error rates at levels α and β, respectively.

Lewis et al. (2007) found optimal Bayesian GS tests of $H_0 : |\theta| \leq \theta_0$ which recommend at each IA the action minimizing the total expected additional loss under the quadratic decision loss function:

$$L(\theta, \text{Accept } H_0) = d_0\theta^2$$

$$L(\theta, \text{Declare superiority}) = \begin{cases} d_1(\theta - \theta_0)^2 & \text{if } \theta < \theta_0 \\ 0 & \text{if } \theta \geq \theta_0 \end{cases}$$

$$L(\theta, \text{Declare inferiority}) = \begin{cases} 0 & \text{if } \theta \leq -\theta_0 \\ d_1(\theta - \theta_0)^2 & \text{if } \theta > -\theta_0 \end{cases}$$

Utilities d_0 and d_1 are calibrated to ensure the Bayesian procedure has certain frequentist type I and type II error rates. Murray et al. (2018) propose GS tests for ordinal data in a heterogeneous population which can accommodate treatment-by-subgroup interactions. Levels of the ordinal outcome are assigned utilities. Then, at each IA k, $k = 1, \ldots, K$, for each subgroup we calculate the posterior probability that the mean utility on the experimental therapy exceeds that on control: if this is larger than $(1 - p_k)$, recruitment into the subgroup is halted. Here, $p_k = \Phi(c_k)$, the nominal p-value corresponding to the stage-k critical value of an error-spending maximum duration GS design (Jennison and Turnbull, Section 7.2.3, 2000). Murray et al. (2018) use Monte Carlo simulation to verify appropriate control of frequentist error rates within each subgroup.

10.3.3 Non-Normal or Non-Asymptotic Endpoints

As described in Section 10.3.1, most applications of GS designs are based on the canonical normal distribution (10.1). There are a few examples of popular tests for which the canonical normal distribution does not apply asymptotically, for example, weighted log-rank test statistics used in survival analysis (Jennison and Turnbull (1999), Chapter 13) or tests based on generalized estimating equations in generalized linear models (i.e., overdispersed Poisson and logistic regression models) (Shoben and Emerson, 2014). In many cases where the asymptotic canonical distribution does not apply for a considered method, it can be replaced by another method for which it does apply in the GS context (see e.g. Mütze et al. (2018)). For some specific GS test statistics, researchers have derived an asymptotic multivariate normal distribution with a covariance structure which is different from the canonical form. Chapter 11 of Jennison and Turnbull (1999) is a pointer to some of these results.

When sample sizes are small or the original data distributions are far from the normal distribution, we might also be interested in the non-asymptotic joint distributions of repeated GS test statistics. Some work in this regard has been done for small sample sizes of normally distributed data with unknown variance and inference based on t-statistics (Stein, 1945; Timmesfeld et al., 2007).

The practically most relevant case in the clinical trial context is that of GS comparisons of event rates. In such cases, the endpoint of the clinical trial is the number of responders to treatment (e.g., the number of patients who are symptom-free after 2 weeks of treatment) which is usually assumed to be binomially distributed. In oncology, single-arm trials testing whether the objective response rate (ORR) is above some fixed clinically meaningful threshold are quite common. Otherwise, the aim of the study is usually the comparison of responder rates between a treatment and a control group, possibly adjusting for additional covariates. Several methods are available, such as Fisher's exact test and its generalizations for stratified contingency tables, Wald or likelihood-ratio-χ^2-tests and various versions of parameter tests in logistic regression models.

Closer examinations of the non-asymptotic properties of GS methods are warranted in this context for two reasons:

1. Since the response from every individual patient is only a "success" or a "failure", convergence of the rate estimate's distribution to a normal distribution is slow.

2. Since only a discrete number of outcomes is possible (hence, the true distribution of the test statistic is also discrete), the type I error rate cannot be fully exhausted in most cases[1]. This "unused" type I error rate can be redistributed to allow testing at additional IAs.

The development of methods dealing with discreteness in GS trials has a long history (see e.g. Jennison and Turnbull (1999), chapter 12) and was investigated for single-arm trials by Mander and Thompson (2010) and Englert and Kieser (2012) more recently. Improved computing power could, in the future, be used to extend the basic ideas of exact inference to more complex situations such as the repeated comparisons of two treatments with a Cochran-Mantel-Haenszel test. Ristl et al. (2018) investigated such a search for optimal rejection regions in discrete multivariate sample spaces, albeit their concrete application is in the slightly different context of multiple testing of two binary endpoints.

10.3.4 GS Tests for Delayed Responses and Leveraging Short-Term Readouts

In many clinical trials, a patient's primary endpoint will be measured after they have been treated or followed up for some period of time, in order to capture whether the treatment has a sustained effect on response. When there is a delay in response, at each IA of a GS test, there will be patients who have been randomized and started treatment but have not yet been followed up long enough to provide their primary endpoint. In some scenarios, it will be ethical to continue to monitor these patients and their responses after the decision has been taken to halt recruitment at the interim. These "pipeline" responses (Hampson and Jennison, 2013) can complicate the interpretation of the GS test, particularly if early stopping is triggered by a positive trend in the interim data only for there to be some dilution of the trend once we incorporate the pipeline responses. However, such test statistic sample paths are entirely possible, particularly if pipeline responses contribute a substantial amount of information for θ. For simplicity, in what follows we will restrict attention to endpoints where the delay is fixed and known ahead of time. For example, glycated hemoglobin (HbA_{1c}) at week 24 is an accepted primary endpoint in diabetes trials; change from baseline in the APCC score after 5 years of treatment was one of the primary endpoints of the clinical trial in AD described in Section 10.2. However, a similar challenge can also arise with survival endpoints subject to independent adjudication, such as time to cause-specific mortality.

Whitehead (1992) proposed an ad-hoc approach for calculating a final p-value after a GS test has overrun which is based on the stagewise ordering of the sample space of the GS test defined in Section 10.3.6. The IA that prompted the decision to halt accrual is effectively deleted from the records, and we assume the final analysis was always intended to occur at the information level incorporating the pipeline data. Hall and Ding (2008) proposed combining stagewise p-values based on the data collected before and after the

[1]Unless "randomized decisions" are allowed which, however, are strongly discouraged by health authorities in clinical trials.

final IA. Sooriyarachchi et al. (2003) use simulation to compare operating characteristics of GS tests using either the deletion or p-value combination approaches to accommodate overrunning. However, neither approach enables us to plan ahead for the pipeline data we know will accrue if the trial terminates early. Hampson and Jennison (2013) proposed delayed response GS tests (DR GSTs) of $H_0 : \theta \leq 0$. DR GSTs distinguish between IAs, where we make decisions about whether or not to stop recruitment, from the decision analysis, where we make a decision about whether or not to reject H_0.

Suppose that at IA k, we use all available data to calculate the standardized test statistic Z_k associated with Fisher information $\mathcal{I}_k$. If recruitment is stopped, we wait for the pipeline data to be collected before updating the test statistic to $\tilde{Z}_k$ with information $\tilde{\mathcal{I}}_k$. The GS test is defined by boundaries $\{\ell_1, \ldots, \ell_{K-1}\}$ and $\{u_1, \ldots, u_{K-1}\}$, which are applied at the $(K-1)$ IAs, and $\{c_1, \ldots, c_K\}$ which are applied at the decision analyses. Then the stopping rule of the DR GST is of the form:

At IA $k = 1, \ldots, K-1$:

 If $Z_k \geq u_k$ or $Z_k \leq \ell_k$, stop accrual and proceed to decision analysis k.

 Otherwise, continue to IA $(k+1)$.

At decision analysis $k = 1, \ldots, K$:

 If $\tilde{Z}_k \geq c_k$, reject H_0

 If $\tilde{Z}_k < c_k$, do not reject H_0.

We define $u_{K+1} = \ell_{K+1}$ to ensure that in the absence of early stopping, we must make a final hypothesis decision at decision analysis K. The stopping boundaries of a two-stage test are illustrated in Figure 10.2. Hampson and Jennison derive error-spending versions of these DR GSTs and optimal versions of boundaries minimizing the average expected number of patients recruited on termination of the test while controlling the type I error rate at level α when $\theta = 0$ and the type II error rate at level β when $\theta = \delta$. It is common for designs to stipulate $c_k < u_k$ so that some dilution of the interim trend is permitted before the test stops without rejecting H_0.

Hampson and Jennison noted that sequences of test statistics of the form $\{Z_1, \ldots, Z_k, \tilde{Z}_k\}$ are based on nested subsets of the data accumulated at decision analysis k, meaning that asymptotically they will follow canonical joint distribution (10.1) given information levels $\mathcal{I}_1, \ldots, \mathcal{I}_k, \tilde{\mathcal{I}}_k$ when test statistics are based on MLEs of parameters in a general parametric model. This means that the boundaries and operating characteristics of DR GSTs can be calculated using standard software such as the mvtnorm package in R (Genz et al., 2019). As the length of the delay in response increases, the savings in the expected recruited sample size achieved by an optimal DR GST (versus the fixed sample test) decrease.

When there is a delay in the primary endpoint, short-term correlated intermediate endpoints are often also measured. For example, we measure change from baseline at 12 and 24 weeks. Several authors have proposed leveraging these short-term data to support interim decisions, including the cases of binary responses (Marschner and Becker, 2001; Niewczas et al., 2019; Whitehead et al., 2008) and continuous responses (Galbraith and Marschner, 2003; Hampson and Jennison, 2013; Wüst and Kieser, 2003). Suppose a single short-term response is measured, and label X_{SDi} (X_{LDi}) as the short-term (long-term) response of the ith patient on treatment $D \in \{C, E\}$. Furthermore, assume pairs of measurements on the same patient follow a bivariate normal distribution so that:

$$\begin{pmatrix} X_{SDi} \\ X_{LDi} \end{pmatrix} \sim N\left(\begin{pmatrix} \mu_{SD} \\ \mu_{LD} \end{pmatrix}, \begin{pmatrix} \sigma_S^2 & \tau\sigma_S\sigma_L \\ \tau\sigma_S\sigma_L & \sigma_L^2 \end{pmatrix} \right) \quad \text{for } i = 1, 2, \ldots;\ D \in \{C, E\}. \tag{10.3}$$

At an IA, stopping decisions are driven by the MLE of the long-term treatment effect $\theta_L = \mu_{EL} - \mu_{CL}$ derived from the joint likelihood of all short- and long-term responses. This MLE is unbiased and is only a function of θ_L, meaning that it is sufficient to control the type I error rate of the GS test under $\theta_L = 0$. Although Equation (10.3) implies the variance matrix is identical across treatment arms, it is straightforward to relax this assumption. Incorporating short-term data increases the information available for θ_L at each IA, enabling more informed decision-making. When responses are continuous and normally distributed, the gain in information depends on the delay in the short-term endpoint (relative to the long-term endpoint) and the within-patient correlation of responses. When parameters of the variance matrix in (10.3) are unknown, Hampson and Jennison (2013) show that an information monitoring approach using estimates in place of the variance parameters provides adequate error rate control, although larger deviations from the nominal type I error rate are seen under small sample sizes.

10.3.5 Illustrative Example Designing GS Tests

To illustrate and compare different GS designs, we follow Section 10.2 and consider a RCT of a new therapy for AD. For illustration, we modify the example slightly to imagine how that GS trial would have been designed to compare high-dose therapy (H) versus standard of care (C) if the primary endpoint had been change from baseline at 18 months in global function, measured by the Alzheimer's Disease Cooperative Study-Clinical global impression of change (ADCS-CGIC). Suppose patient outcomes are normally distributed, with $X_{iH} \sim N(\mu_H, \sigma^2)$ on the high dose and $X_{iC} \sim N(\mu_C, \sigma^2)$ on control, for $i = 1, \ldots, N$. Defining $\theta = \mu_H - \mu_C$, we design a two-stage GS test of $H_0 : \theta \leq 0$ vs $H_1 : \theta > 0$ with type I error rate $\alpha = 0.025$ at $\theta = 0$ and power 0.9 at $\theta = 0.4$, assuming $\sigma = 1.54$. These settings appear reasonable in light of the design assumptions made for the LEADe Study (Jones et al., 2008). The target effect corresponds to a 40% difference in change between H and C, assuming an average decrease of 1.0 on C.

Assuming recruitment proceeds at a rate of 12 patients per month, Figure 10.1 plots the boundaries of four different two-stage GS tests, as listed below. All tests are designed with a nonbinding futility boundary:

1. Haybittle-Peto design permitting early stopping for efficacy with p-value < 0.001 and with a beta-spending futility boundary defined by the spending function $f(t) = \min\{1 - \Phi(z_\beta/\sqrt{t}), \beta\}$, for $t \geq 0$.

2. Error-spending test with O'Brien-Fleming shaped boundaries spending type I and type II error probabilities according to the functions $g(t) = \min\{2 - 2\Phi(z_{\alpha/2}/\sqrt{t}), \alpha\}$ and $f(t) = \min\{1 - \Phi(z_\beta/\sqrt{t}, \beta\}$, respectively, where $z_\alpha = \Phi^{-1}(1 - \alpha)$.

3. Error spending test defined by the ρ-family spending functions (with $\rho = 1$) $g(t) = \alpha \min\{t, 1\}$ and $f(t) = \beta \min\{t, 1\}$.

4. Error spending test defined by the ρ-family spending functions (with $\rho = 2$) $g(t) = \alpha \min\{t^2, 1\}$ and $f(t) = \beta \min\{t^2, 1\}$.

Let us focus on Design 2, which recruits a maximum of 640 patients and schedules an IA once the primary endpoint has been measured on $n_1 = 320$. Given the 18 month delay in the primary endpoint and assuming a recruitment rate of 12 patients per month, $\tilde{n}_1 = n_1 + (18 \times 12) = 536$ patients will have been recruited by the time of the IA and will be used for an efficacy test if reached. If recruitment is not stopped at the IA, a final decision will be based on responses from $\tilde{n}_2 = 640$ patients. We can derive an error-spending DR GST which

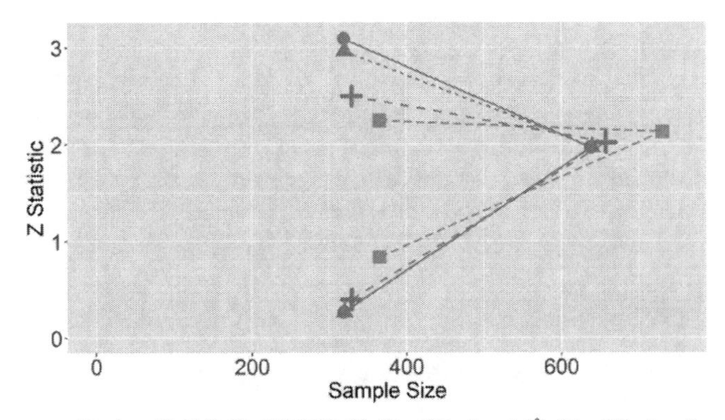

FIGURE 10.1
Critical boundaries defining four K = 2-stage GS tests of $H_0 : \theta \leq 0$. Tests have type I error rate $\alpha = 0.025$ at $\theta = 0$ and power $= 0.9$ at $\theta = 0.4$. Boundaries are for monitoring standardized test statistics.

acknowledges that these additional pipeline responses will accrue after a decision is taken to stop the trial early. We follow Method 1 presented in Hampson and Jennison (2013) to design the error spending DR GST for the information sequence $\{\mathcal{I}_1 = 33.7, \tilde{\mathcal{I}}_1 = 56.5, \tilde{\mathcal{I}}_2 = 67.5\}$, finding the interim boundaries (ℓ_1, u_1) to satisfy:

$$Pr\{Z_1 \geq u_1; \theta = 0\} = g(\mathcal{I}_1/\mathcal{I}_{max}) \quad \text{and} \quad Pr\{Z_1 \leq \ell_1; \theta = \delta\} = f(\mathcal{I}_1/\mathcal{I}_{max}), \qquad (10.4)$$

where g and f are as defined for Design 2 above. The critical values, c_1 and c_2, at decision analyses $k = 1, 2$ are then found as the solutions to:

$$Pr\{Z_1 \leq \ell_1, \tilde{Z}_1 \geq c_1; \theta = 0\} = Pr\{Z_1 \geq u_1, \tilde{Z}_1 \leq c_1; \theta = 0\} \qquad (10.5)$$

$$Pr\{Z_1 \in (\ell_1, u_1), \tilde{Z}_2 \geq c_2; \theta = 0\} = \alpha - g(\mathcal{I}_1/\mathcal{I}_{max}). \qquad (10.6)$$

Choosing c_1 to satisfy Equation (10.5) implies the type I error rate spent in stage 1, given by $Pr\{Z_1 \geq u_1, \tilde{Z}_1 \geq c_1; \theta = 0\} + Pr\{Z_1 \leq \ell_1, \tilde{Z}_1 \geq c_1; \theta = 0\}$ simplifies to $Pr\{Z_1 \geq u_1; \theta = 0\}$, which is equal to $g(\mathcal{I}_1/\mathcal{I}_{max})$ by design, due to our choice of u_1 specified in (10.4). Finding c_2 as the solution to Equation (10.6) ensures that the cumulative type I error rate spent by decision analysis K is α. We label this DR GST as "Test 1" and its boundaries are plotted in Figure 10.2. It has power 0.901 and the expected number of patients recruited on termination

$$F = \int E(N; \theta) = 592.$$

To see the impact that leveraging information on a short-term endpoint can have, suppose that an endpoint can be measured after 6 months which is correlated with the primary endpoint with coefficient $\rho = 0.9$. Incorporating this information at the first IA decreases the information in the pipeline at the IA, so that we can design a DR GST for the information sequence $\{\mathcal{I}_1 = 45.1, \tilde{\mathcal{I}}_1 = 56.5, \tilde{\mathcal{I}}_2 = 67.5\}$. The boundaries of this test are also shown in Figure 10.2: the test has power $= 0.896$ and expected sample size $F = 569$.

10.3.6 Point Estimation in GS Designs

Upon termination of a clinical trial with a GS or an adaptive design, it is usually required to provide point estimates and confidence intervals of the treatment effect. In practice, the

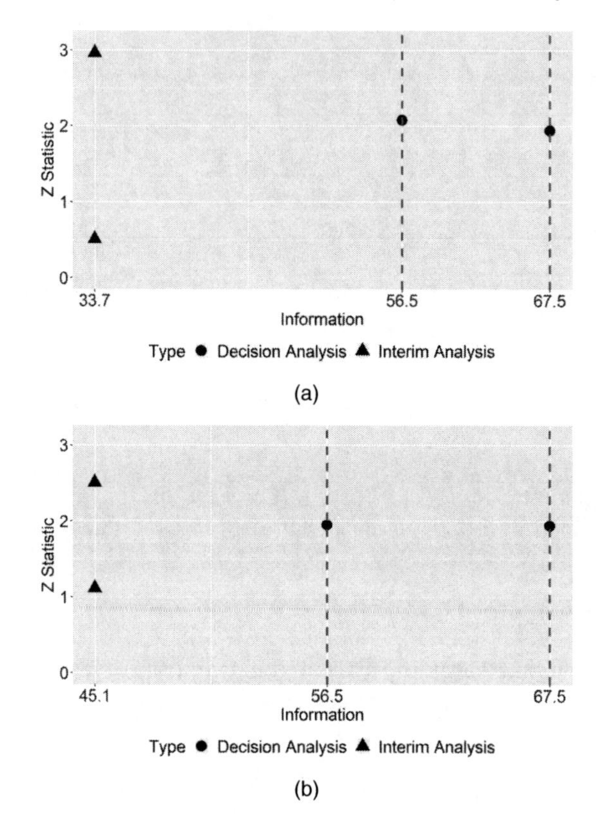

FIGURE 10.2
Critical boundaries defining $K = 2$-stage DR GSTs of $H_0 : \theta \leq 0$ leveraging: (a) information on the long-term primary endpoint measured at 18 months; and (b) information on the primary endpoint and a highly correlated short-term endpoint observed after 6 months, to support interim decision making. Both tests have type I error rate $\alpha = 0.025$ at $\theta = 0$. Critical boundaries are for monitoring standardized test statistics.

most popular point estimate is the "usual" unadjusted estimate ignoring that data were obtained in the framework of a GS trial. To illustrate this, consider a GS trial comparing an experimental treatment T with the standard-of-care C on a continuous endpoint X with $k \geq 2$ stages. GS testing is done on the null hypothesis $H_0 : \theta \leq 0$ vs $H_A : \theta > 0$ where θ is the difference in expected responses on T and C. For simplicity, assume that we have a sample

$$x_{ijk} \sim N(\theta_j, \sigma^2) \tag{10.7}$$

with a known σ^2 from patient i in treatment group j and trial stage k with $\bar{\bar{x}}_{jk} = \frac{1}{n_{j1}+\cdots+n_{jk}} \sum_{l=1}^{k} \sum_{i=1}^{n_l} x_{ijl}$. The unadjusted estimate of θ would then be $\hat{\theta} = \bar{\bar{x}}_{TS} - \bar{\bar{x}}_{CS}$ where S denotes the stage at which the trial stopped. It is easy to see that $\hat{\theta}$ is the maximum-likelihood estimate (MLE) of θ in this situation (Chang, 1989).

The estimate $\hat{\theta}$ is biased. Assume that there is only one IA and stopping for efficacy is permitted when $\hat{\theta}_1 = \bar{x}_{T1} - \bar{x}_{C1} > c_1$. If $\hat{\theta}_1 > c_1$, there will be a *reporting bias*:

$$E\left(\hat{\theta}_1 \middle| \hat{\theta}_1 > c_1\right) > E(\hat{\theta}_1) = \theta.$$

There is also a reporting bias in $\hat{\theta}_2 = \bar{\bar{x}}_{T2} - \bar{\bar{x}}_{C2}$, since

$$E\left(\hat{\theta}_2 \middle| \hat{\theta}_1 \leq c_1\right) < E\left(\hat{\theta}_2\right) = \theta.$$

In this simple situation, $\left(\hat{\theta}_1 \middle| \hat{\theta}_1 > c_1\right)$ has a truncated normal distribution and $\left(\hat{\theta}_2 \middle| \hat{\theta}_1 \leq c_1\right)$ hence has a distribution with probability density function (pdf)

$$\frac{\int_{z=-\infty}^{c_1} f_{\hat{\theta}_2|\hat{\theta}_1}\left(\hat{\theta}_2 \middle| \hat{\theta}_1 = z\right) \cdot f_{\hat{\theta}_1}(z)dz}{P(\hat{\theta}_1 \leq c_1)}$$

where $f_{\hat{\theta}_2|\hat{\theta}_1}\left(\hat{\theta}_2 \middle| \hat{\theta}_1 = z\right)$ is the pdf of the conditional normal distribution $N\left(\sqrt{\mathcal{I}_2}\left(1 - \frac{\mathcal{I}_1}{\mathcal{I}_2}\right)\right.$ $\left.\theta + \sqrt{\frac{\mathcal{I}_1}{\mathcal{I}_2}}z, \sigma^2 \cdot \left(1 - \frac{\mathcal{I}_1}{\mathcal{I}_2}\right)\right)$ of $\left(\hat{\theta}_2 \middle| \hat{\theta}_1 = z\right)$ (cf. the canonical distribution (10.1)) and $f_{\hat{\theta}_1}(z)$ the pdf of $\hat{\theta}_1$.

Emerson and Fleming (1990) used this idea to obtain an analytical expression for the bias in the normal two-sample case. Unfortunately, the bias depends on the unknown true value of θ: It is largest when $\theta = c_1$ and converges toward 0 as the true θ moves away from c_1. For example the bias is

$$bias(\theta) = \frac{\sigma}{\sqrt{n_1}}\left(1 - \frac{n_1}{n_2}\right)\phi(\sqrt{n_1}(c_1 - \theta)) \tag{10.8}$$

in the case above with two equally sized groups tested in two stages where $\phi(\cdot)$ denotes the pdf of $N(0,1)$ and $n_k = n_{k1} + n_{k2} = 2n_{k1}$ is the total sample size of stage $k = 1,2$.

In GS studies that stop early for efficacy, this bias and the resulting "exaggerated" treatment effect estimation very often raise concerns from health authorities and healthcare providers. Unfortunately, an unbiased estimate of θ is surprisingly hard to obtain. Liu and Hall (1999) proved that a uniformly minimum variance unbiased estimator does not exist. They instead suggest a so-called "truncation-adaptable" unbiased estimator which conditions on the stage S at the end of which the trial was stopped, ignoring the reasons for why the stopping occurred and if and how the trial would have continued had the stopping not occurred. While this concept of conditional unbiasedness allows for an elegant solution to the problem of minimum-variance unbiased estimation, it remains debatable to what extent it offers help in practice. In the abovementioned example, if the trial stops at IA 1, the truncation-adaptable unbiased estimator becomes $\hat{\theta}_1$ - the very same number which usually triggers the discussion about an "exaggerated treatment effect estimate". Therefore, the notion of "truncation-adaptable unbiasedness" has not found acceptance in the community of statisticians working in clinical trials. In addition, the resulting estimate also performs worse than the simple MLE $\hat{\theta}$ in terms of its mean-squared error.

"Unbiasedness" seems an elusive concept in this regard. A more detailed discussion about different sources of bias (in particular about *reporting* and *selection* bias) can be found in Bauer et al. (2010).

A potentially more practical approach was suggested by Whitehead (1986). He suggested to estimate $bias(\theta)$ from the data by plugging $\hat{\theta}$ into (10.8) and then correct the original estimate $\hat{\theta}$ by subtracting the estimated bias. Since $\hat{\theta}$ itself is biased, the bias estimate is biased too, but by iterating the plug-in procedure, the bias of $\hat{\theta}$ can be substantially reduced - though not completely removed. Wassmer and Brannath (2016), Section 4.2, discuss this estimator and some related estimators in more detail.

The reporting of an unadjusted estimate such as the MLE $\hat{\theta}$ in a GS trial is often dismissively labeled as being "naive". It should be acknowledged, however, that all "non-naive" estimates suggested so far in the literature have one or several drawbacks. They do not entirely remove the bias, or they are only "unbiased" in some specific sense or they have a large MSE.

10.3.7 Interval Estimation in GS Designs

The selection bias discussed in Section 10.3.6 also affects interval estimation: A conventional $(1-\alpha)$-confidence interval (CI) for the treatment effect θ, calculated at the point in time when the GS trial stopped, will not cover the true θ with probability $1-\alpha$. For example, in the normal distribution case (10.7), Armitage et al. (1969) calculate the true coverage probability of nominal 95%-CIs for θ as 86% in a 5-look GS design with equally spaced IAs.

There are two solutions to this challenge:

- *Repeated-confidence intervals* (RCI; see Jennison and Turnbull (1999), chapter 9): RCIs $\{I_k\}$, $k = 1, \ldots, K$, are a sequence of intervals which fulfill

$$P_\theta\left(\theta \in I_k\right) \geq 1-\alpha \text{ for all } k \tag{10.9}$$

  where $P_\theta(x)$ denotes the probability of event x if the true treatment effect is θ. Obviously, (10.9) is a stricter condition than

$$P_\theta\left(\theta \in I_k\right) \geq 1-\alpha \text{ for the } k \text{ at which the trial stopped.}$$

  Hence, an RCI automatically defines a (conservative) $(1-\alpha)$-CI for θ.

  In practice, a particularly easy way of obtaining an RCI is based on the CI-construction method of inverting a two-sided test for θ. The basic idea is that the "acceptance region" $\{\theta_0 : \psi_{\theta_0}(X) = 0\}$ of a family of tests $\psi_{\theta_0}(X) \in \{0, 1\}$ on data X for $H_0(\theta_0) : \theta = \theta_0$ defines a $1-\alpha$ confidence set. Here, $\psi_{\theta_0}(X) = 1$ denotes rejection of $H_0(\theta_0)$ and $\psi_{\theta_0}(X) = 0$ non-rejection (which we will call "acceptance" in this context for simplicity). The RCI at stopping time k is simply the ordinary, unadjusted $(1-\alpha_k)$-*CI* at this time, where α_k is the "GS" alpha-level of the GS test used at time k, cf. Section 10.3.1. For example, in a GS trial using O'Brien-Fleming-type Lan-DeMets boundaries (Lan and DeMets, 1983) for two equally spaced interim and a final analysis and the canonical distribution (10.1), I_1 would be a 99.98%-CI, I_2 a 98.80%-CI and I_3 a 95.37%-CI.

  Jennison and Turnbull (1999), Chapter 9, discuss the RCI approach in detail. Mehta et al. (2007) have extended it to adaptive designs. While its simplicity is attractive, its conservativeness is a disadvantage. In particular, if a trial is stopped early for efficacy, it appears unsatisfactory that very wide confidence intervals should arise: Early stopping means a low sample size, and this is compounded by a substantially increased confidence level $1-\alpha_k$. This has led many researchers to look for alternatives.

- *Confidence intervals based on stagewise ordering:* To derive an alternative to the repeated CIs, we must provide an order relation $(\preceq)$ on the sample space of $(T, \hat{\theta})$. Here, T is the random variable with support $\{1, \ldots, k\}$ which denotes the time at which the trial was stopped and $\hat{\theta}$ is the estimate of the parameter estimated in the GS trial. The confidence region is then a set $\left\{\theta_0 : (k, \hat{\theta}) \in A(\theta_0)\right\}$ where $(k, \hat{\theta})$ are the stopping stage k and the estimate of θ obtained from the data and $A(\theta_0)$ is the acceptance region of a test for $H_0 : \theta = \theta_0$ which uses the order relation. For example, it could be the set of all θ_0 for which $P_{\theta_0}\left((k_l(\theta_0), z_l(\theta_0)) \preceq (k, \hat{\theta}) \preceq (k_u(\theta_0), z_u(\theta_0))\right) \geq \alpha$ for critical values $k_l(\theta_0), z_l(\theta_0), k_u(\theta_0), z_u(\theta_0)$ calculated such that they yield a level α-test for $H_0 : \theta = \theta_0$.

For a two-sided test used in a GS trial without stopping for futility and a stop for efficacy if $|\hat{\theta}| > c_{\theta_0}$ for some critical value c_{θ_0}, stagewise ordering is defined as follows: $(k_1, \hat{\theta}_1) \succ (k_2, \hat{\theta}_2)$ if

- $k_1 < k_2$ or
- $k_1 = k_2$ and $|\hat{\theta}_1| > |\hat{\theta}_2|$.

In any case, the concrete calculation of CIs derived from orderings of the sample space requires numerical methods.

Owing to limited space, we have only scratched the surface of CI construction for GS designs. There are many additional complications related to the monotonicity of the order relation, the desire to match test decisions with confidence intervals, and questions regarding the possibility of the point estimate being outside of the CI. Jennison and Turnbull (1999), chapter 8, and Wassmer and Brannath (2016), Chapter 4, give thorough discussions of these issues, as well as of other orderings. They conclude that stagewise ordering with inversion of two one-sided tests is the most appropriate approach because it avoids problems such as confidence regions which are not intervals and because stagewise ordering does not "mix" outcomes from different trial stages and thus makes the approach amenable to GS designs with error spending and to adaptive designs.

Like point estimation, the choice of the most appropriate interval estimation method remains a challenge for the practice of clinical trials. Up to now, there is no entirely satisfactory solution to the problem. RCIs guarantee correct coverage (asymptotically if parameter estimates are asymptotically normally distributed), but are conservative, especially when trials are stopped early. Adjusted CIs based on sample space ordering are technically complicated and not readily available in standard software. More importantly, however, they rely on fixed stopping rules and do not necessarily retain the correct coverage if these rules are violated, which sometimes happens in practice (e.g., when a data monitoring committee decides to continue with a trial after an IA although the stopping criteria are formally fulfilled). In practice, usually only unadjusted CIs are reported with a caveat that they have to be interpreted in an exploratory sense, or they are reported together with RCIs.

10.4 Adaptive Designs

10.4.1 Introduction to Adaptive Designs

An adaptive trial design is a procedure which "... allows for prospectively planned modifications to one or more aspects of the design based on accumulating data from subjects in the trial" (Food and Drug Administration, 2019). Following the seminal paper by Bauer (1989), a huge variety of adaptive procedures have been proposed for early phase dose-finding trials as well as late-stage confirmatory studies (Bauer et al., 2016; Bhatt and Mehta, 2016; Pallmann et al., 2018). Adaptations of clinical trials are typically restricted to the following features (apart from early stopping as in GS trials):

- Blinded or unblinded sample-size reevaluation
- Termination of recruitment into one or several of the treatment arms
- Selection of a subpopulation of patients from the original trial population.

From a practical point of view, we find it helpful to distinguish between designs basing adaptations on unblinded and blinded data. For example, designs allowing sample size modifications based on blinded estimates of a nuisance parameter may be less challenging to implement than those basing adaptations on unblinded effect estimates, due to the lower risk of information leakage (and thus bias) at an IA. Relatively often, only a single IA allows design adaptations. In addition, there may be regular safety reviews which are not considered as being a part of the adaptive design, i.e., they are not formally considered as an "IA". Seamless phase II/III trials, which use phase II data to select which treatment(s) or population(s) should be taken forward to phase III, can be thought of as two-stage adaptive designs (although both stages can proceed group-sequentially). Multi-arm multi-stage (MAMS) trials, which begin comparing several treatment strategies and allow trial arms to be dropped at subsequent IAs, are examples of multistage procedures.

Adaptive designs have found their way into practice. In a review of 123 scientific advice letters issued between 2007 to mid-2012 by the Committee for Human Medicinal Products (CHMP) of the European Medicines Agency (EMA) in response to requests from pharmaceutical sponsors, 59 addressed questions on adaptive designs for phase II or phase III trials (Elsässer et al., 2014). The most commonly planned adaptations used unblinded interim data either to reassess the target sample size or decide whether or not to drop a treatment arm or enrich the patient population. More recent reviews of adaptive designs have found seamless phase II/III designs to be a commonly implemented adaptive approach (Cerqueira et al., 2020; Collignon et al., 2018).

In the context of adaptive designs for confirmatory trials, attention has focused on finding frequentist designs which either preserve the type I error rate or, if answering more than one scientific question, maintain strong control of the FWER. The FWER is defined as the probability of rejecting one or more true null hypotheses (Hochberg and Tamhane, 1987). Care is needed to prevent inflation of the type I error rate. To illustrate this, suppose we run a two-stage trial which begins randomizing patients in a 1:1 ratio between an experimental therapy and control, and second stage group sizes are allowed to depend on the unblinded stage 1 data. If equal allocation must continue in stage 2, the maximum type I error rate is increased from 0.025 to 0.06 by always choosing the second stage group sizes to maximize the conditional type I error rate under $\theta = 0$ (Proschan and Hunsberger, 1995). When different proportional changes are allowed to the second stage group sizes on each arm (e.g., a doubling of the planned number on experimental but no increase on control), the maximum type I error rate increases from 0.025 to 0.11 (Graf and Bauer, 2011). Graf et al. (2014) extend these results to the case that in addition, stage 1 begins comparing k treatments with control and the best therapy is always selected at the interim. Then, controlling for the multiplicity (but not the adaptivity) of the design by performing the final hypothesis test using an α-level Dunnett critical value based on the original 1:1 randomization ratio, the maximum type I error rate is 0.09. If the multiplicity is ignored and an unadjusted critical value of $\Phi^{-1}(1 - 0.025)$ used, the maximum type I error rate is 0.15 (Graf et al., 2014).

Trials following master protocols are designed to address several scientific questions simultaneously and are becoming increasing popular with the advance of precision medicine (Food and Drug Administration, 2018; Woodcock and LaVange, 2017). They comprise three main types of design: basket trials, evaluating a single targeted therapy in target-positive patients with different diseases; umbrella trials, evaluating different drugs in different biomarker-defined subgroups with the same disease; and platform trials which permit trialists to add in new treatments (to an umbrella trial) or new diseases (to a basket trial) over time (Collignon et al., 2020). Since trials attempt to answer multiple questions, adjustments for multiplicity may be necessary. In addition, all of the master protocol trials mentioned above can and often do incorporate elements of adaptivity, such as response adaptive randomization, adding or dropping treatment arms, as well as GS testing. The BATTLE (Kim et al., 2011) and I-SPY 2 (Barker et al., 2009) trials are high-profile examples of umbrella and platform trials, respectively, using flexible Bayesian designs calibrated to control the overall type I error rate. The STAMPEDE (Carthon and Antonarakis, 2016) trial is an example of a MAMS trial following a frequentist design, where the aim is to control the type I error rate for each pairwise comparison with control rather than maintain strong FWER control.

Inflation of the overall type I error rate will also occur in a seamless phase II/III trial even if we only select one from J treatments or subgroups at the first IA but fail to adjust the final analysis for this selection process. This is because selecting, for example, the best out of J treatments, introduces selection bias and the estimate $\hat{\theta}_{1i^\star}$ comparing the selected treatment $i^\star$ with control based on stage 1 data, will be biased away from the null. The size of this bias will increase with J and depend on the precise configuration of true effects $\theta_1, \ldots, \theta_J$. However, as long as adaptation decisions are based only on the long-term primary endpoint, the increment in the score statistics $\sqrt{\mathcal{I}_{2,i^\star}} Z_{2,i^\star} - \sqrt{\mathcal{I}_{1,i^\star}} Z_{1,i^\star}$ will be independent of $\sqrt{\mathcal{I}_{1,i^\star}} Z_{1,i^\star}$ and follow the canonical joint distribution (Stallard and Todd, 2003). The situation is more complex when information on patients who contribute to the increment $\sqrt{\mathcal{I}_{2,i^\star}} Z_{2,i^\star} - \sqrt{\mathcal{I}_{1,i^\star}} Z_{1,i^\star}$ is used to make the adaptation decision (Bauer and Posch, 2004), that is, when data from a correlated short-term endpoint are used to inform selection decisions. Jenkins et al. (2011) consider using progression-free survival (PFS) to support an interim subgroup selection decision but base final inferences on overall survival (OS). Then, there may be patients who have progressed at the interim but have not yet died: these patients will contribute to both the adaptation and the final OS log-rank statistic. Therefore, the increment in the OS log-rank statistic (final minus the interim statistic) in the selected subgroup will not have the usual null distribution due to selection bias. To overcome this, Jenkins et al. proposed partitioning OS data by stage of recruitment, so that patients recruited before the IA contribute to the stage 1 OS test statistic, and they prespecify the maximum duration of OS follow-up so that this is not influenced by correlated short-term data. Stallard (2010) formulates seamless designs for continuous endpoints using short-term data to inform adaptations.

When discussing adaptive designs, we draw a distinction between: (a) "flexible" adaptive designs permitting unplanned design modifications; (b) preplanned adaptive designs; and (c) GS tests. GS tests allow decisions about the continuation of recruitment to be determined by unblinded accumulating data usually summarized by a sufficient statistic for θ. However, other important aspects of the design, such as the maximum information level, the information timing of future IAs, and the shape of stopping boundaries (i.e. error spending function) are prespecified and should not depend on observed treatment effect estimates (Bretz et al., 2009). Preplanned adaptive designs can accommodate more general adaptations, such as treatment selection or sample size reestimation in response to interim estimates of θ, although adaptation decisions must follow a prespecified rule. Deviations from this rule result in inflation of the type I error rate. Since all aspects of the trial design are prespecified, the (asymptotic) joint distribution of the sufficient test statistics under the global null can be found, and classical designs controlling the type I error rate calculated. While preplanned adaptive designs sacrifice some flexibility, they have several advantages; they are also of interest theoretically, providing useful benchmarks for comparison with flexible adaptive designs. Flexible and preplanned adaptive designs are discussed further in Sections 10.4.2 and 10.4.4 below.

10.4.2 Flexible Adaptive Designs

10.4.2.1 Scope of Flexible Adaptive Designs

Bauer (1989), Bauer and Köhne (1994) based their approach to adaptive designs on the combination of p-values. This idea dates back to Fisher (1925) who noticed that $-2\log(p_1 \cdot$

$p_2)$ has a $\chi^2(4)$-distribution if p_1 and p_2 are uniformly $U(0,1)$-distributed, independent p-values. This fact can be used to construct tests for a hypothesis which was tested on two independent datasets or, more generally, with two stochastically independent tests. From the 1990s onward, this idea was generalized in many directions leading to a large number of suggestions for combining stages of a multistage clinical trial, either by combining p-values or test statistics or calculating conditional critical values for subsequent stages of a multistage trial. Proschan and Hunsberger (1995), Lehmacher and Wassmer (1999), and Müller and Schäfer (2001, 2004) among many others made important contributions. Bretz et al. (2009), and Wassmer and Brannath (2016) give an overview.

One of the key ideas in this regard is given by the conditional rejection principle (CRP) (Müller and Schäfer, 2004) which we will very briefly describe in an informal way: Assume that a clinical trial is planned in two stages aiming at the test of a treatment effect described by a parameter θ. Assume further that in the original plan for the trial, there are two test statistics t_1 and t_2 for the two stages, both testing the same hypothesis about θ at IA $j = 1, 2$. Finally, assume that after IA 1, stage 2 of the trial is modified and that this modification results in a test statistic t_2^* which is different from the originally planned t_2. If we can establish that the distribution of $(t_2^* \,|\text{stage-1-data})$ under the hypothesis is the same as that of $(t_2 \,|\text{stage-1-data})$ for all realizations of t_1, then we are allowed to test the hypothesis using t_2^* instead of t_2. (A conservative test arises if $(t_2^* \,|\text{stage-1-data})$ is stochastically smaller than $(t_2 \,|\text{stage-1-data})$ under the hypothesis.) This general principle is very powerful and encompasses many possible adaptive tests such as combination p-value approaches and tests based on the conditional error function (CEF). One of the most popular methods is the inverse normal p-value combination approach (Lehmacher and Wassmer, 1999): At IA 1 of the two-stage trial, the p-value p_1 is obtained. H_0 is rejected if $p_1 \leq \alpha_1$, or equivalently if $\Phi^{-1}(1-p_1) \geq \Phi^{-1}(1-\alpha_1)$. If H_0 cannot be rejected at IA 1, stage 2 data are collected and a test that uses only these stage 2 data is used to produce p-value p_2 which is stochastically independent of the stage 1 data. Before the trial was started, a weight $w \in [0,1]$ had been fixed. This is used to calculate the combination statistic

$$w \cdot \Phi^{-1}(1-p_1) + \sqrt{1-w^2} \cdot \Phi^{-1}(1-p_2) \tag{10.10}$$

which under H_0 follows an $N(0,1)$-distribution. Because of this latter fact, we can now test the hypothesis again with critical value c_2 which fulfills $1 - \Phi_w\left(\Phi^{-1}(1-\alpha_1), c_2\right) = \alpha$ where $\Phi_{w_1}(x,y)$ denotes the cdf of the bivariate Normal distribution with means 0, variances 1 and correlation w. Note that c_2 is also the critical value which would arise in a two-stage GS design with normally distributed test statistic where early stopping for efficacy is done at IA 1 when $p_1 \leq \alpha_1$ and the correlation between interim and final test statistic is w.

Section 10.4.3 discusses the application of flexible adaptive designs in more detail.

10.4.2.2 Combination Approaches for Implementing Flexible Designs

In theory, the combination approaches described in Section 10.4.2.1 allow a complete re-design of an entire study after the IA. For example, the p-values p_1 and p_2 from level-α-tests on the first- and second-stage data, respectively, can be combined into a single α-level test by Fisher's combination function $-2\log(p_1 \cdot p_2)$, as long as they are stochastically independent under H_0. The concrete statistical tests from which they arise are irrelevant[2]. Hence, one

[2] All that needs to be established is that the two p-values are stochastically independent. Oftentimes, this is achieved by deriving them from a stage 2 test statistic t_2 which conditional on stage 1 data always has the same distribution (typically $N(0,1)$) and thus is also unconditionally stochastically independent of the stage 1 data. The p-value p_2 is then usually simply considered to be $p_2 = 1 - \Phi(t_2)$ which means that it has a uniform distribution on $(0,1)$ if $t_2 \,|\, \text{stage 1 data} \sim N(0,1)$. The details of this approach can be tricky, however. For example, a stage 2 t-test in an adaptive design where the sample size is changed after

could in theory redesign the entire study, change the primary endpoint, remove treatment arms or introduce entirely new treatment arms, switch from a parametric to a nonparametric test, introduce previously unplanned IAs and the like.

In practice, however, regulatory, operational, and interpretational considerations force trialists to obey several limitations. For example, health authorities demand a concrete time limit for trial termination and a final analysis. They usually also insist on a limited number of IAs to be prespecified and put high hurdles on the introduction of additional IAs into an ongoing trial (even if the data remains blinded). In addition, upfront planning of a trial must make provisions for the handling of drug supply, preparing sites to participate in the trial, database locking and data cleaning required before an IA, etc. Finally, the rejection of a generic null hypothesis such as "the new treatment has no advantage over the standard of care" is unsatisfactory if "advantage" means different things in two stages of a clinical trial.

10.4.3 Example

The trial introduced in Section 10.2 had two primary endpoints, but for the sake of simplicity, we will focus on one of them here, a cognitive ability score called APCC. An IA was planned after $n_1 = 300$ patients had completed 5 years of study. At this point in time, safety data on the high dose H would be used to make a decision about stage 2 of the trial. This can be viewed as a sample size re-estimation where the number of patients randomized to dose H may be reduced to 0 and that on the low dose L increased correspondingly. More concretely, without any safety concerns about dose H, another $n_{H2} = n_{C2} = 680$ patients would be studied for 5 years on dose H and control treatment C, respectively, and another $n_{L2} = 340$ on dose L. In case of safety concerns, no further high-dose patients would be studied in stage 2 of the trial, but recruitment into the dose L arm would be intensified to yield 740 patients on dose L in stage 2. Ultimately, the study would have $n_{HF} : n_{LF} : n_{CF} = 800 : 400 : 800$ patients $(H : L : C)$ in case of no safety concerns and $n^\star_{LF} : n^\star_{CF} = 800 : 800$ patients $(L : C)$ if safety concerns arise on dose H. In the latter case, no statistical test for efficacy of dose H would be done.

It is not acceptable to condition on the "observed" sample sizes. The sample sizes are modified due to outcomes (safety data in this case) which may be related to the primary endpoint. Only if we were sure that the safety results are stochastically independent of efficacy results, it would be permissible to condition on them, i.e. treat the modified sample sizes as if they had been fixed in advance.

Under the assumption that APCC asymptotically has a normal distribution, the joint distribution of the test statistics at the IA is

$$\begin{pmatrix} Z_{H1} \\ Z_{L1} \end{pmatrix} \sim N\left(\begin{pmatrix} f_{H1}\theta_H \\ f_{L1}\theta_L \end{pmatrix}, \begin{pmatrix} 1 & \sqrt{\frac{n_{H1}}{n_{H1}+n_{C1}}\frac{n_{L1}}{n_{L1}+n_{C1}}} \\ \sqrt{\frac{n_{H1}}{n_{H1}+n_{C1}}\frac{n_{L1}}{n_{L1}+n_{C1}}} & 1 \end{pmatrix} \right) \tag{10.11}$$

where $Z_{jk} = f_{jk} \cdot \frac{\bar{x}_{jk} - \bar{x}_{Ck}}{\sigma}$ is the test statistic for comparing dose $j = H, L$ with the control group C (in stage $k = 1, 2$ or with all data across the two stages if $k = F$; 'F' for "final"); $n_{j1}, j = H, L, C$ is the interim sample size of group j; θ_j is the true treatment difference between group $j = H, L$ and C and

$$f_{jk} = \sqrt{\frac{n_{jk} \cdot n_{Ck}}{n_{jk} + n_{Ck}}} \text{ for } k = 1, 2, \text{ or } F.$$

stage 1 is only asymptotically independent of the stage 1 data, because its degrees of freedom depend on the modified sample size which in turn depends on stage 1 observations.

By construction the final test statistic is $Z_{jF} = f_{jF} \cdot \frac{\bar{x}_{jF} - \bar{x}_{CF}}{\sigma} = wZ_{j1} + \sqrt{1-w^2} Z_{j2}$ with $w = \sqrt{\frac{n_{C1}}{n_{CF}}}$ (the same w for all three groups).

If the high dose appears safe at the IA, we would proceed with the trial as planned. At the final analysis, there would be two final Z-statistics Z_{HF} and Z_{LF} based on n_{CF}, n_{HF} and n_{LF} patients, respectively. The joint distribution of the test statistics is given by

$$\begin{pmatrix} Z_{H1} \\ Z_{L1} \\ Z_{HF} \\ Z_{LF} \end{pmatrix} \sim N \left(\begin{pmatrix} f_{H1}\theta_H \\ f_{L1}\theta_L \\ f_{HF}\theta_H \\ f_{LF}\theta_H \end{pmatrix}, \begin{pmatrix} 1 & \rho_{HL} & w & w\rho_{HL} \\ \rho_{HL} & 1 & w\rho_{HL} & w \\ w & w\rho_{HL} & 1 & \rho_{HL} \\ w\rho_{HL} & w & \rho_{HL} & 1 \end{pmatrix} \right) \tag{10.12}$$

where $\rho_{HL} = \sqrt{\frac{n_{H1}}{n_{H1}+n_{C1}}} \sqrt{\frac{n_{L1}}{n_{L1}+n_{C1}}}$ is the correlation between Z_{H1} and Z_{L1} and $w = \sqrt{\frac{n_{H1}}{n_{HF}}}$ is the correlation between Z_{H1} and Z_{HF}. Note that ρ_{HL} is also the correlation between Z_{HF} and Z_{LF} and between Z_{H2} and Z_{L2} and w is also the correlation of Z_{L1} and Z_{LF}.

We can equivalently write (10.12) in terms of the independent increments as

$$\begin{pmatrix} Z_{H1} \\ Z_{L1} \\ Z_{H2} \\ Z_{L2} \end{pmatrix} \sim N \left(\begin{pmatrix} f_{H1}\theta_H \\ f_{L1}\theta_L \\ f_{H2}\theta_H \\ f_{L2}\theta_L \end{pmatrix}, \begin{pmatrix} 1 & \rho_{HL} & 0 & 0 \\ \rho_{HL} & 1 & 0 & 0 \\ 0 & 0 & 1 & \rho_{HL} \\ 0 & 0 & \rho_{HL} & 1 \end{pmatrix} \right) \tag{10.13}$$

There are several options for implementing the adaptive trial in this case. All of them use the closed test procedure (see Chapter 1 of this book): First, the intersection hypothesis $H_0 = H_{0L} \cap H_{0H}$ is tested at level α. If and only if H_0 is rejected, $H_{0L} : \theta_L \leq 0$ and $H_{0H} : \theta_H \leq 0$ are tested at level α. H_0 is rejected if $\max(Z_{L1}, Z_{H1}) \geq c_1$ where c_1 is such that $1 - \Phi_{\rho_{HL}}(c_1, c_1) = \alpha_1$ where $\alpha_1 < \alpha$ is the "α spent" at the interim. Let $p_1 = 1 - \Phi_{\rho_{HL}}(\max(Z_{L1}, Z_{H1}), \max(Z_{L1}, Z_{H1}))$ be the corresponding p-value for H_0. Then an equivalent way of stating the rejection criterion for H_0 is

$$\Phi^{-1}(1 - p_1) \leq \Phi^{-1}(1 - \alpha_1).$$

If this is fulfilled, then H_{0j}, $j = L, H$ will be tested using a standard Z-test which rejects H_{0j} if

$$Z_{j1} \geq \Phi^{-1}(1 - \alpha_1) \Leftrightarrow p_{j1} = 1 - \Phi(Z_{j1}) \leq \alpha_1.$$

It is easy to show that $c_1 \geq \Phi^{-1}(1 - \alpha_1)$ and that therefore rejection of H_0 implies rejection of either H_{0L} or H_{0H}, i.e. this closed test procedure is consonant.

If H_0 cannot be rejected at the interim, and modifications to the preplanned design are contemplated, the following options can be considered for the final analysis:

1. Inverse normal p-value combination: Before start of the trial, we fix weights $0 \leq w_0 \leq 1$ and $0 \leq w_j \leq 1$, $j = H, L$ for stage 1 of the trial. At the final, we use the test statistic

$$t^* = w_0 \cdot \Phi^{-1}(1 - p_1) + \sqrt{1 - w_0^2} \cdot \Phi^{-1}(1 - p_2^*)$$

   to test for H_0. Here $p_2^* = 1 - \Phi_{\rho^*_{HL}}(\max(Z^*_{L2}, Z^*_{H2}), \max(Z^*_{L2}, Z^*_{H2}))$ if dose H is not dropped. If dose H is dropped at the interim, we use $p_2^* = p^*_{L2} = 1 - \Phi(Z^*_{L2})$. Rejection of H_0 occurs if $t^* > c$ where c is calculated from $1 - \Phi_{w_0}(\Phi^{-1}(1 - \alpha_1), c) = \alpha$. The notation Z^*_{jk} indicates that we allow the stage 2 sample size to change from the original plan from n_{j2} to n^*_{j2}. The correlation ρ^*_{HL} only changes if the sample size ratios $H : L : C$ change between the two stages.

We note that $t^* \neq \max(Z_{LF}, Z_{HF})$; not even if the original design remains unchanged at the interim (no dropping of dose H, no sample size changes in stage 2). Thus, this method does not exactly recreate the usual Dunnett test statistic at the final analysis.

If H_0 can be rejected, H_{0j} will be tested using $Z_j^* = w_j \cdot \Phi^{-1}(1-p_{j1}) + \sqrt{1-w_j^2} \cdot \Phi^{-1}(1-p_{j2}^*)$ with the "group sequential" critical value c_j that fulfills $1 - \Phi_{w_j}\left(\Phi^{-1}(1-\alpha_1), c_j\right) = \alpha$. If $w_j = w = \sqrt{\frac{n_{C1}}{n_{CF}}}$ is chosen and $\frac{n_{j1}}{n_{jF}} = \frac{n_{C1}}{n_{CF}}$, then Z_j^* is the usual z-test statistic at the final analysis, if no changes to the preplanned sample sizes were made in groups j and C.

2. Conditional error function approach: To test for H_0, we calculate the conditional error

$$A(\text{stage-1-data}) = P_{H_0}(\max(Z_{LF}, Z_{HF}) \geq c\,|\text{stage-1-data}) \tag{10.14}$$

where c is obtained from (10.12) as

$$P_{H0}(\max(Z_{L1}, Z_{H1}) < c_1, \max(Z_{LF}, Z_{HF}) \geq c) = \alpha - \alpha_1.$$

If $p_2^* < A(\text{stage-1-data})$, H_0 is rejected. Here again, $p_2^* = 1 - \Phi_{\rho_{HL}^*}(\max(Z_{L2}^*, Z_{H2}^*), \max(Z_{L2}^*, Z_{H2}^*))$ is used if dose H is not dropped and $p_2^* = p_{L2}^* = 1 - \Phi(Z_{L2}^*)$ if dose H is dropped. If H_0 can be rejected, H_{0j} is tested correspondingly. It can be shown that for H_0, this procedure can also be represented as an inverse normal p-value combination (Posch and Bauer, 1999).

This description is not unique. The stage 1 data we are conditioning on could be (Z_{L1}, Z_{H1}) or it could only be $\max(Z_{L1}, Z_{H1})$. These two approaches lead to different inferences on H_0. It is easier to work with (Z_{L1}, Z_{H1}) because then $A(Z_{L1}, Z_{H1}) = 1 - \Phi_{\rho_{HL}}\left(\frac{c - wZ_{L1}}{\sqrt{1-w^2}}, \frac{c - wZ_{H1}}{\sqrt{1-w^2}}\right)$.

3. Recalculation of conditional critical values (cumulative MAMS) approach: Based on the conditional rejection principle (Müller and Schäfer, 2004), Ghosh et al. (2020) suggest to test H_0 with the use of Z-test statistics Z_{jF}^* which are calculated from all data across the two stages as if no modification after stage 1 had been done. If dose H is dropped, $t^* = Z_{LF}^*$ is used as the test statistic for H_0. If dose H is not dropped, $t^* = \max(Z_{LF}^*, Z_{HF}^*)$ is used. The critical values are calculated by matching:

$$P_{H_0}(t^* \geq c^*|\text{stage-1-data}) = P_{H_0}(\max(Z_{LF}, Z_{HF}) \geq c\,|\text{stage-1-data}). \tag{10.15}$$

Here, c is obtained from (10.12) since

$$P_{H_0}(\max(Z_{L,1}, Z_{H1}) < c_1, \max(Z_{LF}, Z_{HF}) \geq c) = \alpha - \alpha_1,$$

just like in the CEF approach. Equation (10.15) must be solved numerically for c^*. This is feasible if $n_{C1}/n_{C2} = n_{j1}/n_{j2} = n_{j1}^*/n_{j2}^*$ for all nondropped treatment arms j, i.e. if the randomization ratios between the treatment arms which stay in the study are not changed across the stages; only the total sample size of stage 2 being subject to changes.

Like with the CEF approach, the stage 1 data to condition on can be (Z_{L1}, Z_{H1}) or $\max(Z_{L1}, Z_{H1})$ giving different versions of this approach. Unlike the inverse Normal and the CEF approach, this method does not combine separate test statistics or p-values from the two stages. Hence, if no modifications are made relative to the original plan, the decision rule for H_0 is exactly the same as with the corresponding GS test.

If H_0 is rejected, H_{0j} is tested in the same way as with the other two approaches.

The inverse normal combination has the advantage of simplicity. The inverse normal combination and the CEF approach can be used even if the sample size ratios n_{Lk}/n_{Ck} change between stages $k = 1$ and 2. According to Ghosh et al. (2020) there is, however, a power loss relative to the cumulative MAMS approach.

Since in this example, it is desired to change the sample size in the case where the high dose must be dropped for safety concerns, we illustrate the approach using the inverse Normal p-value combination. It is not foreseen that the study stops at the interim, so we use $\alpha_1 = 0$ (in practice, often a very small α_1 is used, e.g. $\alpha_1 = 10^{-6}$, which can be ignored for power calculations or the final analysis). At the IA, the decision is to either continue as planned with both doses or drop the high dose and adjust sample sizes in the remaining low dose arm as described above. We could potentially also consider changing the sample size based on the interim results in some other way or to stop the study altogether for futility. As described in Section 10.3.1, this futility stop is considered to be "nonbinding".

To implement the combination p-value approach, we need to specify weights for the two stages of the three tests for H_0, H_{0H} and H_{0L}. For H_{0H}, the choice is obvious: If dose H makes it to stage 2, there will be n_{jk}, $k = 1, 2$ patients in stage k on both groups $j = C, H$. Hence, the optimal weight is $w_H = \sqrt{n_{H1}/n_{HF}} = \sqrt{120/800} = 0.387$. Since the final Z-test statistic calculated from 1,600 patients in groups H and C at the final analysis can be written as $Z_{HF} = w_H \cdot Z_{H1} + \sqrt{1 - w_H^2} \cdot Z_{H2}$, this simplifies to the usual two-group Z-test.

For H_{0L}, the optimal weight also is $w = \sqrt{n_{C1}/n_{CF}} = w_H$ if the trial moves on in an unmodified way to the final analysis. However, if the high dose is safe, it is anticipated that it will have much better efficacy, hence a test of the low dose is of lesser importance in this situation. For this reason, it was decided to use weights which focus on a high power of testing L versus C when dose H is dropped. In this case, stage 1 has a total of $n_{C1} + n_{L1} = 180$ patients who are relevant to the comparison of L and C and stage 2 has $n_{C2}^* + n_{L2}^* = 680 + 740 = 1420$ patients. We therefore pick weights

$$w_L = \sqrt{\frac{\frac{1}{n_{L2}^*} + \frac{1}{n_{C2}^*}}{\frac{1}{n_{L1}} + \frac{1}{n_{C1}} + \frac{1}{n_{L2}^*} + \frac{1}{n_{C2}^*}}} = 0.318$$

for testing H_{0L} at the final analysis. The weight w_L corresponds to the sample-size factor f_{LF} in a two-sample Z-test with n_{LF}^* and n_{CF}^* patients, respectively (see formula (10.12)). However, since the randomization ratio between C and L changes from stage 1 to stage 2, this approach does not reproduce the unmodified usual two-sample-Z-test statistic.

Regarding the test of the intersection hypothesis H_0, we use the same weights as for the high dose ($w_0 = w_H$), that is, we use the test statistic

$$t^* = w_H \cdot \Phi^{-1}(1 - p_1) + \sqrt{1 - w_H^2} \cdot \Phi^{-1}(1 - p_2^*)$$

where $p_1 = 1 - \Phi_{\rho_{HL}}(\max(Z_{L1}, Z_{H1}), \max(Z_{L1}, Z_{H1}))$ and $p_2^* = 1 - \Phi_{\rho_{HL}}(\max(Z_{L2}, Z_{H2}), \max(Z_{L2}, Z_{H2}))$ if dose H is continued and $p_2^* = 1 - \Phi(Z_{L2}^*)$ if it is discontinued.

To give a numerical example, let us assume that after stage 1 of the trial, we obtain $Z_{L1} = 0.78$ and $Z_{H1} = 1.42$, corresponding to $p_{L1} = 0.218$ and $p_{H1} = 0.078$. Hence, $\max(Z_{H1}, Z_{L1}) = 1.42$ which corresponds to the p-value $p_1 = 1 - \Phi_{\rho_{HL}=0.408}(1.42, 1.42) = 0.137$.

We go through both cases of dose H discontinued or not:

1. *Dose H is not discontinued*: Recruitment continues as planned and results in stage 2-only-observations of $Z_{L2} = 1.80$ and $Z_{H2} = 3.29$. In this case, we thus have $Z_{LF} = w_H \cdot Z_{L1} + \sqrt{1 - w_H^2} \cdot Z_{L2} = 1.96$ and likewise $Z_{HF} = 3.58$. As we are using the p-value-combination approach, $Z_{HF} = \max(Z_{LF}, Z_{HF})$ is not the test statistic for the

intersection hypothesis H_0. This is rather $t = w_H \cdot \Phi^{-1}(1-p_1) + \sqrt{1-w_H^2} \cdot \Phi^{-1}(1-p_2) = 3.275$ with a corresponding p-value of 0.000528. Hence, H_0 can be rejected. If there had been no IA, the test statistic would have been 3.58, but the p-value calculation would also have to be modified to $p = 1 - \Phi_{\rho_{HL}}(3.58, 3.58) = 0.000337$. Obviously, there must be a power loss from using the combination, but here as well as in many other examples, it is not very severe.

Sample sizes n_{H2} and n_{C2} were realized as preplanned, therefore the test for H_{0H} is the ordinary Z-test at the final analysis with $Z_{HF} = 3.58$ and a corresponding p-value of 0.00017. Thus, by the closed test principle, it is concluded that the high dose provides a statistically significant benefit over the control treatment.

Regarding H_{0L}, it was decided upfront to use the weight w_L for stage 1. Hence, the test statistic becomes $Z_L = w_L \cdot Z_{L1} + \sqrt{1-w_L^2} \cdot Z_{L2} = 1.955$ with a p-value of 0.02531. Thus, H_{0L} can just not be rejected at a level of $\alpha = 0.025$. In contrast, had it been planned from the start to use the weights $w = w_H$ for Z_L as well, the p-value from the ordinary Z-test would have been 0.02490 - just below $\alpha = 0.025$.

2. *Dose H is discontinued*: No additional patients are recruited into treatment arm H. Instead, the stage 2 sample size for dose L is increased to $n_{L2}^* = 740$. Assume that we observe $Z_{L2}^* = 2.03$. To test H_0, we use the combination test statistic $t^* = w \cdot \Phi^{-1}(1-p_1) + \sqrt{1-w^2} \cdot Z_{L2}^* = 2.295$ with p-value 0.011 such that H_0 is rejected. H_{0L} is now tested using $Z_L^* = w_L \cdot Z_{L1} + \sqrt{1-w_L^2} \cdot Z_{L2}^* = 2.173$ (p-value 0.0149). Hence, H_{0L} is rejected. Again, there is a small efficiency loss from the combination: If the two stages had been preplanned with total sample sizes $n_{L1}+n_{L2}^*$ and $n_{C1}+n_{C2}$, respectively, the resulting Z-test would have given a test statistic $Z_L = 2.267$ (p-value 0.0117). As mentioned above, the p-value combination approach cannot reproduce this test exactly.

10.4.4 Preplanned Adaptive Designs

Inferences for a "flexible" adaptive design permitting, for example, sample size modifications, are commonly based on a weighted combination of stagewise test statistics. The combination weights must be prespecified, typically on the basis of planned stagewise sample sizes, meaning that when the target sample size is adapted, the weight assigned to any stagewise statistic may not be proportional to the information it actually represents. The consequence of using a test statistic which does not weight all observations equally is that inference is based on a nonsufficient statistic for θ (Jennison and Turnbull, 2003).

However, sufficiency of the test statistic is only one factor in determining the operating characteristics of an adaptive clinical trial. Other design choices will play a role, such as the interim adaptation rule and the shape of the stopping boundaries. Preplanned adaptive designs base inferences on sufficient statistics and can be further optimized and used as benchmarks for comparison with flexible adaptive designs. For example, Jennison and Turnbull (2006a) quantify the impact of adaptivity on efficiency by comparing the average expected sample sizes of the optimal preplanned adaptive design permitting unblinded sample size re-estimation and the optimal non-adaptive GS test. Comparing the sample size re-assessment rule for the $(k+1)$th group under the optimal adaptive design ($n_{\text{opt}}(\hat{\theta}_k)$) with that of the rule targeting a specified conditional power ($n_{\text{CP}}(\hat{\theta}_k)$), the authors find that $n_{\text{opt}}(\hat{\theta}_k)$ is an umbrella shaped function of $\hat{\theta}_k$ while $n_{\text{CP}}(\hat{\theta}_k)$ recommends larger increases in sample size as $\hat{\theta}_k$ decreases. Optimal preplanned adaptive designs achieve average expected sample sizes which are smaller than the minima achieved by non-adaptive GS tests, although savings are small (Jennison and Turnbull, 2006a).

As mentioned in Section 10.4.1, seamless phase II/III trials are often run as preplanned adaptive designs. A single overarching protocol is written ahead of time which prespecifies all aspects of the rule that will be used at the end of phase II by the IDMC to adapt the phase III design. Having this in place reduces the logistical white space that normally separates the phases. However, it also reduces a sponsor's flexibility to respond to unexpected safety signals in phase II or changes in the standard of care (Cuffe et al., 2014). This means seamless confirmatory designs are most appropriate for programs with an already substantial safety database or therapeutic areas with a stable standard of care. As with GS tests, a quickly available primary endpoint, or a short-term surrogate which is highly correlated with the primary endpoint, is also essential to ensure timely adaptations are possible. Stallard and Todd (2003) propose a seamless GS phase II/III design where the best of K treatments, labeled $i^\star$, is selected at the first IA. Since the selection rule is prespecified, one can derive the joint distribution of $\{Z_{1,i^\star}, Z_{2,i^\star}, \ldots\}$ and compute GS boundaries which control the FWER under the global null hypothesis. Jennison and Turnbull (2006b) note that this procedure can be thought of as a closed test which maintains strong FWER control. Magnusson and Turnbull (2013) propose a preplanned seamless procedure which identifies responsive subgroups at the first IA and subsequently GS tests the one-sided null hypothesis of no benefit of treatment in the union of the selected subgroups. Boundaries are designed to control the FWER under the global null, which implies strong FWER control only if effects in different subgroups cannot lie in opposite directions (Magnusson and Turnbull, 2013).

Hampson and Jennison (2015) find optimal frequentist procedures for inferentially seamless phase II/III trials as admissible Bayes tests. They consider trials that begin randomizing n_1 patients to each of J active doses and control. At the end of phase II, one selects the best performing dose so long as it is associated with a positive trend, that is, $\hat{\theta}_{1i^\star} = \max_{j=1,\ldots,J}\{\hat{\theta}_{1j}\} > 0$, otherwise the trial is stopped for futility. A further n_2 patients are then randomized to dose $i^\star$ and control in phase III. In this setting, it is unclear which data should be used to test $H_{0i^\star}$ at the end of phase III. For instance, should we use only $\hat{\theta}_{2i^\star}$, the treatment effect estimate from phase III, which is not subject to any selection bias? Or should we leverage data on treatment $i^\star$ and control from both stages, summarized by $\hat{\theta}_{1i^\star}$ and $\hat{\theta}_{2i^\star}$? Or should we use instead all data from phase II and III, including effect estimates for non-selected treatments since these provide context for the selection decision made at the end of phase II? Hampson and Jennison find optimal procedures which have maximum power, defined as $Pr\{\text{Select the most effective treatment and reject } H_{0i^\star}; \boldsymbol{\theta}\}$, under effect configurations of the form $\boldsymbol{\theta} = (\gamma, \ldots, \gamma, 1)\delta$, in the class of tests controlling $Pr\{\text{Reject any true } H_0; \boldsymbol{\theta} = (0, \ldots, 0)\}$ at level α. The authors then check that weak control of the FWER ensures strong control and while in many cases it does, this is not always the case.

For the seamless testing problem considered by Hampson and Jennison (2015), a uniformly most powerful test of $H_{0i^\star}$ does not exist because the form of the optimal test depends on the configuration of $\boldsymbol{\theta}$ at which we set power. However, the authors find that the Thall, Simon and Ellenberg procedure (TSE, (Thall et al., 1988)), which for a normal response bases a final test of $H_{0i^\star}$ on the statistic shown below, has robust efficiency:

$$\sqrt{n_1/(n_1+n_2)}Z_{1,i^\star} + \sqrt{n_2/(n_1+n_2)}Z_{2,i^\star} > c_1(J, n_1, n_2),$$

where $c_1(J, n_1, n_2)$, is calibrated to ensure the FWER is α when $\boldsymbol{\theta} = (0, \ldots, 0)$, which is sufficient to maintain strong FWER control (Jennison and Turnbull, 2006b).

10.4.5 Estimation in Adaptive Designs

In adaptive designs, there is more flexibility with respect to possible modifications than in GS designs. Hence, the topic of point estimation following the termination of an adaptive design is a lot broader, and different approaches are required for different types of adaptations.

In the prespecified adaptive designs of Section 10.4.4, the concept of unbiasedness can often be rendered more tangible than in GS designs. The cases that have received most attention are those of treatment arm selection and of sample size re-estimation at the IA. The situations considered are characterized by the following features:

- The trial will continue for a predetermined number of stages. There is no early stopping of the entire trial.
- The rules for trial modification are clearly defined before the start of the study. Two important examples are:
  - "drop-the-loser" or "keep-the-winner" rules (e.g. after stage 1, select the two treatments with the highest average response to continue into stage 2)
  - a concrete formula for sample size recalculation depending on the observed interim data (e.g. calculate the stage 2 sample size such that the conditional probability of rejecting the null hypothesis given the interim results is 80%).

In this situation, there is a preplanned final stage of the trial (K, say). If x_{ijk} is the response from patient i in treatment group j and trial stage k and the x_{ijk} are mutually stochastically independent (usually because every patient is under observation only during a single stage j of the trial), then the "last-stage-only" estimate $\bar{x}_{jK} = \frac{1}{n_{jK}} \sum_i x_{ijK}$ is an unbiased estimate of $E(x_{ijk}) = \theta_j$ for those treatments which remain in the study till the final stage K. Hence, $\bar{x}_{jK}$ can be used as the basis for an improved unbiased estimator of the selected treatments using the Rao-Blackwell theorem (Bowden and Glimm, 2008; Cohen and Sackrowitz, 1989; Stallard and Kimani, 2018). This also requires conditioning on the selection rules, but the conditioning is more plausible than for the "truncation-adaptable" estimator which requires users to pretend that they are ignorant of the reasons for why the study stopped after stage S - in the face of the fact that this is something we precisely know in GS trials. Like in the case of GS point estimates, the conditionally unbiased estimates have a larger MSE than the MLE. Hence, a number of other methods (Bebu et al., 2013, 2010; Bowden et al., 2014; Bowden and Glimm, 2014; Carreras and Brannath, 2013; Whitehead, 1986) has been suggested to provide compromises between strictly unbiased estimates and the MLE. For example, the estimator by Carreras and Brannath (2013) is less biased and has a smaller MSE than the MLE.

A disadvantage of these methods is that they only apply to fixed prespecified adaptation rules. Among flexible adaptive designs (see Section 10.4.2), sample size reestimation has received most attention. Assume that in the case where individual observations follow (10.7), the originally planned sample size n_{j2} of treatment arm j in stage 2 of the trial might be modified to $n^{\star}_{j2}$, using $\bar{x}_{j1}$, but without a concrete algorithm for how this sample size modification is calculated. Similar to treatment arm selection and early stopping in GS designs, the MLE is ignorant of such modifications. It is

$$\hat{\theta}_j = \frac{n_{j1}\bar{x}_{j1} + n^{\star}_{j2}\bar{x}_{j2}}{n_{j1} + n^{\star}_{j2}}$$

and is again biased, but the bias can now no longer be precisely quantified, not even if θ_j are assumed to be known. A relatively simple alternative to the MLE $\hat{\theta}_j$ is the median-unbiased estimator (MEUE). It is given by

$$\tilde{\theta}_j = \frac{n_{j1}\bar{x}_{j1} + \sqrt{n_{j2}n^{\star}_{j2}}\bar{x}_{j2}}{n_{j1} + \sqrt{n_{j2}n^{\star}_{j2}}}.$$

The MEUE is not (mean-)unbiased, but less biased than the MLE and performs well in terms of MSE (Bretz et al., 2009).

Regarding interval estimation, generalizations of the methods presented in Section 10.3.7 are applied. We just give a flavor of how these methods work in a simple case. Assume that the inverse normal p-value combination resulted in uniformly $U[0,1]$-distributed, stochastically independent p-values p_1 and p_2 in a two-stage design with sample size reestimation after stage 1. Let $Z_k = \Phi^{-1}(1-p_k)$ be the corresponding test statistic from stage k alone. Assume further that Z_k is related to a parameter θ and a test of $H_0 : \theta = \theta_0$. The quintessential example would be the test of the difference between the two means from two independent normal samples. To be more precise, we assume that $Z_k = \frac{\hat{\theta}_k}{\sigma/\sqrt{\mathcal{I}_k}}$ where σ^2 is a known variance parameter and $\mathcal{I}_k$ the information available from stage $k = 1,2$ such that Z_k is (asymptotically) $N(0,1)$-distributed under H_0. Furthermore Z_2 and $\mathcal{I}_2$ will be indexed with a $*$ where necessary to indicate that they are calculated with a possible sample size modification after stage 1. An (asymptotic) repeated confidence interval for θ is then given by

$$\left[\sigma \cdot \frac{Z_1}{\sqrt{\mathcal{I}_1}} \pm \Phi^{-1}(1-\alpha_1/2) \cdot \sigma/\sqrt{\mathcal{I}_1}\right] \tag{10.16}$$

if the trial is stopped after stage 1 and

$$\left[\sigma \cdot \frac{w \cdot Z_1 + \sqrt{1-w^2} Z_2^*}{w\sqrt{\mathcal{I}_1} + \sqrt{1-w^2} \cdot \sqrt{\mathcal{I}_2^*}} \pm c \cdot \sigma / \left(w\sqrt{\mathcal{I}_1} + \sqrt{1-w^2}\sqrt{\mathcal{I}_2^*}\right)\right] \tag{10.17}$$

if it finishes after stage 2 (Bretz et al. (2009), section 6.4). Here, w is the pre-defined weight for stage 1, $\mathcal{I}_2$ is the originally planned and $\mathcal{I}_2^*$ the actually observed information for stage 2 after the sample size modification; $c = \Phi^{-1}(1-\alpha_2/2)$ is the critical value for the combination test statistic $w \cdot Z_1 + \sqrt{1-w^2} \cdot Z_2^*$ calculated as in Section 10.3.7, i.e. the normal quantile that corresponds to the "local" test level α_2 used at the end of stage 2. Furthermore, if $w = \sqrt{\frac{\mathcal{I}_1}{\mathcal{I}_1+\mathcal{I}_2}}$ is used (where $\mathcal{I}_2$ denotes the originally planned information of stage 2), (10.16) is the usual $1-\alpha_1$-confidence interval based on the normal distribution. If in addition $\mathcal{I}_2 = \mathcal{I}_2^*$, then (10.17) is equal to the repeated confidence interval from the corresponding GS test without a sample size reestimation. We also note in passing that the midpoint of confidence interval (10.17) is the MEUE.

Chapter 8 of the book by Wassmer and Brannath (2016) gives an extensive discussion of confidence intervals for adaptive designs including confidence intervals based on stagewise ordering and a discussion about one- and two-sided intervals. In practice, oftentimes ordinary unadjusted confidence intervals are used for the same reasons that were discussed in Section 10.3.7: Repeated confidence intervals are very conservative, in particular when early stopping arises. Confidence intervals based on stagewise ordering require complicated numerical integrations and one may argue about the appropriateness of stagewise ordering.

10.4.6 Future Developments

We have focussed in this chapter on GS and adaptive clinical trials where type I error control is required. Typically, this requirement is mandatory for pivotal (phase III) clinical trials which are submitted to health authorities to obtain approval for a new treatment. These trials focus on *confirming* a beneficial treatment effect. Earlier phases of clinical development are usually *exploratory* in nature. Type I error rate control is less important and therefore, methods not (strictly) controlling type I errors are often applied. Examples of such methods are response-adaptive designs (Atkinson and Biswas, 2019) and biased coin and urn designs (Antognini and Giovagnoli, 2019). In such designs, randomization probabilities are continuously modified, depending on responses from previously treated patients in an attempt to "zero in" on the best of many treatment options. Bayesian methods are often applied to do the corresponding calculations.

This chapter does not deal with these methods. There is, however, a recent trend toward confirmatory clinical trials which allow more flexibility than even the adaptive designs discussed here. These may contain elements from confirmatory adaptive designs, response-adaptive designs, and other modifications of an ongoing trial. As briefly mentioned in Section 10.4.1, such trials have recently been discussed under the label "master protocols", see Woodcock and LaVange (2017) for an overview. The increased research interest has been fueled by the US 21st Century Cures Act and the need for speed in reaction to the Ebola epidemic (Dodd et al., 2016) and the COVID-19 pandemic.

It is uncontroversial that for master protocols, multiplicity and the corresponding probability of erroneous decisions regarding treatment efficacy are important concerns. Currently, however, important questions surrounding the precise quantification of type I errors, and the requirements on the limitations of type I error probabilities are not resolved. The scientific community and health authorities will have to find a new consensus on these questions in the near future.

10.5 Discussion

This chapter presents GS and adaptive designs applied in confirmatory clinical trials. The use of GS methods in clinical trials dates back to the early 1980s and initially met with some resistance from health authorities, but by now these methods are firmly established in clinical trials and routinely applied in many therapeutic areas. In oncology, for example, GS trials are by now the norm rather than the exception.

Adaptive designs have also become more widespread and found their way into applications (e.g. Martin et al. (2017), Barnes et al. (2010) for just two examples). Selection among prespecified subpopulations (for example, patients with or without genetic mutations suspected to influence cancer tumor growth) and selection of treatment regimens are quite common. Still, adaptive designs are not nearly as ubiquitous as GS designs. In spite of the fact that the statistical methodology for adaptive designs is well developed and implemented in readily-available software, e.g. EAST (Cytel, 2016) and the R package RPACT (Wassmer and Pahlke, 2019), concerns about trial integrity and operational feasibility sometimes "kill" adaptive designs. In big, complex and expensive phase III programs, they can sometimes offer substantial benefits.

There are many related topics which we are not covering here. We already mentioned response-adaptive and biased-coin designs, but we also do not discuss blinded sample size reevaluation (Kieser and Friede, 2003), the so-called minimization designs (Pocock and Simon, 1975; Proschan et al., 2011) or the use of resampling tests in adaptive designs (Proschan, 2017; Proschan et al., 2014).

Acknowledgments

We would like to thank the editors and the reviewers for their comments on an earlier version of this chapter. Their comments helped improving the chapter.

Bibliography

Anderson, K. (2016). gsdesign (version 3.0-1).

Antognini, A. and A. Giovagnoli (2019). *Adaptive Designs for Sequential Treatment Allocation.* Chapman & Hall / Boca Raton, FL: CRC Press.

Atkinson, A. and A. Biswas (2019). *Randomised Response-Adaptive Designs in Clinical Trials.* Chapman & Hall / Boca Raton, FL: CRC Press.

Barber, S. and C. Jennison (2002). Optimal asymmetric one-sided group sequential tests. *Biometrika 89*(1), 49–60.

Barker, A., C. Sigman, G. Kelloff, N. Hylton, D. Berry, and L. Esserman (2009). I-SPY 2: An adaptive breast cancer trial design in the setting of neoadjuvant chemotherapy. *Clinical Pharmacology and Therapeutics 86*(1), 97–100.

Barnes, P., S. Pocock, H. Magnussen, A. Iqbal, B. Kramer, M. Higgins, and D. Lawrence (2010). Integrating indacaterol dose selection in a clinical study in COPD using an adaptive seamless design. *Pulmonary Pharmacology & Therapeutics 23*(3), 165–171.

Baud, O., L. Maury, F. Lebail, D. Ramful, F. El Moussawi, C. Nicaise, V. Zupan-Simunek, A. Couresol, A. Beuchè, P. Bolot, P. Andrini, D. Mohamed, and C. Alberti (2016). Effect of low-dose hydrocortisone on survival without bronchopulomary dysplasia in extremely preterm infants (PREMILOC): a double-blind, placebo-controlled, multicentre, randomised trial. *Lancet 387*, 1827–1836.

Bauer, P. (1989). Multistage testing with adaptive designs. *Biometrie und Informatik in Medizin und Biologie 20*, 130–148.

Bauer, P., B. Bretz, V. Dragalin, F. König, and G. Wassmer (2016). Twenty-five years of confirmatory adaptive designs: opportunities and pitfalls. *Statistics in Medicine 35*(3), 325–347.

Bauer, P. and K. Köhne (1994). Evaluation of experiments with adaptive interim analyses. *Biometrics 50*(4), 1029–1041.

Bauer, P., F. König, W. Brannath, and M. Posch (2010). Selection and bias – two hostile brothers. *Statistics in Medicine 29*(1), 1–13.

Bauer, P. and M. Posch (2004). Modifcation of the sample size and the schedule of interim analyses in survival trials based on data inspections. *Statistics in Medicine 23*(8), 1333–1335.

Bebu, I., V. Dragalin, and G. Luta (2013). Confidence intervals for confirmatory adaptive two-stage designs with treatment selection. *Biometrical Journal 55*(3), 294–309.

Bebu, I., G. Luta, and V. Dragalin (2010). Likelihood inference for a two-stage design with treatment selection. *Biometrical Journal 52*(6), 811–822.

Bhatt, D. and C. Mehta (2016). Adaptive designs for clinical trials. *The New England Journal of Medicine 375*, 65–74.

Bowden, J., W. Brannath, and E. Glimm (2014). Empirical Bayes estimation of the selected treatment mean for two-stage drop-the-loser trials: a meta-analytic approach. *Statistics in Medicine 33*(3), 388–400.

Bowden, J. and E. Glimm (2008). Unbiased estimation of selected treatment means in two-stage trials. *Biometrical Journal 50*(4), 515–527.

Bowden, J. and E. Glimm (2014). Conditionally unbiased and near unbiased estimation of the selected treatment mean for multistage drop-the-losers trials. *Biometrical Journal 56*(2), 332–349.

Bretz, F., F. König, W. Brannath, E. Glimm, and M. Posch (2009). Adaptive designs for confirmatory clinical trials. *Statistics in Medicine 28*(8), 1181–1217.

Carreras, M. and W. Brannath (2013). Shrinkage estimation in two-stage adaptive designs with mid-trial treatment selection. *Statistics in Medicine 32*(10), 1677–1690.

Carthon, B. and E. Antonarakis (2016). The STAMPEDE trial: paradigm-changing data through innovative trial design. *Translational Cancer Research 5*(3), S485–S490.

Cerqueira, F., A. Jesus, and M. Cotrim (2020). Adaptive design: a review of the technical, statistical, and regulatory aspects of implementation in a clinical trial. *Therapeutic Innovation and Regulatory Science 54*(1), 246–258.

Chang, M. (1989). Confidence intervals for a normal mean following a group-sequential test. *Biometrics 45*(1), 247–254.

Cohen, A. and H. Sackrowitz (1989). Two stage conditionally unbiased estimators of the selected mean. *Statistics and Probability Letters 8*(3), 273–278.

Collignon, O., C. Gartner, A. Haidich, R. Hemmings, B. Hofner, F. Pétavy, M. Posch, K. Rantell, K. Roes, and A. Schiel (2020). Current statistical considerations and regulatory perspectives on the planning of confirmatory basket, umbrella, and platform trials. *Clinical Pharmacology & Therapeutics 107*(5), 1059–1067.

Collignon, O., F. Koenig, A. Koch, R. Hemmings, F. Pétavy, A. Saint-Raymond, M. Papaluca-Amati, and M. Posch (2018). Adaptive designs in clinical trials: from scientific advice to marketing authorisation to the European Medicine Agency. *Trials 19*, 642–656.

Cox, D. and D. Hinkley (1974). *Theoretical Statistics*. Boca Raton, FL: Chapman and Hall.

Cuffe, R., D. Lawrence, D. Stone, and M. Vandemeulebroecke (2014). When is a seamless study desirable? Case studies from different pharmaceutical sponsors. *Pharmaceutical Statistics 13*(4), 229–237.

Cytel (2016). East 6 (version 6.4).

Dodd, L., M. Proschan, J. Neuhaus, J. Koopmeiners, J. Neaton, J. Beigel, K. Barrett, H. Lane, and R. Davey Jr (2016). Design of a randomized controlled trial for Ebola virus disease medical countermeasures: PREVAIL II, the Ebola MCM Study. *Journal of Infectious Diseases 213*(12), 1906–1913.

Dodge, H. and H. Romig (1929). A method for sampling inspection. *Bell System Technical Journal 8*(4), 613–631.

Eales, J. (1995). Optimal two-sided group sequential tests. *Sequential Analysis 14*(4), 273–286.

Eales, J. and C. Jennison (1992). An improved method for deriving optimal one-sided group sequential tests. *Biometrika 79*(1), 13–24.

Elsässer, A., J. Regnstrom, T. Vetter, F. Koenig, R. Hemmings, M. Greco, M. Papaluca-Amati, and M. Posch (2014). Adaptive clinical trial designs for European marketing authorization: a survey of scientific advice letters from the European Medicines Agency. *Trials 15*, 383.

Emerson, S. and T. Fleming (1990). Parameter estimation following group-sequential hypothesis testing. *Biometrika 77*(3), 875–892.

Englert, S. and M. Kieser (2012). Improving the flexibility and efficiency of phase II designs for oncology trials. *Biometrics 68*(3), 886–892.

European Medicines Agency (2007). *Reflection paper on methodological issues in confirmatory clinical trials planned with an adaptive design.* Amsterdam: European Medicines Agency.

Fisher, R. (1925). *Statistical methods for research workers.* Oliver and Boyd, Edinburgh.

Food and Drug Administration (2018). *Master Protocols: Efficient Clinical Trial Design Strategies to Expedite Development of Oncology Drugs and Biologics–Guidance for Industry.* U.S. Department of Health and Human Services.

Food and Drug Administration (2019). *Adaptive Designs for Clinical Trials of Drugs and Biologics – Guidance for Industry.* U.S. Department of Health and Human Services.

Galbraith, S. and I. Marschner (2003). Interim analysis of continuous long-term endpoints in clinical trials with longitudinal outcomes. *Statistics in Medicine 22*(11), 1787–1805.

Gallo, P., L. Mao, and V. Shih (2014). Alternative views on setting clinical trial futility criteria. *Journal of Biopharmaceutical Statistics 24*(5), 976–993.

Genz, A., F. Bretz, T. Miwa, X. Mi, F. Leisch, F. Scheipl, and T. Hothorn (2019). mvtnorm: Multivariate Normal and t distributions. R package version 1.0-11.

Gerber, F. and T. Gsponer (2019). gsbdesign (version 1.0.1).

Ghosh, B. (1991). A brief history of sequential analysis. In B. Ghosh and P. Sen (Eds.), *Handbook of Sequential Analysis*, pp. 1–19. Marcel Dekker, New York, NY.

Ghosh, P., L. Liu, and C. Mehta (2020). Adaptive multi-arm multi-stage clinical trials. *Statistics in Medicine 39*(8), 1084–1102.

Glimm, E., M. Bezuidenhoudt, A. Caputo, and W. Maurer (2018). A testing strategy with adaptive dose selection and two endpoints. *Statistics in Biopharmaceutical Research 10*(3), 196–203.

Goodman, S. (2007). Stopping at nothing? Some dilemmas of data monitoring in clinical trials. *Annals of Internal Medicine 146*(12), 882–887.

Graf, A. and P. Bauer (2011). Maximum inflation of the type 1 error rate when sample size and allocation rate are adapted in a pre-planned interim look. *Statistics in Medicine 30*(14), 1637–1647.

Graf, A., P. Bauer, E. Glimm, and F. Koenig (2014). Maximum type 1 error rate inflation in multiarmed clinical trials with adaptive interim sample size modifications. *Biometrical Journal 56*(4), 614–630.

Grossman, J., M. Parmar, D. Spiegelhalter, and L. Freedman (1994). A unified method for monitoring and analyzing controlled trials. *Statistics in Medicine 13*(18), 1815–1826.

Gsponer, T., F. Gerber, B. Bornkamp, D. Ohlssen, M. Vandemeulebroecke, and H. Schmidli (2014). A practical guide to Bayesian group sequential designs. *Pharmaceutical Statistics 13*(1), 71–80.

Hall, W. and K. Ding (2008). Sequential tests and estimates after overrunning based on p-value combination. In B. Clarke and S. Ghosal (Eds.), *Pushing the limits of Contemporary Statistics: Contributions in Honor of Jayanta K Ghosh*, Volume 3, pp. 33–45. Beachwood: Institute of Mathematical Statistics, New York, NY.

Hampson, L. and C. Jennison (2013). Group sequential tests for delayed responses (with discussion). *Journal of the Royal Statistical Society, Series B 75*(1), 3–54.

Hampson, L. and C. Jennison (2015). Optimizing the data combination rule for seamless phase II/III clinical trials. *Statistics in Medicine 34*(1), 39–58.

Hochberg, Y. and A. Tamhane (1987). *Multiple Comparison Procedures.* New York, NY: John Wiley & Sons, Inc.

ICON (2020). Addplan: Adaptive designs - plans and analyses.

Jenkins, M., A. Stone, and C. Jennison (2011). An adaptive seamless phase II/III design for oncology trials with subpopulation selection using correlated survival endpoints. *Pharmaceutical Statistics 10*(4), 347–356.

Jennison, C. and B. Turnbull (1989). Interim analyses: the repeated confidence interval approach (with discussion). *Journal of the Royal Statistical Society, Series B 51*(3), 305–361.

Jennison, C. and B. Turnbull (1993). Group sequential tests for bivariate response: interim analyses of clinical trials with both efficacy and safety endpoints. *Biometrics 49*(3), 741–752.

Jennison, C. and B. Turnbull (1997). Group-sequential analysis incorporating covariate information. *Journal of the American Statistical Association 92*(405), 1330–1341.

Jennison, C. and B. Turnbull (1999). *Group Sequential Methods with Applications to Clinical Trials.* Boca Raton, FL: CRC Press.

Jennison, C. and B. Turnbull (2003). Mid-course sample size modification in clinical trials based on the observed treatment effect. *Statistics in Medicine 22*(6), 971–993.

Jennison, C. and B. Turnbull (2006a). Adaptive and nonadaptive group sequential tests. *Biometrika 93*(1), 1–21.

Jennison, C. and B. Turnbull (2006b). Confirmatory seamless phase II/III clinical trials with hypotheses selection at interim: opportunities and limitations. *Biometrical Journal 48*(4), 650–655.

Jones, R., M. Kivipelto, H. Feldman, L. Sparks, R. Doody, D. Waters, J. Hey-Hadavi, A. Breazna, R. Schindler, H. Ramos, LEADe investigators (2008). The Atorvastatin/Donepezil in Alzheimer's Disease Study (LEADe): design and baseline characteristics. *Alzheimer's & Dementia 4*(2), 145–153.

Kieser, M. and T. Friede (2003). Simple procedures for blinded sample size adjustment that do not affect the type I error rate. *Statistics in Medicine 22*(23), 3571–3581.

Kim, E., R. Herbst, I. Wistuba, J. Lee, G. Blumenschein Jr, A. Tsao, D. Stewart, M. Hicks, J. Erasmus Jr, S. Gupta, C. Alden, S. Liu, X. Tang, F. Khuri, H. Tran, B. Johnson, J. Heymach, L. Mao, F. Fossella, M. Kies, V. Papadimitrakopoulou, S. David, S. Lippman, W. Hong (2011). The BATTLE trial: personalizing therapy for lung cancer. *Cancer Discovery 1*(1), 44–53.

Korn, E., B. Freidlin, and M. Mooney (2009). Stopping or reporting early for positive results in randomized clinical trials: The National Cancer Institute Cooperative Group experience from 1990 to 2005. *Journal of Clinical Oncology 27*(10), 1712–1721.

Kosorok, M., Y. Shi, and D. DeMets (2004). Design and analysis of group sequential clinical trials with multiple primary endpoints. *Biometrics 60*(1), 134–145.

Lachin, J. (2005). A review of methods for futility stopping based on conditional power. *Statistics in Medicine 24*(18), 2747–2764.

Lan, K. and D. DeMets (1983). Discrete sequential boundaries for clinical trials. *Biometrika 70*(3), 659–663.

Lehmacher, W. and G. Wassmer (1999). Adaptive sample size calculations in group sequential trials. *Biometrics 55*, 1286–1290.

Lewis, R., A. Lipsky, and D. Berry (2007). Bayesian decision-theoretic group sequential clinical trial design based on a quadratic loss function: a frequentist evaluation. *Clinical Trials 4*(1), 5–14.

Liu, A. and W. Hall (1999). Unbiased estimation following a group-sequential test. *Biometrika 86*(1), 71–78.

Magnusson, B. and B. Turnbull (2013). Group sequential enrichment design incorporating subgroup selection. *Statistics in Medicine 32*(16), 2695–2714.

Mander, A. and S. Thompson (2010). Two-stage designs optimal under the alternative hypothesis for phase II cancer clinical trials. *Contemporary Clinical Trials 31*(6), 572–578.

Marschner, I. and S. Becker (2001). Interim monitoring of clinical trials based on long-term binary endpoints. *Statistics in Medicine 20*(2), 177–192.

Martín, M., A. Chan, L. Dirix, J. O'Shaughnessy, R. Hegg, A. Manikhas, M. Shtivelband, P. Krivorotko, and et al (2017). A randomized adaptive phase II/III study of buparlisib, a pan-class I-PI3K inhibitor, combined with paclitaxel for the treatment of HER2-advanced breast cancer(BELLE-4). *Annals of Oncology 28*(2), 313–320.

Mehta, C., P. Bauer, M. Posch, and W. Brannath (2007). Repeated confidence intervals for adaptive group sequential trials. *Statistics in Medicine 26*, 5422–5433.

Mehta, C. and A. Tsiatis (2007). Flexible sample size considerations using information-based interim monitoring. *Therapeutic Innovation and Regulatory Science 35*(4), 1095–1112.

Müller, H. and H. Schäfer (2001). Adaptive group sequential designs for clinical trials: Combining the advantages of adaptive and of classical group sequential approaches. *Statistics in Medicine 57*, 886–891.

Müller, H. and H. Schäfer (2004). A general statistical principle for changing a design any time during the course of a trial. *Statistics in Medicine 23*, 2497–2508.

Murray, T., Y. Yuan, P. Thall, J. Elizondo, and W. Hofstetter (2018). A utility-based design for randomized comparative trials with ordinal outcomes and prognostic subgroups. *Biometrics 74*, 1095–1103.

Mütze, T., E. Glimm, H. Schmidli, and T. Friede (2018). Group sequential designs for negative binomial outcomes. *Statistical Methods in Medical Research 26*, 5422–5433.

Niewczas, J., C. Kunz, and F. König (2019). Interim analysis incorporating short- and long-term binary endpoints. *Biometrical Journal 61*, 665–687.

O'Brien, P. and T. Fleming (1979). A multiple testing procedure for clinical trials. *Biometrics 35*, 549–556.

Öhrn, F. (2011). *Group sequential and adaptive methods–topics with applications for clinical trials*. Ph. D. thesis, University of Bath.

Öhrn, F. and C. Jennison (2010). Optimal group-sequential designs for simultaneous testing of superiority and non-inferiority. *Statistics in Medicine 29*, 743–759.

Pallmann, P., A. Bedding, B. Choodari-Oskooei, M. Dimairo, L. Flight, L. Hampson, J. Holmes, A. Mander, L. Odondi, M. Sydes, S. Villar, J. Wason, C. Weir, G. Wheeler, C. Yap, and T. Jaki (2018). Adaptive designs in clinical trials: why use them, and how to run and report them. *Trials 16*, 29.

Pocock, S. (1977). Group sequential methods in the design and analysis of clinical trials. *Biometrika 64*, 191–199.

Pocock, S. and R. Simon (1975). Sequential treatment assignment with balancing for prognostic factors in the controlled clinical trial. *Biometrics 31*, 103–115.

Posch, M. and P. Bauer (1999). Adaptive two stage designs and the conditional error function. *Biometrical Journal 41*(6), 689–696.

Proschan, M. (2017). Re-randomization tests for unplanned changes in clinical trials. *Clinical Trials 14*, 425–431.

Proschan, M., E. Brittain, and L. Kammerman (2011). Minimize the use of minimization with unequal allocation. *Biometrics 67*(6), 1135–1141.

Proschan, M., E. Glimm, and M. Posch (2014). Connections between permutation and t-tests: relevance to adaptive methods. *Statistics in Medicine 33*, 4734–4742.

Proschan, M. and S. Hunsberger (1995). Designed extension of studies based on conditional power. *Biometrics 51*, 1315–1324.

Proschan, M., K. Lan, and J. Turk Wittes (2006). *Statistical Monitoring of Clinical Trials.* Springer, New York, NY.

Ristl, R., D. Xi, E. Glimm, and M. Posch (2018). Optimal exact tests for multiple binary endpoints. *Computational Statistics and Data Analysis 122*, 1–17.

Rosenblatt, M. (2017). The large pharmaceutical company perspective. *The New England Journal of Medicine 376*, 52–60.

SAS (2016). SAS (version 9.4).

Scharfstein, D., A. Tsiatis, and J. Robins (1997). Semiparametric efficiency and its implication on the design and analysis of group-sequential studies. *Journal of the American Statistical Association 92*(405), 1342–1350.

Schmidli, H., S. Gsteiger, S. Roychoudhury, A. O'Hagan, D. Spiegelhalter, and B. Neuenschwander (2014). Robust meta-analytic-predictive priors in clinical trials with historical control information. *Biometrics 70*, 1023–1032.

Schulz, K. and D. Grimes (2005a). Multiplicity in randomised trials ii: subgroup and interim analyses. *Lancet 365*, 1657–1661.

Schulz, K. and D. Grimes (2005b). Randomized trials stopped early for benefit–a systematic review. *Journal of American Medical Association 294*, 2203–2209.

Shoben, A. and S. Emerson (2014). Violations of the independent increment assumption when using generalized estimating equation in longitudinal group sequential trials. *Statistics in Medicine 33*(8), 5041–5056.

Sooriyarachchi, M., J. Whitehead, T. Matsushita, K. Bolland, and A. Whitehead (2003). Incorporating data received after a sequential trial has stopped into the final analysis: implementation and comparison of methods. *Biometrics 59*, 701–709.

Stallard, N. (2010). A confirmatory seamless phase II/III clinical trial design incorporating short-term endpoint information. *Statistics in Medicine 29*, 959–971.

Stallard, N. and P. Kimani (2018). Uniformly minimum variance conditionally unbiased estimation in multi-arm multi-stage clinical trials. *Biometrika 105*(2), 495–501.

Stallard, N. and S. Todd (2003). Sequential designs for phase III clinical trials incorporating treatment selection. *Statistics in Medicine 22*, 689–703.

Stein, C. (1945). A two-sample test for a linear hypothesis whose power is independent of the variance. *nnals of Mathematical Statistics 16*, 243–258.

Thall, P., R. Simon, and S. Ellenberg (1988). Two-stage selection and testing designs for comparative clinical trials. *Biometrika 75*, 303–310.

Timmesfeld, N., H. Schäfer, and H. Müller (2007). Increasing the sample size during clinical trials with t-distributed test statistics without inflating the type I error rate. *Statistics in Medicine 26*(12), 2449–2464.

Todd, S. (2007). A 25-year review of sequential methodology in clinical studies. *Statistics in Medicine 26*, 237–252.

Wassmer, G. and W. Brannath (2016). *Group Sequential and Confirmatory Adaptive Designs in Clinical Trials.* Springer, New York, NY.

Wassmer, G. and F. Pahlke (2019). Rpact (version 2.0.6).

Whitehead, A., M. Sooriyarachchi, J. Whitehead, and K. Bolland (2008). Incorporating intermediate binary responses into interim analyses of clinical trials: a comparison of four methods. *Statistics in Medicine 27*, 1646–1666.

Whitehead, J. (1986). On the bias of maximum likelihood estimation following a sequential test. *Biometrics*, 573–581.

Whitehead, J. (1992). Overrunning and underrunning in sequential clinical trials. *Controlled Clinical Trials 13*, 106–121.

Whitehead, J. (1997). *The Design and Analysis of Sequential Clinical Trials.* Chichester: John Wiley & Sons, Chichester.

Woodcock, J. and L. LaVange (2017). Master protocols to study multiple therapies, multiple diseases, or both. *New England Journal of Medicine 377*(1), 62–70.

Wüst, K. and M. Kieser (2003). Blinded sample size recalculation for normally distributed outcomes using long- and short-term data. *Biometrical Journal 8*, 915–930.

11

Multiple Testing for Dose Finding

Frank Bretz
Novartis AG

Dong Xi
Novartis Pharmaceuticals

Björn Bornkamp
Novartis AG

CONTENTS

11.1 Introduction

A good understanding and characterization of the dose-response relationship is a fundamental step in the investigation of any new compound, be it a medicinal drug, a herbicide or fertilizer, a molecular entity, an environmental toxin, or an industrial chemical (Ruberg, 1995a). Already 500 years ago, the Swiss humanist, physician and chemist Paracelsus pointed out the fundamental problem that any compound, even the most harmless one, is potentially toxic if administered at high enough doses, as illustrated by the well-known quote (Letter, 2000, p. 162):

> All things are poison and nothing is without poison,
> only the dose permits something not to be poison.

DOI: 10.1201/9780429030888-13

That is to say, compounds typically considered toxic can be harmless if administered at small enough doses. Conversely, all compounds can be lethal at sufficiently high doses. This also holds for compounds typically considered to be nontoxic or even essential for living such as water and table salt, which become lethal for humans at doses of 10 ℓ and 300 g, respectively (Unkelbach and Wolf, 1985); see Olson (2004) and Noakes et al. (1985) among many others for further examples and medical case reports of drug overdose.

The basic difficulty in getting the right dose is the trade-off between wanted and unwanted effects, especially in the context of drug development. In view of selecting a dose for confirmatory Phase III trials and potential market authorization, a solid characterization of the dose-response relationships with respect to both efficacy and safety are fundamentally important. Failing to do so could lead to incorrect dose recommendations for subsequent trials, possibly leading to a failed development program. For example, if the dose is set too high, safety and tolerability problems are likely to arise while selecting too low a dose makes it difficult to establish adequate efficacy in the confirmatory phase. Incorrect dosing has been recognized as one of the major reasons for the delay in developing new drugs. Even after market authorization, dose adjustments in the drug label continue to be required with some frequency (Cross et al., 2002; Heerdink et al., 2002; Sacks et al., 2014).

Even if it is generally agreed that understanding the dose-response relationships for a new therapeutic drug is important, the objective setting for an actual trial may be subject to much debate. For example, Ruberg (1995a,b) formulated four rather diverse objectives commonly posed in dose-response trials; see also Chuang-Stein and Agresti (1997) and Bretz et al. (2008).

(i) *Is there any evidence of a drug effect?*
The detection of a dose-response signal is often related to the determination of proof-of-concept in a Phase II development program. This is a critical decision point because a positive proof-of-concept outcome, coupled with a subsequent commitment to go into full development, leads to substantial investments. In Section 11.2.1, we review trend tests, which are designed to detect significant dose-response signals under a suitable order restriction.

(ii) *Which doses are (relevantly) different from control?*
This question is closely related to the estimation of a minimum effective dose, i.e., "the smallest dose with a discernible useful effect" (ICH, 1994) and other target doses of interest. If confirmatory pairwise comparisons with a control are of main interest (such as in Phase III trials), multiple comparison procedures as described in Section 11.2.2 may be appropriate to answer this question.

(iii) *What is the dose-response relationship?*
This question is broader than the previous ones in the sense that it asks for a complete functional description of the dose-response relationship. If this is of main interest, modeling approaches may be appropriate to take full advantage of the observed data. However, because the underlying dose-response models are typically unknown in clinical practice, model uncertainty has to be taken into account. In Section 11.3, we describe the MCP-Mod approach that maintains the flexibility of modeling for dose-response estimation while preserving robustness to model misspecification.

(iv) *What is the optimal dose?*
Although very natural, this question is likely to be the most difficult to answer. In practice, this question may not even be well defined in the sense that different stakeholders may have a different understanding of what 'optimal dose' means. The trade-off between efficacy and safety might be perceived differently between different stakeholders. For example, pharmaceutical companies tend to be interested in marketing

sufficiently high doses to ensure good efficacy, whereas regulatory agencies may focus on lower doses to ensure the safety of a new drug. In all circumstances, any answer to this question will be a trade-off between efficacy considerations, safety issues, and regimen convenience. Statistical methods like the MCP-Mod approach described in Section 11.3 offer the quantitative basis for such discussions.

In view of this variety of clinical dose-response trial objectives, it becomes evident that tailored designs and analysis methods are necessary. This chapter contains a brief overview of dose-finding problems encountered at different stages of drug development, explains the common objectives of clinical dose-response trials, and describes key statistical methodologies, which can be applied to such trials. This includes multiple comparison procedures, modeling techniques, and hybrid approaches combining multiple comparisons with modeling.

Accordingly, this chapter is organized as follows. In Section 11.1.1 we introduce a dose-response trial, which will be used later to illustrate the concepts and results developed in this chapter. Sections 11.2 and 11.3 contain the core material as we provide an overview of the main statistical approaches frequently applied in drug development. We first review multiple comparison procedures applied to dose-response testing and pairwise comparisons with a control. Moving on toward exploratory analyses, we then discuss recently introduced hybrid dose-finding methods that combine principles of multiple comparisons with modeling techniques. We conclude this chapter with further discussions in Section 11.4. For general reading about dose finding in drug development, we refer to the edited books by Ting (2006), Chevret (2006), Krishna (2006) and O'Quigley et al. (2017), and the references therein.

11.1.1 Clinical Trial Example

Irritable bowel syndrome (IBS) is a common functional gastrointestinal disorder. Its symptom-based diagnosis is based on chronic abdominal pain, discomfort, bloating, and alteration of bowel habits. Consider the following Phase II dose-response trial on a new drug for the treatment of IBS (Biesheuvel and Hothorn, 2002). Patients were randomized to either placebo or one of four active dose levels, corresponding to doses 0, 1, 2, 3, and 4. Note that the original dose levels were chosen log-equidistantly but have been blinded here for confidentiality, although this does not change the utility of the example for this chapter.

Improvement in abdominal pain is an important efficacy endpoint in evaluating the treatment of IBS. In our case study, the intensity of the abdominal pain was assessed on a daily basis by the patients on a five-point scale, ranging from none (0) to incapacitating (4). The endpoint of interest is the baseline adjusted average daily abdominal pain during the last week of treatment. A larger reduction in pain score as compared to baseline means a clinical benefit.

In total, 369 patients completed the trial, with nearly balanced allocation across the doses; see Table 11.1 for the summary data. The complete data are available with, for example, the `IBScovars` data set in the `R` package `DoseFinding` (Bornkamp et al., 2019). For this chapter, we only use the two variables `dose` and `resp` and ignore gender and other covariate information.

Figure 11.1 displays the boxplots at each of the five doses. We conclude empirically that it is reasonable to assume equal variances in each dose group and the data to be normally distributed. Following the discussion in the Introduction, we pose the following questions of clinical interest:

(i) Is there any evidence of a dose-related drug effect?

(ii) Does any of the investigated doses respond significantly different from the placebo response?

TABLE 11.1
Summary Data of the IBS Dose-Response Trial from Biesheuvel and Hothorn (2002)

			95% Confidence Interval	
Dose	**Estimate**	**Standard Error**	**Lower**	**Upper**
0	0.21691	0.09052	0.0389	0.3949
1	0.50155	0.08637	0.3317	0.6714
2	0.51383	0.08808	0.3406	0.6870
3	0.56766	0.08989	0.3909	0.7444
4	0.56475	0.08928	0.3892	0.7403

(iii) What is the functional form of the dose-response relationship?

(iv) Which is the smallest dose that achieves a clinically relevant improvement of at least 0.2 units (on the five-point scale) over the placebo response?

From Figure 11.1, one could conclude that there is a dose-related trend because the treatment effect seems to increase with higher doses. However, it is not clear whether such a trend is statistically significant and whether any of the investigated doses is significantly better than placebo. Proper statistical analysis methods can be applied to give a more precise answer than simply looking at the plot, with the possibility of including clinical relevance considerations as well. Either multiple comparison procedures or modeling

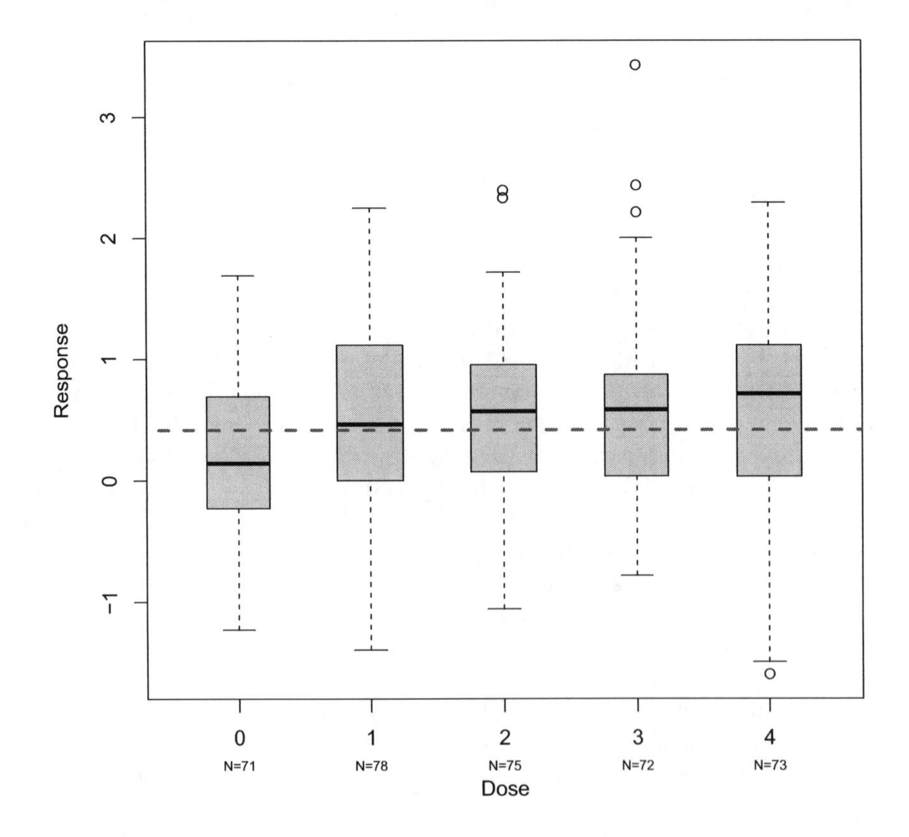

FIGURE 11.1
Boxplots for the `IBScovars` data. Dashed line: Improvement of 0.2 units over the mean placebo response.

techniques can be used to answer this question, depending on, for example, whether the study is intended to be confirmatory or not. Accordingly, we will revisit this example in later sections and answer the above questions as we gradually introduce various methods.

11.2 Hypothesis Testing Approaches

Hypothesis testing based on multiple comparison procedures regard the dose as a qualitative factor. Such approaches can be used for detecting an overall dose-response trend and for pairwise comparisons with a control. They are robust in the sense that they make a few, relatively weak assumptions about the underlying dose-response relationships. However, they are not designed for interpolation (or even extrapolation) of information beyond the observed dose levels. Inference is thus confined to the dose levels under investigation. In the sequel, we follow the outline from Bretz et al. (2009) and review multiple comparison procedures to the extent that they are used in dose-response trials.

11.2.1 Trend Tests

Consider the one-way analysis-of-variance (ANOVA) model

$$y_{ij} = \mu_j + \varepsilon_{ij}, \tag{11.1}$$

where y_{ij} denotes the observation on the i-th patient at dose level j for $i = 1, \ldots, n_j$, $j = 1, \ldots, k$, and $k \geq 2$. Here, μ_j denotes the mean treatment effect at dose $j = 1, \ldots, k$ and ε_{ij} denote the independent normal distributed random errors with mean 0 and variance σ^2. The index $j = 1$ often denotes the placebo group. Further, let $\bar{y}_j$ and $\bar{y}$ denote the arithmetic means in group $j = 1, \ldots, k$, and the overall mean, respectively. That is,

$$\bar{y}_j = \frac{1}{n_j} \sum_{i=1}^{n_j} y_{ij} \quad \text{and} \quad \bar{y} = \frac{1}{n} \sum_{j=1}^{k} n_j \bar{y}_j,$$

where $n = \sum_{j=1}^{k} n_j$ denotes the total number of patients. Also, let

$$s^2 = \frac{1}{\nu} \sum_{j=1}^{k} \sum_{i=1}^{n_j} \left(y_{ij} - \bar{y}_j\right)^2.$$

denote the pooled variance estimate with $\nu = n - k$ degrees of freedom.

We are interested in testing the null hypothesis

$$H\colon \mu_1 = \cdots = \mu_k \tag{11.2}$$

of no differential effects among the k dose groups against the restricted alternative

$$K\colon \mu_1 \leq \cdots \leq \mu_k \text{ with } \mu_1 < \mu_k. \tag{11.3}$$

Note that the formulation of the alternative hypothesis (11.3) explicitly assumes the treatment means μ_i to be monotonically ordered. That is, if we can reject H then we conclude in favor of a significant dose-response signal in form of an increasing trend. Situations with decreasing trends are handled similarly.

In what follows, we review common trend tests under the order restriction (11.3) that are more powerful than tests that do not exploit this restriction. Well-known trend tests developed under the classical ANOVA assumptions (11.1) include the likelihood ratio test for homogeneity of normal means under the total order restriction (Bartholomew, 1961), the modified two-sample t test of Williams (1971) and its modified version investigated by Marcus et al. (1976). These and other approaches, however, have limited use in practice because critical values are often available only for equal group sample sizes and the methods cannot be used in the presence of covariates. In the following, we instead focus on contrast tests, as first introduced by Abelson and Tukey (1963) and Schaafsma and Smid (1966) in the context of dose-response testing. These are powerful methods to detect an overall dose-response trend and can be applied to a variety of different statistical models, including general linear models allowing for covariates and/or factorial treatment structures.

Let $c_1, \ldots, c_k$ denote fixed constants (known as contrast coefficients) such that $\sum_{j=1}^k c_j = 0$. A single contrast test is based on the test statistic

$$t = \frac{\sum_{j=1}^{k} c_j \bar{y}_j}{s\sqrt{\sum_{j=1}^{k} \frac{c_j^2}{n_j}}}.$$

By construction, the test statistic t follows a central t distribution with ν degrees of freedom under the null hypothesis H from (11.2). When H is not true, t follows a non-central t distribution with non-centrality parameter

$$\tau = \frac{\sum_{j=1}^{k} c_j \mu_j}{\sigma\sqrt{\sum_{j=1}^{k} \frac{c_j^2}{n_j}}}.$$

Numerous proposals for the choice of the contrast coefficients have been made, and we refer to Tamhane et al. (1996) for some examples. However, for fixed sample sizes n_j, the non-centrality parameter τ depends only on the contrast coefficients $c_1, \ldots, c_k$. Consequently, in order to maximize the chance of rejecting H, we select the coefficients such that the non-centrality parameter $\tau = \tau(c_1, \ldots, c_k)$ is maximized. It can be shown that in the one-way ANOVA model, (11.1) the optimal contrast coefficients are proportional to

$$c_j = n_j(\mu_j - \bar{\mu}), \quad j = 1, \ldots, k, \tag{11.4}$$

where $\bar{\mu} = \sum_{j=1}^k n_j \mu_j / \sum_{j=1}^k n_j$ denotes the overall standardized mean value (Bretz et al., 2005; Casella and Berger, 1990). Because contrast tests are invariant to scalar changes, a unique representation of the optimal contrast can be obtained conveniently by imposing the regularity condition $\sum_{i=1}^k c_i^2 = 1$. Therefore, under model (11.1), the optimal contrast coefficients depend only on the group sample sizes n_j and the expected mean responses μ_j. In the balanced case $n_1 = \ldots = n_k$, the optimal contrast coefficients do not even depend on the sample sizes and the computation is further simplified.

In practice, the values of μ_j required in Equation (11.4) are unknown. Best guesses (so-called guesstimates) of these values have to be used in order to calculate optimal contrast coefficients. However, if the guesstimates do not reflect well, the true values of μ_j, the power of rejecting H could be much smaller than anticipated. Figure 11.2 illustrates how the choice of contrast coefficients impacts the power of a single contrast test. Suppose that

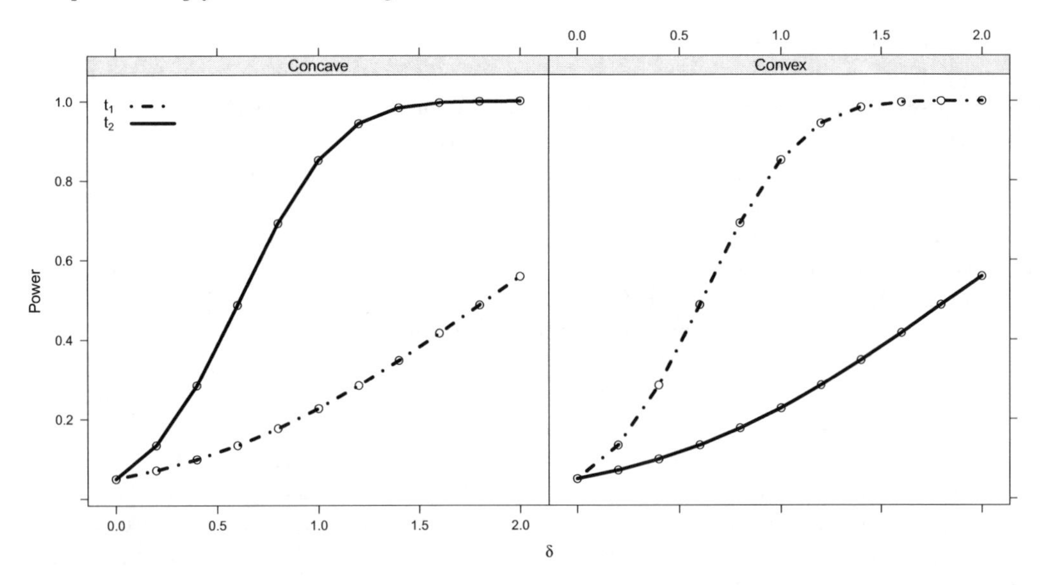

FIGURE 11.2
Power of t_1 (dotted line) and t_2 (solid line) with $n_i = 10$, $\sigma = 1$, $\alpha = 0.05$ for the concave and convex mean configurations. Definition of t_1 and t_2 are given in the text.

we compare $k - 1 = 3$ dose levels of a new drug with placebo. We investigate two sets of contrast coefficients. Let the first contrast test t_1 use the coefficients $c_1 = c_2 = c_3 = -1$, $c_4 = 3$ and the second contrast test t_2 use $c_1 = -3$, $c_2 = c_3 = c_4 = 1$. Each of these two sets of contrast coefficients reflects the order restriction (11.3) by choosing increasing coefficients $c_i \leq c_{i+1}, i = 1, 2, 3$. We compute the power at the two configurations $\mu_1 = 0$, $\mu_2 = \mu_3 = \mu_4 = \delta$ (concave shape) and $\mu_1 = \mu_2 = \mu_3 = 0$, $\mu_4 = \delta$ (convex shape) for varying values of the shift parameter $\delta \in [0, 2]$. The results in Figure 11.2 indicate that the power of single contrast tests depends substantially on the extent to which the contrast coefficients reflect the pattern of the μ_j's. In extreme cases, the loss in power can be as high as 60%.

Multiple contrast tests have been introduced to achieve more robustness with respect to misspecification of the contrast coefficients. The key idea is to identify "... a set of vectors that are 'strategically' located within the alternative region" (Mukerjee et al., 1987). Consider $q \geq 2$ contrast vectors $\mathbf{c}_\ell = (c_{\ell 1}, \ldots, c_{\ell k})'$, $\ell = 1, \ldots, q$. Let $t_1, \ldots, t_q$ denote the corresponding single contrast test statistics. A multiple contrast test then uses the maximum test statistic

$$t_{\max} = \max\{t_1, \ldots, t_q\}. \tag{11.5}$$

Assuming the normal model (11.1), it can be shown that under the null hypothesis H the test statistics $t_1, \ldots, t_q$ are jointly multivariate t distributed with ν degrees of freedom and correlation matrix $\mathbf{R} = (\rho_{\ell\ell'})$, where

$$\rho_{\ell\ell'} = \frac{\sum_{j=1}^{k} \frac{c_{\ell j} c_{\ell' j}}{n_j}}{\sqrt{\left(\sum_{j=1}^{k} \frac{c_{\ell j}^2}{n_j}\right)\left(\sum_{j=1}^{k} \frac{c_{\ell' j}^2}{n_j}\right)}}, \quad 1 \leq \ell, \ell' \leq q. \tag{11.6}$$

Let $t_{1-\alpha}(q, \nu, \mathbf{R})$ denote the upper α equicoordinate quantile of this distribution, i.e. $P(t_1 \leq t, \ldots, t_q \leq t) = P(\max_i t_i \leq t) = 1 - \alpha$ for $t = t_{1-\alpha}(q, \nu, \mathbf{R})$. If $t_{\max} > t_{1-\alpha}(q, \nu, \mathbf{R})$, then the multiple contrast test rejects H and we can conclude in favor of a significant dose-response trend. Alternatively, we can calculate multiplicity adjusted p values and compare their minimum with the significance level α. We refer to Genz and Bretz (2009) for details on computing multivariate t probabilities.

The class of multiple contrast tests is very broad and includes many common trend tests that can be formulated as multiple contrast tests, such as the Williams test and its modification by Marcus (Bretz, 2006). In Section 11.3, we describe an approach to derive optimal (in a certain sense) contrast coefficients for a given candidate set of possible dose-response models. Other multiple comparison procedures (not necessarily restricted to trend tests), which can be formulated as multiple contrast tests, include many-to-one comparisons (Dunnett, 1955), all-pair comparisons (Tukey, 1953), and multiple comparisons with the best (Hsu, 1984); see Bretz et al. (2001) for a comprehensive overview.

We conclude this section with a remark about the monotonicity assumption (11.3). It is a strong assumption and may be violated in practice. It has been shown that already small deviations from this assumption may lead to invalid trend tests (Bauer, 1997). For example, in situations with an effect reversal at higher doses, they do not control the probability of incorrectly declaring a dose to be effective when it is not. Irrespective of which trend test is going to be applied, the decision for its use should always consider the plausibility of the order restriction (11.3). Tests for dose-response signals subject to effect reversals at higher doses were investigated by Simpson and Margolin (1990), Pan and Wolfe (1996) and Bretz and Hothorn (2001) among others.

11.2.2 Pairwise Comparisons

In confirmatory trials, we are interested in comparing the individual dose levels against placebo. This results in testing the $k - 1$ one-sided null hypotheses

$$H_i\colon \mu_{i+1} \leq \mu_1, \quad i = 1, \ldots, k-1, \tag{11.7}$$

assuming that larger values indicate a treatment benefit. The null hypothesis H_i thus reflects that the mean effect for the placebo group ($i = 1$) is higher than that for dose i. Accordingly, the alternative hypotheses are given by

$$K_i\colon \mu_{i+1} > \mu_1, \quad i = 1, \ldots, k-1.$$

Rejecting any of the $k-1$ null hypotheses H_i thus ensures that the new drug is better than placebo for at least one of the $k - 1$ doses under investigation. This is the classical many-to-one problem of comparing several treatments with a common control, which is different from the problem of detecting an overall dose-response trend by testing the null hypothesis (11.2) against the restricted alternative (11.3).

Following conventional regulatory standards, a strong control of the family-wise error rate (FWER) is necessary when multiple hypotheses are simultaneously tested in confirmatory clinical trials (EMA, 2017; FDA, 2017) at a given significance level $\alpha \in (0, 1)$. The standard multiple test procedure to address the many-to-one problem is the Dunnett test (Dunnett, 1955), which compares each dose group with the control group. One can conceptually think of Dunnett's method as a set of $k - 1$ two-sample t tests, adjusted for multiplicity. More specifically, the Dunnett test is a multiple contrast test with contrast vectors $\mathbf{c}_\ell$, $\ell = 1, \ldots, k-1$, where $c_{\ell,j} = -1$ for $j = 1$, $c_{\ell,j} = 1$ for $j = \ell + 1$ and $c_{\ell,j} = 0$ otherwise. Should the $k - 1$ pairwise dose-control comparisons be performed for more than one endpoint or population, more advanced procedures may be required to address the resulting multiple test problem, such as the graphical approaches introduced in Bretz et al. (2009) and Burman et al. (2009).

One possibility to improve the Dunnett test is to use the closed test procedure (Marcus et al., 1976). More specifically, the Dunnett test is applied to each intersection hypothesis $H_I = \bigcap_{i \in I} H_i$ for $I \subseteq \{1, \ldots, k-1\}$ and $H_I \neq \emptyset$, where $H_i, i = 1, \ldots, k-1$, are given in (11.7). This results in the step-down Dunnett test (Dunnett and Tamhane, 1991) for the pairwise doses-control comparisons, which by construction rejects all hypotheses rejected by the original Dunnett test and possibly others. In Section 11.3.3.2, we use similar ideas to replace the Dunnett test through more powerful MCP-Mod trend tests within the closed testing framework (König, 2015).

11.2.3 Clinical Trial Example Revisited

We now revisit the IBS case study from Section 11.1.1 to illustrate the methods introduced in Sections 11.2.1 and 11.2.2. The `multcomp` package in `R` provides a convenient implementation of these methods (Bretz et al., 2010; Hothorn et al., 2019). Readers should not expect to get the same results as stated in this chapter. For example, differences in random number generating seeds or truncating the number of significant digits in the results may result in slight differences to the output shown here.

11.2.3.1 Trend Tests

For illustration, we consider in the following the Williams test in its modification using contrast tests (Bretz, 2006). In what follows, we use the `aov` function to fit the one-way ANOVA model (11.1). Then we apply the `glht` function from the `multcomp` package to perform the Williams contrast test by specifying the `mcp(dose = "Williams")` option to the `linfct` argument.

```
> library(multcomp)
> data("IBScovars", package = "DoseFinding")
> IBScovars$dose  <- as.factor(IBScovars$dose)
> ibs.aov         <- aov(resp ~ dose, data = IBScovars)
> ibs.mc          <- glht(ibs.aov, alternative = "greater",
+                              linfct = mcp(dose = "Williams"))
```

In the previous call, `resp` denotes the response and `dose` contains the dose assignments for each patient. The `alternative = "greater"` specifies an increasing trend according to (11.3).

At this stage, we can apply the `summary` method to summarize and display the results returned from the `glht` function. Alternatively, we can save the information provided by the `summary` method into the object

```
> ibs.res         <- summary(ibs.mc)
```

The contrast coefficients for the active doses can be then retrieved by

```
> print(ibs.res$linfct[,2:5])
       dose1     dose2     dose3     dose4
C 1 0.000000 0.0000000 0.0000000 1.0000000
C 2 0.000000 0.0000000 0.4965517 0.5034483
C 3 0.000000 0.3409091 0.3272727 0.3318182
C 4 0.261745 0.2516779 0.2416107 0.2449664
```

Furthermore, we can extract the numerical information from the resulting list element `ibs.res$test` for further analyses. For example,

```
> round(min(ibs.res$test$pvalues),4)
[1] 0.0016
```

returns the minimum adjusted p value from the previous analysis. Because of the small p value, we can reject the null hypothesis (11.2) and conclude in favor of a significant dose-response signal.

11.2.3.2 Pairwise Comparisons

We now consider the Dunnett test (Dunnett, 1955) introduced in Section 11.2.2. In analogy to the previous `glht` call, we can invoke

```
> ibs.mc2          <- glht(ibs.aov, alternative = "greater",
+ linfct = mcp(dose = "Dunnett"))
> ibs.res2         <- summary(ibs.mc2)
```

and extract the multiplicity adjusted p values

```
> round(ibs.res2$test$pvalues,4)
[1] 0.0386 0.0320 0.0111 0.0115
```

to compare the four active dose levels with placebo. We conclude at the significance level $\alpha = 0.025$ that the two higher doses are better than placebo, but not the two lower doses. Note that the minimum adjusted p value is 0.0111 and therefore larger than the 0.0016 obtained for the Williams contrast test. This indicates that trend tests tend to be more powerful than the pairwise comparisons from Dunnett for testing (11.2).

11.3 Approaches Combining Multiple Comparisons with Modeling

In Section 11.2.1, we introduced trend tests using multiple contrasts, but without specifying how to choose the contrast coefficients. In this section, we review hybrid methods that embed dose-response modeling techniques within the class of multiple contrast tests. This allows us to define contrast coefficients based on clinical and pharmacological considerations. An early reference is Tukey et al. (1985), who recognized that the power of standard hypothesis tests to detect a dose-response signal depends critically on the unknown dose-response relationship. To account for this uncertainty, they proposed to simultaneously use several trend tests and subsequently adjust the resulting p values for multiplicity. The MCP-Mod approach is an extension that maintains the flexibility of modeling for dose-response estimation while preserving robustness to model misspecification (Bretz et al., 2004, 2005). It has gained considerable attention over the past years, as evidenced by the Qualification Opinion of the European Medicines Agency (CHMP, 2014) and the Fit-for-Purpose Determination of the U.S. Food and Drug Administration (FDA, 2016).

11.3.1 MCP-Mod for Normal Distributed Data

The MCP-Mod approach relies on modeling the dose-response relationship and thus makes better use of the available information than hypothesis testing approaches in the sense that they consider dose as a continuous variable and interpolate information across doses instead of treating every dose separately. The drawback of modeling approaches is their dependence on the assumed model: any inference (e.g., target dose estimation) depends on the employed dose-response model and can be highly sensitive to its choice. Because the dose-response model and its parameters are unknown before a clinical trial, one is faced with model uncertainty, a problem that is often ignored. A common approach is to fit several dose-response models once the data have been observed and select the best fitting model. However, such a naive approach does not account for model uncertainty and can lead to undesirable effects due to data dredging, such as overfitting, biased treatment effect estimates, and over-optimistic analysis results; see Chatfield (1995), Draper (1995), and Hoeting et al. (1999) among others. It is also important to have the flexibility to a-priori exclude dose-response models that are considered implausible due to biological or clinical reasons.

MCP-Mod is a hybrid approach that combines the advantages of the multiple comparison procedures (MCP) and modeling approaches (Mod). At the trial design stage, a suitable set of candidate models is identified through clinical team discussions and statistical considerations, which also impact the decisions on the number of doses, required sample sizes, patient allocations, etc. The MCP step performed at the trial analysis stage focuses on detecting the existence of a dose-response signal using suitable contrast tests derived from the candidate model set and adjusting for the fact that multiple candidate models are considered. Once a dose-response signal is established, one proceeds to the Mod step where the best model out of the prespecified candidate model set or a model averaging approach is used for dose-response and target-dose estimation. In the following, we describe the individual steps in more detail.

Different to Section 11.2.1, we now treat 'dose' as a quantitative variable, denoted as d. We consider the regression model

$$y_{ij} = f(d_j, \boldsymbol{\theta}) + \varepsilon_{ij}, \tag{11.8}$$

where as before y_{ij} denotes the response and ε_{ij} the error term for patient $i = 1, \ldots, n_j$ within dose group $j = 1, \ldots, k$. As before, $j = 1$ denotes the placebo group. Moreover, $f(.)$ is parameterized through the vector $\boldsymbol{\theta}$. Most dose-response models used in practice can be expressed in terms of a standardized model f^0 such that $f(d, \boldsymbol{\theta}) = \theta_0 + \theta_1 f^0(d, \boldsymbol{\theta}^0)$. This decomposition is often useful in practice; for example, in fitting nonlinear regression models, starting estimates are only required for the standardized model parameters $\boldsymbol{\theta}^0$.

Since the true underlying regression model f is unknown at the trial design stage, the MCP-Mod approach postulates a set of M candidate dose-response models. Table 11.2 summarizes commonly used dose-response models, together with their standardized versions; see also Pinheiro et al. (2006). Note that the Sigmoid *Emax* model in Table 11.2 is sometimes referred to as four-parameter logistic model (Ritz and Streibig, 2005). Besides choosing an expression for $f^0(d, \boldsymbol{\theta}^0)$, we also need initial guesses for the standardized

TABLE 11.2
Commonly Used Dose-Response Models and Their Standardized Versions. For the beta model $B(\delta_1, \delta_2) = (\delta_1 + \delta_2)^{\delta_1+\delta_2}/(\delta_1{}^{\delta_1}\delta_2{}^{\delta_2})$.

Model	$f(d, \boldsymbol{\theta})$	$f^0(d, \boldsymbol{\theta}^0)$
Emax	$E_0 + E_{\max} d/(ED_{50} + d)$	$d/(ED_{50} + d)$
Sigmoid *Emax*	$E_0 + E_{\max} d^h/(ED_{50}^h + d^h)$	$d^h/(ED_{50}^h + d^h)$
Exponential	$E_0 + E_1[\exp(d/\delta) - 1]$	$\exp(d/\delta) - 1$
Power	$E_0 + E_1 d^\alpha$	d^α
Linear	$E_0 + \delta d$	d
Linear log-dose	$E_0 + \delta \log(d + c)$	$\log(d + c)$
Quadratic	$E_0 + \beta_1 d + \beta_2 d^2$	$d + (\beta_2/\lvert\beta_1\rvert)d^2$
Beta	$E_0 + E_{\max} B(\delta_1, \delta_2)(d/D)^{\delta_1}(1 - d/D)^{\delta_2}$	$(d/D)^{\delta_1}(1 - d/D)^{\delta_2}$

parameter vector $\boldsymbol{\theta}^0$, called guesstimates, to calculate the optimal contrast coefficients further below. For example, they can be derived from some initial knowledge of the expected percentage of the maximum response for a given dose. Each of these candidate shapes ultimately produces a mean response vector $\boldsymbol{\mu}_m = (\mu_{m1}, \ldots, \mu_{mk})'$, where $\mu_{mi} = f_m^0(d_i, \boldsymbol{\theta}_m^0)$ depends on both dose d_i and the parameter vector $\boldsymbol{\theta}^0$ of the standardized model, $m = 1, \ldots, M$.

For any given candidate shape, we use suitably chosen multiple contrast tests to asses the dose-response signal. Assuming model (11.8), we consider the single contrast $\mathbf{c}'\boldsymbol{\mu}$, where $\mathbf{c}$ denotes a vector of contrast coefficients such that $\sum_{i=1}^k c_i = 0$ and $\boldsymbol{\mu} = (\mu_1, ..., \mu_k)'$ the vector of mean responses across the dose levels. Given an alternative shape $\boldsymbol{\mu}$, optimal contrast coefficients $\mathbf{c}_m^{\text{opt}} = (c_{m1}^{\text{opt}}, \ldots, c_{mk}^{\text{opt}})'$ can be determined from equation (11.4). Recall that contrast tests are invariant to any shift and scale change on the mean response vector, thus independent of intercept and slope of the dose-response model $f(d, \boldsymbol{\theta})$. As a result, we only need to consider standardized dose-response models when choosing candidate shapes.

With the MCP-Mod approach, we calculate for each candidate shape $\boldsymbol{\mu}_m$ the optimal contrast $\mathbf{c}_m^{\text{opt}}$, such that the power of the resulting single contrast test is maximized when the true underlying mean response equals $\boldsymbol{\mu}_m$. For example, for the linear regression model, the resulting linear contrast coefficients for equally spaced doses and balanced patient allocation are such that the difference between any two adjacent contrast coefficients is a constant, resulting in a powerful test to detect the linear trend. Similarly, any dose-response relationship characterized through $\boldsymbol{\mu}_m$ can be tested by determining the corresponding optimal contrast test whose coefficients are defined in dependence of the assumed $\boldsymbol{\mu}_m$.

The resulting multiple contrast test uses the maximum test statistic (11.5). Since the joint distribution of contrast test statistics $t_1(\mathbf{c}_1^{\text{opt}}), \ldots, t_M(\mathbf{c}_M^{\text{opt}})$ is known to be multivariate t, probabilities can be calculated using the `mvtnorm` package in R (Genz et al., 2020). Here, we use the notation $t_m(\mathbf{c}_m^{\text{opt}})$ to emphasize the dependence of the contrast test statistics (11.4) on $\mathbf{c}_m^{\text{opt}}$, $m = 1, \ldots, M$. As described in Section 11.2.1, if at the final analysis the observed multiple contrast test statistic is larger than the associated multiplicity-adjusted critical value, we reject the null hypothesis (11.2) and conclude in favor of a significant dose-response signal. Alternatively, we can calculate multiplicity-adjusted p values and compare their minimum with the significance level α. Every single contrast test ultimately translates into a decision whether a candidate dose-response curve is significant given the observed data while controlling the Type I error rate of incorrectly declaring a significant dose-response signal at a prespecified significance level α. If no candidate shape is statistically significant, the MCP-Mod procedure stops and indicates that a dose-response signal cannot be established from the observed data. In practice, one would still proceed with the modeling step in order to estimate, for example, the responses at the investigated dose levels.

Once an overall dose-response signal is established, we can select a single model, which could be the one with the most significant contrast test statistic, or based on other model selection criteria such as the Akaike information criteria (AIC). Alternatively, multiple models can be selected if model averaging is preferred (Buckland et al., 1997). We refer to Schorning et al. (2016) and Verrier et al. (2014) for further discussions on the use of model selection versus model averaging strategies in dose-finding studies. Finally, fitting the selected model to the data and estimating adequately the target dose(s) of interest can be achieved using standard nonlinear regression analysis.

11.3.2 Clinical Trial Example Revisited

The `DoseFinding` R package implements the design and analysis of dose finding trials (Bornkamp et al., 2019); see also Xun and Bretz (2017). In the following, we illustrate

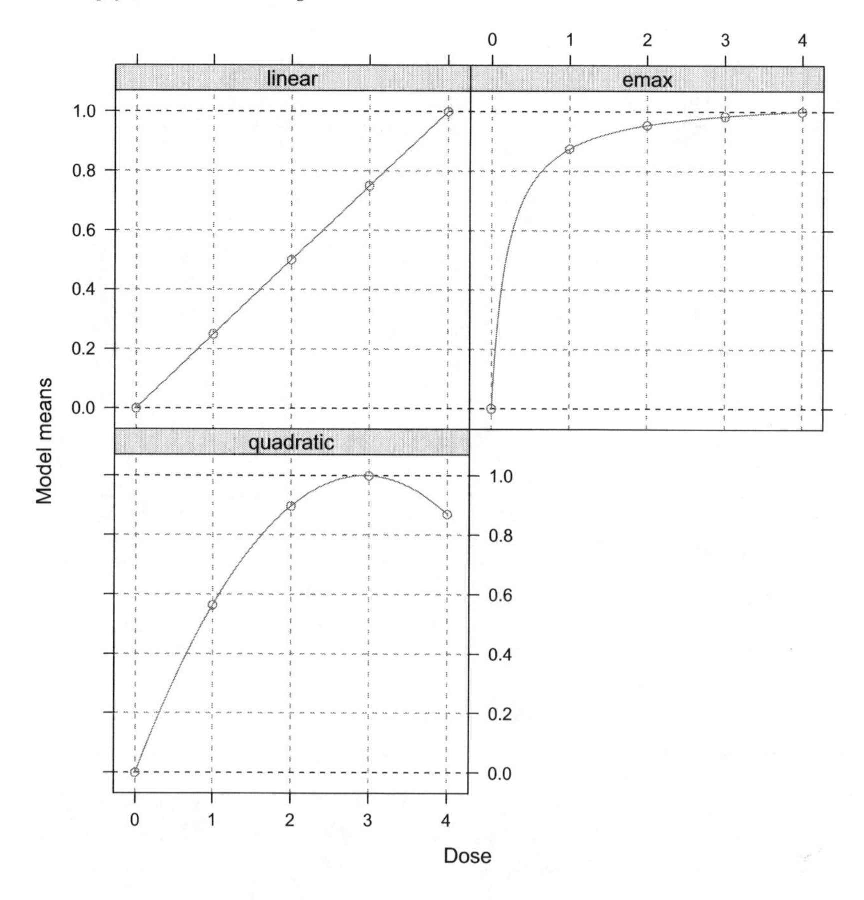

FIGURE 11.3
Candidate dose-response shapes for the IBS trial.

the application of the MCP-Mod approach with the IBS trial from Section 11.1.1. Recall that the data is available with the `IBScovars` data set in the `DoseFinding` package.

```
> library(DoseFinding)
> data(IBScovars)
```

Assume that at the trial design stage, the three shapes plotted in Figure 11.3 are included in the candidate model set. Using the standardized versions of the dose-response models, these shapes are as follows: an *Emax* model with $ED_{50} = 0.2$, the linear model, and a quadratic model with $\delta = -0.17$.

These candidate shapes can be defined with the `Mods` function and visualized afterward. The output object `models` will be used as an input object for other functions to extract the mean response or target doses.

```
> mods <- Mods(linear=NULL, emax=0.2, quadratic=-0.17, doses=0:4)
> plot(mods)
```

For the MCP step at the analysis stage, the multiple contrast test is performed using the `MCTtest` function. By specifying `models=mods`, optimal contrasts are derived automatically (Table 11.3). In this example, all models result in small p values indicating a strong evidence for the existence of a dose-response relationship (Table 11.4).

TABLE 11.3
Optimal Contrast Using MCP-Mod for the `IBScovars` Data

Dose	Linear	Emax	Quadratic
0	−0.617	−0.889	−0.815
1	−0.338	0.135	−0.140
2	0.002	0.227	0.294
3	0.315	0.253	0.408
4	0.637	0.275	0.252

TABLE 11.4
Test Statistics, Adjusted p Values and AIC Values Using MCP-Mod for the `IBScovars` Data

	Linear	Emax	Quadratic
Test statistics	2.645	3.215	3.092
Adjusted p values	0.008	0.001	0.002
AIC values	851.8201	850.392	851.2303

```
> test <- MCTtest(dose, resp, data=IBScovars, models=mods)
> print(test)
```

For the Mod step, we can fit the individual models using `fitMod` and select the best model according to the AIC. For example, we can use the following call to fit an *Emax* model and calculate the corresponding AIC value.

```
> fitEmax <- fitMod(dose, resp, data=IBScovars,
+                   model="emax", bnds=c(0.01,5))
> print(fitEmax)
> AIC(fitEmax)
```

Other models can be fit similarly. The *Emax* model has the smallest AIC value (Table 11.4) and is therefore selected for further analyses. The fitted *Emax* model based on the `fitEmax` object is $0.217 + 0.377d/(0.363 + d)$, for $d \in [0, 4]$. We can use the `plot` function to generate the plot of the fitted *Emax* curve shown in Figure 11.4, which could be useful in reporting the analysis results.

```
> plot(fitEmax, CI=TRUE, plotData="meansCI", level=0.95)
```

The fitted model is plotted together with the 95% confidence intervals. Based on this model fit, we can estimate the dose that achieves a clinically relevant improvement over placebo using the `TD` function. For the `IBScovars` data, where increasing response is beneficial, the dose achieving an effect of 0.2 over placebo is 0.409, obtained with the following call.

```
> TD(fitEmax, Delta=0.2, direction="increasing")
```

Alternatively, we can use the `ED` function to estimate a dose ED_p that achieves a certain percentage p of the full effect size over placebo within the observed dose range (Bretz et al., 2008). For example, for $p = 0.9$ we obtain $ED_{90} = 1.712$, at which 90% of the maximum effect size is achieved, through the call

```
> ED(fitEmax, p=0.9)
```

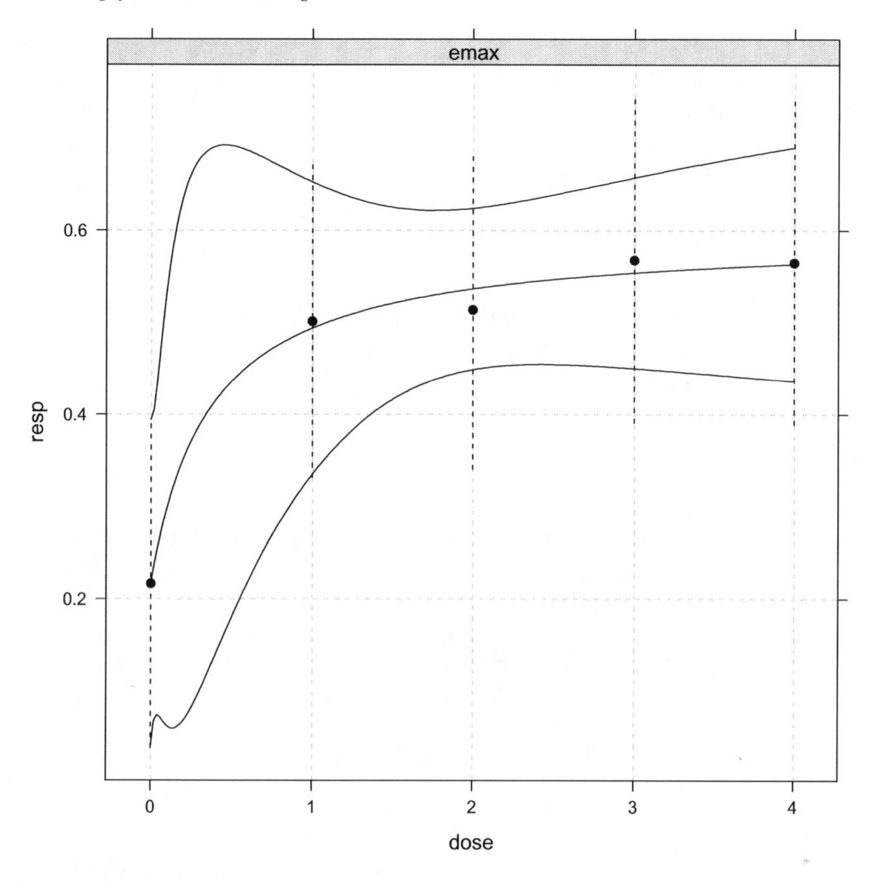

FIGURE 11.4
Estimated *Emax* curve for the `IBScovars` data.

11.3.3 Extensions

The MCP-Mod approach was originally developed for a normal distributed, homoscedastic response variable measured at a single timepoint in parallel group study designs. In the following, we describe various extensions thereof. In Section 11.3.3.1 we describe an extension to general parametric models that greatly broadens the utility of MCP-Mod. In Section 11.3.3.2 we describe an application of closed testing to obtain adjusted p values for the MCP-Mod trend test as well as for the pairwise dose-control comparisons. We briefly review further extensions in Section 11.3.3.3.

11.3.3.1 MCP-Mod for General Parametric Models

Practical problems often involve non-normal response variables such as count, binary, or time to event data. In addition, the final analysis has usually to be adjusted for relevant covariates (e.g., region, age, ...), patients measurements are often recorded over time (necessitating the use of longitudinal models), and patients might receive more than one treatment (such as in cross-over or incomplete block designs). Specific extensions of the MCP-Mod approach to binary data were considered in Klingenberg (2009) and Benda (2010). In this section, we describe the generalization of MCP-Mod developed by Pinheiro et al. (2014) for general parametric models. The authors developed an overarching methodology to perform efficient multiple comparisons and modeling for dose finding, under uncertainty about the dose-response shape, using general parametric models. Their framework is quite broad and can be utilized in situations involving, for example, generalized nonlinear models, linear and nonlinear mixed effects models, and Cox proportional hazards models.

Following Xun and Bretz (2017), let $\mathbf{y}$ denote the response of a patient receiving dose d, which could be a vector in specific applications. We consider the model

$$\mathbf{y} \sim F(\mu(d), \mathbf{z}, \boldsymbol{\eta}), \tag{11.9}$$

where $\mu(d)$ denotes the dose-response parameter, $\mathbf{z}$ are possible covariates, and $\boldsymbol{\eta}$ contains the nuisance parameters. This formulation generalizes the notation from Section 11.3.1 for a normal distributed response variable to any parametric distribution. If F is a normal distribution, $\mu(d)$ is then the mean response at dose d, and $\boldsymbol{\eta} = \sigma^2$ is a nuisance parameter. The main idea of the generalized MCP-Mod approach is to extract dose-response parameters $\mu(d)$ from model (11.9) so that contrast tests and dose-response model fitting can be performed on these parameters. Because all dose-response information is contained in $\mu(d)$, this parameter should be easily interpretable in order to communicate with clinical teams, choose candidate dose-response shapes, and specify clinically relevant effects. We refer to Pinheiro et al. (2014) for examples on how to specify $\mu(d)$ in practical settings.

At the design stage, we specify a set of candidate dose-response shapes $\boldsymbol{\mu}_1, ..., \boldsymbol{\mu}_M$. Each of the candidate shapes determines an optimal contrast to evaluate dose response. Consider a dose-response mean vector $\boldsymbol{\mu}_m = (\mu_m(d_1), \ldots, \mu_m(d_k))'$, for k doses $d_1, ..., d_k$, including placebo. The optimal contrast for this shape then satisfies

$$\mathbf{c}_m^{\text{opt}} \propto \mathbf{S}^{-1} \left(\boldsymbol{\mu}_m - \frac{\boldsymbol{\mu}_m \mathbf{S}^{-1}\mathbf{1}}{\mathbf{1}'\mathbf{S}^{-1}\mathbf{1}} \right), \text{ for } m = 1, \ldots, M,$$

where $\mathbf{1}$ is a vector of 1's of proper length and $\mathbf{S}$ is the covariance matrix of the estimated dose-response parameters $\widehat{\boldsymbol{\mu}}$ (Bornkamp, 2006; Pinheiro et al., 2014).

At the analysis stage, we proceed in a similar way as Hothorn et al. (2008), who extended certain simultaneous inference procedures to general parametric models. To be more precise, we fit an ANOVA-type model with dose as a factor, for any distribution $F(\cdot)$, and extract estimated both, the dose-response parameter estimate $\widehat{\boldsymbol{\mu}}$ and the covariance matrix estimate $\widehat{\mathbf{S}}$, using appropriate estimation methods such as maximum likelihood, generalized estimating equations or partial likelihood. The contrast test statistics is then

$$z_m = \frac{(\mathbf{c}_m^{opt})' \widehat{\boldsymbol{\mu}}}{\sqrt{(\mathbf{c}_m^{opt})' \widehat{\mathbf{S}} (\mathbf{c}_m^{opt})}}, \text{ for } m = 1, \ldots, M.$$

In many parametric estimation problems, $\widehat{\boldsymbol{\mu}}$ is asymptotically multivariate normal distributed with covariance matrix $\mathbf{S}$, such as in generalized linear models, parametric time-to-event models and mixed-effect models. When this condition is met, the joint distribution of the contrast statistics $z_1, \ldots, z_M$ is also multivariate normal. P values are then calculated via the joint distribution of $z_1, \ldots, z_m$ under the null hypothesis of no dose response while controlling the Type I error rate at a prespecified significance level α.

Once a dose-response signal is established, we proceed to the Mod step by fitting the dose-response profile and estimating target doses based on all models with significant contrast test statistics in the MCP step. There are many ways to fit the dose-response models to the observed data, including approaches based on maximizing the likelihood or the restricted likelihood. Alternatively, Pinheiro et al. (2014) suggested a two-stage approach based on generalized least squares $[\widehat{\boldsymbol{\mu}} - \boldsymbol{\mu}(d)]' \widehat{\mathbf{S}}^{-1} [\widehat{\boldsymbol{\mu}} - \boldsymbol{\mu}(d)]$. Although this approach relies on asymptotic results, it has the appeal of being a general-purpose application, as it depends only on $\widehat{\boldsymbol{\mu}}$ and $\widehat{\mathbf{S}}$. This two-stage approach is implemented in the `DoseFinding` package (Bornkamp et al., 2019). We refer to Pinheiro et al. (2014) and Xun and Bretz (2017) for further details and examples.

11.3.3.2 MCP-Mod under Closed Testing

The standard MCP-Mod approach only allows to test for a dose-response trend but does not make inferences about the individual dose-control comparisons. In this section, we describe an extension using closed testing. Following König (2015), a suitable MCP-Mod trend test is applied to each intersection hypothesis $H_I = \bigcap_{i \in I} H_i$ for $I \subseteq \{1, \ldots, k-1\}$ and $H_I \neq \emptyset$, where H_i are defined in (11.7). In doing so, we can calculate multiplicity-adjusted p values for the $k-1$ dose-control comparisons. We describe below this closed version of the MCP-Mod approach for a single, normal distributed endpoint following Section 11.3.1. Extensions to the general parametric models introduced in Section 11.3.3.1 are straightforward. Simulations comparing the closed MCP-Mod approach with various common multiple comparison procedures were reported in Bretz et al. (2017).

Using closed testing, we need to define suitable α-level tests for each intersection hypothesis H_I. Following König (2015), we can apply MCP-Mod as long as only those doses are considered for the construction of the contrast test that are part of the intersection hypothesis. That is, we can use the same set of prespecified dose-response models for all intersection hypotheses but need to calculate the optimal contrast based on the set of available doses. Note that for the elementary hypotheses H_i all contrast tests reduce to pairwise comparisons. More formally, the closed MCP-Mod approach is defined as follows; see also Bretz et al. (2017):

(i) let $H_1, \ldots, H_{k-1}$ denote the $k-1$ elementary hypotheses defined in (11.7);

(ii) construct all intersection hypotheses $H_I = \bigcap_{i \in I} H_i$, $I \subseteq \{1, \ldots, k-1\}$;

(iii) define a set $\mathcal{M}$ of pres-pecified candidate models;

(iv) test each intersection hypothesis H_I using the MCP-Mod contrast tests by calculating the optimal contrasts for each model, based on the doses d_j, $j \in J = 1 \cup \{I+1\}$ and where $j = 1$ denotes the placebo group;

(v) reject an elementary hypothesis H_i, if all hypotheses H_I with $i \in I$ are rejected by their α-level tests.

In other words, we perform a multiple contrast test for each intersection hypothesis. The single contrast test for detecting the m-th model shape uses only the data of the doses d_j with $j \in J = 1 \cup \{I+1\}$, where the notation $\{I+1\}$ shall indicate that all indices contained in I are incremented by 1. Thus, we use the modified test statistics $t_{m,I}$ based on the contrasts $\mathbf{c}_{m,I} = (c_{m1,I}, \ldots, c_{mk,I})'$ with coefficients $c_{mj,I}, j \in 1 \cup \{I+1\}$, that are the optimal contrasts for testing model shape m using only the active dose levels $j \in \{I+1\}$ and the placebo group $j = 1$. For any index $j \notin 1 \cup \{I+1\}$, we set the coefficient $c_{mj,I} = 0$. Note that we used to index I to emphasize the dependence of the test statistics and contrast coefficients on it.

The final test statistic for the intersection hypothesis H_I is the maximum $t_{\max,I} = \max\{t_{1,I}, \ldots, t_{M,I}\}$ of all single contrast tests. An intersection hypothesis H_I can hence be rejected if $t_{\max,I}$ is larger than or equal to the multiplicity adjusted critical value from the multivariate t distribution, which can be computed using the `mvtnorm` package in R (Genz et al., 2020). Alternatively, we can calculate multiplicity adjusted p values and compare them with the significance level α.

Note that the test for the global null hypothesis $H_{1,\ldots,k-1} = \bigcap_{i=1}^{k-1} H_i$ is exactly the MCP-Mod contrast test to assess an overall dose-response signal. As mentioned above, for the elementary null hypothesis H_i, $i = 1, \ldots, k-1$, the contrast coefficients of all model shapes $m = 1, \ldots, M$ reduce to the pairwise dose-control comparisons. Finally, we note that the optimal contrasts need to be further constrained by requiring that the sign of the contrast coefficients of the active doses and placebo need to be different. They are available with the `DoseFinding` package using the `type = "constrained"` option from the `optContr` function. We refer to König (2015) for further discussions on this topic, including a detailed investigation of the Type I error rate control.

11.3.3.3 Further Extensions

The original MCP-Mod approach from Bretz et al. (2004, 2005) has been subject to various other extensions than described so far. In the following, we provide a brief literature review.

Practical considerations regarding the implementation of MCP-Mod were discussed by Pinheiro et al. (2006), including sample size calculations for the MCP part. Likewise, Xun and Bretz (2017) provided further recommendations when implementing MCP-Mod, including general considerations on when to apply MCP-Mod as well as specific considerations applicable at the trial design and analysis stages.

Response-adaptive extensions of MCP-Mod with one or multiple interim analyses are possible and often advisable as they may result in, for example, power gains to detect a dose-response signal or in higher precision to estimate the dose-response curve or a target dose of interest (Bornkamp et al., 2011; Dragalin et al., 2010; Franchetti et al., 2013; Miller, 2010). Conversely, such designs also allow the possibility to stop a trial early in case of lack of efficacy; see Mielke and Dragalin (2017) for a recent review of two-stage designs in dose finding.

Several authors have extended MCP-Mod using likelihood ratio tests (Baayen et al., 2015; Dette et al., 2015; Gutjahr and Bornkamp, 2017). While general purpose software is not yet available for these methods, these extensions have the appeal that they do not require the specification of guesstimates, which simplifies the use of MCP-Mod in practice.

Instead of employing maximum test statistics, Ma and McDermott (2020) suggested the use of different statistics for combining dependent p values, such as Fisher's combination method or the inverse normal combination method. The authors proposed the application of generalized multiple contrast tests, which can also be used for model selection and dosage selection by employing a closed testing procedure.

Further extension of MCP-Mod include the use of Bayesian methods for estimating or selecting the dose-response curve from a sparse dose design (Thomas, 2006; Wakana et al., 2007) and its application to joint efficacy-safety modeling (Tao et al., 2015), subgroup analyses (Thomas et al., 2020), and multiregional clinical trials (Yamaguchi and Sugitani, 2020).

11.4 Conclusions

In this chapter, we reviewed multiple comparison procedures for clinical trials comparing several doses of a new compound with placebo. We briefly discussed the importance of dose finding in drug development and summarized the main objectives of related clinical trials in Phases II and III. Sections 11.2 and 11.3 contained the core material as we provided an overview of the main statistical approaches frequently applied in drug development. We first reviewed multiple comparison procedures applied to dose-response testing and pairwise comparisons with a control. Moving on toward exploratory analyses, we then discussed hybrid dose-finding methods that combine principles of multiple comparisons with modeling techniques. We used a clinical trial example to illustrate the methods and available software implementations in R.

As in many other areas of drug development, dose finding is not a problem that can be resolved by statisticians alone. While there exist many statistical methods that provide satisfactory solutions, in practice, the problem of selecting the right dose requires a collaborative effort that includes statisticians, pharmacometricians, and their clinical partners. It is critical that important inputs are taken from every stakeholder before a decision is made on the trial design or data analysis method to address the dose-finding problem at hand.

Bibliography

Abelson, R. and J. Tukey (1963). Efficient utilisation of non-numerical information in quantitative analysis: General theory and the case of simple order. *Annals of Mathematical Statistics 34*, 1347–1369.

Baayen, C., P. Hougaard, and C. Pipper (2015). Testing effect of a drug using multiple nested models for the dose–response. *Biometrics 71*(2), 417–427.

Bartholomew, D. J. (1961). Ordered tests in the analysis of variance. *Biometrika 48*, 325–332.

Bauer, P. (1997). A note on multiple testing procedures in dose finding. *Biometrics 53*, 1125–1128.

Benda, N. (2010). Model-based approaches for time-dependent dose finding with repeated binary data. *Statistics in Medicine 29*(10), 1096–1106.

Biesheuvel, E. and L. A. Hothorn (2002). Many-to-one comparisons in stratified designs. *Biometrical Journal 44*, 101–116.

Bornkamp, B. (2006). *Comparison of model-based and model-free approaches for the analysis of dose-response studies.* Fachbereich Statistik, Universität Dortmund, Germany. Diploma thesis.

Bornkamp, B., F. Bretz, H. Dette, and J. Pinheiro (2011). Response-adaptive dose-finding under model uncertainty. *Annals of Applied Statistics 5*, 1611–1631.

Bornkamp, B., J. Pinheiro, and F. Bretz (2019). *DoseFinding: Planning and Analyzing Dose Finding experiments.* R package version 0.9–17.

Bretz, F. (2006). An extension of the Williams trend test to general unbalanced linear models. *Computational Statistics & Data Analysis 50*, 1735–1748.

Bretz, F., A. Genz, and L. A. Hothorn (2001). On the numerical availability of multiple comparison procedures. *Biometrical Journal 43*, 645–656.

Bretz, F. and L. A. Hothorn (2001). Testing dose-response relationships with a priori unknown, possibly non-monotonic shapes. *Journal of Biopharmaceutical Statistics 11*, 193–207.

Bretz, F., T. Hothorn, and P. Westfall (2010). *Multiple comparisons using R.* Boca Raton, FL: Chapman and Hall/CRC.

Bretz, F., J. Hsu, J. Pinheiro, and Y. Liu (2008). Dose finding - a challenge in statistics. *Biometrical Journal 50*, 480–504.

Bretz, F., F. König, and B. Bornkamp (2017). Multiple test strategies for comparing several doses with a control in confirmatory trials. In O. J, I. A, and B. B (Eds.), *Handbook of Methods for Designing, Monitoring and Analyzing Dose-finding Trials*, pp. 279–290. CRC Press, Boca Raton, FL.

Bretz, F., W. Maurer, W. Brannath, and M. Posch (2009). A graphical approach to sequentially rejective multiple test procedures. *Statistics in Medicine 28*, 586–604.

Bretz, F., J. Pinheiro, and M. Branson (2004). On a hybrid method in dose-finding studies. *Methods of Information in Medicine 43*, 457–460.

Bretz, F., J. Pinheiro, and M. Branson (2005). Combining multiple comparisons and modeling techniques in dose-response studies. *Biometrics 61*, 738–748.

Bretz, F., A. C. Tamhane, and J. Pinheiro (2009). Multiple testing in dose response problems. In A. C. T. A. Dmitrienko and F. Bretz (Eds.), *Multiple testing problems in pharmaceutical statistics*, Boca Raton, pp. 99–130. Chapman & Hall/CRC Biostatistics Series.

Buckland, S. T., K. P. Burnham, and N. H. Augustin (1997). Model selection: An integral part of inference. *Biometrics 53*, 603–618.

Burman, C.-F., C. Sonesson, and O. Guilbaud (2009). A recycling framework for the construction of bonferroni-based multiple tests. *Statistics in Medicine 28*, 739–761.

Casella, G. and R. Berger (1990). *Statistical Inference.* Belmont, Calif.: Duxbury Press.

Chatfield, C. (1995). Model uncertainty, data mining and statistical inference (with discussion). *Journal of the Royal Statistical Society Series A 158*, 419–466.

Chevret, S. (2006). *Statistical Methods for Dose Finding Experiments.* New York: Wiley.

CHMP (2014). *Qualification Opinion of MCP-Mod as an efficient statistical methodology for model-based design and analysis of Phase II dose finding studies under model uncertainty.* Committee for Medical Product for Human Use.

Chuang-Stein, C. and A. Agresti (1997). A review of tests for detecting a monotone dose-response relationship with ordinal response data. *Statistics in Medicine 16*, 2599–2618.

Cross, J., H. Lee, A. Westelinck, J. Nelson, C. Grudzinkas, , and C. Peck (2002). Postmarketing drug dosage changes of 499 fda-approved new molecular entities, 1980−1999. *Pharmacoepidemiology and Drug Safety 11*, 439–446.

Dette, H., S. Titoff, S. Volgushev, and F. Bretz (2015). Dose response signal detection under model uncertainty. *Biometrics 71*(4), 996–1008.

Dragalin, V., B. Bornkamp, F. Bretz, F. Miller, S. Padmanabhan, N. Patel, I. Perevozskaya, J. Pinheiro, and J. Smith (2010). A simulation study to compare new adaptive dose-ranging designs. *Statistics in Biopharmaceutical Research 2*, 487–512.

Draper, D. (1995). Assessment and propagation of model uncertainty. *Journal of the Royal Statistical Society Series B 57*, 45–97.

Dunnett, C. W. (1955). A multiple comparison procedure for comparing several treatments with a control. *Journal of the American Statistical Association 50*, 1096–1121.

Dunnett, C. W. and A. C. Tamhane (1991). Step-down multiple tests for comparing treatments with a control in unbalanced one-way layouts. *Statistics in Medicine 10*(6), 939–947.

EMA (2017). *Guideline on multiplicity issues in clinical trials (Draft).* European Medicines Agency. Available at http://www.ema.europa.eu/docs/en_GB/document_library/Scientific_guideline/2017/03/WC500224998.pdf.

FDA (2016). *Fit-for-Purpose of MCP-Mod.* Rockville, MD: U.S. Food and Drug Administration.

FDA (2017). *Guidance for Industry Multiple Endpoints in Clinical Trials (Draft).* Food and Drug Administration. Available at https://www.fda.gov/regulatory-information/search-fda-guidance-documents/multiple-endpoints-clinical-trials-guidance-industry.

Franchetti, Y., S. J. Anderson, and A. R. Sampson (2013). An adaptive two-stage dose-response design method for establishing proof of concept. *Journal of Biopharmaceutical Statistics 23*(5), 1124–1154.

Genz, A. and F. Bretz (2009). *Computation of Multivariate Normal and t Probabilities,* Volume 195 of *Lecture Note Series.* Heidelberg: Springer.

Genz, A., F. Bretz, T. Miwa, X. Mi, F. Leisch, F. Scheipl, B. Bornkamp, M. Maechler, T. Hothorn, and M. T. Hothorn (2020). Package 'mvtnorm'. *Journal of Computational and Graphical Statistics 11*, 950–971.

Gutjahr, G. and B. Bornkamp (2017). Likelihood ratio tests for a dose-response effect using multiple nonlinear regression models. *Biometrics 73*(1), 197–205.

Heerdink, E., J. Urquhart, and H. Leufkens (2002). Changes in prescribed dose after market introduction. *Pharmacoepidemiology and Drug Safety 11*, 447–53.

Hoeting, J., D. Madigan, A. Raftery, and C. Volinsky (1999). Bayesian model averaging: a tutorial (with discussion). *Statistical Science 14*, 382–417.

Hothorn, T., F. Bretz, and P. Westfall (2008). Simultaneous inference in general parametric models. *Biometrical Journal 50*, 346–363.

Hothorn, T., F. Bretz, P. Westfall, R. M. Heiberger, A. Schuetzenmeister, and S. Scheibe (2019). *Multcomp: simultaneous inference in general parametric models.* R package version 1.4-12. Available at https://cran.r-project.org/web/packages/multcomp.

Hsu, J. (1984). Constrained simultaneous confidence intervals for multiple comparisons with the best. *Annals of Statistics 12*, 1136–1144.

ICH (1994). *ICH Topic E4: Dose-response information to support drug registration.* International Conference on Harmonization.

Klingenberg, B. (2009). Proof of concept and dose estimation with binary responses under model uncertainty. *Statistics in Medicine 28*(2), 274–292.

König, F. (2015). Confirmatory testing for a beneficial treatment effect in dose-response studies using mcp-mod and an adaptive interim analysis. Presented at the Isaac Newton Institute, http://www.ideal.rwth-aachen.de/wp-content/uploads/2014/02/201507092.pdf.

Krishna, R. (2006). *Dose Optimization in Drug Development.* New York: Informa Healthcare.

Letter, P. (2000). *Paracelsus.* Klein Königsförde, Germany: Königsfurt Verlag.

Ma, S. and M. P. McDermott (2020). Generalized multiple contrast tests in dose-response studies. *Statistics in Medicine 39*(6), 757—772.

Marcus, R., E. Peritz, and K. R. Gabriel (1976). On closed testing procedures with special reference to ordered analysis of variance. *Biometrika 63*(3), 655–660.

Mielke, T. and V. Dragalin (2017). Two-stage designs in dose finding. In J. O'Quigley, A. Iasonos, and B. Bornkamp (Eds.), *Handbook of Methods for Designing, Monitoring, and Analyzing Dose-Finding Trials*, Chapter 14, pp. 247–265. Boca Raton, FL: CRC Press.

Miller, F. (2010). Adaptive dose-finding: Proof of concept with type i error control. *Biometrical journal 52*(5), 577–589.

Mukerjee, H., T. Roberston, and F. Wright (1987). Comparison of several treatments with a control using multiple contrasts. *Journal of the American Statistical Association 82*, 902–910.

Noakes, T. D., N. Goodwin, and B. L. Rayner (1985). Water intoxication: a possible complication during endurance exercise. *Med Sci Sports Exerc 17*, 370–375.

Olson, K. C. (2004). *Poisoning & drug overdose*. New York: Lange Medical Mooks/McGraw-Hill.

O'Quigley, J., A. Iasonos, and B. Bornkamp (2017). *Handbook of Methods for Designing, Monitoring, and Analyzing Dose-Finding Trials*. Boca Raton, FL: CRC Press.

Pan, G. and D. Wolfe (1996). Comparing groups with umbrella orderings. *Journal of the American Statistical Association 91*, 311–317.

Pinheiro, J., B. Bornkamp, and F. Bretz (2006). Design and analysis of dose finding studies combining multiple comparisons and modeling procedures. *Journal of Biopharmaceutical Statistics 16*, 639–656.

Pinheiro, J., B. Bornkamp, E. Glimm, and F. Bretz (2014). Model-based dose finding under model uncertainty using general parametric models. *Statistics in Medicine 33*(10), 1646–1661.

Pinheiro, J., F. Bretz, and M. Branson (2006). Analysis of dose-response studies: Modeling approaches. In N. Ting (Ed.), *Dose Finding in Drug Development.*, pp. 146–171. Springer Verlag, New York, NY.

Ritz, C. and J. C. Streibig (2005). Bioassay analysis using r. *Journal of Statistical Software 12*(5), 1–22.

Ruberg, S. J. (1995a). Dose-response studies. I. Some design considerations. *Journal of Biopharmaceutical Statistics 5*, 1–14.

Ruberg, S. J. (1995b). Dose-response studies. II. Analysis and interpretation. *Journal of Biopharmaceutical Statistics 5*, 15–42.

Sacks, L., H. Shamsuddin, Y. Yasinskaya, K. Bouri, M. Lanthier, and R. Sherman (2014). Scientific and regulatory reasons for delay and denial of fda approval of initial applications for new drugs, 2000-2012. *Journal of the American Medical Association 311*(4), 378–384.

Schaafsma, W. and L. Smid (1966). Most stringent somewhere most powerful tests against alternatives restricted by a number of linear inequalities. *Annals of Mathematical Statistics 37*, 1161–1172.

Schorning, K., B. Bornkamp, F. Bretz, and H. Dette (2016). Model selection versus model averaging in dose finding studies. *Statistics in Medicine 35*(22), 4021–4040.

Simpson, D. and B. Margolin (1990). Nonparametric testing for dose-response curves subject to downturns: Asymptotic power considerations. *Annals of Statistics 18*, 373–390.

Tamhane, A. C., Y. Hochberg, and C. W. Dunnett (1996). Multiple test procedures for dose finding. *Biometrics 52*, 21–37.

Tao, A., Y. Lin, J. Pinheiro, and W. J. Shih (2015). Dose finding method in joint modeling of efficacy and safety endpoints in phase ii studies. *International Journal of Statistics and Probability 4*(1), 33.

Thomas, M., B. Bornkamp, M. Posch, and F. König (2020). A multiple comparison procedure for dose-finding trials with subpopulations. *Biometrical Journal 62*(1), 53–68.

Thomas, N. (2006). Hypothesis testing and Bayesian estimation using a sigmoid Emax model applied to sparse dose designs. *Journal of Biopharmaceutical Statistics 16*, 657–677.

Ting, N. (2006). *Dose Finding in Drug Development.* New York: Springer.

Tukey, J. W. (1953). The problem of multiple comparisons. Unpublished manuscript reprinted in: *The Collected Works of John W. Tukey, Volume 8,* 1994, H. I. Braun (Ed.), Chapman and Hall, New York.

Tukey, J. W., J. L. Ciminera, and J. F. Heyse (1985). Testing the statistical certainty of a response to increasing doses of a drug. *Biometrics 41*, 295–301.

Unkelbach, H. and T. Wolf (1985). *Qualitative Dosis-Wirkungs-Analysen.* Stuttgart: Gustav Fischer Verlag.

Verrier, D., S. Sivapregassam, and A.-C. Solente (2014). Dose-finding studies, mcp-mod, model selection, and model averaging: Two applications in the real world. *Clinical Trials 11*(4), 476–484.

Wakana, A., I. Yoshimura, and C. Hamada (2007). A method for therapeutic dose selection in a phase ii clinical trial using contrast statistics. *Statistics in Medicine 26*, 498–511.

Williams, D. A. (1971). A test for difference between treatment means when several dose levels are compared with a zero dose control. *Biometrics 27*, 103–117.

Xun, X. and F. Bretz (2017). The MCP-Mod methodology: Practical considerations and the DoseFinding R package. In O'Quigley J, Iasonos A, and Bornkamp B (Eds.), *Handbook of Methods for Designing, Monitoring and Analyzing Dose-finding Trials*, pp. 205–227. CRC Press, Boca Raton, FL.

Yamaguchi, Y. and T. Sugitani (2020). Sample size allocation in multiregional dose-finding study using MCP-Mod. *Statistics in Biopharmaceutical Research.* (in press).

12

Multiple Endpoints

Bushi Wang

Boehringer Ingelheim Pharmaceuticals, Inc.

CONTENTS

12.1 Overview

Multiple endpoints in clinical trials are important source of multiplicity. In 2017, US Food and Drug Administration (2017) published a comprehensive draft guidance on multiple endpoints, which reviewed the general background of multiplicity and principles to address multiplicity issue for different types of multiple endpoints problems. It also provided a tutorial on several common multiple testing procedures that can be used in the multiple endpoints setting. Many of the points and discussion are referenced but not repeated in this chapter.

This chapter aims to provide a simple and clear understanding of what types of multiple endpoints are most commonly seen in clinical trials and the statistical considerations associated with such types of multiple endpoints. The chapter allows the readers to choose the appropriate types of multiple endpoints for their own clinical trial design and understand the implication and associated advantages or risks for each choice. There are also sufficient references provided for complex statistical procedures to deal with particular multiple endpoint scenarios.

DOI: 10.1201/9780429030888-14

The structure of this chapter will help the reader to navigate through different topics in multiple endpoints. First of all, Section 12.2 is the most important section with a detailed explanation of the difference between sequentially ordered endpoints, co-primary endpoints, alternative primary endpoints, and composite endpoints. It also includes some basic statistical theories such as why multiplicity adjustment is not required for sequentially ordered endpoints and co-primary endpoints. After reading Section 12.2, it should be clear whether co-primary endpoints in particular trial design should be preferred or avoided, and what kind of multiplicity adjustment might need to be prespecified for composite endpoint and its components.

Section 12.3 and 12.4 are written as a reference that readers can dive into more details for relevant topics of interest. The co-primary endpoint topic will be discussed in further detail in Section 12.3, where several proposals of upward adjustment of type I error control are reviewed. However, the fundamental question why family-wise error rate needs to be controlled strongly in clinical trials is explained, for readers to understand why certain upward adjustment of type I error control is clearly not acceptable in regulatory judgment. Section 12.4 focused on the key fact that family-wise error control gets much more complicated when there is another source of multiplicity. In particular, when there are multiple endpoints in clinical trials investigating multiple doses of a drug, or with multiple interim analyses or multiple subgroups of interest, the multiplicity control require some additional attention. Many of the more advanced multiple testing procedures that were reviewed in the FDA draft guidance are developed in the context of additional sources of multiplicity besides multiple endpoints. These discussion will hopefully help the reader to understand the logic restrictions when applying closed testing-based procedures for more than one sources of multiplicity. For example, logic restrictions when sequentially ordered endpoints (primary and secondary) are tested for different doses of an experimental drug. Further references are provided in respective sections for more details of specific methods.

12.2 Types of Multiple Endpoints

12.2.1 Sequentially Ordered Endpoints

It is common that clinical endpoints are *ordered sequentially* according to their importance or likelihood of showing effect, where the terms of *primary*, *secondary* and *tertiary* endpoints are used. The order of the endpoints should align with the objective of the study and it also has major implication on drug label from regulatory perspective. Sometimes the term key secondary endpoint and other secondary endpoints are used to differentiate the importance among secondary endpoints, then followed by a group of exploratory/further endpoints. In the FDA draft guidance on multiple endpoints (US Food and Drug Administration, 2017), it was suggested that primary endpoints are essential to establish effectiveness for approval, and the secondary endpoints may be used to support the primary endpoint(s) and/or demonstrate additional effects.

However, the determination of sequentially ordered primary and secondary endpoints can sometimes be more complicated. For example, when global registration is considered, different regions/health authorities might accept/request different definitions of the primary endpoint. In this case, a careful evaluation of the sequence is of critical importance to the overall success of the clinical development. In particular, defining *co-primary endpoints* as detailed in Section 12.2.2 might not be the best option due to the reduction in study power and unnecessary hurdle for regions that do not require co-primary endpoints in the target indication. More discussion around this topic can be found in Section 12.3.

As for multiplicity control for sequentially ordered endpoints, each endpoint hypothesis will be tested at level α following the predetermined order, and no additional adjustment is required. The reason is that when null hypotheses H_{01} and H_{02} are tested in sequence, H_{02} only gets to be tested if H_{01} is already rejected. Therefore, the two hypotheses being tested are in fact H_{01} and $H_{a1} \cap H_{02}$, where H_{a1} is the alternative hypothesis of H_{01}. For simplicity, the same notation is used here to denote both the hypothesis and its parameter space, and the alternative hypothesis space is defined as the complement of the null space. Because H_{01} and $H_{a1} \cap H_{02}$ are disjoint, only one of them can be true and multiplicity issue is not applicable. In other words, multiplicity issue exists only when multiple null hypotheses can be true at the same time. This is the *principle of partitioning*, where parameter space is partitioned into disjoint regions. As guaranteed by the partitioning principle, no multiplicity adjustment is required as long as each partition is tested at level α. At the same time, since $H_{a1} \cap H_{02}$ is a subset of H_{02}, a level α test for H_{02} is also a level α test for $H_{a1} \cap H_{02}$, the procedure simplifies to finding level α test for the original ordered null hypotheses H_{01} and H_{02}. The formal proof that multiplicity adjustment is not necessary for sequentially ordered hypotheses was first written down by Hsu and Berger (1999) in the context of stepwise confidence intervals.

As seen in the proof, following the predefined sequence is mandatory. Any failure to reject an earlier endpoint will break the chain of the testing and no further endpoint hypotheses can be rejected. In general, it is required by many regulatory agencies as well as scientific communities that even nominal p-values of further endpoints may not be displayed when the chain breaks to eliminate the possibility of misleading claim.

Sequentially ordered endpoints is a natural case of *fixed-sequence multiple testing procedure.*

12.2.2 Co-primary Endpoints

Co-primary endpoints refers to the situation where multiple primary endpoints are equally weighted in importance and an intervention is deemed efficacious only if it improves on all of them. No multiplicity adjustment is necessary to test co-primary endpoints with similar reason shown in Section 12.2.1. Since all co-primary endpoints must be rejected to make a claim, they can be tested in any order.

The major difference between sequentially ordered endpoints and co-primary endpoints is that while success is already achieved after rejecting the first hypothesis for sequential endpoints, it is only possible after rejecting all hypotheses for co-primary endpoints case. In some sense, the definition of "success" determines how to power a study with either sequentially ordered endpoints or co-primary endpoints. See Section 12.3 for details.

Although there is no limitation on how many co-primary endpoints could be used in one clinical study, normally no more than two endpoints would be recommended. The reason is straightforward, the more co-primary endpoints to consider, the more difficult to power the study to show significant result for all of them. The top recommendation from the Multiple Endpoints Expert Team of the Pharmaceutical Research and Manufacturers of America (PhRMA) is that a single primary endpoint should be identified whenever possible or a composite developed from the medical perspective (Offen et al., 2007).

The FDA draft guidance (US Food and Drug Administration, 2017) provided two circumstances where co-primary endpoints are required. The first circumstance is that for some disorders, there are two or more different features that are so critically important to the disease under study that a drug will not be considered effective without demonstration of an effect on all of these features. One of the examples is to consider pain and an individually specified most bothersome second feature (among photophobia, phonophobia, and nausea) are both shown to be improved by the treatment for migraine headaches. The second kind of circumstance is when there is a single identified critical feature of the disorder, but uncertainty as to whether an effect on the endpoint alone is clinically meaningful. In these cases, two endpoints are often used. One is specific for the disease feature intended to be affected by the drug but not readily interpretable as to the clinical meaning, and the second is clinically interpretable but may be less specific for the intended action of the test drug. The provided example is in Alzheimer's disease where a measure of cognition (e.g., the Alzheimer's Disease Assessment Scale-Cognitive Component, ADAS-Cog) and a clinically interpretable measure of function (such as a clinician's global assessment or an Activities of Daily Living Assessment, iADL) are used as co-primary endpoints.

It is worth noting that considerations involving global development and strategies for registration with multiple regulatory agencies are beyond the scope of the FDA guidance but should not be neglected. It is highly recommended to discuss with different regulatory agencies before finalizing pivotal clinical trial endpoints in complex situations of multiple endpoints. For example, if endpoints A and B by themselves are considered most important by different health authorities in different regions of the world. Designating A and B as co-primary endpoints has a disadvantage, that is, if any one of them fails, the trial is considered a failure, and no registration is possible with any health authorities. On the other hand, designating the endpoint with potentially larger effect size as a primary endpoint and the other as secondary will have a higher chance of success. If both endpoints are efficacious, the product can be registered globally. If the primary endpoint is efficacious, at least the product can be registered at regions where this endpoint is considered the most important. If the primary endpoint fails, no registration is possible, which would have been the same result as if co-primary endpoints were used. Overall, the probability of success for global registration is unchanged but the chance to seek regional registrations is improved by using sequentially ordered endpoints instead of co-primary endpoints.

Since rejecting multiple null hypotheses is harder than rejecting a single null hypothesis, the impact of multiplicity in co-primary endpoints situation is mainly from inflated type II error rate. In other words, the power to reject multiple co-primary endpoints hypotheses is lower than the individual power for each single hypothesis. A simple remedy is to increase the sample size such that each individual hypothesis is powered at a higher level so that overall power is maintained at, for example 80%. It is not a simple topic to properly maintain the overall power and keep the sample size within reasonable range. We provide additional discussion in Section 12.3.4.

12.2.3 Alternative Primary Endpoints

When the demonstration of a treatment effect on at least one of several primary endpoints is sufficient, it is termed as *alternative primary endpoint* scenario for lack of a better term. Here "alternative" indicates that each primary endpoint is an alternative to other primary endpoints in determining the treatment effect. The more endpoints to include in this case, the easier to demonstrate effect on at least one of them. Therefore, alternative primary endpoint scenario introduces a typical multiplicity problem and a proper multiplicity adjustment is critical. However, it is fairly straightforward to control the error rate in alternative primary endpoints scenarios with common multiple testing procedures. Please refer to the other chapters of this book for details on common multiple testing procedures.

12.2.4 Composite Endpoint

Composite endpoint is a single endpoint composed by several component clinical outcomes, where all of them are expected to be affected by the treatment. When the components correspond to distinct events, composite endpoints are often assessed as the time to first occurrence of any one of the components. The main rationale to use composite endpoint is to increase the number of events from different components, thereby accelerating clinical trial conduct and reducing the sample size to a reasonable and feasible range. If individual component endpoints are considered separately with proper multiplicity adjustment, the trial could end up with prohibitively large sample size requirement or becoming extremely underpowered. A comprehensive review of using composite endpoint can be found in Sankoh et al. (2014, 2017).

The major concerns regarding composite endpoint are mostly related to the consistency of results analyzing the composite endpoint and each component separately. First of all, as shown by Sankoh et al. (2017), highly correlated components do not add trial efficiency regarding gain in the overall event rate or endpoint sensitivity compared with disparate or independent components. Therefore, loosely correlated components are common in practice and inconsistent result do happen frequently. A good example of composite endpoint is the three-point major adverse cardiovascular event (3P-MACE), where cardiovascular (CV) death, nonfatal myocardial infarction (MI), and non-fatal stroke are components of this composite endpoint. Sometimes, an alternative of four-point MACE (4P-MACE) adding the hospitalization for unstable angina (HUA) was considered or debated as the primary endpoint. The inclusion of HUA certainly increases the number of events and shortens the time to study completion. However, as comprehensively reviewed by Marx et al. (2017), it also comes with noteworthy disadvantages including the clinical subjectivity in diagnosis of unstable angina, and its low prognostic relevance compared with CV death, non-fatal MI, or non-fatal stroke. Marx et al. (2017) provided a detailed review of several cardiovascular outcome trials involving type 2 diabetes from 1977 to 2013 and focused on the discussion whether and when components such as unstable angina and heart failure should be considered in such studies.

There are many large clinical trials using composite endpoints, which showed interesting and controversial results, because of inconsistency among components. Two very representational examples are the ARISE trial (Tardif et al., 2008) and LIFE trial (Dahlöf et al., 2002).

The ARISE trial is a phase III multi-center, double-blind, parallel group, and placebo-controlled trial to assess the effects of the antioxidant succinobucol (AGI-1067) on cardiovascular outcomes in patients with recent acute coronary syndromes already managed with conventional treatments. The composite primary endpoint was a MACE-like endpoint of time to first occurrence of cardiovascular death, resuscitated cardiac arrest, myocardial infarction, stroke, hospitalization due to unstable angina, or hospitalization due to coronary revascularization. The last two components, though less serious than the other four, had far more frequent occurrences. Despite enrolling 6,144 patients, the trial completed within merely 37 months after the enrollment start and achieved the prespecified number of primary outcome events. However, the primary composite endpoint shows no effect of succinobucol versus placebo (530 events in the succinobucol group vs 529 in the placebo group; hazard ratio 1.00, 95% CI 0.89–1.13, $p = 0.96$). The composite secondary endpoint of cardiovascular death, cardiac arrest, myocardial infarction, or stroke occurred in fewer patients in the succinobucol group than in the placebo group (207 vs. 252 events; hazard ratio 0.81, 95% CI 0.68–0.98). Nevertheless, since the trial failed on the primary endpoint, the result of secondary and further endpoints cannot be statistically significant due to multiplicity. It seems that succinobucol reduced the frequency of the first four components of their composite but result in greater frequency of hospitalization. The strategy to include less serious and more frequent components in the primary composite endpoint seems backfired in this study.

The Losartan Intervention For Endpoint reduction in hypertension (LIFE) study enrolled 9193 patients with the primary composite endpoint was 3P-MACE. The primary composite endpoint occurred in 508 losartan and 588 atenolol patients (relative risk 0.87, 95% Cl 0.77–0.98, $p = 0.021$), and therefore, the interpretation in the original publication Dahlöf et al. (2002) states that losartan prevents more cardiovascular morbidity and death than atenolol. However, further analysis of the study shows 204 losartan and 234 atenolol patients died from cardiovascular disease (0.89, 0.73–1.07, $p = 0.206$); 232 and 309, respectively, had fatal or nonfatal stroke (0.75, 0.63–0.89, $p = 0.001$); and myocardial infarction (nonfatal and fatal) occurred in 198 and 188, respectively (1.07, 0.88–1.31, $p = 0.491$). The statistical significant finding with the primary composite endpoint was deemed to be driven by the softest of the three component, stroke, with little or no positive contribution from the more clinically important components (cardiovascular death and MI). A wrong interpretation of the composite endpoint result may cause potential widespread distribution of misleading conclusion. In the LIFE study case, the U.S. Food and Drug Administration correctly restricted the label of losartan for reduction of nonfatal stroke as opposed to the original primary composite endpoint.

To address the issue that heterogeneity may be observed in components of a composite endpoint, usually it is required that all components of the composite endpoint at least trend positively in favor of the treatment or noninferior compared to the control. It is also expected that the overall effect of the composite endpoint is not driven by the clinically less important component. Regarding composite endpoint, the FDA draft guidance on multiple endpoints (US Food and Drug Administration, 2017) states that only findings on prespecified endpoints that are statistically significant, with adjustment for multiplicity, are considered demonstrated effects of a drug. All other findings are considered descriptive and would require further study to demonstrate that they are true effects of the drug. To demonstrate an effect on a specific component or components of a composite endpoint, the component or components should be included prospectively as a secondary endpoint for the study or possibly as an additional primary endpoint, with appropriate type I error rate control. The FDA draft guidance also recommended to preplan analysis of the contribution of each component of a composite endpoint and report such findings together with the main result.

Statistical considerations and specific multiple testing methods are proposed to allocate certain amount of significance level to the most important component of a composite endpoint such that in the event that composite endpoint failed to demonstrate statistical significant benefit, further testing on the component or components are still possible. Readers can refer to Huque et al. (2011) for discussion on statistical testing in the context of composite endpoint.

12.3 Co-primary Endpoints

We dedicate this section to discuss important topics pertinent to co-primary endpoints. Sometimes the term *reverse* multiplicity problem was used (for example, by Offen et al. (2007)) to describe situations where no multiplicity adjustment is needed to control the family wise error rate. However, it is a misleading term originated from the wrong impression that requiring all co-primary endpoints to be rejected at 5% level is "too conservative" and one should "reverse" from the strong control of family-wise error rate and test each hypothesis at level higher than 5% (two-sided). Therefore, we will clearly describe situations where no multiplicity adjustment is required and avoid using the wrong phrase of "reverse multiplicity".

Why does family-wise error rate need to be controlled strongly in drug development? In the long run, regulatory agencies would like to make sure no more than 2.5% of the approved drugs are not efficacious (one-sided). Drug companies are not going to test the same compound in the same indication over and over again. Different drugs are developed for different diseases by different companies. The fundamental theory behind controlling the family-wise error rate is the law of large numbers for *independent but not identically distributed* samples. In particular, Kolmogorov's strong law of large numbers (see e.g. Sen and Singer (1993), Theorem 2.3.10) states that $\bar{X}_n - \mathrm{E}[\bar{X}_n]$ converges almost surely to 0, provided that each X_i has finite second moment and $\sum_{k\geq 1} k^{-2}\mathrm{Var}[X_k] < \infty$. X_i is the indicator if a family-wise error is made for the i-th clinical trial. The strong control of family-wise error rate guarantees that under *any* configuration of true and nontrue null hypotheses as well as any specified and unspecified nuisance parameters, the probability of making at least one type I error is no more than α. The expected value $\mathrm{E}[\bar{X}_n]$ is no more than α if strong control of the family-wise error rate is satisfied for every clinical trial. However, if for some trials, the family-wise error rate is not controlled strongly, the upper bound for the strong law convergence value cannot be obtained.

Consistent with the interpretation of the non-identically distributed version of law of large numbers, the FDA draft guidance on multiple endpoints stated that the relaxation of statistical testing criteria for each co-primary endpoint is generally not acceptable.

Nevertheless, it was proposed in Offen et al. (2007) that in some cases the increase of sample size to keep sufficient power for co-primary endpoints could become substantial. Some statistical approaches were proposed to relax the strong control of family-wise error rate with co-primary endpoints. In general, these proposals fall under one of the three directions, which will be briefly reviewed in the following subsections.

12.3.1 Restricted Null Space

The strong control of family-wise error requires the maximum false-positive rate over the entire null space to be at most α. In the scenario that there is no treatment effect on one of the co-primary endpoints, the type I error is essentially controlled at the least favorable condition where the treatment has no effect on one endpoint and infinite large effect on the other endpoint. The type I error control in this scenario is equivalent to testing a single null hypothesis at level α. This relation can be expressed in the following equation, where $T_2 > c$ has probability 1 when μ_2 is approaching ∞.

$$P_{\mu_1=0,\mu_2=\infty}(T_1 > c, T_2 > c) = P_{\mu_1=0,\mu_2=\infty}(T_1 > c) = P_{\mu_1=0}(T_1 > c) \qquad (12.1)$$

Restricted null space approaches argue that it is unrealistic to assume one of the endpoints could have infinite effect and propose to limit the null space to certain level M. Therefore, to control

$$P_{\mu_1=0,\mu_2=M}(T_1 > c, T_2 > c) \leq \alpha. \qquad (12.2)$$

The critical constant c can be calculated given the joint distribution of T_1 and T_2 with known correlation.

The restricted null space suits favorably for the situation where one of the endpoints is expected to display a strong effect while the other endpoint might involve a rare event or otherwise expected to be underpowered. It is desirable to consider a slightly larger significance level for the other endpoint to confirm a positive trend given one endpoint is clinically critical and significantly improved by the treatment.

However, the upward adjustment of the significance level is generally limited using restricted null space. As shown by simulation in Chuang-Stein et al. (2007), when M or correlation between endpoints increase, or when sample size is large enough, the upward adjustment diminishes quickly. The simulation shows that the restricted null space is not overly liberal. However, the restricted null space and using the estimated correlation of the endpoints to replace the true correlation result in the loss of strong control for family-wise error. Offen et al. (2007) provided a brief review of other earlier version of the restricted null space approach, which will not be repeated here.

12.3.2 Average Type I Error

The second direction of approaches is to control average type I error instead of family-wise error. As mentioned earlier, the strong control of family-wise error requires the maximum error under all three configurations of the null space $(0,0)$, $(0,\infty)$ and $(\infty,0)$ is no more than α (two endpoints). The average type I error proposal is to control some weighted average of type I error under the three configurations, which forms an upper bound to the average over the entire null space.

Offen et al. (2007) and Chuang-Stein and Li (2017) discussed the equally weighted case and Snapinn and Sarkar (1996) considered a data-dependent weighting. In the simple equal weighting case, one can control the individual type I error at α^* for $(0,\infty)$ and $(\infty,0)$. The type I error under $(0,0)$ depends on the correlation between the endpoints and equal to $(\alpha^*)^2$ for independent endpoints. The value of α^* can be solved by equating $((\alpha^*)^2+2\alpha^*)/3$ to α.

The concern over average type I error approaches is the significant deviation from the strong control of family-wise error rate, which was required in almost all regulatory applications. Average type I error brings more upward adjustment to the individual hypothesis testing level because the complete null configuration $(0,0)$ will bring down the average type I error much more than restricted null space can achieve by limiting on $(0,\infty)$. As a consequence, the family-wise error rate over the entire null space can increase from the required 0.025 to as high as 0.036 with two endpoints and a small correlation (Chuang-Stein and Li, 2017). As commented by Hung and Wang (2009) and Hung and Wang (2010), the situation lead to possibility that p-values for both endpoints are above 0.025 but still conclude a treatment to be efficacious, which deviates from the current regulatory requirement and could render confusion to the consumers if p-values are included in the label. Average type I error is putting more weight on the **weak** control of family-wise error rate and less weight on the strong control. It is not well accepted by regulatory consideration.

12.3.3 Balanced Adjustment Method

Kordzakhia et al. (2010) proposed a balanced adjustment method where a strong treatment benefit in one of the co-primary endpoints can compensate a weaker treatment benefit in the other co-primary endpoint, under the totality consideration of treatment effect. Denote by p_1 and p_2 the p-values associated with the two co-primary endpoints, and let α_L be a real number that $\alpha < \alpha_L < 2\alpha$. The rejection region of the two co-primary endpoints hypotheses are defined as

$$R(\alpha_L) = \{(p_1, p_2) \in (0, \alpha_L)^2 : p_1 + p_2 + cp_1p_2 < \alpha_L\} \tag{12.3}$$

where $c = (\alpha_L - 2\alpha)/\alpha^2$ is a constant depending on α_L and α. It can be shown that if one of the p-values is above α, the other p-value must be smaller than α to fall into the rejection region. This feature made the balanced adjustment method more appealing to the average type I error approach in some sense.

Kordzakhia et al. (2010) proposed a few considerations to select a proper α_L. For example, to consider the strong control of family-wise error rate under restricted null space and calculate the appropriate α_L level. Alternatively, α_L can also be selected to control the average type I error at certain level. Chuang-Stein and Li (2017) pointed out that when α_L

is selected this way, it corresponds to the adjusted significance level in Chuang-Stein et al. (2007).

The balanced adjustment method is a monotone test, where if (p_1, p_2) result in rejection of both co-primary endpoints hypotheses, then for any $p_1' < p_1$ and $p_2' < p_2$, (p_1', p_2') should also result in rejection of the hypotheses. A non-monotone test may, for example, reject both co-primary endpoints with p-values $(0.03, 0.06)$ but fail to reject with $(0.01, 0.06)$. Although monotone is a desired property, some proposals to deal with co-primary endpoints can be non-monotone. For example, Gibson and Overall (1989) proposed to compare the larger p-value with a critical value depends on the level of the smaller p-value. Snapinn and Sarkar (1996)'s Bayesian approach with data-dependent weighting to control average type I error as well as the new approach in Chuang-Stein and Li (2017) are all non-monotone.

As mentioned in Section 12.2.2, the FDA draft guidance on multiple endpoints (US Food and Drug Administration, 2017) provided two circumstances require co-primary endpoints, when there are two or more different features that are so critically important to the disease under study that a drug will not be considered effective without demonstration of an effect on all of these features, and when there is a single identified critical feature of the disorder, but uncertainty as to whether an effect on the endpoint alone is clinically meaningful. In practice, if a strong effect in one of the endpoints can clinically compensate a weaker effect on another endpoint needs to be discussed with regulatory agencies. Most likely only one direction of compensation is reasonable. Clearly, the balanced adjustment method does not control family-wise error rate strongly, but it touches the key question whether co-primary endpoints are the right way to set up the hypotheses.

12.3.4 Discussion on Sample Size and Power Consideration with Co-primary Endpoints

In general, any regulatory application would require rigorous multiplicity adjustment to strongly control the family-wise error rate. With co-primary endpoints, any of the upward adjustment of the significance level at individual hypothesis as discussed above must be discussed with regulatory agencies before being implemented in pivotal trials. Clinical trialists should understand the statistical implication and consequence when deciding if co-primary endpoints or primary-secondary endpoints should be used in the trial. Simulation and sample size calculation might help the determination whether to use co-primary endpoints or primary-secondary endpoints.

As argued above, while relaxation of individual hypothesis significance level is questionable, increasing sample size to maintain proper study power is always acceptable. In co-primary endpoints situation, the power of interest is to reject both co-primary null hypotheses (assuming two co-primary endpoints without loss of generality). Suppose 80% power is desired, which is an overall type II error rate of 0.2. A simple Bonferroni adjustment tells us that we can power each individual hypothesis at 90%, or individual type II error $0.2/2 = 0.1$ without any assumption on correlation between endpoints. In reality, the effect size of the two endpoints is most likely different; therefore, one might want to split the overall type II error rate unequally to avoid over-powering one hypothesis and under-powering the other. Sample size can be calculated based on this simple Bonferroni split of type II errors.

On the other hand, if there is no regulatory requirement to use co-primary endpoints, one can also calculate sample size based on a single primary endpoint with 90% power. The final decision whether to use co-primary or primary-secondary endpoints can be based on the sample size calculation and which scenario feels more comfortable to the project team. It should also consider how much assumptions are made in the calculation and the likelihood and consequence of wrong assumptions (e.g., wrong assumption on the sequential order of endpoints based on their likelihood to demonstrate drug effect).

There are other possibilities to define study power in different situations such as alternative endpoints, sequentially ordered endpoints, or dose-response study. For a specific power definition of interest, the first step is to select a multiple testing procedure (MTP) that strongly control the family-wise error rate and figure out what would be the α level for each individual hypothesis following the MTP. Then sample size can be calculated based on a proper split of type II errors among the null hypotheses and its assigned α level. For detailed methods on proper sample size calculation, one can refer to Wang and Ting (2016) where different power definitions and splitting type II error among individual hypothesis were explored based on different study objectives.

12.4 Multiple Endpoints (Sequentially Ordered) with Another Source of Multiplicity

A major reason that multiple endpoint is an important topic that deserves its own chapter in this book, and the FDA published draft guidance on this topic is because multiple endpoints are likely not the only source of multiplicity to deal with in a clinical trial. It can be considerably more complicated when more than two sources of multiplicity present in a study. To be more specific, while comparing two doses of an experimental drug with a control for a single primary endpoint is a typical multiple testing problem, it becomes more complex when co-primary endpoints or secondary endpoints need to be considered at the same time. Sometimes, the naive application of multiple testing procedure, which was developed under the framework of single source of multiplicity will not control the family-wise error rate as intended. As a result, multiple endpoints with another source of multiplicity expand to a large collection of research topics. In the following sections, we review the situations where the second source of multiplicity comes from either multiple experimental arms (doses), multiple interim analyses, or from subgroups. Without loss of generality, sequentially ordered primary and secondary endpoints will be discussed in this section. The co-primary endpoints case will not be re-discussed in this section because no multiplicity adjustment is actually needed for the co-primary endpoints and the focus would have been on the other single source of multiplicity in a straightforward setting.

12.4.1 Multiple Endpoints and Multiple Doses

Multiplicity is a typical topic in dose-response studies, and various multiple testing procedures can be applied in practice. When there are multiple endpoints in a dose-response study, the usual closed testing procedure fails because each intersection hypothesis was considered equally without any logical restrictions regarding that efficacy in the secondary endpoint would not be relevant before efficacy in the primary endpoint was shown. The simple Bonferroni adjustment and fixed-sequence procedure will not be recommended in this setting because they are either too conservative and significantly reduce the power or weighing too much on a guess of the likelihood to show effect for each endpoint and dose combination, respectively. In the following, we focus on two classes of popular approaches, differing in how to take in the logical restrictions on the sequential order of multiple endpoints.

A large collection of multiple testing strategies were centered around the gatekeeping procedure of Dmitrienko et al. (2003), which was proposed to address exactly the issue in the analysis of clinical trials with multiple endpoints and multiple dose levels. The foundation of the *gatekeeping procedure* is built on grouping hypotheses into sequential families, and each family serves as a gatekeeper for the subsequent families, where the primary family of hypotheses comprising the primary objectives of the study and the secondary and tertiary objectives forms other families are tested only if one or more gatekeeper hypotheses have been rejected. The term parallel gatekeeping is used to describe this particular procedure. Important further research in this area include the generalization of serial and parallel gatekeeping procedures to the tree-structured gatekeeping procedures (Dmitrienko et al., 2008, 2006, 2007), and building multistage parallel gatekeeping procedures based on tests that are more powerful than Bonferroni (Dmitrienko et al., 2008). A combination of more powerful separable truncated multiple testing procedures and application to scenarios beyond the parallel gatekeeping restriction is proposed by Dmitrienko and Tamhane (2011).

Another thread of development around the same topic focused on general *graphical representation* of sequentially rejective multiple testing procedures (Bretz et al., 2009). In this framework, most of the Bonferroni-based multiple testing procedures, including the parallel gatekeeping procedure can be presented by simple, interactive graphs which can be easily communicated to clinical trialists with nonstatistical background. An extension of the graphical approach to weighted Bonferroni and non-Bonferroni-based procedures was later published by Bretz et al. (2011). It is noteworthy that there is an R package 'gMCP' for researchers to easily implement and customize graphical multiple testing procedures with a graphical user interface (GUI).

Logical restriction among hypotheses of multiple endpoints and multiple doses is one of the most important considerations and debate in these type of clinical trials. As stipulated by Hung and Wang (2009, 2010), when the clinical decision involves several dose-endpoint pair hypotheses, then the line between primary and secondary endpoints may be unclear. Here two distinct approaches should be made clear. Gatekeeping procedures by Dmitrienko et al. (2003, 2007) emphasized on the '*independence condition*' where a decision to reject a null hypothesis in the primary family is independent of decision made for hypotheses in subsequent families. Alternatively, the '*decision paths principle*' represented by partition testing (Liu and Hsu, 2009) emphasized that, **for a given dose**, efficacy in the secondary endpoint is relevant only if efficacy in the primary endpoint has been shown. Two principles disagree with each other when implemented in testing procedures, which is illustrated in Figure 12.1.

A graphical representation of the parallel gatekeeping procedure can be found on the left panel of Figure 12.1 which follows the independence condition. Here H_1 and H_2 denote the null hypotheses in the primary family which are primary endpoint for high dose (H_1) and low dose (H_2). Similarly, H_3 and H_4 denote the null hypotheses in the secondary family which are secondary endpoint for high dose (H_3) and low dose (H_4). We assume the readers are familiar with the graphical approach notations; otherwise, please refer to Bretz et al. (2009) or the R package 'gMCP' manual. The procedure represented on the right panel of Figure 12.1 is an alternative procedure by Maurer et al. (2011), which follows the decision paths principle.

The distinction is well represented graphically. While in the gatekeeping procedure, there is no arrow from H_3 to H_2, or from H_4 to H_1, indicating that the decision on primary endpoint (e.g., H_1 high-dose primary endpoint) does not depend on the rejection of secondary endpoint (e.g., H_4 low-dose secondary endpoint), regardless of dose levels. However, it is possible that low dose could fail primary endpoint (H_2) but still be tested for secondary endpoint (H_4) because H_4 could be tested as long as H_1 is rejected, regardless of H_2 result. Therefore, the gatekeeping procedure could potentially claim that one of the doses is efficacious in the secondary endpoint but not in the primary endpoint, a rather invalid result from a regulatory perspective.

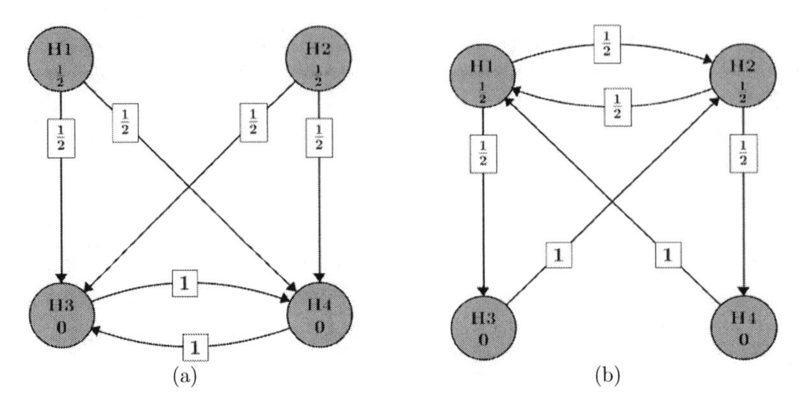

FIGURE 12.1
The graphical representation of parallel gatekeeping by Dmitrienko et al. (2003) and alternative multiple testing procedure by Maurer et al. (2011) following the decision paths principle.

The decision paths principle will not allow testing secondary endpoint (e.g., H_4) if the primary endpoint hypothesis for the same dose (e.g., H_2) is not rejected. Depicted by no arrows pointing to H_3 and H_4 without passing through their corresponding same-dose primary endpoints H_1 and H_2.

The independence condition of Dmitrienko et al. (2003) requires that testing of H_2 (primary endpoint low dose) must not be influenced by the result of testing H_3 (secondary endpoint high dose) is artificial and irrelevant when multiple doses are considered. By insisting on this artificial independence condition, the testing procedure is prone to violate a more important principle that efficacy of certain dose on the secondary endpoint cannot be claimed without rejecting the null hypothesis of the primary endpoint (of the same dose).

Note that by removing the arrow from H_1 to H_4 and H_2 to H_3 on the left panel of Figure 12.1 will not solve the problem, and it can still result in the invalid conclusion potentially. As long as there is a path connecting H_1 with H_4 (through H_1 to $H3$ and H_3 to H_4) without passing through H_2, the decision paths principle is violated and potentially H_4 can be rejected without rejecting H_2.

In general, decision paths principle of Liu and Hsu (2009) should be applied in designing multiple testing procedures for multiple doses and multiple endpoints. Practitioners should exercise with caution when independence condition (Dmitrienko et al., 2003) is implemented and expect to explain why it is preferred in particular situations and how invalid result can be avoided. As a conclusion, the more flexible graphical approach is recommended with consideration of decision paths principle to handle multiplicity in clinical trial with multiple endpoints and multiple doses.

12.4.2 Multiple Endpoints in Adaptive Design

In this section, we discuss the situation where the second source of multiplicity is introduced by testing repeatedly the same hypothesis at different time points, namely, at interim and at final analysis. Group sequential trials and adaptive design is a common practice when interim analyses are needed for various reasons in clinical study. When multiple endpoints, such as hierarchically ordered primary and secondary endpoints, are considered in a group sequential trial, several considerations are offered in the literature.

Denote H_p and H_s as the **one-sided** primary and secondary endpoints. For simplicity, one-sided hypothesis is considered to avoid the complication of directional error (type III).

Without loss of generality, consider clinical trials with one interim analysis and final analysis. Let Z_{pi} and Z_{si} denote the test statistics for H_p and H_s at the interim analysis ($i = 1$) and final analysis ($i = 2$).

One of the common stagewise strategies is to test the primary endpoint hypothesis H_p at the interim with critical value c_{p1}. If $Z_{p1} \geq c_{p1}$, reject H_p, stop the trial and test the secondary endpoint hypothesis H_s with critical constant c_{s1}. Otherwise if $Z_{p1} < c_{p1}$, the trial continues and the primary hypothesis H_p is tested again at final analysis with critical value c_{p2}. If $Z_{p2} > c_{p2}$, the primary endpoint hypothesis is rejected at the final analysis and H_s is then tested with critical value c_{s2}. The family-wise error rate under H_s (regardless if H_p is true or false) can be written as

$$P(Z_{p1} \geq c_{p1}, Z_{s1} \geq c_{s1}) + P(Z_{p1} < c_{p1}, Z_{p2} \geq c_{p2}, Z_{s2} \geq c_{s2}) \tag{12.4}$$

Several articles discussed this strategy explicitly Glimm et al. (2010); Hung et al. (2007); Tamhane et al. (2010). Common α-spending function can be selected separately for primary and secondary endpoints. For primary endpoint, if O'Brien-Fleming boundary is used (O'Brien and Fleming, 1979), $c_{p2} = \sqrt{t}c_{p1}$, where t is the information fraction. Similarly, if Pocock boundary is used (Pocock, 1977), then $c_{p2} = c_{p1}$.

The same or different α-spending function can be used for the secondary endpoint. With O'Brien-Fleming boundary, $c_{s2} = \sqrt{t}c_{s1}$ and for Pocock boundary, $c_{s2} = c_{s1}$.

Specifically, if $c_{s1} = c_{s2} = z_\alpha$, Hung et al. (2007) first pointed out that this naive strategy motivated by the fixed sequence test procedure for classical trial design without interim analysis does not maintain the family-wise error rate at level α. In this set up, the secondary endpoint H_s is tested at level α with a Pocock boundary whenever the primary endpoint is significant. Tamhane et al. (2010) tabulated the numerical value of c_{s1} for several primary boundary and secondary boundary combinations for different correlation level of the endpoints. They also proved analytically that the naive strategy does not control family-wise error rate for all correlations between endpoints and effect size of the primary endpoint.

In addition, Tamhane et al. (2010) explored the selection of α-spending boundaries where (a) $c_{p1} = c_{s1}$, $c_{p2} = c_{s2}$; (b) $c_{p1} > c_{s1}$, $c_{p2} < c_{s2}$; and 3) $c_{p1} < c_{s1}$, $c_{p2} > c_{s2}$. Where the maximum family-wise error rate will be achieved for different correlations and effect size was discussed. The power of the study was also evaluated for the three situations. In their conclusion, choosing O'Brien-Fleming boundary for the primary endpoint is uniformly more powerful than the Pocock boundary in terms of primary power, which is defined as the probability that the primary endpoint is rejected either at the interim analysis or at the final analysis when the null hypothesis H_p is false. Secondary power is defined as the probability where secondary endpoint is either rejected at interim analysis after rejecting the primary endpoint or rejected at the final analysis after primary endpoint. As the secondary power is a function of correlation ρ and the effect size δ_p and δ_s, it is more complicated to describe the scenarios when different boundaries were selected. However, in general, Tamhane et al. (2010) concluded that using O'Brien-Fleming boundary for the primary endpoint and Pocock or ad hoc boundary for the secondary endpoint will result in the preferable secondary power in most common combinations of ρ and δ's.

Glimm et al. (2010) independently evaluated this testing strategy and reached the same conclusion. In particular, Glimm et al. (2010) pointed out that in practice, the choice of the spending function is guided by the wish to stop the trial early unless there is very strong evidence in favor of the new treatment. Therefore, allocating small significance level at the interim analysis (O'Brien-Fleming boundary) is much more popular than the balanced type (Pocock boundary). For secondary endpoint, using Pocock boundary is usually more powerful than using Bonforroni test, which is, in turn, more powerful than O'Brien-Fleming boundary. In addition, Glimm et al. (2010) also discussed other testing strategies which might be of interest in particular scenarios. For example, following the stagewise strategy but do not stop the trial at interim if only the primary endpoint is rejected. Instead, test the secondary endpoint again at final analysis. As pointed out by Glimm et al. (2010), some of those strategies can control the family-wise error rate strongly and some can only control type I error at individual endpoint level. However, there are still places for such strategies to be considered and discussed with the clinical team.

12.4.3 Multiple Endpoints and Multiple Populations

Multiplicity considerations in *subgroup* analysis is an important topic itself. Dmitrienko et al. (2017) and Dane et al. (2019) gave a comprehensive summary on the multiplicity issue in subgroup analysis. In general, subgroup analysis is considered either exploratory or confirmatory. Different strategies for exploratory or confirmatory purpose are reviewed in the abovementioned articles. Multiplicity in subgroup analysis is already a complicated topic with many ongoing discussions. Currently, there is not enough consideration to handle multiple endpoints in multiple subgroups or multiple populations. In this section, a brief high-level discussion is provided in this area.

One particular scenario is that treatment effect was evaluated in a confirmatory setting for both overall population and a target subpopulation with primary and secondary endpoints. It can be handled in a similar way as multiple endpoints with multiple dose groups. The decision paths principle still works in this situation where secondary endpoint of a particular population is only relevant if the benefit from primary endpoint is already shown for that population. Sometimes, there is also incentives to consider the primary endpoint for both overall population and the targeted subgroup before moving on with secondary endpoint. The graphical approach on the right panel of Figure 12.1 can be a good option. Another possibility is to consider truncated multiple testing procedure for primary endpoint in the two population, then apply to a multistage gatekeeping procedure as described in Dmitrienko et al. (2008). Readers can also refer to Qiu et al. (2018) for discussion on graphical representation using truncated procedures.

However, we would like to point out a common trap to apply a fixed sequence procedure with prespecified order of null hypotheses to avoid multiplicity adjustment (See Section 12.2.1). In particular, fixed sequence procedure should not be recommended unless the user is fairly confident that the hypothesis in the first order has higher power than the hypothesis in the second order. Dmitrienko et al. (2017) provided example of phase III clinical trial that the fixed sequence procedure prevented the sponsor from making valid claims because of the unfortunate specification of testing sequence. Sometimes, there is an illusion that fixed sequence procedure results in the highest power with given sample size, while the reality is that the higher power is a bet on the user correctly guessed the right testing sequence. When the bet is wrong, the high power disappears. Therefore, although other more flexible multiple testing procedures might require larger sample size, it can be more robust to the uncertainty of observed effect size and significance level.

12.5 Discussion

In this chapter, we discussed multiplicity topics related to multiple endpoints and situations when another source of multiplicity present besides multiple endpoints. A few principles in multiple testing are reviewed to help explaining some of the handling in multiple endpoints topics.

The fundamental principle of why multiplicity adjustment is required for drug regulatory consideration is revisited in Section 12.3. More specifically, the requirement for family-wise error rate to be controlled strongly is rooted from the independent but not identically distributed version of law of large numbers.

Another important principle briefly covered in this chapter is the partitioning principle (see also Chapter 4 of this book for more detailed introduction on partitioning principle and decision paths). Using partitioning principle, it is straightforward to understand why multiplicity adjustment is not needed for sequentially ordered endpoints and co-primary endpoints.

When multiple endpoints and multiple doses are both presented in a single study, it is recommended to follow the decision paths principle of partitioning such that secondary endpoint hypothesis can be rejected only if the primary endpoint hypothesis of the same dose has already been rejected. It is pointed out in Section 12.4.1 that the independence condition required by gatekeeping procedures may result in the type of invalid conclusion to claim secondary endpoint efficacious for a dose but fail to show efficacy in the primary endpoint.

Bibliography

Bretz, F., W. Maurer, W. Brannath, and M. Posch (2009). A graphical approach to sequentially rejective multiple test procedures. *Statistics in Medicine 28*(4), 586–604.

Bretz, F., M. Posch, E. Glimm, F. Klinglmueller, W. Maurer, and K. Rohmeyer (2011). Graphical approaches for multiple comparison procedures using weighted bonferroni, simes, or parametric tests. *Biometrical Journal 53*(6), 894–913.

Chuang-Stein, C. and J. D. Li (2017). Changes are still needed on multiple co-primary endpoints. *Statistics in Medicine 36*(28), 4427–4436.

Chuang-Stein, C., P. Stryszak, A. Dmitrienko, and W. Offen (2007). Challenge of multiple co-primary endpoints: a new approach. *Statistics in Medicine 26*(6), 1181–1192.

Dahlöf, B., R. B. Devereux, S. E. Kjeldsen, S. Julius, G. Beevers, U. de Faire, F. Fyhrquist, H. Ibsen, K. Kristiansson, O. Lederballe-Pedersen, L. H. Lindholm, M. S. Nieminen, P. Omvik, S. Oparil, and H. Wedel (2002). Cardiovascular morbidity and mortality in the losartan intervention for endpoint reduction in hypertension study (life): a randomised trial against atenolol. *The Lancet 359*(9311), 995 – 1003.

Dane, A., A. Spencer, G. Rosenkranz, I. Lipkovich, T. Parke, and on behalf of the PSI/EFSPI Working Group on Subgroup Analysis (2019). Subgroup analysis and interpretation for phase 3 confirmatory trials: White paper of the efspi/psi working group on subgroup analysis. *Pharmaceutical Statistics 18*(2), 126–139.

Dmitrienko, A., B. Millen, and I. Lipkovich (2017). Multiplicity considerations in subgroup analysis. *Statistics in Medicine 36*(28), 4446–4454.

Dmitrienko, A., W. W. Offen, and P. H. Westfall (2003). Gatekeeping strategies for clinical trials that do not require all primary effects to be significant. *Statistics in Medicine 22*(15), 2387–2400.

Dmitrienko, A. and A. C. Tamhane (2011). Mixtures of multiple testing procedures for gatekeeping applications in clinical trials. *Statistics in Medicine 30*(13), 1473–1488.

Dmitrienko, A., A. C. Tamhane, L. Liu, and B. L. Wiens (2008). A note on tree gatekeeping procedures in clinical trials. *Statistics in Medicine 27*(17), 3446–3451.

Dmitrienko, A., A. C. Tamhane, X. Wang, and X. Chen (2006). Stepwise gatekeeping procedures in clinical trial applications. *Biometrical Journal 48*(6), 984–991.

Dmitrienko, A., A. C. Tamhane, and B. L. Wiens (2008). General multistage gatekeeping procedures. *Biometrical Journal 50*(5), 667–677.

Dmitrienko, A., B. L. Wiens, A. C. Tamhane, and X. Wang (2007). Tree-structured gatekeeping tests in clinical trials with hierarchically ordered multiple objectives. *Statistics in Medicine 26*(12), 2465–2478.

Gibson, J. M. and J. E. Overall (1989). The superiority of a drug combination over each of its components. *Statistics in Medicine 8*(12), 1479–1484.

Glimm, E., W. Maurer, and F. Bretz (2010). Hierarchical testing of multiple endpoints in group-sequential trials. *Statistics in Medicine 29*(2), 219–228.

Hsu, J. C. and R. L. Berger (1999). Stepwise confidence intervals without multiplicity adjustment for dose-response and toxicity studies. *Journal of the American Statistical Association 94*(446), 468–482.

Hung, H. M. J. and S.-J. Wang (2009). Some controversial multiple testing problems in regulatory applications. *Journal of Biopharmaceutical Statistics 19*(1), 1–11. PMID: 19127460.

Hung, H. M. J., S.-J. Wang, and R. O'Neill (2007). Statistical considerations for testing multiple endpoints in group sequential or adaptive clinical trials. *Journal of Biopharmaceutical Statistics 17*(6), 1201–1210. PMID: 18027226.

Huque, M. F., M. Alosh, and R. Bhore (2011). Addressing multiplicity issues of a composite endpoint and its components in clinical trials. *Journal of Biopharmaceutical Statistics 21*(4), 610–634. PMID: 21516560.

James Hung, H. M. and S.-J. Wang (2010). Challenges to multiple testing in clinical trials. *Biometrical Journal 52*(6), 747–756.

Kordzakhia, G., O. Siddiqui, and M. F. Huque (2010). Method of balanced adjustment in testing co-primary endpoints. *Statistics in Medicine 29*(19), 2055–2066.

Liu, Y. and Hsu, J. C. (2009). Testing for efficacy in primary and secondary endpoints by partitioning decision paths. *Journal of the American Statistical Association 104*(488), 1661–1670.

Marx, N., D. K. McGuire, V. Perkovic, H.-J. Woerle, U. C. Broedl, M. von Eynatten, J. T. George, and J. Rosenstock (2017). Composite primary end points in cardiovascular outcomes trials involving type 2 diabetes patients: Should unstable angina be included in the primary end point? *Diabetes Care 40*(9), 1144–1151.

Maurer, W., E. Glimm, and F. Bretz (2011). Multiple and repeated testing of primary, coprimary, and secondary hypotheses. *Statistics in Biopharmaceutical Research 3*(2), 336–352.

O'Brien, P. C. and T. R. Fleming (1979). A multiple testing procedure for clinical trials. *Biometrics 35*(3), 549–556.

Offen, W., C. Chuang-Stein, A. Dmitrienko, G. Littman, J. Maca, L. Meyerson, R. Muirhead, P. Stryszak, A. Baddy, K. Chen, K. Copley-Merriman, W. Dere, S. Givens, D. Hall, D. Henry, J. D. Jackson, A. Krishen, T. Liu, S. Ryder, A. J. Sankoh, J. Wang, and C.-H. Yeh (2007). Multiple co-primary endpoints: Medical and statistical solutions: A report from the multiple endpoints expert team of the pharmaceutical research and manufacturers of america. *Drug Information Journal 41*(1), 31–46.

Pocock, S. J. (1977). Group sequential methods in the design and analysis of clinical trials. *Biometrika 64*(2), 191–199.

Qiu, Z., L. Yu, and W. Guo (2018). A family-based graphical approach for testing hierarchically ordered families of hypotheses.

Sankoh, A. J., H. Li, and R. B. D'Agostino Sr (2014). Use of composite endpoints in clinical trials. *Statistics in Medicine 33*(27), 4709–4714.

Sankoh, A. J., H. Li, and R. B. D'Agostino Sr. (2017). Composite and multicomponent end points in clinical trials. *Statistics in Medicine 36*(28), 4437–4440.

Sen, P. K. and J. M. Singer (1993). *Large Sample Methods in Statistics.* London: Chapman & Hall, Inc.

Snapinn, S. M. and S. K. Sarkar (1996). A note on assessing the superiority of a combination drug with a specific alternative. *Journal of Biopharmaceutical Statistics 6*(3), 241–251. PMID: 8854229.

Stefansson, G., W. Kim, and J. C. Hsu (1988). On confidence sets in multiple comparisons. In S. S. Gupta and J. O. Berger (Eds.), *Statistical Decision Theory and Related Topics IV*, Volume 2, pp. 89–104. Springer-Verlag, New York, NY.

Tamhane, A. C., C. R. Mehta, and L. Liu (2010). Testing a primary and a secondary endpoint in a group sequential design. *Biometrics 66*(4), 1174–1184.

Tardif, J.-C., J. J. McMurray, E. Klug, R. Small, J. Schumi, J. Choi, J. Cooper, R. Scott, E. F. Lewis, P. L. L'Allier, and M. A. Pfeffer (2008). Effects of succinobucol (agi-1067) after an acute coronary syndrome: a randomised, double-blind, placebo-controlled trial. *The Lancet 371*(9626), 1761 – 1768.

US Food and Drug Administration (2017, Jan). Multiple endpoints in clinical trials. *Guidance for Industry.*

Wang, B. and N. Ting (2016). Sample size determination with familywise control of both type i and type ii errors in clinical trials. *Journal of Biopharmaceutical Statistics 26*(5), 951–965. PMID: 26881972.

13

Subgroups Analysis for Personalized and Precision Medicine Development

Yi Liu

Nektar Therapeutics

Hong Tian

BeiGene

Jason C. Hsu

The Ohio State University

CONTENTS

DOI: 10.1201/9780429030888-15

Subgroup analysis occurs in diverse areas such as personalized medicine and web analytics. This chapter describes them in the setting of a randomized controlled trials (RCTs) for personalized/precision medicine development. To personalize medicine is to compare the efficacy of treatment versus control in subgroups and their mixtures. There are natural relationships among efficacy in subgroups and their mixtures. This chapter provides a guide to subgroup analysis that respects such logical relationships.

For binary and time-to-event outcomes, there has been an oversight in the analyses of efficacy stratified on a biomarker, in the sense that they do not reflect logical relationships among efficacy in a mixture with efficacy in the subgroups. Causes of the illogical analyses are as follows: (a) the use of efficacy measures such as odds ratio and hazard ratio (HR), which are not collapsible and therefore not logic-respecting and (b) incorrect mixing of efficacy measure such as relative response (RR) even when they are logic-respecting. We will explain RR and ratio of median (RoM) survival times are logic-respecting (which implies they are collapsible) in RCTs. We will further explain that, for binary and time-to-event outcomes, mixing efficacy in subgroups by prevalence will lead to illogical results, in general, that efficacy should be mixed by the prognostic effect instead. Finally, we show that the path to achieve confident logical inference on efficacy in subgroups and their mixtures is (1) choose a logic-respecting efficacy measure, (2) model the data and adjust for imbalance using the Least Squares means technique, and (3) apply the subgroup mixable estimation (SME) principle to infer efficacy in subgroups and their mixtures.

13.1 Targeted Therapy and Personalized/Precision Medicine

Targeted therapies, which as Woodcock (2015) states are sometimes called "personalized medicine" or "precision medicine", target specific pathways. For example, pembrolizumab (Keytruda®) and nivolumab (Opdivo®) are medicines that target PD-1, the so-called Programmed cell Death protein 1 on immune T cells. By blocking PD-1, these targeted therapies boost the immune response against cancer cells, which can shrink some tumors or slow their growth.

In personalized/precision medicine, we are concerned with finding whether there are subgroups of an overall patient population that exhibit a differential response to treatment. Any subgroup with a significantly better response to treatment could be identified for tailoring with appropriate labeling language and reimbursement considerations in the market. Conversely, subgroups with a worse response to treatment could be appropriately contraindicated in labeling.

Subgroups can be defined by biomarkers or by other characteristics such as countries or regions. In the former case, decision-making involves assessing efficacy in the subgroups and their mixtures. In the latter case, the typical practice is to adjust for baseline differences in the subgroups in assessing a presumed common efficacy across the subgroups.[1] This chapter focuses on the former situation.

Targeted therapies make use of blood chemistry tests, genotyping, imaging, immunohistochemistry (IHC), or other technology to measure each subject's biomarker value or values. These biomarker values can then be used to determine who are more likely to benefit from a treatment.

[1]In the analysis, there is no interaction term between region and treatment, but there is an interaction term between biomarker and treatment.

We will focus on the situation where there is a "treatment" and a "control", abbreviated as Rx and C respectively. Our subgroup analysis discussion will be mainly in the setting of an RCT.

13.2 Respecting Logical Relationships between Subgroups and Their Mixtures

In any study, it is important to have confidence that it is the new *treatment* that causes patients to have better outcome.

An RCT is a scientific study where subjects are randomly allocated to one or other of the different treatments under study. It is assumed that there is no differential propensity in treatment assignment. Random assignment of subjects to treatments then reduces imbalance of subject characteristics across treatments if the sample size is large (i.e., prevalence of each subgroup is about the same under Rx and under C), reducing the likelihood of spurious causality.

Let $\mu^{Rx}(x)$ and $\mu^{C}(x)$ denote the true effect of Rx and C at each biomarker value x. Let $p(x)$ be the density of patient biomarker values in the population which, in our RCT setting, is the same for Rx and C. Suppose a biomarker cut-point value c divides the entire population into two subgroups, the marker-negative $g^- = \{x < c\}$ subgroup, and the marker-positive $g^+ = \{x \geq c\}$ subgroup.

Denote the true (unknown) efficacy in g^-, g^+, and all-comers $\{g^-, g^+\}$ by $\eta_{g^-}, \eta_{g^+}, \eta_{\{g^-,g^+\}}$ respectively. Since all-comers is a mixture of g^+ and g^-, it is desirable for efficacy measures to meet the criterion that efficacy for all-comers lies between the efficacy of the complementary subgroups:

$$\text{Definition: An efficacy measure is } \textit{logic-respecting} \text{ if } \eta_{\{g^-,g^+\}} \in [\eta_{g^-}, \eta_{g^+}] \tag{13.1}$$

13.2.1 Three Causes for Efficacy Assessment to Be Illogical

Efficacy measures and their properties (such as being collapsible and logic-respecting) are defined at the population level (i.e., in the parameter space, with infinite sample size).

In the literature, an efficacy measure is said to be *collapsible* if g^- and g^+ patients deriving the same efficacy ($= 3$ say) implies all-comers $\{g^-, g^+\}$ derive the same efficacy ($= 3$) as well. Therefore, logic-respecting implies collapsibility.

Non-collapsibility is taken to be an indication of non-causality. Rubin (1978) showed that conducting a study as an RCT is sufficient to avoid non-causality (when the sample size is large). His proof of *strong ignorability* applies if efficacy is measured as a *difference* of means, but not necessarily if efficacy is measured as a ratio (as it implicitly assumes efficacy in the subgroups determine efficacy in the overall population).

Estimated efficacy in finite samples may exhibit illogic behavior due to

1. Using a *not* logic-respecting efficacy measure (including assuming efficacy in a mixture can be determined by efficacy in the subgroups)
2. Not adjusting for imbalance in the data (over reliance on efficacy measure being logic-respecting)
3. Over-extension of Least Squares means (LSmeans) for continuous outcome to binary and time-to-event outcomes in computer packages (even when efficacy measure is logic-respecting)

This chapter shows how each pitfall can be avoided.

Specifically, we will explain why difference of means, relative response (RR), and ratio of median (RoM) survival times are logic-respecting. We provide a (balanced data infinite sample size) example that proves odds ratio is not collapsible (and therefore not logic-respecting). A counterexample in the literature will be cited that proves HRs is not collapsible.

We will also demonstrate by examples the danger of not adjusting for imbalance in the data, even if efficacy measure is logic-respecting.

Surprisingly, currently computer packages give misleading subgroup analysis results even when the efficacy measure is logic-respecting, and the data is perfectly balanced. Ironically, current computer package implementations can mask the fact that efficacy measures such as HR are not collapsible.

It is possible to give non-misleading subgroup analysis results. In this chapter, we will explain how the new SME principle, working in concert with the LSmeans technique, confidently produces logical inference on subgroups and their mixtures.

13.3 Prognostic and Predictive Biomarkers

For a therapy to target a subgroup, it is important to have confidence that patients with the targeting biomarker value indeed benefit more, that is, the biomarker is not merely *prognostic* but *predictive*, in the following sense.

The Merriam-Webster dictionary definition of "prognostic" is "something that foretells". We say a biomarker is *treatment-effect* prognostic if its value has some ability to foretell the outcome for a patient given that treatment. A biomarker is thus **not** *treatment-effect* prognostic if its value has no such ability, that is, patients form a single population under that treatment.

There are other definitions of a prognostic biomarker. For example, BEST (2016) defines a prognostic biomarker as one which predicts the increased likelihood of an event *without an intervention.* Those biomarkers can be called *disease-progression* prognostic biomarkers.

For brevity, in the RCT setting of this chapter, a prognostic biomarker refers to a *treatment-effect* prognostic biomarker. Our definition of a "prognostic" biomarker is treatment arm specific, to distinguish between the situation where the marker is not prognostic in one arm but is prognostic in the other (as in Figure 13.1a), and the situation where the marker is equally prognostic in both arms (as in Figure 13.1b).

A biomarker is **predictive** if its value has some ability to differentiate between the effect of Rx from the effect of C (i.e., it has some ability to foretell the *efficacy* of Rx vs. C). We might say the biomarker is *purely predictive* in the case of Figure 13.1a, while we say the biomarker is *purely prognostic* in the case of Figure 13.1b.

Overlooked in the literature is that, even in an RCT, efficacy in $\{g^-, g^+\}$ involves the prognostic effect if efficacy is measured as a *ratio*, be it odds ratio, relative response, HR, or ratio of medians (RoM). This oversight plays a role in incorrect subgroup analyses in current computer packages.

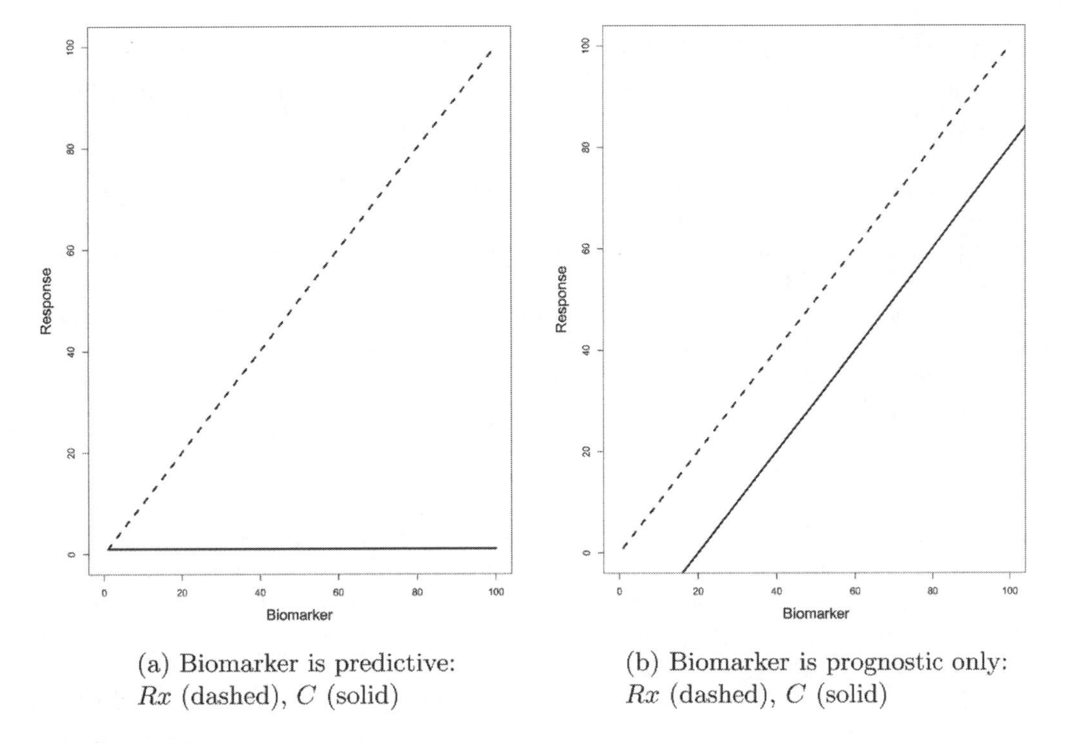

(a) Biomarker is predictive: Rx (dashed), C (solid)

(b) Biomarker is prognostic only: Rx (dashed), C (solid)

FIGURE 13.1
Predictive marker (a) vs. non-predictive marker (b)

13.4 Logic-Respecting Efficacy Measures

Denote by $\mu_{g^+}^{Rx}$, $\mu_{g^-}^{Rx}$, $\mu_{g^+}^{C}$, $\mu_{g^-}^{C}$ the true *expected* outcomes in the g^+ and g^- subgroups for each treatment arm, and denote by μ^{Rx} and μ^{C} the true *expected* outcome over the entire patient population if the entire population had received Rx or C, respectively.

13.4.1 Difference of Means is Logic-Respecting

In therapeutic areas such as Type 2 diabetes and Alzheimer's Disease with continuous outcome measures, traditionally efficacy of Rx vs. C is measured by the *difference* of *mean* treatment and control effects, so

$$\eta_{g^+} = \mu_{g^+}^{Rx} - \mu_{g^+}^{C} \text{ and } \eta_{g^-} = \mu_{g^-}^{Rx} - \mu_{g^-}^{C}$$

represent efficacy of Rx vs. C in the g^+ and g^- subgroups. In our RCT setting, with population prevalence of the g^+ subgroup being γ^+,

$$\mu^{Rx} = \gamma^+ \times \mu_{g^+}^{Rx} + (1-\gamma^+) \times \mu_{g^-}^{Rx}, \tag{13.2}$$

$$\mu^{C} = \gamma^+ \times \mu_{g^+}^{C} + (1-\gamma^+) \times \mu_{g^-}^{C}. \tag{13.3}$$

Therefore, in the case of efficacy being a difference of means, efficacy in the combined population is

$$\eta_{\{g^-,g^+\}} = \mu^{Rx} - \mu^{C} = \gamma^+ \times \eta_{g^+} + (1-\gamma^+) \times \eta_{g^-}, \tag{13.4}$$

and is therefore logic-respecting.

13.4.2 Relative Response is Logic-Respecting in a Logistic Model

A fundamental truth is that, in general, efficacy in all-comers $\{g^-, g^+\}$ cannot be determined merely by Rx versus C efficacies in g^- and g^+ (the two vertical down arrows in Figure 13.2) because it depends on efficacy of Rx in g^- versus C in g^+ and efficacy of Rx in g^+ versus C in g^- (the two diagonal arrows in the left panel of Figure 13.2). Knowing the *prognostic* effect of Rx in g^+ vs. C in g^- (represented by the bottom horizontal arrow) allows us to deduce the two missing efficacies from efficacies in g^- and g^+ , as illustrated in part by the right panel of Figure 13.2.

It so happens that, if efficacy is measured as a *difference of means*, then knowledge of the prognostic effect is not needed (i.e., one does not need to go through the bottom solid arrow) because addition and subtraction can be done in any order. Adding the top row and subtracting the left column let us in effect account for the solid diagonal arrow. However, ratio efficacies are affected by the prognostic effect (bottom solid arrow) because addition and division have to be done in the proper sequence.

Law of nature dictates how response probabilities mix. If, under Rx, the response probability in g^- is 25% and the response probability in g^+ is 75%, and the entire population consists of a 50/50 mix of g^- and g^+, then naturally the response probability under Rx in $\{g^-, g^+\}$ is 50%. That is, response probabilities naturally mix within each arm, weighted by prevalence of the g^- and g^+ patients.[2]

Therefore, if we operate in the proper sequence, adding response probabilities within each treatment arm first, dividing the combined response probabilities second, then no

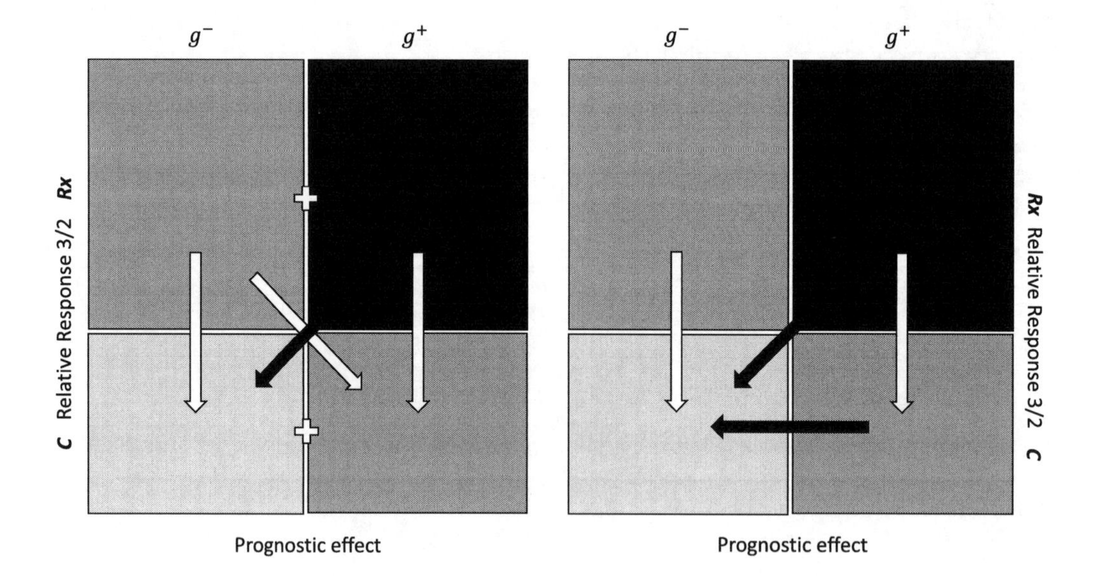

FIGURE 13.2
In general, Rx versus C efficacy in $\{g^-, g^+\}$ depends not only on efficacy in g^- and g^+ (the two vertical down arrows), but also on efficacy of Rx in g^+ versus C in g^- and Rx in g^- versus C in g^+ (the two diagonal arrows in (a)). In the case of Relative Response, population efficacy depends on the prognostic effect (horizontal left arrow in (b)) in addition to efficacy within g^- and g^+.

[2]Mixing logarithms of probabilities by prevalence and then exponentiating results in 0.4330127 which is incorrect because it violates the law of nature.

knowledge of the prognostic effect is need to obtain the correct relative response (RR) in $\{g^-, g^+\}$. This is the natural path taken by the SME principle, to be described in 13.6.

On the other hand, if we divide response probabilities within the g^- and g^+ subgroups first, then in order to combine them, we need to know whether each column's RR is a ratio of two big numbers or two small numbers relative to the other column's RR (information contained in the two diagonal arrows), and that information can be deduced from the prognostic effect (bottom solid arrow). The prognostic factor is the proper coefficient for mixing RR, not the prevalence, as we will demonstrate explicitly.

Let RR_{g^+}, RR_{g^-} and $\overline{RR}$ denote the relative response for the g^+, g^- subpopulations and the mixture $\{g^+, g^-\}$ all-comers population respectively. Interestingly, $\overline{RR}$ is not a mixture of RR_{g^+} and RR_{g^-} weighted by Prevalence, the population proportion of the subgroups. Rather, $\overline{RR}$ is a mixture of RR_{g^+} and RR_{g^-} weighted by the population proportion of responders **under** C who are g^+ and g^- respectively, weights which Lin et al. (2019) call the prognostic factor.

Denote by RR_{g^+}, RR_{g^-} and $\overline{RR}$ the relative response for the g^+, g^- subpopulations and for the mixture $\{g^+, g^-\}$ all-comers population respectively so, in terms of the marginal responder probabilities in Table 13.1,

$$RR_{g^+} = \frac{p_{g^+}^{Rx}}{p_{g^+}^{C}}, \quad RR_{g^-} = \frac{p_{g^-}^{Rx}}{p_{g^-}^{C}}, \quad \overline{RR} = \frac{p^{Rx}}{p^{C}}. \tag{13.5}$$

Note intuitively and crucially that natural mixing is in terms of responder probabilities **within each arm**. With population prevalence of the g^+ subgroup being γ^+, the responder rates in the mixture $\{g^+, g^-\}$ population are

$$p^{Rx} = \gamma^+ \times p_{g^+}^{Rx} + (1-\gamma^+) \times (p_{g^-}^{Rx}) \tag{13.6}$$

$$p^{C} = \gamma^+ \times p_{g^+}^{C} + (1-\gamma^+) \times (p_{g^-}^{C}) \tag{13.7}$$

But (13.5), (13.6) and (13.7) from the marginal probabilities in Table 13.1 are insufficient to reveal the relationships among RR_{g^+}, RR_{g^-} and $\overline{RR}$. For that, one needs Table 13.2, which gives *in the total population* the logical relationship between responder probabilities in the g^+ and g^- subpopulations under Rx and C and their combined probabilities in the mixture $\{g^+, g^-\}$ population. In terms of Table 13.2,

$$RR_{g^+} = \frac{p_{g^+}^{Rx}(R)/(\gamma^+\tau^{Rx})}{p_{g^+}^{C}(R)/(\gamma^+\tau_{g^+}^{C})}, \quad RR_{g^-} = \frac{p_{g^-}^{Rx}(R)/(\gamma^-\tau_{g^-}^{Rx})}{p_{g^-}^{C}(R)/(\gamma^-\tau_{g^-}^{C})}, \quad \overline{RR} = \frac{p^{Rx}(R)/\tau^{Rx}}{p^{C}(R)/\tau^{C}}. \tag{13.8}$$

Since

$$\frac{p_{g^+}^{C}(R)}{p_{g^+}^{C}(R)+p_{g^-}^{C}(R)}\frac{p_{g^+}^{Rx}(R)/\tau^{Rx}}{p_{g^+}^{C}(R)/\tau^{C}} + \frac{p_{g^-}^{C}(R)}{p_{g^+}^{C}(R)+p_{g^-}^{C}(R)}\frac{p_{g^-}^{Rx}(R)/\tau^{Rx}}{p_{g^-}^{C}(R)/\tau^{C}} = \frac{p^{Rx}(R)}{p^{C}(R)}\frac{\tau^{C}}{\tau^{Rx}}, \tag{13.9}$$

TABLE 13.1
Conditional Response Probability Given Treatment Rx or C for Patients in the g^+ and g^- Biomarker Subgroups and Marginal Probability in the All-Comers Population

	g^+ Subpopulation			**g^- Subpopulation**			**Population**		
	R	**NR**		**R**	**NR**		**R**	**NR**	
Rx	$p_{g^+}^{Rx}$	$1-p_{g^+}^{Rx}$	1	$p_{g^-}^{Rx}$	$1-p_{g^-}^{Rx}$	1	p^{Rx}	$1-p^{Rx}$	1
C	$p_{g^+}^{C}$	$1-p_{g^+}^{C}$	1	$p_{g^-}^{C}$	$1-p_{g^-}^{C}$	1	p^{C}	$1-p^{C}$	1
	p_{g^+}	$1-p_{g^+}$	1	p_{g^-}	$1-p_{g^-}$	1	p	$1-p$	1

TABLE 13.2
Joint Probabilities of Response (R) or Non-Response (NR) in the Total Population, with Prevalence γ^+ and $\gamma^-(= 1 - \gamma^+)$ for the $g^+$and g^- Subgroups

	g^+ **Subpopulation**				g^- **Subpopulation**				**Population**		
	R	**NR**			**R**	**NR**			**R**	**NR**	
Rx	$p^{Rx}_{g^+}(R)$	$p^{Rx}_{g^+}(NR)$	$\gamma^+\tau^{Rx}$	+	$p^{Rx}_{g^-}(R)$	$p^{Rx}_{g^-}(NR)$	$\gamma^-\tau^{Rx}$	=	$p^{Rx}(R)$	$p^{Rx}(NR)$	τ^{Rx}
C	$p^{C}_{g^+}(R)$	$p^{C}_{g^+}(NR)$	$\gamma^+\tau^{C}$	+	$p^{C}_{g^-}(R)$	$p^{C}_{g^-}(NR)$	$\gamma^-\tau^{C}$	=	$p^{C}(R)$	$p^{C}(NR)$	τ^{C}
	$p_{g^+}(R)$	$p_{g^+}(NR)$	γ^+	+	$p_{g^-}(R)$	$p_{g^-}(NR)$	γ^-	=	$p(R)$	$p(NR)$	1

The table on the right displays the correct marginal probabilities when the g^+ and g^- subgroups are combined so that the sum of the probabilities in corresponding cells of the two tables at the left equals the probability in the corresponding cell of the right-hand table.

the true mixture relative response $\overline{RR}$ can be represented as

$$\overline{RR} = \frac{p^C_{g^+}(R)}{p^C(R)} \times RR_{g^+} + \frac{p^C_{g^-}(R)}{p^C(R)} \times RR_{g^-}. \tag{13.10}$$

So $\overline{RR}$ is in fact a mixture of RR_{g^+} and RR_{g^-} weighted by $\frac{p^C_{g^+}(R)}{p^C(R)}$ and $\frac{p^C_{g^-}(R)}{p^C(R)}$, the population proportion of responders **under** C who are g^+ and g^- respectively. Therefore, note importantly, the efficacy measure relative response RR *is* logic-respecting.

If the biomarker is not prognostic, then the (joint) responder rates under C and in g^+ or g^- (i.e., $p^C_{g^+}(R)$ and $p^C_{g^-}(R)$) would be proportional to the overall responder rate under C, therefore $p^C_{g^+}(R) = \gamma^+ \times p^C(R)$ and $p^C_{g^-}(R) = (1 - \gamma^+) \times p^C(R)$, in which case

$$\overline{RR} = \gamma^+ \times RR_{g^+} + (1 - \gamma^+) \times RR_{g^-}. \tag{13.11}$$

This illustrates, in general, linearly mixing *logarithms* of efficacies (that happen to be coefficients in models linearized for computational purpose) and then exponentiating violates the law of nature. This is one of the oversights in current computer packages that will be demonstrated in Sections 13.5.4 and 13.5.5.

Further, if the biomarker has a prognostic effect, then (13.11) would not equal (13.10), and $\overline{RR}$ in the all-comers population *cannot* be determined by RR in the g^+ and g^- subpopulations and the prevalence γ^+. Another oversight in current computer packages is to assume in general efficacy in the all-comers population *can* be determined by efficacies in the g^+ and g^- subpopulations and the prevalence γ^+, an oversight that will be demonstrated in Sections 13.5.4 and 13.5.5 as well.

In contrast, the SME principle that will be described in Section 13.6 follows a path that adheres to the law of nature so that these pitfalls do not appear.

13.4.3 Ratio of Survival Times is Logic-Respecting in a Weibull Model

Median survival times are often of interest in oncology trials with time-to-event outcomes. The ratio of the median survival times between Rx and C provides direct information on the relative treatment effects. For example, if the median survival time for patients randomized to Rx is 18 months and the median survival time for patients randomized to C is 12 months. Then Rx median survival time is 1.5 times (=18/12) that of C. Following Ding (2016) et al., we show that, under a Weibull model (a special case of the Cox Proportional Hazard model), that ratio of median survival times is logic-respecting (and therefore collapsible). That is, the efficacy of the mixture stays within the interval of the subgroups' efficacy.

Proposition 13.1 *Assume the time-to-event data fit the following Cox Proportional Hazard (PH) model:*

$$h(t|Trt, M) = h_0(t)\exp\{\beta_1 Trt + \beta_2 M + \beta_3 Trt \times M\}, \tag{13.12}$$

where $Trt = 0$ (C) *or* $Trt = 1$ (Rx), $M = 0$ (g^-) *or* $M = 1$ (g^+), *and* $h_0(t) = h(t|C, g^-)$ *is the hazard function for the* g^- *subgroup receiving* C. *Further assume that the survival function* $S_0(t)$ *for* C, g^- *is from a Weibull distribution with scale* λ *and shape* k, *i.e.,*

$$S_0(t) (= S^C_{g^-}(t)) = e^{-(t/\lambda)^k}, \quad t \geq 0.$$

If efficacy is defined as the ratio of median survival times (between Rx *and* C*), then the efficacy of* g^-, g^+, *and their mixture can all be represented by a function of the five model parameters* $(\lambda, k, \beta_1, \beta_2, \beta_3)$. *More importantly, the efficacy of the mixture is always guaranteed to stay within the interval of the subgroups' efficacy.*

Proof 2 *Denote by ν^{Rx} and ν^{C} the true median survival times over the entire patient population (randomized to Rx and C respectively). Denote by $\nu^{Rx}_{g^+}$, $\nu^{Rx}_{g^-}$, $\nu^{C}_{g^+}$, $\nu^{C}_{g^-}$ the corresponding median survival times in the g^+ and g^- subgroups. Denote $\theta_1 = e^{\beta_1}$, $\theta_2 = e^{\beta_2}$ and $\theta_3 = e^{\beta_3}$. Note that $\theta_1, \theta_2, \theta_3$ all > 0.*

By the PH property, the survival function for each of the subgroups has the following form

$$S^{C}_{g^-}(t) = e^{-(t/\lambda)^k}, \quad S^{Rx}_{g^-}(t) = e^{-\theta_1 (t/\lambda)^k},$$
$$S^{C}_{g^+}(t) = e^{-\theta_2 (t/\lambda)^k}, \quad S^{Rx}_{g^+}(t) = e^{-\theta_1\theta_2\theta_3 (t/\lambda)^k}.$$

Straightforward calculation gives the median survival time for each subgroup as follows

$$\nu^{Rx}_{g^+} = \lambda(\frac{\log 2}{\theta_1\theta_2\theta_3})^{1/k}, \quad \nu^{C}_{g^+} = \lambda(\frac{\log 2}{\theta_2})^{1/k}, \quad \nu^{Rx}_{g^-} = \lambda(\frac{\log 2}{\theta_1})^{1/k}, \quad \nu^{C}_{g^-} = \lambda(\log 2)^{1/k}. \tag{13.13}$$

Then the ratios of median for g^+ and g^- are

$$r_{g^+} = (\theta_1\theta_3)^{-1/k} \quad \text{and} \quad r_{g^-} = \theta_1^{-1/k}, \tag{13.14}$$

which are functions of (k, θ_1, θ_3).

For the mixture of g^+ and g^-, according to the law of nature, survival functions mix within each treatment arm because survival probabilities are probabilities. Therefore, for the mixture of g^+ and g^-, the median survival times for Rx and C are the solutions for the following two equations respectively.

$$t = \nu^{Rx}: \ (1-\gamma^+)e^{-\theta_1(t/\lambda)^k} + \gamma^+ e^{-\theta_1\theta_2\theta_3(t/\lambda)^k} = 0.5, \tag{13.15}$$
$$t = \nu^{C}: \ (1-\gamma^+)e^{-(t/\lambda)^k} + \gamma^+ e^{-\theta_2(t/\lambda)^k} = 0.5. \tag{13.16}$$

Then the ratio of median for the mixture group $\overline{r} \equiv \nu^{Rx}/\nu^{C}$ is an implicit function of $(\lambda, k, \theta_1, \theta_2, \theta_3)$. Notice that θ_2, the prognostic effect of the biomarker, is involved.

Now, we show that $\overline{r}$ is between r_{g^-} and r_{g^+}. Let $t = \nu^C r_{g^-} = \nu^C \theta_1^{-1/k}$ and plug into the left side of Equation (13.15), we have

$$(1-\gamma^+)e^{-\theta_1(\nu^C\theta_1^{-1/k}/\lambda)^k} + \gamma^+ e^{-\theta_1\theta_2\theta_3(\nu^C\theta_1^{-1/k}/\lambda)^k} \tag{13.17}$$
$$= (1-\gamma^+)e^{-(\nu^C/\lambda)^k} + \gamma^+ e^{-\theta_2\theta_3(\nu^C/\lambda)^k}. \tag{13.18}$$

The first term in Equation (13.18) equals the first term on the left side of (13.16) with ν^C plugged in. Therefore, whether (13.18) > 0.5 *or* < 0.5 *depends on whether $\theta_3 < 1$ or > 1. Without loss of generosity, assume $\theta_3 > 1$. Then by the property that the all survival functions are non-increasing functions, comparing (13.15) with (13.17), we have*

$$\nu^{Rx} > \nu^C \theta_1^{-1/k} = \nu^C r_{g^-}.$$

Thus, $\overline{r} = \nu^{Rx}/\nu^C < r_{g^-}$. With a similar argument, we can show that $\overline{r} > r_{g^+}$ (if $\theta_3 > 1$). Hence, we have shown that the ratio of median survival time for the mixture population is within the interval of the ratios for the subgroups and each ratio can be represented by a function of $(\lambda, k, \theta_1, \theta_2, \theta_3)$ either explicitly or implicitly.

13.5 Adjusting for Imbalance in the Data Even in an RCT

An unstratified RCT just randomly assigns subjects to Rx and C. Randomization does not ensure perfect balance. Imbalance in the data on covariates such as baseline measurements and blocking factors such as region are routinely adjusted for in stratified *analyses* using the Least Squares means (LSmeans) technique, to be described in Section 13.5.2.

Some RCTs are stratified by *design*, stratified on known or anticipated predictive factor such as the subject's biomarker value in the drug's targeted pathway. A stratified RCT would randomly assign subjects to Rx and C within each stratum of the predictive factor. Contrary to the belief by some, the principal purpose of stratifying the design is not to achieve balance in the data, because imbalance can be taken care by LSmeans. Rather, a stratified design may sharpen the Rx vs. C comparison, if patients are relatively homogeneous within each stratum. Increasingly, stratified designs are used to ensure an adequate sample size of patients in a subgroup of potential interest. To assess the efficacy in the overall population, the analysis of such a study then readjusts the prevalence of patient subgroups.

It may be impractical to execute an RCT that stratifies on every possible factor though. An oncology study may stratify on the subjects' status in the gene that the therapy targets (e.g., the MET gene) so that MET+ patients are randomized to Rx and C, and separately MET− patients are randomized to Rx and C. But it might be impractical to further stratify the study on the subjects' status in another potentially predictive gene (e.g., the EGFR gene). Therefore, with not very large samples, there may be an imbalance between Rx and C in the unstratified factor's subject status, which can potentially skew the result. To avoid biased result, it is important that statistical analysis employs a technique that adjusts for imbalance in subjects status across the treatments for factors that might affect the outcome.

There are (at least) two parallel approaches to adjusting for imbalance in the data. One is the imputation technique of Little and Rubin (1987). The other technique, which we describe in some detail in this chapter, is least squares means.

13.5.1 Analyses Stratified on Biomarker Subgroups Should Include a Rx:C × Biomarker Interaction Term

For Alzheimer's Disease (AD), *change* in Alzheimer's Disease Assessment Scale-Cognitive Subscale (ADAS-Cog) from baseline ADAS-Cog is a common measure of a treatment's effect. For Type 2 diabetes (T2DM), *change* in hemoglobin A1c from baseline A1c is the usual clinical measure of a treatment's effect. For schizophrenia, *change* in Positive and Negative Syndrome Scale (PANSS) from baseline PANSS is a typical clinical measure of a treatment's effect.

Randomization does not achieve perfect balance. Having healthier patients in one treatment arm and sicker patients in the other arm biases the result. *Baseline* is often included in the model as a (continuously valued) covariate to adjust for imbalance in the severity of illness of patients when they are initially assigned to Rx and C. In AD, T2DM, and schizophrenia studies, with the assumption that baseline measurement affects Rx and C in the same way, *baseline* is included in the model as a main effect, without a *baseline* and Rx:C interaction term (i.e., *baseline* has the same slope under Rx and C).

Clinical trials across multiple regions of the world have become common practice. Having *Region* as a blocking factor allows inference on a common efficacy even if measurements in the European Union are systematically higher (or lower) than measurements in the U.S., for instance. With the assumption that the systemic difference affect Rx and C in the same way, *Region* is often included in the model as a (categorical) main effect, without a *Region* and Rx:C interaction term. The purpose of such modeling is to utilize all the data to infer on a presumed common Rx vs. C efficacy while adjusting for a systemic effect.

This notion of a *common* efficacy is well-defined provided the *differential* between Rx and C remains constant across baseline values and/or the blocking factor's levels, at the population level (Hsu 1996, pp. 182–1833). As with any modeling, this no-interaction assumption should be based on domain knowledge and checked against actual data. When such a model is appropriate, multiple comparisons as described in Chapter 7 of Hsu (1996) based on Least Squares means (LSMmeans) are unbiased. See Chapter 7 of Hsu (1996) for a detailed guideline of LSmeans analysis in a model that does not include an interaction term between Rx:C and covariates and/or the blocking factors.

The situation with a biomarker for potential patient targeting is different. A marker such that Rx vs. C efficacy remains constant across its values, a **purely prognostic** marker, is not useful for patient targeting. We are interested in **predictive** biomarkers, which interact with Rx:C. Many of the targeted therapies for Alzheimer's Disease that have been tried target the clearance of beta amyloid in patients, for instance. The ApoE gene is postulated to be involved in the clearance of beta amyloids. Therefore, it is reasonable for the analyses of such studies to take into account the patients' ApoE status, and the model should include an ApoE$\times Rx$:C interaction term in addition to an ApoE main effect term.

For analyzing time-to-event data, fitting a Cox PH (proportional hazard) model to compare Rx vs. C, if one puts in a `strata(biomarker)` statement for a categorical biomarker, then HR is assumed to be constant across the subgroups defined by the biomarker, which would be inappropriate because for any useful biomarker HR would not be constant. The log-Rank test should not be used either, stratified or not, because its error rate control is extremely "weak". Section 13.7 shows dramatically that Type I error rate control of the stratified log-Rank test does not control the rate of making incorrect clinical decisions.

Description of LSmeans below is for LSmeans analysis in a model that includes an interaction term between Rx:C and g^+:g^-.

13.5.2 Least Squares Means

For Alzheimer's Disease (AD), *change* in Alzheimer's Disease Assessment Scale-Cognitive Subscale (ADAS-Cog) from baseline ADAS-Cog is a common measure of a treatment's outcome. For Type 2 diabetes (T2DM), *change* in hemoglobin A1c from baseline A1c is the usual clinical measure of a treatment's outcome. These outcomes are *continuous* in nature, and efficacy is typically defined as the mean difference between Rx and C.

Consider the data in Table 13.3, where a larger (more positive) outcome is better.

The *marginal* means estimate of Rx vs. C efficacy in the combined population $\{g^-, g^+\}$,

$$\hat{\theta}_2^{MG} = \frac{1.96 + 2.18 + 4.86}{3} - \frac{1.16 + 4.67 + 4.35}{3} = -0.393 < 0,$$

suggests that the Rx treatment is harmful. This estimate of broad efficacy has a negative bias because the imbalance in Rx vs. C sample sizes between the two subgroups is unfavorable to Rx. In other words, imbalance in the data can cause Simpson's Paradox phenomenon.

For continuous outcome modeled linearly, computer packages apply the Gauss-Markov theorem to adjust for imbalance in the data. This implementation is commonly referred to as least squares means (LSmeans):

> Simply put, they are estimates of the class or subclass arithmetic means that would be expected had equal subclass numbers been obtainable (Goodnight and Harvey (1978)).
>
> LS-means are predicted population margins - that is, they estimate the marginal means over a balanced population (SAS manual).

TABLE 13.3
Imbalance in Data Lead to Different Least Squares Means and Marginal Means

Subgroup	g^-	g^+
Treatment (Rx) observed outcomes	1.96, 2.18	4.86
Control (C) observed outcomes	1.16	4.67, 4.35

Data in Table 13.3 indicate treatment (Rx) is better than control (C) within each subgroup. For a model that includes indicators for Rx/C and g^-/g^+ and their interaction, LSmeans for the 2 treatment-by-subgroup combinations are just the cell means. Assuming prevalence of each subgroup is 50%, the *Least Squares* means estimate of Rx over C efficacy in the combined population $\{g^-, g^+\}$ is

$$\hat{\theta}_2^{LS} = (0.5 \times \frac{1.96 + 2.18}{2} + 0.5 \times 4.86) - (0.5 \times 1.16 + 0.5 \times \frac{4.67 + 4.35}{2}) = 0.630 > 0.$$

Unlike marginal means, the Least Squares means estimate correctly suggests a beneficial treatment effect.

The Means statement in Proc GLM of SAS compares treatments based on Marginal means, and therefore should not be used.

13.5.3 LSmeans Subgroup Analysis in Computer Packages Are Correct for Continuous Outcomes

Consider a (perfectly) balanced population, with the prevalence of each of the g_1, g_2, g_3 subgroups being $\frac{1}{3}$, and within each subgroup half of the subjects are given Rx while the other half given C, as depicted in Table 13.4.
True difference of the Rx and C effects is exactly zero.

Now consider an (artificial) unbalanced data set from this balanced population as depicted in Table 13.5. This imbalance can be from stratifying the *design* purposely allocating patients to $\{g_1, g_2\}$ and g_3) in the 10:4 ratio, with retrospective genotyping of g_1 and g_2 turning up an imbalance between them in the Rx and C arms or simply because the sample size is small.

LSmeans will unbiasedly estimate Rx versus C efficacy for the balanced population in Table 13.4. If the intended patient population is not balanced but with unequal prevalence between the subgroups, then one can use the ESTIMATE statement in SAS to unbiasedly

TABLE 13.4
Each Number in the Table Represents One Million Copies (n_∞ in the millions), so There Are One Million Copies of 7.5 in g_2 Given C for Example

	g_1	g_2	g_3	**Average Effect**
n_∞	3×10^6	3×10^6	3×10^6	
Rx	5.5	5	5	$\frac{5.5+5+5}{3} = 5.167$
C	3	7.5	5	$\frac{3+7.5+5}{3} = 5.167$
$Rx - C$	2.5	−2.5	0	0

TABLE 13.5

Least Squares Means Unbiasedly Estimate Means for a *Balanced* **Population** from Unbalanced Data *Regardless of Whether Design of the Study is Stratified or Not*, but Marginal Means Do Not. This statement holds whether each number in the table represents one number (n_{sm} just a few numbers), so there is one 7 and one 8 in g_2 given C for example, or one thousand copies (n_{lg} in the thousands), so there are one thousand copies of 7 and one thousand copies of 8 in g_2 given C for example.

	g_1	g_2	g_3	Marginal Means	LSmeans
n_{sm}	5	5	4		
n_{lg}	5×10^3	5×10^3	4×10^3		
Rx	5,6	3,5,7	6,4	$\frac{5+6+3+5+7+6+4}{7} = 5.143$	$\frac{\frac{5+6}{2}+\frac{3+5+7}{3}+\frac{6+4}{2}}{3} = 5.167$
C	3,3,3	7,8	4,6	$\frac{3+3+3+7+8+4+6}{7} = 4.857$	$\frac{\frac{3+3+3}{3}+\frac{7+8}{2}+\frac{4+6}{2}}{3} = 5.167$
$Rx - C$	2.5	-2.5	0	0.286	0

estimate Rx versus C efficacy for the intended population by specifying the prevalence as the coefficients.

Our explanation and demonstration of LSmeans should remove the surprising ignorance of the distinct purposes between a stratified *design* and a stratified *analysis*:

- Stratifying a *design* is primarily to avoid sparsity, and/or for enrichment. If the subjects are relatively homogeneous within the strata, then there is a power gain as well.
- A stratified *analysis* uses LSemeans to adjust for data imbalance (sample size and covariate value imbalance, in the specified stratification factors). A *stratified design* is *not needed* for LSmeans to produce unbiased estimates.

You might wonder why, instead of estimating effects in a balanced population, LSmeans does not estimate effects weighted by prevalence? (Weighing by observed marginal prevalence is the OM *option* in LSmeans in SAS.) However, when computer first became powerful enough to compute LSmeans, efficacy in mixture of subgroups was hardly discussed. While it is true that today targeted therapies are common, we do not know in the future what the most pressing problem will be. The problem-neutral default of estimating effects in a balanced population is a safe and sensible choice. One can specify other mixing coefficients using the ESTIMATE statement in SAS or a vector of coefficients in R.

Data in Table 13.5 can be analyzed using SAS codes such as

```
proc glm;
class Group Trt;
/* Incorrect Marginal means model */
model Y= Trt;
lsmeans Trt;

proc glm;
class Group Trt;
/* Correct LSmeans model with both main effects and interaction */
model Y= Trt Group Group*Trt;
lsmeans Trt;
```

to illustrate how the LSmeans statement in Proc GLM or Proc Mixed in SAS applies the Gauss-Markov theorem to correctly estimate efficacy in treatments for continuous outcome modeled linearly with Normally distributed errors.

13.5.4 LSmeans Subgroup Analysis in Computer Packages Are Misleading for Binary Outcomes

The Gauss-Markov theorem applies to linear models. Therefore, to avoid imbalance in the data biasing results, data with binary outcomes are routinely linearized by fitting a logistic or a log-linear model, with parameters in the model estimated by LSmeans. However, contrary to the implication in Hothorn et al. (2008), parameters in such models should *not* be mixed as if they were in a linear model for continuous outcomes.

Consider the balanced population in Table 13.6. Suppose efficacy is measured by Relative Response (RR), the ratio of response probability between Rx and C. If, under Rx, the response probability in g^- is 25% and the response probability in g^+ is 75%, and the entire population consists of a 50/50 mix of g^- and g^+, then naturally the response probability under Rx in $\{g^-, g^+\}$ is 50%. Table 13.6 is computed using this law of nature.

Let RR_{g^+}, RR_{g^-} and $\overline{RR}$ denote the relative response for the g^+, g^- subpopulations and the mixture $\{g^-, g^+\}$ all-comers population respectively. As explained in Section 13.4.2, $\overline{RR}$ is not a mixture of RR_{g^+} and RR_{g^-} weighted by Prevalence, the population proportion of the subgroups. Rather, $\overline{RR}$ is a mixture of RR_{g^+} and RR_{g^-} weighted by the population proportion of responders **under** C who are g^+ and g^- respectively, weights which Lin et al. (2019) call the prognostic factor.

From Table 13.6, we see that for a balanced population, there are two ways to arrive at $\overline{RR}$, the correct RR for the combined $\{g^-, g^+\}$ population. One way is to mix the response rates *within each arm* by prevalence first, and then compute $\overline{RR}$ for $\{g^-, g^+\}$. This is what SME to be described in Section 13.6 does. The other way is to compute RR_{g^-} and RR_{g^+} separately for g^- and g^+ first and then mix them by the prognostic factor.

SAS codes such as

```
proc genmod;
class Trt;
/* model for Marginal means */
model Response/nTotal= Trt / dist=binomial link=log;
LSmeans Trt;


proc genmod;
class Subroup Trt;
/* model for LSmeans means stratified on subgroup */
model Response/nTotal= Subroup Trt Subroup*Trt / dist=binomial link=log;
LSmeans Trt;
```

produce Marginal means and LSmeans results displayed in Table 13.7 from computer packages that indicate there are two mistakes in LSmeans estimation of $\overline{RR}$ in current packages, even for a balanced population (a balanced data set with $n \to \infty$).

The first, easy to spot, issue is LSmeans in current computer packages linearly mix whatever parameters are in the model which in the case of a logistic or log-linear model are on a *logarithmic* scale rather than the probability scale. In our example, the stratified LSmeans 0.6609 for $\log(\overline{RR})$ is strictly larger than the true $\log(\overline{RR})$ of 0.5109.

The second, more fundamental, issue is mixture of LSmeans estimates for subgroups are weighed by *prevalences* in current computer packages, $\frac{1}{2}$ in the case of Table 13.6, instead of the proper prognostic factor $\frac{0.10}{0.10+0.50}$ and $\frac{0.50}{0.10+0.50}$. For our example, had LSmeans mixed

TABLE 13.6
Response Rates and Relative Responses in a *Balanced* ($n \to \infty$) Population

	Rx (n = 20000)	C (n = 20000)
g^- $(n = 20000)$	$\frac{2500}{10000} = 0.25$	$\frac{1000}{10000} = 0.10$
g^+ $(n = 20000)$	$\frac{7500}{10000} = 0.75$	$\frac{5000}{10000} = 0.50$
$\{g^-, g^+\}$	$\frac{10000}{20000} = 0.50$	$\frac{6000}{20000} = 0.30$
$\overline{RR} :=$ True RR in $\{g^-, g^+\}$	$\frac{0.50}{0.30} = \frac{5}{3} = 1.6667 = e^{0.5109}$	
Mixing RR by prevalence	$\frac{1}{2} \times 2.5 + \frac{1}{2} \times 1.5 = 2 \neq \frac{5}{3}$	
Mixing $\log(RR)$ by prevalence	$\frac{1}{2} \times \log(2.5) + \frac{1}{2} \times \log(1.5) = 0.6609 \neq 0.5109$	
Mixing RR by the prognostic factor	$\frac{0.10}{0.10+0.50} \times 2.5 + \frac{0.50}{0.10+0.50} \times 1.5 = \frac{5}{3}$	

TABLE 13.7
Law of Nature Mixes Probabilities (Not Logarithms) within Each Treatment Arm, While Computer Packages Currently Mix Parameters in Models Parameterized by Human

	Marginal Means	LSmeans in Computer Packages
Rx	$-0.6931 = log\left(\frac{0.25+0.75}{2}\right)$	$-0.8370 = \frac{\log(0.25)+\log(0.75)}{2}$
C	$-1.2040 = log\left(\frac{0.10+0.50}{2}\right)$	$-1.4979 = \frac{\log(0.10)+\log(0.50)}{2}$
$Rx - C$	$-0.6931 - (-1.204) = 0.5109 = \log(\frac{5}{3})$	$-0.8370 - (-1.4979) = 0.6609 \neq \log(\frac{5}{3})$

RR (not its logarithm) by prevalence, the result would have been 2 which again is strictly larger than the true $\overline{RR}$ of $\frac{5}{3}$.

Mixing response rates within each arm by prevalence first, SME in Lin et al. (2019) does not have these issues.

Marginal mean happens to correctly estimate $\overline{RR}$ for a **perfectly** balanced data set (perfect balance between Rx and C in sample sizes across subgroups, and values of additional covariates if they are present), because ignoring the subgroup label in effect mixes the responders within each arm. However, Marginal mean incorrectly estimates $\overline{RR}$ for Rx vs. C in the combined $\{g^-, g^+\}$ population if the data is imbalanced, and therefore should not be used.

13.5.5 LSmeans Subgroup Analysis in Computer Packages Are Misleading for Time-to-Event Outcomes

Similar to the binary outcome case, one fundamental issue in the over-extension of LSmeans to time-to-event outcomes is it linearly mixes whatever parameters are in a model linearized for adjusting for imbalance in the data by LSmeans. While these model parameters are equivalent to the efficacy parameters of interest, the scales on which they are measured (such as the logarithmic scale) often make them unsuitable for linear mixing.

An even more fundamental issue of the over-extension of LSmeans is it assumes efficacy in a mixture is a function of efficacies in the subgroups and the *prevalence*. This is generally a false assumption for binary and time-to-event outcomes. While logic-respecting efficacy such as RoM is perfectly well-defined and computable for mixtures, they are functions of efficacies in the subgroups and the *prognostic* effect, not prevalence.

Time-to-event data are typically fitted to a Cox proportional hazard model, which, for LSmeans purpose, is parameterized as a log-linear model. A Weibull model is a special case of this, and RoM is logic-respecting in such a model as we showed in Section 13.4.3. On the other hand, HR is known to be *not* collapsible. See Aalen et al. (2015). So, to be clear, HR should not be used to measure efficacy when there are subgroups because it is not logic-respecting, and using it can lead to illogical decision-making.

Nevertheless, HR currently is still used as an efficacy measure in subgroup analysis. What we show below is that thinking

1. One can always find some function to represent efficacy in $\{g^+, g^-\}$ as a linear combination of (that function of) efficacies in g^+ and g^-

2. with the mixing coefficient in this linear combination being the *prevalence*

have resulted in current computer packages masking the fact that efficacy measures such as HR is not collapsible (and thus not logic-respecting).

Consider a trial with time-to-event data (e.g., progression free survival) from either treatment (Rx) or control (C) arm with 1:1 randomization ratio and there exists a subgroup effect (g^+ or g^-) with prevalence 50%. To give insight into the over-extension, we use the simplest Cox model, one in which each of the $2 \times 2 = 4$ combinations of $Rx : C$ and $g^+ : g^-$ subgroups has an exponential distribution. Suppose within each treatment and subgroup combination, the data follows an exponential distribution with specified medians and we focus on the efficacy measure HR, with prognostic effect defined as the RoM between g^+ and g^- in the control arm C. For such a simple model, quantities such RoM in $\{g^+, g^-\}$ can be computed without resorting to simulation.

Table 13.8 gives an example where there is no prognostic effect (i.e., median for C is the same 10 months for the two subgroups) but differential efficacy in two subgroups (i.e., HR for $g^+ = 0.5$ and $g^- = 2$).

Table 13.9 gives an example where there is a prognostic effect of 6 in C in terms of RoM but efficacy for the subgroups are the same (i.e. HR=0.5 for g^- and g^+).

When there is a prognostic effect or when there is differential efficacy between subgroups, HR between Rx and C for the overall population $\{g^+, g^-\}$ is not well defined as it depends on time t (see Figure 13.3). Nonetheless, computer packages (e.g. Proc PHreg in SAS) will provide "HR" estimates in these cases upon user demand. Proposition 13.2 below shows the computed HR estimates an "average" HR in some sense.

Proposition 13.2 *In the absence of censoring, the HR estimator from a marginal Cox model (i.e. with treatment indicator as the only predictor) converges in probability to the "average" HR defined in (13.19)*

$$\log(\text{“}HR\text{”}) = -\int_0^\infty \log(HR(t))dS(t) \tag{13.19}$$

where $HR(t)$ is the ratio of hazard functions between Rx and C for the overall population $\{g^+, g^-\}$ and $S(t)$ is the survival function for the overall population $\{g^+, g^-\}$ combining Rx and C arm patients, both of which are functions of time t.

Proof 3 *See Xu and O'Quigley (2000).*

RoM is always well-defined. It is logic-respecting, and RoM in $\{g^-, g^+\}$ can be computed according to the SME principle using software accompanying Ding *et al.* However, suppose one calculates RoM in $\{g^-, g^+\}$ by linearly mixing RoM in g^- and g^+ weighted by prevalence, then probably no one would be surprised that such mixing does not produce

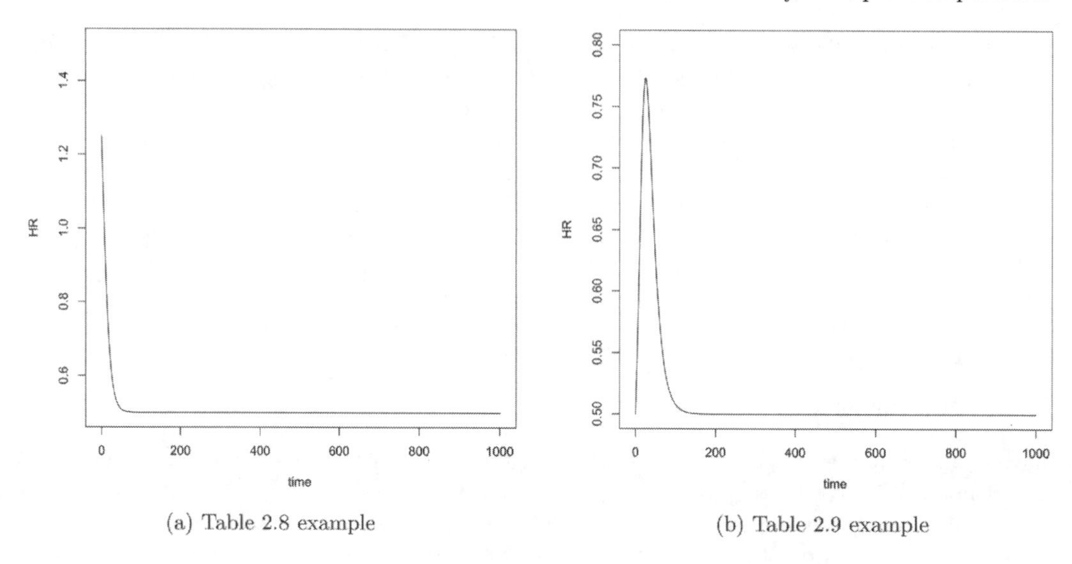

(a) Table 2.8 example (b) Table 2.9 example

FIGURE 13.3
Hazard Ratio (HR) in the overall population $\{g^-, g^+\}$ depends on time t. (a) Table 13.8 example, (b) Table 13.9 example.

the correct RoM in $\{g^-, g^+\}$, and one might in fact wonder "What are they thinking?" Yet, mixing HR or logarithms of HR by prevalence is similar in nature. Indeed, as shown in Tables 13.8 and 13.9, mixing HR or $\log(HR)$ by prevalence does not lead to the theoretical "average" HR.

To understand what computer packages currently do, we generate 200,000 patients' time-to-event data based on the setting in Tables 13.8 and 13.9 with perfectly balanced data and 50% prevalence for subgroup g^+. Then for each setting, we obtain the LSmeans estimate of the HR for Rx vs C from three Cox models: marginal model (treatment indicator only), model without interaction (treatment and subgroup indicators only), and model with interaction (treatment and subgroup indicators and their interaction) with the following SAS codes:

TABLE 13.8
An Example of Theoretical HR and RoM in g^-, g^+ and Overall $\{g^+, g^-\}$ When There Is No Prognostic Effect

	***Rx* Median**	***C* Median**	**Prognostic Effect**	***Rx*:*C* HR, RoM**
g^-	5	10	$\frac{C \text{ median } g^+}{C \text{ median } g^-} = 1$	HR=2, RoM=0.5
g^+	20	10		HR=0.5, RoM=2
$\{g^-, g^+\}$	9.3	10		HR(t) Figure 13.3a RoM=0.93
"Average" HR in $\{g^-, g^+\}$	$\exp\{-\int_0^\infty \log(HR(t))dS(t)\} = \exp(-0.17) = 0.84$			
Mixing HR by prevalence	$\frac{1}{2} \times 2 + \frac{1}{2} \times 0.5 = 1.25 \neq 0.84$			
Mixing $\log(HR)$ by prevalence	$\frac{1}{2} \times \log(2) + \frac{1}{2} \times \log(0.5) = 0 \neq -0.17 = \log(0.84)$			

TABLE 13.9

An Example of Theoretical HR and RoM in g^-, g^+ and Overall $\{g^+, g^-\}$ When This Is a Prognostic Effect

	Rx Median	*C* Median	Prognostic Effect	*Rx:C* HR	*Rx:C* RoM
g^-	10	5	$\frac{C \text{ median } g^+}{C \text{ median } g^-} = 6.0$	0.5	2
g^+	60	30		0.5	2
$\{g^-, g^+\}$	21.7	10.9		HR(t) Fig. 13.3b	2
"Average" HR in $\{g^-, g^+\}$	$\exp\{-\int_0^\infty \log(HR(t))dS(t)\} = 0.60 \neq 0.5$				
Mixing HR by prevalence	$\frac{1}{2} \times 0.5 + \frac{1}{2} \times 0.5 = 0.5 \neq 0.60$				
Mixing $\log(HR)$ by prevalence	$\frac{1}{2} \times \log(0.5) + \frac{1}{2} \times \log(0.5) = -0.69 \neq -0.51 = \log(0.60)$				

```
/* Fit a marginal model */
proc phreg;
class trt(ref='0') / param=glm;
model time*event(0)= trt ;
hazardratio 'H1' trt / diff=all cl=both;
lsmeans trt;
run;

/* Fit model without interaction term */
proc phreg;
class trt(ref='0') subgroup (ref='0') / param=glm;
model time*event(0)= trt subgroup;
hazardratio 'H1' trt / diff=all cl=both;
lsmeans trt;
run;

/* Fit model with interaction term */
proc phreg;
class trt(ref='0') subgroup (ref='0') / param=glm;
model time*event(0)= trt subgroup trt*subgroup;
hazardratio 'H1' trt / diff=all cl=both;
lsmeans trt;
run;
```

Results from SAS are shown in Tables 13.10 and 13.11. They confirmed that the HR computed by fitting a marginal Cox model (without any biomarker term in the model) estimates the "average" HR defined in (13.19) (for a balanced data set without censoring with sample size approaching infinity).

The example in Table 13.8 has no prognostic effect, but there is differential efficacy between the subgroups. It is not surprising that the model without interaction does not give correct estimation of "average" HR since the model is incorrectly specified. However, even when an interaction term is added, LSmeans in computer packages currently still does not lead to correct estimation of the logarithm of "average" HR because it is not an average of the logarithms of the HRs in the subgroups.

For the example in Table 13.9, models with and without interaction estimate (correctly) $\log(HR)$ in the g^- and g^+ subgroups as $\log(0.5)$. The LSmeans option produces an estimate of $\log(0.5)$ for the "average" HR which might look correct but actually is not. The correct

TABLE 13.10
Computer Packages Currently Combine Unequal Hazard Ratios in Subgroups in a Way Inconsistent with (13.19) Even When There Is No Prognostic Effect (Table 13.8 Example), Based on a Total of 200,000 Patients with 1:1 Randomization Ratio and a Prevalence of 0.5

Marginal model	log ("average" HR) for $\{g^-, g^+\} = -0.16 \approx \log(0.84)$
LSmeans without interaction	$\log(HR)$ for $\{g^-, g^+\} = 0.01 \neq \log(0.84)$
LSmeans with interaction	$\log(HR)$ for $g^- = 0.702 \approx \log(2)$
	$\log(HR)$ for $g^+ = -0.705 \approx \log(0.5)$
	$\log(HR)$ for $\{g^-, g^+\} \approx \frac{\log(2)+\log(0.5)}{2} \neq \log(0.84)$

TABLE 13.11
Computer Packages Currently Do Not Combine Equal HR in Subgroups Sensibly When There Is a Prognostic Effect (Table 13.9 Example), Based on a Total of 200,000 Patients with 1:1 Randomization Ratio and a Prevalence of 0.5

Marginal model	log ("average" HR) for $\{g^-, g^+\} = -0.5 \approx \log(0.60)$
LSmeans without interaction	$\log(\text{HR})$ for $\{g^-, g^+\} = -0.697 \approx \log(0.5) \neq \log(0.60)$
LSmeans with interaction	$\log(\text{HR})$ for $g^- = -0.694 \approx \log(0.5)$
	$\log(\text{HR})$ for $g^+ = -0.701 \approx \log(0.5)$
	$\log(\text{HR})$ for $\{g^-, g^+\} \approx \frac{\log(0.5)+\log(0.5)}{2} \neq \log(0.60)$

"average" HR is log(0.52) because of the prognostic effect. Ironically, the seemingly logical results given by computer packages currently might give the illusion that HR is collapsible (while HR is not collapsible and not logic-respecting).

13.6 The Subgroup Mixable Estimation Principle

SME in Ding et al. and Lin et al. is a principled approach that produces logic-respecting inferences for efficacy in the mixture by mixing g^+ and g^- within each arm first, then compare Rx with C. In a three-step process, SME takes the LSmeans estimates for the (canonical) parameters in models appropriate for the RCT data (eg., logistic, log-linear, Weibull) to whatever space appropriate for mixing (e.g., responder probability or survival probability) within each arm, and then calculate efficacy in g^+, g^-, and $\{g^+, g^-\}$:

1. Fit a model for the clinical outcome, obtain LSmeans estimates for the model parameters and their estimated variance-covariance matrix.

2. *Within each of the Rx and C arms*, estimate the effects μ_{g+}^{Rx}, μ_{g-}^{Rx}, μ_{g+}^{C}, μ_{g-}^{C} in the g^+ and g^- subgroups as appropriate functions of the model parameters. In addition, estimate the effects μ^{Rx} and μ^{C} (be it response probability or median survival time) in $\{g^+, g^-\}$ *within each of the Rx and C arms*, mixing in accordance to prevalence in the intended patient population.[3] Obtain estimated variance-covariance matrices for the estimates of $(\mu_{g+}^{Rx}, \mu_{g-}^{Rx}, \mu^{Rx})$ and $(\mu_{g+}^{C}, \mu_{g-}^{C}, \mu^{C})$ by the delta method.

[3]Some stratified studies are "enriched" so that the proportion of g^+ patients in the study is γ_E^+ instead of the prevalence γ^+ in the intended patient population. For such studies, estimation of the effects in Step 2 of SME should be based on γ^+ in the intended patient population, not γ_E^+.

3. Estimate efficacy in g^+, g^- subgroups and in all-comers $\{g^+, g^-\}$ by comparing Rx with C, deriving the estimated variance-covariance matrix of these estimates by the delta method.

An app demonstrating SME for analyzing time-to-event data is available at `https://jchsustatsci.shinyapps.io/Ratio_of_Median_survival_times`.

While SME naturally takes the prognostic effect into account, it will not magically transform a non-collapsible efficacy measure into a logic-respecting one. One should start with a logic-respecting efficacy measure and then apply the SME principle to it.

13.6.1 Implication toward Causal Inference

In general, causal inference terms, an (association) measure is collapsible if "collapsed" conditional measures equals the marginal measure. (See Greenland and Robins 2009, for example.)

In the language of Section 13.2.1, an efficacy measure is collapsible if g^- and g^+ patients deriving the same efficacy (e.g. $RR = 3$) implies all-comers derive the same efficacy as well ($RR = 3$).[4] A logic-respecting efficacy measure is automatically a collapsible efficacy measure, because it pinches the efficacy in all-comers $\{g^-, g^+\}$ between the (potentially different) efficacies in g^- and g^+ patients.

Efficacy measures are defined in the population space, not the sample space. As stated in Section 13.2.1, Rubin (1987) proved that difference of means is collapsible in a RCT, *in the population space*. What Sections 13.5.2 and 13.5.3 showed is that, provided continuous outcome data from an RCT goes through LSmeans adjustment by linear modeling, estimating means in a *balanced population*, efficacy assessment of difference of means has no confounding issue because difference of means is logic-respecting.

Similarly, we showed in Section 13.4.2 that, provided binary data from an RCT goes through LSmeans adjustment by logistic (or log-linear) modeling, SME efficacy assessment of RR has no confounding issue because RR is logic-respecting.

We also showed in Section 13.4.3 that, provided time-to-event data from an RCT goes through LSmeans adjustment by Weibull modeling, SME efficacy assessment of RoM has no confounding issue because RoM is logic-respecting.

On the other hand, an example in Section 13.4.2 showed odds ratio is not collapsible, even in a RCT. And an example in Aalen et al. (2015) and the example in Table 13.9 in Section 13.5.5 showed that HR is not collapsible (even) under a Weibull model, a special case of the Cox PH model.

Taken together, we see that contrary to what is stated in some literature, collapsibility is *not* a model property but an efficacy measure property. We elaborate in the next section.

13.6.1.1 Collapsibility is *not* a Model Property

As seen in Section 13.5.2, RCT data need to go through least square means analysis to avoid Simpson's Paradox behavior. Least squares means is a linear technique, so continuous data go through linear modeling, binary data go through logistic (or log-linear) modeling, and time-to-event data go through Cox or Weibull regression modeling.

While the interaction parameters in a linear model is a difference of means, the interaction parameter in a logistic model is the log of the odds ratio, and the interaction parameter in a Cox model is the log of the HR.

Observing that odds ratio and HR are not collapsible, but ratio of time is under a Weibull model, some literature have phrased collapsibility as a model property, as in "the

[4] This is termed *strict* collapsibility in Greenland et al. (1999).

logistic model is not collapsible" and "the Cox model is not collapsible" but "the Weibull model is collapsible" (e.g., in Aalen et al. 2015).

However, using a linearized model for LSmeans purpose does not obligate one to measure efficacy using whatever happens to be the interaction parameter in that model. For example, one can model binary data using a logistic model and still assess efficacy by the logic-respecting measure RR. One can model time-to-event data using a Weibull model and choose to assess efficacy by the HR (instead of a ratio of time).

SME in Lin et al. (2019) can fit binary data to either a logistic model or a log-linear model and assesses efficacy using the logic-respecting measure RR by applying a sequence of delta methods.

On the other hand, if one fits time-to-event data to a Weibull model and chooses to assess efficacy by the HR (instead of a ratio of time), then Simpson's Paradox might result because HR is not collapsible even under a Weibull model.

Therefore, collapsibility is an efficacy measure property, not a model property. There is no need to discard a tried-and-true model just because the parameter which corresponds to its interaction term happens to not be collapsible, as one can likely transform parameters in such a model to another efficacy measure which is logic-respecting. Choice of efficacy measure should be made medically, logically, but not for mathematical convenience.

13.7 Log-Rank Test Does Not Control Incorrect Decision Rate

A curious practice in the statistical analysis of survival data from clinical trials is to report confidence intervals for HR from the Wald test in a Cox Proportional Hazard (PH) model, but report p-values from the log-Rank test.

The null hypothesis being tested by a log-Rank test is the survival functions under Rx and C are exactly equal at *all* time points. It can be thought of as testing infinitely many equality nulls (4.12) between Rx and C, that the survival probabilities are exactly equal at all time points or, equivalently, that the (population) survival times are exactly equal for all quantiles.

We show that Type I error rate control by the log-Rank test testing this very restrictive null hypothesis offers no protection against the rate of making incorrect decisions. In contrast, we remind ourselves that, decision-making based on confidence sets automatically controls the incorrect decision rate. See Section 5.2 of Lin et al. (2019).

In multiple comparisons, the null hypothesis being tested by a log-Rank test is called a *Complete* null, that all the individual null hypotheses are true.

Definition 13.1 *The complete null is where all the null hypotheses are true.*[5]

Definition 13.2 *Controlling the Type I error rate under the complete null is termed weak control.*

For outcome measures that are not time-to-event, it has long been recognized that weak control of the Type I error rate is inadequate because it may not translate into control of any incorrect regulatory decision rate.

For example, in a dose-response study, weak control of the Type I error rate testing the null hypotheses that the effect at each dose equals the placebo effect may not control the probability of incorrectly inferring an ineffective dose as effective. The reason for this

[5]The *complete* null is also called the *global* null. See Chapter 1 of this Handbook.

inadequacy of weak control is, for methods that pool information across doses (either in terms of point estimates or the data itself), the scenario that has the highest probability of incorrect decision is *not* when all the doses have no effect (see Hsu and Berger 1999).

As another example, with multiple co-primary endpoints, the scenario that has the highest probability for the standard pairwise method to incorrectly infer that there is efficacy in both endpoints is *not* when there is no efficacy in either endpoint (see Hsu and Berger 1999).

The inference given by rejection of the complete null hypothesis tested by a log-Rank test is just

"the survival functions given Rx and C differ at some time point"

which many of us think is a given, with no statistical testing required to establish it.[6]

It is also not an actionable inference. To be useful, the inference needs to state whether Rx is better or worse than C in some clinical sense, such as RoM is greater than one or LLP is less than one half.

IMvigor211 was a phase 3 randomized trial comparing the anti-PD-L1 atezolizumab against chemotherapy in patients with metastatic urothelial carcinoma. For such immunotherapy, a biomarker is PD-L1 expression level, which is typically measured by immunohistochemistry (IHC). In the case of IMvigor 211, the IHC scores were placed in three categories: IC{0}, IC{1}, and IC{2,3}. Both the design and the analysis of IMvigor 211 were stratified on IHC scores.

Reporting on the analysis of the primary endpoint which was *overall survival* (OS) in IMvigor 211, Powles et al. (2018) stated the decision making process after each step of the pre-determined stepwise testing of efficacy in the nested IC{2,3}, IC{1,2,3}, and the IC{0,1,2,3}=ITT populations to be

> If the estimate of the HR is < 1 and the two-sided p-value corresponding to the stratified log-Rank test is < 0.05, the null hypothesis will be rejected and it will be concluded that atezolizumab **prolongs OS** relative to chemotherapy.

So indeed they take the implication of a rejection of the log-Rank test not to be merely "the survival functions given Rx and C differ at some time point", but that overall survival time is increased or decreased depending on whether estimated HR is < 1 or HR is > 1. We will show that Type I error rate control of the log-Rank test does not control the incorrect decision rate.

HR is not collapsible, as we showed in Section 13.5.5. A presumed common HR for the IC{1} and IC{2,3} subpopulations is not the HR for the combined IC{1,2,3} population, even in a balanced (infinite sample size) population with no censoring, if IHC is prognostic. So it is rather hopeless for the Type I error rate control of the stratified log-Rank test to control the rate of making an incorrect clinical decision, be it Rx prolongs OS or otherwise, if decision is made based on estimated HR.

Instead, since RoM is logic-respecting and therefore collapsible (under a Weibull model), let us consider making decision by first conducting a level-α log-Rank test and, upon rejection, declares Rx has longer median survival time compared to C if the estimated median survival time under Rx is longer than the estimated median survival time under C. We will consider a situation where *one* but *not all* of the equality nulls of expected survival times are true, specifically that the median survival times are equal between Rx and C, and show that Type I error rate control of the stratified log-Rank test fails to control the rate of incorrectly making this clinical decision.

---

[6]The null hypothesis tested by the log-Rank test can be called a Null null hypothesis (as in Tukey 1953), because it cannot be *exactly* true. As Tukey (1993) said, "provided we measure to enough decimal places, no two 'treatments' ever have identically the same long-run value".

Suppose the median survival times in the overall population are the same under Rx and C, and the statistical procedure is, once the statistical test for equality of survival functions in the overall population rejects, whichever treatment arm has the longer estimated median survival time, infer that treatment has longer median survival time than the other treatment for the overall population. Of course, either assertion would constitute a directional error. Suppose there are subgroups, so that for the g^+ subgroup, patients give Rx do better than those given C, but the reverse is true for the g^- patients. We conducted a simulation study to see what is the probability that the decision-making process described above would make incorrect directional decision.

For this simulation, we set the median survival time for overall population to be 8 months under both Rx and C. Data is generated from Weibull distributions, with shape parameter values of 1.05 and 1.20 for the g^- and g^+ subgroups respectively. We generate data sets with sample size 1,000, equally randomized to Rx and C, with a prevalence of 50% for each of the g^- and g^+ subgroups, without censoring. For g^+ patients, the median survival times are 12 months and 6 months given Rx and C respectively. Using the fact that, within each treatment arm and at each time point, the survival probability in the overall population is a mixture of survival probabilities in the g^- and g^+ subgroups, the median survival times in the overall population and in the g^+ subgroup determine the scale parameter values in the g^- subgroup in our simulation. Setting the level of the log-Rank test at 5%, the percent of times it rejects was 304 times out of the 1,000 Weibull data sets simulated.

Truth of the Weibull model we generated data from is that median survival times under Rx and C are the same, so inferring either Rx or C has longer median survival time is a directional error, an incorrect decision. For a 5% 2-sided test based on an equal-tailed 95% confidence interval, this incorrect decision rate is no more than 2.5%. On the contrary, for the log-Rank test, since the sum of the two possible directional error rates is estimated to exceed 30%, at least one of the two-directional error rates exceeds 15%. This is an illustration that controlling the Type I error rate of testing a Null null hypothesis may well be a Null control, in terms of controlling any incorrect decision rate.

The log-Rank test is popular because it is perceived to be more powerful than the Wald test. To us, the concept of "power" is inadequate for any multi-action problem because it includes the probability of *rejecting for wrong reasons*. For example, suppose in truth the median survival time under Rx is *higher* than the median survival time under C, so that inferring the median survival time under Rx is *lower* than the median survival time under C is in fact worse than making no inference, making this latter inference is typically counted positively in the calculation of "power". Thus, for time-to-event outcomes, we urge a fundamental re-assessment of the concept of (regulatory) Type I error rate control, vis-à-vis the log-Rank test.

13.7.1 Permutation Testing for Predictive Effect Will Pick Up Purely Prognostic Biomarkers

Rank-based methods are perceived as "nonparametric", based on the notion that all rankings are equally likely under the Null null of identical distributions under Rx and C. Similarly, permutation methods are perceived as "nonparametric", based on the notion that all permutations are equally likely under some "null".

If decision-making goes beyond stating the Null null of identical distributions is false, then permutation-based methods share with the log-Rank test the issue that weak control of Type I error rate may not control the Incorrect Decision Rate in the sense of having inflated directional error rate. Under the same equal median survival times scenario as we had for the log-Rank test simulation, the percent of times a level-5% permutation version of the Cox model likelihood ratio test rejects was 431 times out of the 1,000 Weibull data

TABLE 13.12
An Example with $k = 3$ Cut-Points

Cut-point	$c_1 = 0$	$c_2 = 17$	$c_3 = 53$
Partition subgroup	$0 \leq x \leq 17$	$17 < x \leq 53$	$53 < x \leq 100$
Rx effect	0.2	0.3	0.4
C effect	0.0	0.1	0.2
Efficacy $Rx - C$	0.2	0.2	0.2
Prevalence	1/3	1/3	1/3
Nested subgroup	$x > 0$	$x > 17$	$x > 53$
Efficacy $Rx - C$	0.2	0.2	0.2
Prevalence	1	2/3	1/3

sets simulated, even more than the log-Rank test. So, for the permutation version of the Cox model likelihood ratio test, at least one of its two-directional error rates exceeds 21% in our simulation scenario.

There are two further issues with permutation methods that illustrate the danger of assessing statistical evidence by calculating under a very restricted null, as follows.

One aspect of subgroup identification is to find predictive biomarkers. Sections 13.4.2 and 13.4.3 showed that the prognostic effect needs to be carefully accounted for, to tease out the predictive effect, if the outcome is binary or time-to-event. Suppose one is interested in testing whether a binary biomarker is predictive under a logistic mode using test statistics which is the maximum likelihood estimate of the interaction term. Values far from zero (where 'far' is defined by a reference distribution for the test statistic when the null hypothesis is true) are strong evidence against the null hypothesis. Kil et al. (2020) showed that calculating the null distribution by permuting the biomarker label will cause purely prognostic markers to be inadvertently picked up. This is because permuting the biomarker label makes both the prognostic effect and the predictive effect null, but one cannot assume the prognostic effect is null. Calculating the null distribution by permuting the treatment label has a similar issue because such permutation makes both the treatment main effect and its interaction with the biomarker null.

Another aspect of subgroup identification is to select a cut-point c^* from a set of cut-point values $c_i, i = 1, \ldots, k$, of a continuously valued biomarker x and target patients with $x > c^*$. Subgroup identification methods such as Jiang et al. (2011) and Liu et al. (2016) test for and compute confidence intervals for efficacy in the k (nested) subgroups of patients with $x > c_i, i = 1, \ldots, k$.

To adjust for multiplicity of the k tests, the Cox modeling likelihood ratio testing approach of Jiang et al. (2011) use permutation to compute the null distribution. However, for permutation multiple tests to control the Type I error rate even weakly, the subtle MDJ (Marginals-Determine-the-Joint) condition needs to hold, as explained in Xu and Hsu (2007) and Kaizar et al. (2011). The word "marginal" in MDJ refers to marginal *hypotheses*. To avoid confusion with the word *marginal* referring to collapsing across the strata in causal inference discussion earlier in this chapter, we change the wording from *marginal* to *conditional*, conditioning (in the *distributional* sense) on patients being in a subgroup and re-word the MDJ condition as the CDJ condition:

Definition 13.3 (CDJ) *The Conditionals-Determine-the-Joint (CDJ) condition is said to hold if the truth of all null hypotheses* conditionally *within each subgroup implies the* joint *distributions of the observations (possibly adjusted for the nulls) are identical under Rx and C across all the subgroups.*

The reason CDJ is necessary for permutation tests to control the Type I error rate even weakly is, while permuting treatment label generates a null distribution assuming the *joint* distributions of the observations across all the subgroups are identical under Rx and C, the complete null specifies only some aspect of the distributions under Rx and C are the same *within* each subgroup.

For example, suppose each null hypothesis states the *means* are the same under Rx and C within each subgroup. Then any difference in higher moments in the joint distributions under Rx and C within and across the subgroups, such as differences in variances or skewness or kurtosis between Rx and C within subgroups, or difference in covariances among subgroups between Rx and C, would violate CDJ.

Take the example in Table 13.12 where outcome is binary and efficacy is a difference of means, the difference in the responder probabilities under Rx and C. Suppose each null hypothesis is $Rx : C$ efficacy is 0.2. Then subtracting 0.2 from each observation under Rx while leaving observations under C unchanged would make the mean difference between Rx and C equal to zero in each of the three subgroups. Suppose the test statistic for each of the nested subgroups is the estimated mean difference of these "re-centered" observations, and the form of the multiple test is a maxT test. The three test statistics are correlated since observations with $x > 17$ include observations with $x > 53$ ans so forth. Therefore, one might be tempted to calculate a null distribution for the maxT statistic by permuting the Rx and C treatment label, recalculating the maxT statistic after each permutation. However, the result of Huang et al. (2006) shows this permutation test would not control Type I error rate even weakly because in this case the variances under Rx and C within each nested subgroup would differ, and the covariances among subgroups would differ between Rx and C.

Instead of using permutation to build a null distribution, Liu et al. (2016) show with suitable modeling one can theoretically and numerically compute the joint distribution of pivotal statistics to provide simultaneous confidence intervals for efficacy in the nested subgroups to facilitate choosing a cut-point.

13.8 Summary and Connection

Instead of giving a list of methods for subgroup analysis, we have shown a systematic to develop confident logical inference on efficacy in subgroups and their mixtures, through the following path

1. Choose a logic-respecting efficacy measure;
2. Model the data and adjust for imbalance using the least squares means technique;
3. Apply the SME principle to infer on efficacy in subgroups and their mixtures.

Methods that result, being confidence interval methods, automatically control the directional incorrect decision rate. On the other hand, we urge caution against subgroup analysis methods based on tests of exact equality nulls, as we have shown by example that they may not control the directional incorrect decision rate.

Finally, we briefly indicate how subgroup analyses arise in online testing. In A/B/n web testing, two or more web designs are compared in terms of Key Performance Indicators (KPIs) which include click-through rate (CTR), average order value (AOV), and customer journey. Having subgroups is referred to as having *segmentation*. Customers in different countries may have different preferences; casual gamers behave differently from addicted gamers.

13.9 Acknowledgments

We thank Haiyan Xu for insightful discussions, particularly about the role of *stratification*, both in *design* and in *analysis*. We would like to thank Sue-Jane Wang and Jim Hung for many interesting exchanges over the years as well.

13.10 Glossary

Disease-progression prognostic biomarker: A biomarker which predicts increased likelihood of an event without any treatment.

Treatment-effect prognostic biomarker: A biomarker whose value has some ability to foretell the outcome for a patient given a particular treatment.

Bibliography

Aalen, O. O., R. J. Cook, and K. Røysland (2015). Does Cox analysis of a randomized survival study yield a causal treatment effect? *Lifetime Data Analysis 21*, 579–593.

Ding, Y., H.-M. Lin, and J. C. Hsu (2016). Subgroup mixable inference on treatment efficacy in mixture populations, with an application to time-to-event outcomes. *Statistics in Medicine 35*, 1580–1594.

FDA-NIH Biomarker Working Group (2016). *BEST (Biomarkers, EndpointS, and other Tools) Resource.* Silver Spring (MD): Food and Drug Administration (US). Internet.

Goodnight, J. H. and W. R. Harvey (1997). Least squares means in the fixed effects general model. Technical report, SAS Institute.

Greenland, S. and J. Robins (2009). Identifiability, exchangeability, and epidemiological confounding, revisited. *Epidemiologic Perspectives and Innovations 6*(4).

Greenland, S., J. M. Robins, and J. Pearl (1999). Confounding and collapsibility in causal inference. *Statistical Science 14*(1), 29–46.

Hothorn, T., F. Bretz, and P. Westfall (2008). Simultaneous inference in general parametric models. *Biometrical Journal 50*, 346–363.

Hsu, J. C. (1996). *Multiple Comparisons: Theory and Methods.* London: Chapman & Hall.

Hsu, J. C. and R. L. Berger (1999). Stepwise confidence intervals without multiplicity adjustment for dose response and toxicity studies. *Journal of the American Statistical Association 94*, 468–482.

Huang, Y., H. Xu, V. Calian, and J. C. Hsu (2006). To permute or not to permute. *Bioinformatics 22*, 2244–2248.

Jiang, W., B. Freidlin, and R. Simon (2007). Biomarker-adaptive threshold design: A procedure for evaluating treatment with possible biomarker-defined subset effect. *Journal of the Natational Cancer Institute 99*, 1036–43.

Kaizar, E. E., Y. Li, and J. C. Hsu (2011). Permutation multiple tests of binary features do not uniformly control error rates. *Journal of the American Statistical Association 106*, 1067–1074.

Kil, S., E. Kaizar, S.-Y. Tang, and J. C. Hsu (2020). *Principles and Practice of Clinical Trials*, Chapter Confident Statistical Inference with Multiple Outcomes, Subgroups, and Other Issues of Multiplicity. Springer.

Lin, H.-M., H. Xu, Y. Ding, and J. C. Hsu (2019). Correct and logical inference on efficacy in subgroups and their mixture for binary outcomes. *Biometrical Journal 61*, 8–26.

Little, R. J. A. and D. B. Rubin (1987). *Statistical Analysis with Missing Data*. New York: John Wiley.

Liu, Y., S.-Y. Tang, M. Man, Y. G. Li, S. J. Ruberg, E. Kaizar, and J. C. Hsu (2016). Thresholding of a continuous companion diagnostic test confident of efficacy in targeted population. *Statistics in Biopharmaceutical Research 8*, 325–333.

Powles, T., I. Durán, M. S. van der Heijden, Y. Loriot, N. J. Vogelzang, U. D. Giorgi, S. Oudard, M. M. Retz, D. Castellano, A. Bamias, A. Fléchon, G. Gravis, S. Hussain, T. Takano, N. Leng, E. E. Kadel, R. Banchereau, P. S. Hegde, S. Mariathasan, N. Cui, X. Shen, C. L. Derleth, M. C. Green, and A. Ravaud (2018). Atezolizumab versus chemotherapy in patients with platinum-treated locally advanced or metastatic urothelial carcinoma (IMvigor211): a multicentre, open-label, phase 3 randomised controlled trial. *The Lancet 391*, 748–757.

Rubin, D. B. (1978). Bayesian inference for causal effects: The role of randomization. *Annals of Statistics 6*, 34–58.

Tukey, J. W. (1953). The Problem of Multiple Comparisons. Dittoed manuscript of 396 pages, Department of Statistics, Princeton University.

Tukey, J. W. (1993). Graphical comparisons of several linked aspects: Alternatives and suggested principles. *Journal of Computational and Graphical Statistics 2*(1), 1–33.

Woodcock, J. (2015). FDA Voice. Posted 03/23/2015.

Xu, H. and J. C. Hsu (2007). Using the partitioning principle to control the generalized family error rate. *Biometrical Journal 49*, 52–67.

Xu, R. and J. O'Quigley (2000, 12). Estimating average regression effect under non-proportional hazards. *Biostatistics 1*(4), 423–439.

14

Exploratory Inference: Localizing Relevant Effects with Confidence

Aldo Solari
University of Milano-Bicocca

Jelle J. Goeman
Leiden University Medical Centre

CONTENTS

14.1 Introduction

Modern data analysis, e.g. in genomics or brain imaging, can be highly exploratory. Rather than a single well-defined research question there are tens of thousands of micro-questions, leading to tens of thousands hypothesis tests and tens of thousands of micro-inferences. These can be aggregated to larger-scale inferences in countless ways. For example, in brain imaging, the brain is partitioned into tens of thousands of little cubes (known as *voxels*), each of which may show activity as a response to stimulus. In genomics, researchers may probe hundreds of thousands of genomic locations for evidence of association with a biological phenotype.

Knowing which individual voxels or genomic markers are confidently active is interesting, but the real research questions often relate to patterns of activity at larger scales of aggregation. Neuroscientists are not interested in voxels, but in brain regions. Geneticists are not interested in single genomic locations but in genes or genetic pathways. A common approach is to formulate a null hypothesis of zero activity per tiny unit (voxel, genomic location), calculate p-values per unit, and to look for potentially interesting patterns in these p-values. The biologically interesting questions relate to groups of units taken together.

DOI: 10.1201/9780429030888-16

Two complications arise with such an approach. First, the "zero effect" null hypothesis is usually not truly of interest: true effects are never exactly zero (see Chapter 4 of the present Handbook and Bowring et al., 2019; Ding et al., 2018). It is more meaningful to look for patterns of effects that are large enough to be biologically or practically important. Information on the size of the effect is discarded if we restrict to p-values for the "zero effect" null hypothesis. Proving that an effect is non-zero does not necessarily make it a significant scientific finding if the effect may still be negligible. Effect size estimates should always be judged in relation to a minimal scientifically relevant effect, and true scientific significance of results should be related not only to an assessment of uncertainty in the estimates but also in connection to its relevance limits. Moreover, the use of effect sizes facilitates the comparison of results from different studies and the evaluation of reproducibility and replicability.

Second, the number of potential ways to aggregate units when looking for patterns is virtually limitless. It is well-known that humans are very good at finding seemingly convincing patterns even in pure noise. How confident can the researcher be about a pattern that has been found, if that pattern has been selected from so many potential patterns? Selection is ubiquitous yet subtle, and unavoidable in large-scale testing problems, in particular in the emerging fields of neuroscience and genomics. Researchers often highlight the most captivating discoveries suggested by the data, but false discoveries are likely to intrude into this selection. Traditionally, the only reliable weapon to combat this problem is to use a multiple testing procedure that keeps the false discoveries in check. However, by doing so, the researcher limits their scientific freedom and puts the multiple testing procedure in charge of operating the selection.

We propose an approach that combines solutions to both of these issues. We recommend per unit to test *interval null hypotheses* of "negligible" effect size, so that small effect sizes are considered part of the null hypothesis rather than part of the alternative. This idea will be combined with the idea of *exploratory multiple testing* (Goeman and Solari, 2011), which takes into account the effect of selection, especially when the selection is made in a truly exploratory fashion. We allow researchers to use the data in a data-driven way, i.e. we do not even require the manner in which the data will be used to be specified before seeing the data.

The exploratory approach reverses the traditional roles of the researcher and the multiple testing procedure: it lets the researcher select interesting outcomes freely, and the multiple testing procedure returns a confidence statement on the amount of false discoveries incurred in the chosen selection. Importantly, these confidence statements are simultaneously valid for every selection of potential interest: this allows the researcher to evaluate multiple selections and choose *post hoc* the most promising one. The exploratory multiple testing approach consists of (i) selecting in the light of the data, and (ii) testing the selection with the same data, and, if necessary, iterate (i) and (ii), the researcher benefits from exploration and hypothesis testing in the same breath. The approach does not allow unlimited cherry-picking, however. The requirement is that the "cherry tree" can not be *random*, i.e. the potential hypotheses (cherries: voxels; genomic locations) should be defined a priori and finite in number. While this cuts out some of the more opportunistic use of the "researcher degrees of freedom" (de Groot, 2014; Gelman and Loken, 2013; Simmons et al., 2011), many problems are structured enough to fit this framework.

Common to both aspects of our approach is the use of domain knowledge. The limits of relevance that determine the null hypothesis are defined by the researcher's domain knowledge in order to ensure that statistically significant is also practically significant; moreover, the exploratory search for patterns may also be guided by domain knowledge as well as by the data. A difference is that we demand that the domain knowledge that determines the null hypothesis is pre-specified, while the domain knowledge used in the exploratory step may be left implicit.

The outline of the Chapter is as follows. We first distinguish between confidence intervals and confidence directions for effect sizes. Next, we illustrate with an example how the exploratory multiple testing approach can be used to find interesting patterns of effects by learning from what the data shows and from what is expected by domain knowledge. We illustrate the concepts and methods first with a toy example and subsequently with a neuroimaging data set.

14.2 Assessing uncertainty: intervals and directions

Suppose that per unit (voxel, genomic location) our research question is about the standardized effect

$$\theta = \frac{\mu}{\sigma} \tag{14.1}$$

where μ and σ are the mean and the standard deviation of a Gaussian distribution, respectively. It is worth noting that in other settings, such as pharmaceutical clinical trials, the parameter of interest θ may not need standardization by σ.

Statisticians have classically asked the question "Is the effect 0?", but for many applications this is the wrong question, since the effect is almost always different from 0 — at least in some remote decimal place (Tukey, 1991). The question that should be addressed instead is "Is the effect large enough to be *meaningfully* different from 0?"

This question requires some assessment of the size of the effect. The classical statistical technique for this is point estimation, i.e.

$$\hat{\theta} = \frac{\bar{x}}{s} \tag{14.2}$$

where $\bar{x}$ is the sample mean and s^2 is the unbiased estimator of the variance based on a random sample of size n. A point estimate gives a hint of the most likely value the data seem to point out or suggest, but by itself this is not enough for inference because it ignores variability and gives no insight into uncertainty. We should embrace variation and accept uncertainty (Gelman, 2016).

Tukey (1960; 1991) advocated that the answer should be provided by *confident conclusions*. Confident conclusions are of two types: *directions* and *intervals*. Confidence intervals address the question "What are the likely values for the effect?" by pointing out a whole interval of possible values, chosen so that there can be high confidence that the true value of the effect is among them. Confidence directions address a simpler question: "What is the direction of the effect?" Is it 'positive', 'negative' or 'irrelevant'? Here, 'positive' and 'negative' imply that the effect is meaningfully large, and 'irrelevant' effects do not need to be exactly zero, just not large enough to be meaningful. In Tukey's (1991) words: "We have to accept explicit uncertainty – initially about whether we are confident about direction, ultimately about the exact value of the effect." Since asking only for direction asks for a cruder form of information, we may expect to have more statistical power to answer the question.

Small effects usually matter very little, i.e. are not considered of practical importance. It is the researcher's duty to have enough subject matter knowledge to know the magnitude of a minimally relevant effect. The researcher can choose a *relevance boundary* Δ such that an effect is considered 'irrelevant' when $-\Delta \leq \theta \leq \Delta$ or 'relevant' otherwise, i.e. 'positive' when $\theta > \Delta$ and 'negative' when $\theta < -\Delta$. These bounds are for the true effect θ. Depending on the amount of information we have about θ, we may be able to infer in which category θ lies, or we may be left with uncertainty, inferring e.g. only that θ is either positive or irrelevant, but not negative.

We briefly revisit how to determine confident conclusions about θ. For illustration purposes only, from here on we take the conventional value of $\Delta = 1/5$ for an effect of small magnitude (Cohen, 1988). In other contexts, such as clinical trials, the choice of Δ for a clinically meaning difference (a term perhaps attributable to Stephen J. Ruberg) is now well-established. As pointed out by an anonymous reviewer, the value of $\Delta = 1/5$ frequently used in bioequivalence trials comes from a back calculation that the U.S. Food and Drug Administration (FDA) did. In making a smooth transition from approving if the p-value is close enough to 1 when testing that the effect is exactly zero to approving if the confidence interval is within $\pm\,\Delta$ (on the log scale), the FDA went back to their past data and figured that previous approvals correspond (roughly) to a Δ of 20%.

For a random sample from the Gaussian distribution, $\sqrt{n}\hat{\theta}$ follows a noncentral Student's T distribution with $n-1$ degrees of freedom and non-centrality parameter $\sqrt{n}\theta$. Then, a confidence interval for θ with level $1-\alpha$ is given by

$$I^{1-\alpha} = [\underline{\theta}, \overline{\theta}] \tag{14.3}$$

with $\underline{\theta}$ and $\overline{\theta}$ such that $T_{n-1,\frac{\underline{\theta}}{\sqrt{n}}}(\sqrt{n}\hat{\theta}) = 1-\alpha/2$ and $T_{n-1,\frac{\overline{\theta}}{\sqrt{n}}}(\sqrt{n}\hat{\theta}) = \alpha/2$, respectively, where $T_{\nu,\lambda}(t)$ denotes the cumulative distribution function evaluated at t of a non-central Student's T distribution with ν degrees of freedom and non-centrality parameter λ. For computing confidence intervals for standardized effects, the reader can refer to Kelley (2007).

In the spirit of the three-sided testing approach (Goeman et al., 2010), we partition the parameter space of θ into three disjoint regions: $(-\infty, -\Delta)$, $[-\Delta, \Delta]$ and (Δ, ∞). Then we simultaneously do the following level α tests:

$$\begin{aligned} &1.\ \text{reject } H^+ : \theta \in (\Delta, \infty) \text{ if } \hat{\theta} < a \\ &2.\ \text{reject } H^- : \theta \in (-\infty, -\Delta) \text{ if } \hat{\theta} > -a \\ &3.\ \text{reject } H^0 : \theta \in [-\Delta, \Delta] \text{ if } \hat{\theta} < -b \text{ or } \hat{\theta} > b \end{aligned} \tag{14.4}$$

with $a = -T^{-1}_{n-1,-\sqrt{n}\Delta}(1-\alpha)/\sqrt{n}$ and $b^2 = F^{-1}_{1,n-1,n\Delta^2}(1-\alpha)/n$ where T^{-1} and F^{-1} denote the quantile functions of non-central T and F distributions, respectively.

Because the null hypotheses H^+, H^- and H^0 are disjoint, the true value of θ lies in one of the disjoint regions and no two hypotheses can be simultaneously true. By the partitioning principle (see Chapter 4 of the present Handbook and Finner and Strassburger, 2002; Goeman et al., 2010; Stefansson et al., 1988), all tests of all hypotheses can be simultaneously performed at level α, while still controlling the familywise error at level α. This results in the following confidence direction for θ:

$$D^{1-\alpha} = \begin{cases} (\Delta, \infty) & \text{(positive)} \quad \text{if } \hat{\theta} > b \\ (-\infty, -\Delta) & \text{(negative)} \quad \text{if } \sqrt{n}\hat{\theta} < -b \\ [-\Delta, \Delta] & \text{(irrelevant)} \quad \text{if } -a < \sqrt{n}\hat{\theta} < a \\ (-\infty, \Delta] & \text{(non-positive)} \quad \text{if } -b < \sqrt{n}\hat{\theta} < \min(-a, a) \\ [-\Delta, \infty) & \text{(non-negative)} \quad \text{if } \max(-a, a) < \sqrt{n}\hat{\theta} < b \\ (-\infty, \infty) & \text{(uncertain)} \quad \text{otherwise} \end{cases} \tag{14.5}$$

with the property that

$$\mathrm{P}_\theta(\theta \in D^{1-\alpha}) \geq 1-\alpha. \tag{14.6}$$

Note that it is not possible to conclude that an effect is 'irrelevant' if $a < 0$, which happens for small values of Δ and/or n. This is because testing the null hypothesis $H^- \cup H^+ : \theta \in (-\infty, -\Delta) \cup (\infty, \Delta)$ is based on the intersection-union principle, which rejects $H^- \cup H^+$ if both H^- and H^+ are rejected by the corresponding one-sided tests (Wellek, 2010).

A simplified confidence direction for θ can be obtained by rejecting H^0 and $H^- \cup H^+$ for large and small values of the test statistic $n\hat{\theta}^2$, respectively, resulting in

$$D^{1-\alpha} = \begin{cases} (\Delta, \infty) & \text{(positive)} & \text{if } \hat{\theta} > b \\ (-\infty, -\Delta) & \text{(negative)} & \text{if } \hat{\theta} < -b \\ [-\Delta, \Delta] & \text{(irrelevant)} & \text{if } -c < \hat{\theta} < c \\ (-\infty, \infty) & \text{(uncertain)} & \text{otherwise} \end{cases} \tag{14.7}$$

with $c^2 = F^{-1}_{1,n-1,n\Delta^2}(\alpha)/n$.

Because $a \leq c$, the confidence direction (14.7) is more powerful than (14.5) for detecting irrelevant effects, but it does not allow for conclusions of non-positive and non-negative effects. Figure (14.1) compares the confidence directions (14.5) and (14.7) as function of the relevance boundary Δ with $\alpha = 5\%$ and $n = 50$. The less detail we need in our inference, the more power we have to make such inference. This observation will be a recurring theme in this Chapter.

Now we discuss the comparison of the confidence interval in (14.3) with the confidence direction in (14.7). Suppose that we have obtained a point estimate $\hat{\theta} = 0.46$ based on a

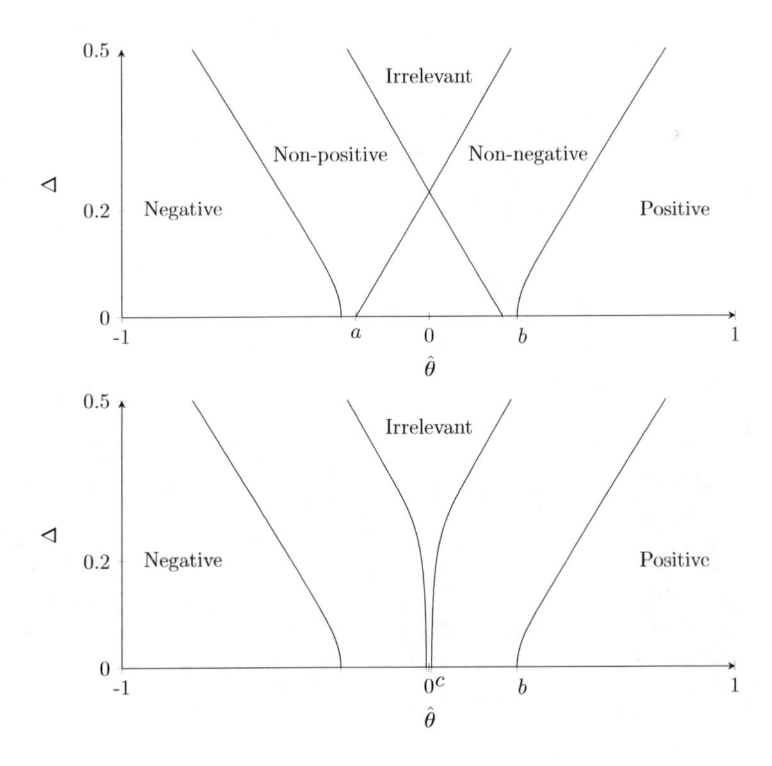

FIGURE 14.1
Comparison of confidence conclusions (14.5) (above) and (14.7) (below) as function of the point estimate $\hat{\theta}$ and the relevance boundary Δ with $\alpha = 5\%$ and $n = 50$. Because $a \leq c$, the confidence direction (14.7) is more powerful than (14.5) for detecting irrelevant effects, but it does not allow for conclusions of non-positive and non-negative effects.

random sample of size $n = 50$. With high confidence, e.g. 95%, we can conclude that the direction is 'positive', i.e. we reject the null hypothesis that $\theta \in (-\infty, \Delta]$ with $\Delta = 0.2$, since $\hat{\theta} > b = 0.446$. The usual confidence interval, however, is $I^{95\%} = [0.1662, 0.7496]$; see Figure 14.2. On the one hand, the interval $[0.1662, 0.7496]$ is clearly more precise than the conclusion that $\theta \in (0.2, \infty)$. On the other hand, this precision comes at a price: $[0.1662, 0.7496]$ does not imply the conclusion that the direction of the effect is 'positive' because the interval contains values below $\Delta = 1/5$. If we are only interested in the tripartite decision regarding relevance and sign of the effect, or more generally if the inference is directed towards some specific subregions of the parameter space, we have more power regarding for the statement of interest, but it may come at the price of a confidence interval of larger length (Benjamini et al., 1998; Goeman et al., 2010; Hayter and Hsu, 1994; Hsu and Berger, 1999).

Figure 14.2 illustrates the example. It shows the possible conclusions about θ (positive, negative or irrelevant) as a function of $\hat{\theta}$ and n at the confidence level 95%. The highlighted regions for concluding about the direction of the effect are larger (i.e. more powerful) than the ones obtained by confidence intervals.

This example illustrates the following principle: it is much easier to detect the direction than to pinpoint the likely values for the effect, or more generally, "the less specific the question is, the more power to answer it". Likewise, as we will see in the next section when considering multiple effects, it is easier to infer the presence or the number of meaningful effects rather than to pinpoint exactly where they are.

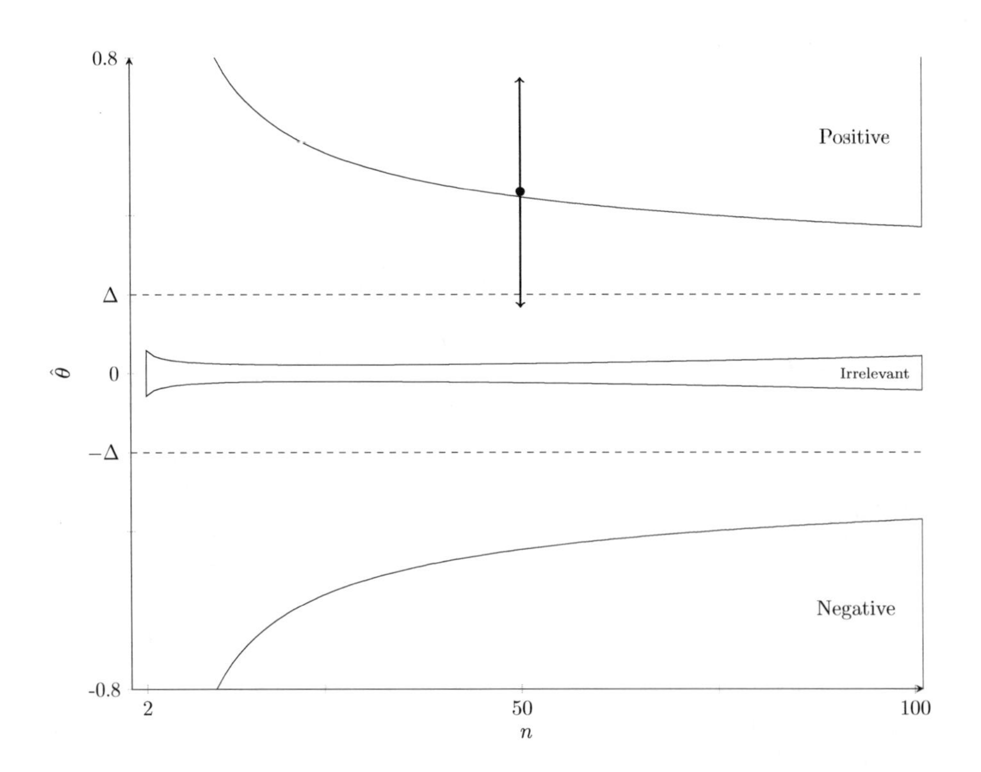

FIGURE 14.2
95% confident conclusions as function of the sample size n and of the point estimate $\hat{\theta}$. For example, with $n = 50$, $\hat{\theta} = 0.46$ (dot) lies in the region of 'positive' direction, with a confidence interval of $I^{95\%} = [0.1662, 0.7496]$.

14.3 Multiplicity and selection

Research questions in genomics and neuroscience typically involve multiple effects $\theta_i = \mu_i/\sigma_i$ for $i = 1,\ldots,m$, with a number of units m in the thousands. Such situations give rise to a collection of questions of the type: "Is the ith effect irrelevant?" Clearly, confident conclusions introduced in the previous section answer each individual question in a straightforward fashion. However, the assessment of uncertainty becomes more difficult when we look at some particular pattern of potential effects, selected in the light of the data (Cox, 1965; Sorić, 1989). Selection renders the statistical inference more complicated: the selected pattern is likely to be the most unexpected one, and as such, it is more likely to be unreal. For this reason, selected confidence intervals tend to have reduced coverage (Benjamini and Yekutieli, 2005).

14.3.1 Simultaneous inference

The emerging field of *selective inference* is concerned with developing statistical methods that account for selection in inference (Benjamini, 2010). Crucial in selective inference is the question what we know about the process of selection. In some situations, the researcher may be willing and able to specify the selection process a priori as a function of the data. This is what classical multiple testing methods such as e.g. Benjamini and Hochberg (1995) assume. In exploratory research, however, the selection process is often haphazard and therefore too complex to model. We may have no idea how the researcher made the selection, or be reluctant to impose any constraints on this creative process. Without any information about the selection strategy, we may still obtain inferential guarantees if we use a method that gives guarantees *simultaneously* for any possible selection (Berk et al., 2013; Blanchard et al., 2020; Genovese and Wasserman, 2006; Goeman and Solari, 2011; Katsevich and Ramdas, 2020; Swanson, 2020).

We now illustrate how the exploratory multiple testing approach Goeman and Solari (2011) can be used to generate new insights, partly from what the data shows and partly from what is expected by the researcher's domain knowledge. We will illustrate this method with a toy problem before moving on to a real data example later. Suppose we have a moderate-scale spatial problem, structured in a $m = 100 \times 100$ square matrix, where for each unit (pixel) we measure the local magnitude of the effect. This set-up is analogous to neural activity detection from 3D brain imaging data (Rosenblatt et al., 2018) that we will come back to later. Figure 14.3 displays the point estimates $\hat{\theta}_i$ based on random sample of size $n = 25$.

Here the scale of the multiplicity is large and the sample size small. We need methods with good power, and taking the lessons of the first part of this Chapter in mind, possibly we should not aim for quantifying effects, but only for assessing whether effects are positive, negative or negligible. Moreover, we imagine that in the substantive theory of our toy data, again in analogy to neuroimaging, it is the shape of the area of non-negligible effects that is of primary interest. Moreover, there may be theory available on the expected shapes and sizes of such regions. Such theory is seldom precise and structured enough to be quantified in terms of Bayesian priors, but is nonetheless crucial for interpretation of results, and should therefore ideally be used in the analysis.

A region S would therefore be of interest of the researcher if it has a high proportion r_S of truly non-negligible effects. How do we count the number of 'relevant' effects in selected regions of interest? For a given region S it is natural to infer that r_S is high using a lower 95%-confidence bound $\underline{r}_S^{95\%}$ on r_S. A lower $(1-\alpha)$-confidence bound should have the property that

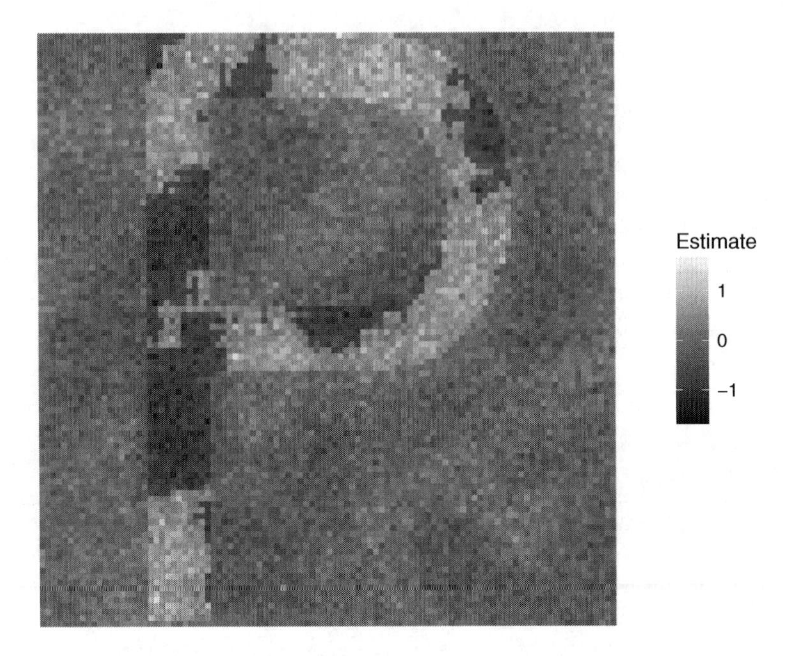

FIGURE 14.3
Example of a moderate-scale spatial problem structured in a 100×100 square matrix, where each pixel corresponds to a point estimate.

$$\mathrm{P}_\theta(r_S \geq \underline{r}_S^{1-\alpha}) \geq 1 - \alpha. \tag{14.8}$$

However, such a confidence bound is useful only if a single set S was of interest, and this set was chosen independently of the data. If several sets S are considered, or if the set S of interest will be chosen on the basis of the data, the confidence bound will have less than $1 - \alpha$ coverage. However, we can make the analysis robust against multiple or post hoc selection by constructing a *simultaneous* lower $(1 - \alpha)$-confidence bound, i.e. a bound with the property that

$$\mathrm{P}_\theta(r_S \geq \underline{r}_S^{1-\alpha} \text{ for all } S) \geq 1 - \alpha, \tag{14.9}$$

where "for all S" means all regions S that the researcher might possibly consider. Since we the researcher will not in practice be able to enumerate all such sets, we simply use all subsets of the 100×100 grid. Simultaneous control over all such S implies that (14.8) also holds for a set S that is selected post hoc as a function of the data. To see why, consider the event E, which happens with probability at least $1 - \alpha$, that $r_S \geq \underline{r}_S^{1-\alpha}$ for all S. If this event happened, then $r_S \geq \underline{r}_S^{1-\alpha}$ for all S so also for the selected S. Consequently, (14.8) holds for the selected S. In fact, the statement also holds simultaneously over multiple selected S, guaranteeing familywise error control if multiple selections are made. Note that these guarantees make no assumptions whatsoever on the selection process: the researcher may use any internal or external information to choose the selected S. She does not have to declare how the set was selected for (14.8) to hold for the selected set, although transparency in the selection process could of course still be desirable for other methodological reasons.

When exploring the data, for every selected region S, an exploratory multiple testing procedure returns an estimate and a lower bound for the true number of relevant effects r_S present in the selected region, i.e. the *post hoc* confidence measures

$$\hat{r}_S \; [\underline{r}_S, |S|] \tag{14.10}$$

where $\hat{r}_S = \underline{r}_S^{50\%}$ and $\underline{r}_S = \underline{r}_S^{95\%}$ are 50% and 95% simultaneous confidence lower bounds for r_S, respectively. The use of multiple levels for simultaneous bounds — the median level 50% and the tail level 95% — is advocated by Tukey (1991) as *multilevel simultaneous*. The 50%-bound plays the role of a point estimate. It is guaranteed to be conservative relative to a median unbiased estimate: it overestimates the amount of true signal in a set S at most 50% of the time. Because this bound is also simultaneous, this property is valid for a selected S.

14.3.2 Closed testing

To construct confidence bounds $\underline{r}_S^{1-\alpha}$ for all S, it has been proposed (Genovese and Wasserman, 2006; Goeman and Solari, 2011) to use closed testing (see Chapter 1 of the present Handbook and Marcus et al., 1976). Moreover, it has been proven that closed testing is optimal for constructing such confidence bounds: every method that is not equivalent to a closed testing procedure can be uniformly improved by one (Goeman et al., 2020).

A closed testing procedure is defined by its local tests, which are α-level tests for the hypotheses $H_S\colon r_S = 0$ for all sets S. The choice of such local tests should be driven by considerations of power and of easy computability. We advocate the use of the Simes test as a local test Goeman et al. (2019), following Hommel (1988). This test rejects H_S if $p_{(i:S)} \leq i\alpha/m_S$ for at least one $1 \leq i \leq m_S$, where $p_{(i:S)}$ is the ith smallest p-value among those in S and $m_S = |S|$ is the cardinality of the set S. $H_\emptyset$ is never rejected. The Simes test is a valid test under the PRDN assumption (Su, 2019), a weaker variant of the PRDS assumptions that is sufficient for the validity of the procedure of Benjamini and Hochberg (1995).

The closed testing procedure corrects all local tests for all S for multiple testing by disregarding the rejections of those S for which a superset S' exists that is not rejected by the local test, and retaining only those rejections for which such supersets do not exist. The remaining rejections will then be all correct rejections with probability at least $1 - \alpha$. To understand why this is true, consider the set T of all truly inactive units. Then H_T is true, and moreover we have that H_S is true if and only if $S \subseteq T$. If H_T is not rejected, which happens with probability at least $1 - \alpha$, then the closed testing procedure does not reject any true H_S.

Starting from the adjusted results of the local tests, the closed testing procedure can be used to calculate the simultaneous confidence bound for r_S as (Goeman and Solari, 2011)

$$\underline{r}_S^{1-\alpha} = \min_V \{|S \setminus V| \colon V \subseteq S, \phi_V = 0\}.$$

where ϕ_V is the indicator of rejection of H_V by the closed testing procedure and $|S|$ denotes the cardinality of a set S. The intuition for the formula is as follows. Let E be the event that the closed testing procedure did not make an error. Then E happens with probability at least $1-\alpha$. Consider the subset of all true hypotheses in S: $W = S \cap T$, so that $|W| = |S| - r_S$. If E happened, then $\phi_W = 0$, since H_W is true. Therefore, $\underline{r}_S^{1-\alpha} \leq |S \setminus W| = r_S$. It follows that $\underline{r}_S^{1-\alpha} \leq r_S$ with probability at least $1 - \alpha$. Moreover, since the event E is the same for all S, we have the desired simultaneous statement (14.13).

We illustrate and explain the working of the closed testing procedure with Simes local tests using a simple example, taken from Goeman et al. (2019), and represented in Figure 14.4. In this example, there are only four potential effects of interest. The p-values for the hypothesis that these effects are negligible are $p_1, \ldots, p_4$, of which $\alpha/2 < p_1 \leq p_2 \leq p_3 \leq 2\alpha/3$ are small, and $p_4 > \alpha$ is large. Here, we see for which sets S the Simes test has rejected the corresponding hypothesis. For example, for $S = \{2, 3, 4\}$ the Simes test rejected H_S because $p_{(2:S)} = p_3 \leq 2\alpha/m_S = 2\alpha/3$. Some hypotheses are rejected by the local test, but not by the closed testing procedure. For example, $H_{\{1\}}$ is not rejected by the closed testing procedure, even though $p_1 \leq \alpha$, because the local test failed to reject $H_{\{1,4\}}$.

To find confidence bounds $\underline{r}_S^{1-\alpha}$, consider as an example the set $S = \{1, 2, 3\}$. We see that the subsets $V \subseteq S$ that are not rejected by the closed testing procedure are the sets $\{1\}$, $\{2\}$ and $\{3\}$, and the empty set. The largest size of unrejected subsets is therefore 1, so that $\underline{r}_S^{1-\alpha} = |S| - 1 = 2$. It is possible that any of the elements of S corresponds to a true null hypothesis since H_1, H_2 and H_3 were not rejected. However, no combination of two units can be simultaneously true, unless the closed testing procedure made an error, since all intersection hypotheses of at least two elements of S were rejected.

The Simes test has a relatively simple structure. One small p-value may immediately imply rejection of the intersection hypotheses corresponding to many sets. For example, in Figure 14.4, the fact that the third ordered p-value $p_3 \leq 2\alpha/3$ immediately implies rejection of the intersection hypothesis for the sets $\{1, 3\}$, $\{2, 3\}$, $\{1, 2, 3\}$, $\{1, 3, 4\}$, $\{2, 3, 4\}$, and $\{1, 2, 3, 4\}$, and since $p_2 \leq p_3$ also $\{1, 2\}$ and $\{1, 2, 4\}$. In fact, in this example $p_3 \leq 2\alpha/3$ implies all rejections made by closed testing. Systematically taking this kind of implications into account, it can be proven Goeman et al. (2019) that with Simes local tests we have

$$\underline{r}_S^{1-\alpha} = \min \left\{ 0 \leq k \leq m_S : \min_{1 \leq i \leq m_S - k} \frac{h}{i} p_{(i+k:S)} > \alpha \right\} \tag{14.11}$$

where m_S is the cardinality of the set S, $p_i = 1 - F_{1,n-1,n\Delta^2}(n\hat{\theta}_i^2)$ is the p-value for testing $H_i^0 \colon \theta_i \in [-\Delta, \Delta]$ against $H_i^- \cup H_i^+ \colon \theta_i \in (-\infty, -\Delta) \cup (\Delta, \infty)$, $p_{(i)}$ is the ith smallest p-value, $p_{(i:S)}$ is the ith smallest p-value among those in S and

$$h = \max \left\{ 0 \leq i \leq m : i p_{(m-i+j)} > j\alpha \text{ for } j = 1, \ldots, i \right\}. \tag{14.12}$$

If the Simes inequality (Simes, 1986) holds for the subset of p-values of true null hypotheses, then for all $\theta = (\theta_1, \ldots, \theta_m)$ we have

$$\mathrm{P}_\theta(r_S \geq \underline{r}_S^{1-\alpha} \text{ for all } S) \geq 1 - \alpha. \tag{14.13}$$

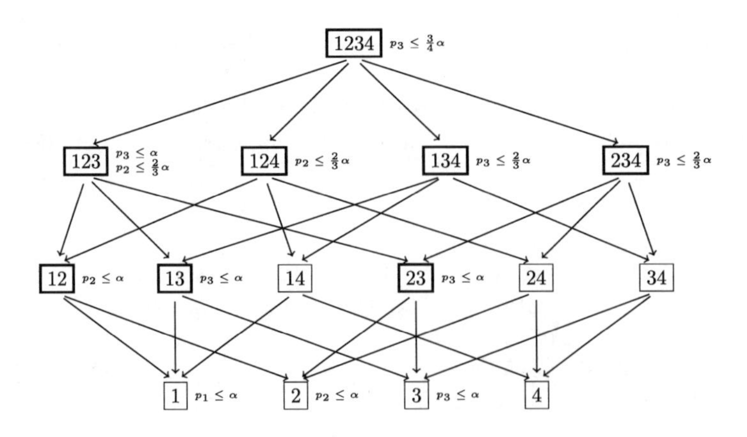

FIGURE 14.4
Example of non-consonant rejections in Hommel's procedure based on $\alpha/2 < p_1 \leq p_2 \leq p_3 \leq 2\alpha/3$, and $p_4 > \alpha$. Nodes in the graph represent intersection hypotheses, labeled by their corresponding index set. Edges represent subset relationships. Hypotheses rejected by the Simes test have the reasons for their rejection given next to them. Hypotheses rejected by the closed testing procedure are marked in bold.

The closed testing procedure as described above can be used in the situation that we have a single hypothesis per spatial unit, for example if we are interested in counting relevant effects, without making distinction between positive and negative. If we want to count positive and negative effects separately, or if we also want to count irrelevant effects $\underline{i}_S^{1-\alpha}$ by replacing p-values p_i with $1 - p_i$ in computing (14.11) and (14.12), we may combine the results of two closed testing procedures performed at level $\alpha/2$. If we want to count positive, negative and irrelevant effects separately, we perform three such procedures at level $\alpha/3$.

For illustration, suppose we have obtained $\underline{r}_S^{1-\alpha/2}$ and $\underline{i}_S^{1-\alpha/2}$ by performing two closed testing procedures at level $\alpha/2$. By using the relationship $r_S = |S| - i_S$, it is possible to derive *simultaneous two-sided confidence limits* for the number of relevant effects, i.e.

$$\mathrm{P}_\theta(\underline{r}_S^{1-\alpha/2} \leq r_S \leq \bar{r}_S^{1-\alpha/2} \text{ for all } S) \geq 1 - \alpha \tag{14.14}$$

where $\bar{r}_S^{1-\alpha/2} = |S| - \underline{i}_S^{1-\alpha/2}$ is the upper bound for the number of relevant effects. Note that (14.14) can be equivalently expressed as two-sided confidence limits for the number of irrelevant effects, i.e.

$$\mathrm{P}_\theta(\underline{i}_S^{1-\alpha/2} \leq i_S \leq \bar{i}_S^{1-\alpha/2} \text{ for all } S) \geq 1 - \alpha$$

where $\bar{i}_S^{1-\alpha/2} = |S| - \underline{r}_S^{1-\alpha/2}$. The construction of two-sided confidence limits described above is similar in spirit to two-sided tests (or intervals) at level α that arise from the combination of two one-sided tests (or upper and lower bounds) at level $\alpha/2$.

14.4 A toy illustration: alphabet theory

The simultaneous approach gives great flexibility to the researcher. This flexibility is a great boon, but also a complication. The confidence bounds give a report on a bewildering 2^m sets, and it would be impossible to report them all. How should a researcher choose sets and bounds to report? In this section we illustrate this on the basis of the toy example of Figure 14.3, illustrating both theory-driven and data-driven ways to choose sets S to report.

Interesting regions may be *data-driven* or *knowledge-based* (or some mix of the two), and we will give examples of both approaches. Usually knowledge-based regions are not uniquely determined a priori: it is likely that many theories have been thrown around with different explanations of which type of effects are to be expected. In our toy example, we assume that "Alphabet theory" conjectures that relevant effects manifest themselves only with the shape of the letters of the alphabet A, B, C, …, Z; alternatively, "Sign theory" conjectures that all relevant effects have either positive or negative sign. Such theories may abound in real data examples as well, where anatomical brain regions may take the role of Alphabet theory. In order to shed light on the plausibility of different explanations, data-driven regions are a natural starting point. In the simultaneous inference framework, we can proceed in stages, asking new questions about the same data at each stage, without incurring increased false positive rates as a result.

A typical example of data-driven regions are clusters of contiguous pixels with corresponding point estimates all passing some threshold. Figure 14.5 displays clusters of at least 50 contiguous pixels (Moore neighborhood) obtained with Low, Medium and High (absolute) thresholds of 0.25, 0.5 and 0.8, respectively. Results for the three thresholds are reported in Table 14.1. As can be seen, the true proportions of signal, 70.8%, 98.1% and 100%, are estimated at about 27.4% [4.8%], 47.8% [8.7%] and 70.1% [7.7%] for the

FIGURE 14.5
Clusters of at least 50 contiguous pixels obtained with Low (left), Medium (centre) and High (right) thresholds. Black and grey pixels indicate positive and negative estimates, respectively.

TABLE 14.1
Post hoc confidence measures: estimate and lower bound for the amount (number and proporti of relevant effects for the overall region, in the unknown regions with only relevant and irreleva effects, and in the regions defined by clusters with Low, Medium and High thresholds and their complementary regions.

	Everything			
	$m = 10000$			
relevant	*number*	*proportion*		
r_S	2633	26.3%		
$\hat{r}_S$ [$\underline{r}_S$]	1045 [171]	10.4% [1.7%]		
$\hat{\dot{r}}_S$ [$\underline{\dot{r}}_S$]	55 [7]	0.5% [0.1%]		
Selection	**All truly relevant effects**		**All truly irrelevant effects**	
	$m_1 = 2633$		$m_0 = 7367$	
relevant	*number*	*proportion*	*number*	*proportion*
r_S	2633	100%	0	0%
$\hat{r}_S$ [$\underline{r}_S$]	942 [171]	35.8% [6.5%]	0 [0]	0% [0%]
$\hat{\dot{r}}_S$ [$\underline{\dot{r}}_S$]	55 [7]	2.1% [0.3%]	0 [0]	0% [0%]
Selection	**Low threshold**		**Low threshold complement**	
	$m_L = 3597$		$m_{L^c} = 6403$	
relevant	*number*	*proportion*	*number*	*proportion*
r_S	2548	70.8%	85	1.3%
$\hat{r}_S$ [$\underline{r}_S$]	986 [171]	27.4% [4.8%]	0 [0]	0% [0%]
$\hat{\dot{r}}_S$ [$\underline{\dot{r}}_S$]	55 [7]	1.5% [0.2%]	0 [0]	0% [0%]
Selection	**Medium threshold**		**Medium threshold complement**	
	$m_M = 1971$		$m_{M^c} = 8029$	
relevant	*number*	*proportion*	*number*	*proportion*
r_S	1934	98.1%	699	8.7%
$\hat{r}_S$ [$\underline{r}_S$]	942 [171]	47.8% [8.7%]	0 [0]	0% [0%]
$\hat{\dot{r}}_S$ [$\underline{\dot{r}}_S$]	55 [7]	2.8% [0.4%]	0 [0]	0% [0%]
Selection	**High threshold**		**High threshold complement**	
	$m_H = 284$		$m_{H^c} = 9716$	
relevant	*number*	*proportion*	*number*	*proportion*
r_S	284	100%	2349	24.2%
$\hat{r}_S$ [$\underline{r}_S$]	199 [22]	70.1% [7.7%]	716 [56]	7.8% [0.6%]
$\hat{\dot{r}}_S$ [$\underline{\dot{r}}_S$]	24 [4]	8.5% [1.4%]	31 [3]	0.3% [0.03%]

Low, Medium and High threshold, respectively. While the High threshold estimate 70.1% is accompanied by a lower bound of only 7.7%, the Medium threshold estimate 47.8% seems more reliable with a lower bound of 8.7%.

If we look at the amount of signal left out of the selection, i.e. at the dual quantity $\hat{r}_{S^c}$ [$\underline{r}_{S^c}$], we find that the numbers 85, 699 and 2349 are estimated by 0 [0], 0 [0] and 716 [56]. There is no evidence of signal outside the selection with Low and Medium thresholds, but strong evidence with the High threshold.

If we select the unknown regions of size $m_1 = 2633$ and $m_0 = 7367$ containing all the signal and all the noise, respectively, the procedure returns an estimated number of relevant effects of 942 [171] and 0 [0]. For the "correct" region containing all the signal, the estimated proportion is only 35.8% [6.5%] compared to 47.8% [8.7%] for the Medium threshold region. This is not surprising, as we should expect tighter bounds for data-driven regions with boosted significance. Note that the procedure gives no evidence of signal outside the correct region, as hoped. If the data-generating process satisfies the assumptions required by the procedure, a correct conclusion will happen in at least 50% and 95% of the times for the estimate and the lower bound, respectively.

Considering everything, i.e. no selection, may provide useful information. The overall number of relevant effects, 2633, is estimated at 1045 [171]. However, as anticipated, it is much easier to conclude on the amount of significant effects rather than to pinpoint where they are. In Table 14.1, $\hat{\dot{r}}_S$ [$\underline{\dot{r}}_S$] denote the number of effects in S that can be pinpointed exactly, i.e. the Hommel rejections at levels 50% and 5%, respectively. Overall, only 55 [7] relevant effects can be perfectly localized.

According to the "Sign theory", relevant effects have either positive or negative sign. The congruence of such theory with the data can be checked by looking at the two regions defined by positive and negative point estimates. For the region of size 5168 identified by positive estimates, the estimated number of relevant effects is 491 [63], whereas for the region of size 4832 identified by negative estimates, the estimated number of relevant effects is 169 [9]. Consequently, we can rule out the "Sign theory" because the results strongly indicate that relevant effects are both with positive and negative sign.

In the light of the results of Table 14.1, the researcher may find a good fit for the P-shaped pattern corresponding to the Medium threshold and recall the "Alphabet theory" as a potential explanation. Uncertain whether the pattern of signal is actually P-shaped or possibly R-shaped as the Low threshold may insinuate, or maybe something else, the researcher can try with all the letters of the alphabet. Table 14.2 reveals that the most likely letter is indeed P. The researcher may also check if a letter's position is well calibrated by considering a circular shift of the letter's position by $1, 2, \ldots, 99$ locations, until it returns at its original position. For the letter P, Figure 14.6 suggests the most likely position is indeed the original.

We have presented a simplistic representation of a "real world" problem in the hope of explaining the proposed approach concisely. In exploratory multiple testing, the researcher uses the data to select promising patterns and (re)uses the same data to obtain valid lower bounds for the amount of relevant effects. All the 50% and 95% lower bounds presented in Table 14.1, Table 14.2 and Figure 14.6 are uniformly correct, although sometimes very conservative. With reference to this matter, statistical power evaluations may help to avoid underpowered studies.

The R code used to generate the toy data example is available in the supplementary information. The analysis was performed by using the `hommel` package (Goeman et al., 2017) for the R software environment (R Core Team, 2018), and customized for fMRI data in R package `ARIbrain`.

TABLE 14.2
Estimate and lower bound for the proportion of relevant effects present in the letters of the alphabet and their complement.

S	P	F	R	E	B	H	D	S	C
r_S	100	84.8	72.8	70.6	63.4	63	56.2	56.2	53.1
$\hat{r}_S$	35.8	27.3	25	23.2	20.4	18.1	17	15	13.9
$\underline{r}_S$	6.5	4.4	4.2	3.6	3	1.9	2.6	2.1	1.8
r_{S^c}	0	9.4	4.4	8.7	7.2	12.9	12.6	15.5	17.6
$\hat{r}_{S^c}$	0	1.8	0.3	1.4	1.2	3.1	2.6	3.8	4.7
$\underline{r}_{S^c}$	0	0	0	0	0	0.3	0.1	0.3	0.4

S	U	G	L	N	K	Z	V	T	O
r_S	49.9	49.5	49.3	46.7	45.5	45.5	43.1	41.7	40.2
$\hat{r}_S$	11.6	13.9	7.9	12.1	11	11.7	7.2	7.7	9.6
$\underline{r}_S$	0.7	1.9	0.2	1.1	1.2	1.8	0.1	1	1.1
r_{S^c}	18.5	16.4	21.9	17.3	19	19.4	21.6	22.7	20.4
$\hat{r}_{S^c}$	5.5	3.8	7.6	4.5	5.5	5.4	7.4	7.8	5.9
$\underline{r}_{S^c}$	0.8	0.2	1.1	0.6	0.6	0.3	1.2	0.8	0.6

S	Q	Y	A	X	J	M	I	W
r_S	37.3	34.9	32.5	31.6	30.7	24.7	23.7	22
$\hat{r}_S$	9.0	4	8.3	5.2	3.5	4.6	2.6	5.3
$\underline{r}_S$	1.0	0.1	0.9	0.2	0.1	0.1	0.1	0.5
r_{S^c}	21.2	24.3	24.3	24.5	25.5	27.4	26.7	28.8
$\hat{r}_{S^c}$	6.1	9	7.2	8.3	9.2	8.8	9.8	8.5
$\underline{r}_{S^c}$	0.6	1.3	0.8	1.1	1.3	1.5	1.4	1.1

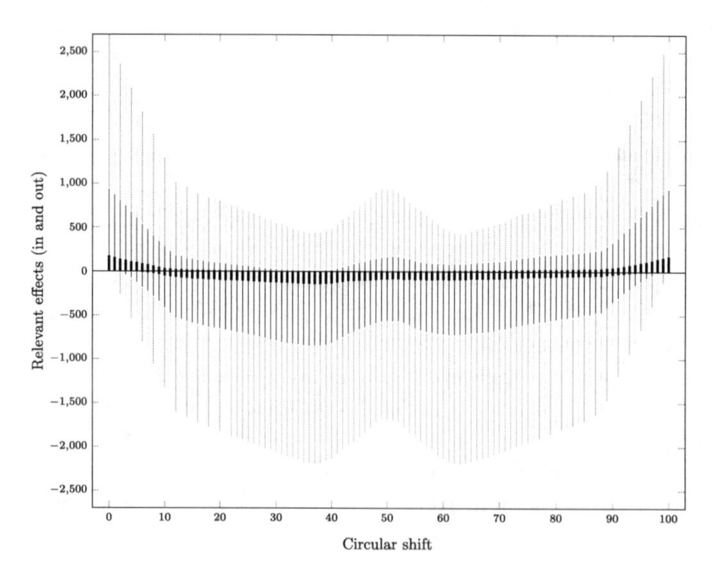

FIGURE 14.6
Amount of signal present in the letter P and outside the letter P (negative sign) as a function of its circular shit of $0, 1, 2, \ldots, 99$ locations, until it returns at its original position $100 = 0$, given twice in the figure. Estimate (thin line) and lower bound (thick line) for the true amount of signal (light line).

14.5 Human Connectome Project Data

Functional magnetic resonance imaging (fMRI) is a modern brain imaging measurement technology used to explore brain activity by obtaining several images of the blood-oxygen-level-dependent (BOLD) signal over time.

For a real-data illustration, we consider a sample of $n = 80$ unrelated subjects from the Human Connectome Project database (HCP, S1200 Release; principal investigators: Bruce Rosen, Arthur W. Toga, Van J. Weeden).

Participants are presented with blocks of trials that either ask them to decide which of two shapes presented at the bottom of the screen match the shape at the top of the screen, or which of two faces presented on the bottom of the screen match the face at the top of the screen, where the faces have either an angry or fearful expression. Information on the task paradigm is given in Hariri et al. (2006). Full details on how the data were acquired and processed is given in Barch et al. (2013) and Glasser et al. (2013).

The focus of our analysis is on the "Face vs Shape" contrast. Subject-specific regression models are fitted, relating the fMRI time series to the design matrix at each voxel separately. An estimate of the contrast coefficient is obtained from the fit. As there is one contrast per voxel, a collection of m voxel-specific estimates is stored as 3D image for each subject. For each voxel, averaging estimates across subjects and dividing by their standard deviation results in the point estimate $\hat{\theta}_i = \bar{x}_i/s_i$ for the 'average' contrast parameter θ_i, $i = 1, \ldots, m$.

We have a large-scale spatial problem, structured in a cube of $91 \times 109 \times 91$ voxels, each voxel of size $2 \times 2 \times 2$ mm. By excluding background voxels, the number of effective voxels is $m = 190794$.

We calculate p-values $p_i = 1 - F_{1,n-1,n\Delta^2}(n\hat{\theta}_i^2)$ for testing $H_i^0 : \theta_i \in [-\Delta, \Delta]$. About the joint distribution of $(p_1, \ldots, p_m)$, we assume that the distribution of $(p_i, i \in I)$ with $I = \{i \in \{1, \ldots, m\} : \theta_i \in [-\Delta, \Delta]\}$ satisfies the Simes' inequality. This assumption is required for the lower bounds $\underline{r}_S^{1-\alpha}$ on the number of relevant effects. Similarly, for p-values $1 - p_i$ testing $H_i^- \cup H_i^+ : \theta_i \in (-\infty, -\Delta) \cup (\Delta, \infty)$, Simes' inequality is assumed for the distribution of $(1 - p_i, i \in R)$ with $R = \{1, \ldots, m\} \setminus I$. This assumption is required for the lower bounds $\underline{i}_S^{1-\alpha}$ on the number of irrelevant effects.

If we are interested in counting either relevant (positive and negative) or irrelevant effects, the closed testing should be performed at level α. If we want to count both relevant and irrelevant effects, we may combine the results of two closed testing procedures performed at level $\alpha/2$.

Conventional value for the relevance boundary Δ are 1/5, 1/2 and 4/5 for an effect of small, medium and large magnitude, respectively (Cohen, 1988). Table 14.3 shows the estimated overall proportion of relevant and/or irrelevant effects as a function of the relevance boundary.

TABLE 14.3
HCP data: estimated overall proportion of relevant and irrelevant effects as a function of the relevance boundary Δ.

	Only relevant	**Only irrelevant**	**Both relevant and irrelevant**	
Δ	$\underline{r}^{50\%}$ [$\underline{r}^{95\%}$]	$\underline{i}^{50\%}$ [$\underline{i}^{95\%}$]	[$\underline{r}^{97.5\%}, \bar{r}^{97.5\%}$]	[$\underline{i}^{97.5\%}, \bar{i}^{97.5\%}$]
0	31.8% [15.3%]	0% [0%]	[13.1%, 100%]	[0%, 86.9%]
1/5	8.7% [5.2%]	1.6% [0%]	[4.7%, 100%]	[0%, 95.3%]
1/2	2.0% [1.3%]	76.5% [56.3%]	[1.2%, 50%]	[50%, 98.2%]
4/5	0.4% [0.2%]	94.4% [89.6%]	[0.2%, 11.8%]	[88.2%, 99.8%]

For the commonly used zero boundary $\Delta = 0$, we see that the estimated proportion of 31.8% of activations is large enough to breed suspicion of undesired 'universal activations' (Bowring et al., 2019) with larger sample sizes. On the other hand, confident conclusions for irrelevant effects can not be established.

At the other extreme, for the large boundary $\Delta = 4/5$, we see that relevant effects are still detectable by Simes tests and that irrelevant effects are ubiquitous among the brain.

As expected, the analysis of both relevant and irrelevant effects given by the formula (14.14) at level $\alpha/2 = 2.5\%$ is slightly less powerful than the individual analyses at level $\alpha = 5\%$. In the following we will explore the intermediate case of the Medium boundary $\Delta = 1/2$ by performing two closed testing at level $\alpha/2 = 5\%$ for counting relevant and irrelevant effects.

14.5.1 Concentration set

A good starting point for a data-driven exploration of regions containing "interesting" voxels is given by the *concentration set* Goeman et al. (2019). The concentration set estimates the locations of relevant effects R and irrelevant effects I by $\hat{R} = \{i \in \{1,\ldots,m\} : p_i < \hat{c}_r\}$ and $\hat{I} = \{i \in \{1,\ldots,m\} : p_i > \hat{c}_i\}$, respectively, where $\hat{c}_r$ and $\hat{c}_i$ are data-dependent p-value thresholds Goeman et al. (2019). All interesting results are "concentrated" in $\hat{R}$ and $\hat{I}$ because the concentration set is the smallest set of voxels with the largest estimate for the number of interesting (relevant or irrelevant) voxels. The residual region $\hat{U} = \{1,\ldots,m\} \setminus \{\hat{R} \cup \hat{I}\}$ represents the uncertainty: it contains voxels for which there is not enough evidence to reach a confident conclusion.

Applied to the HCP data, we obtain $\hat{c}_r = 0.0001$ and $\hat{c}_i = 0.992$. Results for the regions obtained by the concentration set are reported in Table 14.4 and Figure 14.7.

The relevant and irrelevant regions obtained by the concentration set are likely to include mostly relevant and irrelevant voxels, respectively, although $\hat{R}$ may contain at most 13.9% of irrelevant voxels, or $\hat{I}$ at most 11.3% of relevant ones.

The larger region is the irrelevant $\hat{I}$, covering more than half of the brain. The uncertainty region $\hat{U}$ covers about 1/3 of the brain, implying that we can not draw conclusions on 1/3 of voxels. Although $\hat{R}$ covers only 1.5% of the brain, it contains only voxels with positive estimates $\hat{\theta}_i > 0$ and almost half of the relevant voxels in $\hat{R}$ can be localized exactly ($\dot{\underline{r}} = 42.4\%$), compared to hundred thousand found but only one thousand localized irrelevant voxels in $\hat{I}$ ($\dot{\underline{i}} = 1\%$).

14.5.2 Brain atlas

An atlas of the brain provides locations and names of interesting structures and subdivisions of the brain, and this form of domain-knowledge can be used to drive interpretations of the results.

TABLE 14.4
HCP data: relevant, irrelevant and uncertain regions obtained by the concentration set.

Region	**Size**	**% brain**	% of $\hat{\theta}_i > 0$	$[\underline{r}, \bar{r}]$	$[\underline{i}, \bar{i}]$
$\hat{R}$	2876	1.5%	100%	[86.1%, 100%]	[0%, 13.9%]
$\hat{I}$	121201	63.5%	44%	[0%, 11.3%]	[88.7%, 100%]
$\hat{U}$	66715	35%	54%	[0%, 100%]	[0%, 100%]

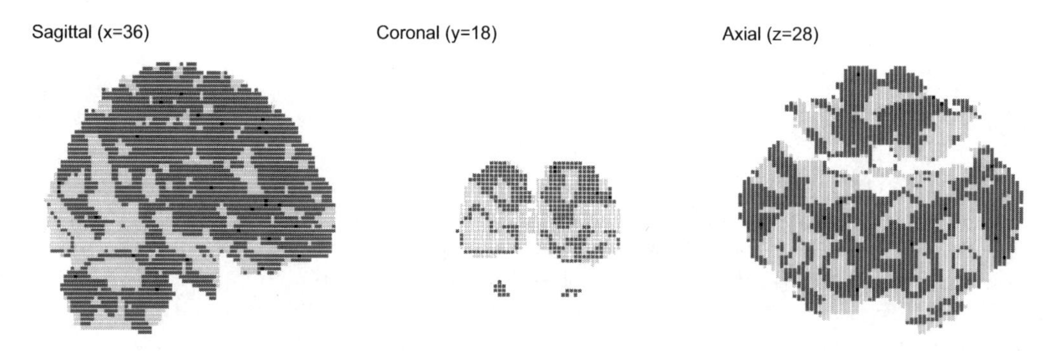

FIGURE 14.7
HCP data: slice views of the concentration set for relevant (dark), irrelevant (light) and uncertain (grey) regions.

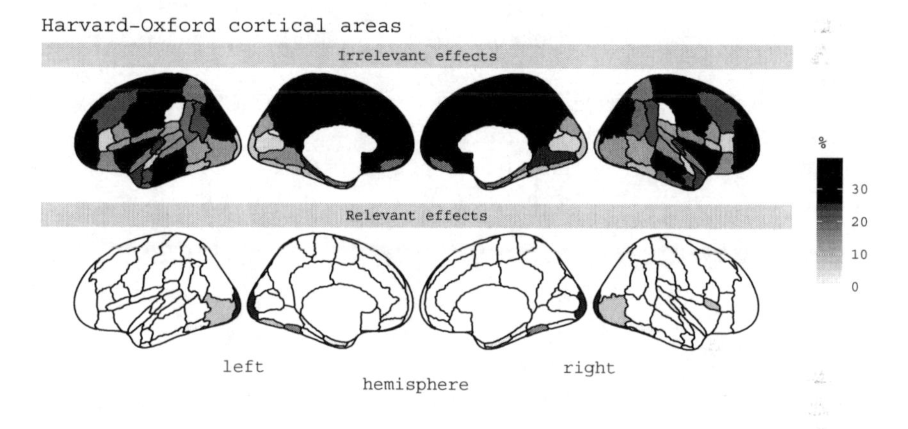

FIGURE 14.8
HCP data: percent of relevant and irrelevant effects found in the Harvard-Oxford cortical areas.

For the present analysis we use the probabilistic Harvard-Oxford Atlas (HOA) at 0% probability threshold, developed at the Center for Morphometric Analysis (CMA) and distributed with the FMRIB Software Library (FSL); see https://identifiers.org/neurovault.image:1704. The HOA contains 48 cortical structural areas covering $m_{\text{HOA}} = 146666$ voxels. Results for each area are displayed in Figure 14.8 and Table 14.4.

The task involving matching shapes and emotionally negative seems to engage a consistent activation in Occipital Pole (793 voxels, 25.2%) and Temporal Occipital Fusiform Cortex (375 voxels, 15.4%), a moderate activation in Lateral Occipital Cortex inferior division (326 voxels, 7.3%) and Occipital Fusiform Gyrus (238 voxels, 7.2%) and low or very low activation in Parahippocampal Gyrus anterior division (27 voxels, 1.3%), Inferior Temporal Gyrus temporooccipital part (14 voxels, 0.9%) and Temporal Fusiform Cortex posterior division (5 voxels, 0.2%).

Areas with large proportions of irrelevant effects are Precentral Gyrus (38.5%), Superior Frontal Gyrus (37.6%), Subcallosal Cortex (37.5%), Middle Temporal Gyrus posterior division (36.4%), Frontal Pole (35.8%), Cingulate Gyrus posterior division (34.9%) and Postcentral Gyrus (34%).

14.5.3 Clusterwise analysis

In this section we illustrate how to find patterns of relevant effects by using both data and domain knowledge. We start with a data-driven clusterwise analysis. Seven clusters are obtained by using a large cluster-forming threshold $\hat{\theta}_i > 0.8$ and requiring a minimum of at least 50 contiguous voxels.

Results are displayed in Table 14.5 and Figure 14.9. We see a very large cluster $\hat{C}_1$ of size 4488 with 67.3% [49.5%] of activations, two smaller clusters $\hat{C}_3$ and $\hat{C}_4$ of sizes 196 and 175 with 35.7% [24.5%] and 36% [26.9%] of activations and the remaining four clusters $\hat{C}_2$, $\hat{C}_5$, $\hat{C}_6$ and $\hat{C}_7$ with very low or no activations. Selecting all the 5568 voxels within the seven clusters, we find evidence for 65.5% [44.3%] of activations; outside these clusters there is evidence for just 2 [1] activations.

We increase the cluster-forming threshold to the concentration set level, i.e. $\hat{\theta}_i > 1.008$, and we selected only the voxels belonging to the five cortical areas that showed at least 25 activations (Occipital Pole, Temporal Occipital Fusiform Cortex, Lateral Occipital Cortex inferior division, Occipital Fusiform Gyrus and Parahippocampal Gyrus anterior division).

This results in 4 subclusters with an increased proportion of activations. The large cluster $\hat{C}_1$ splits into two subclusters $\hat{C}_1'$ and $\hat{C}_1''$ with 97.3% [81%] and 96% [73.5%] of activations spanning Occipital Pole, Temporal Occipital Fusiform Cortex, Lateral Occipital Cortex inferior division and Occipital Fusiform Gyrus.

Drilling-down $\hat{C}_3$ and $\hat{C}_4$ gives the subclusters $\hat{C}_3'$ and $\hat{C}_4'$ with 79.2% [41.7%] and 74.2% [51.6%] of activations within the Parahippocampal Gyrus anterior division.

Results obtained in Barch et al. (2013) seem consistent with the ones provided here, showing bilateral activation of the amygdala and visual cortex including the fusiform.

TABLE 14.5
HCP data: clusters and subclusters of relevant voxels.

Cluster	sub	Size	$\hat{r}$ [$\underline{r}$] number	$\hat{r}$ [$\underline{r}$] percent	$\hat{\dot{r}}$ [$\underline{\dot{r}}$] number	$\hat{\dot{r}}$ [$\underline{\dot{r}}$] percent
$\hat{C}_1$		4488	3020 [2220]	67.3% [49.5%]	1568 [1142]	34.9% [25.4%]
	$\hat{C}_1'$	1457	1417 [1180]	97.3% [81%]	976 [732]	67% [50.3%]
	$\hat{C}_1''$	977	938 [718]	96% [73.5%]	557 [399]	57% [40.8%]
$\hat{C}_2$		378	13 [2]	3.4% [0.5%]	7 [2]	1.9% [0.5%]
$\hat{C}_3$		196	70 [48]	35.7% [24.5%]	52 [39]	26.5% [19.9%]
	$\hat{C}_3'$	24	19 [10]	79.2% [41.7%]	12 [10]	50% [41.7%]
$\hat{C}_4$		175	63 [47]	36% [26.9%]	53 [34]	30.3% [19.4%]
	$\hat{C}_4'$	31	23 [16]	74.2% [51.6%]	21 [14]	67.7% [45.2%]
$\hat{C}_5$		133	0 [0]	0% [0%]	0 [0]	0% [0%]
$\hat{C}_6$		132	2 [0]	1.5% [0%]	2 [0]	1.5% [0%]
$\hat{C}_7$		66	0 [0]	0% [0%]	0 [0]	0% [0%]

TABLE 14.6
HCP data: percent of relevant and irrelevant effects in the Harvard-Oxford cortical structural areas.

Cortical structural area	Size	$\hat{r}$	$\underline{r}$	$\hat{i}$	$\underline{i}$
Occipital Pole	3148	29.4	25.2	17.7	6
Temporal Occipital Fusiform Cortex	2438	20.3	15.4	19.4	7.3
Lateral Occipital Cortex inferior division	4473	11	7.3	26.1	12.1
Occipital Fusiform Gyrus	3322	12.3	7.2	18.3	7
Parahippocampal Gyrus anterior division	2017	2.2	1.3	37.1	17
Inferior Temporal Gyrus temporooccipital part	1628	2.8	0.9	22.1	7.7
Temporal Fusiform Cortex posterior division	2083	1.7	0.2	21.9	6.1
Inferior Frontal Gyrus pars opercularis	1671	0.1	0.1	30.4	12.1
Supramarginal Gyrus anterior division	1905	0	0	4.1	0.2
Temporal Fusiform Cortex anterior division	570	0	0	15.6	0.2
Supracalcarine Cortex	424	0	0	18.2	2.4
Superior Temporal Gyrus anterior division	621	0	0	21.4	4
Intracalcarine Cortex	2211	0.2	0	13.7	4.1
Inferior Frontal Gyrus pars triangularis	1545	0.1	0	15.9	4.3
Parietal Operculum Cortex	1682	0	0	25.7	10.3
Middle Temporal Gyrus temporooccipital part	2458	0	0	29	10.4
Frontal Operculum Cortex	1030	0	0	36	10.5
Cuneal Cortex	1499	0	0	34.5	11.3
Planum Temporale	1424	0	0	32.7	12.1
Central Opercular Cortex	2543	0	0	37.1	13
Lingual Gyrus	4922	0.1	0	29.3	13.3
Superior Temporal Gyrus posterior division	2108	0	0	33.5	13.4
Frontal Orbital Cortex	3325	0	0	34.6	13.7
Planum Polare	1181	0	0	41.5	15.2
Superior Parietal Lobule	2939	0	0	35.7	16.6
Frontal Medial Cortex	1237	0	0	43.4	17.5
Angular Gyrus	2832	0	0	43.4	17.6
Inferior Temporal Gyrus anterior division	674	0	0	50.9	19.7
Supramarginal Gyrus posterior division	2940	0	0	44.7	21.2
Middle Temporal Gyrus anterior division	826	0	0	57.4	21.3
Middle Frontal Gyrus	5645	0.1	0	44.8	21.6
Heschl's Gyrus	786	0	0	58.4	23.3
Temporal Pole	2668	0.1	0	46.4	23.4
Parahippocampal Gyrus posterior division	1954	0	0	56.1	24.8
Lateral Occipital Cortex superior division	8079	0	0	50.1	26.4
Inferior Temporal Gyrus posterior division	2092	0	0	55.2	26.9
Paracingulate Gyrus	4042	0	0	54.5	28.3
Cingulate Gyrus anterior division	4144	0	0	55.6	28.9
Insular Cortex	3389	0	0	57.2	29
Juxtapositional Lobule Cortex	1830	0	0	62.5	29.3
Precuneous Cortex	6805	0	0	57.3	30.2
Postcentral Gyrus	6614	0	0	59.5	34
Superior Frontal Gyrus	5525	0	0	63.1	34.9
Subcallosal Cortex	1594	0	0	65.4	35.8
Middle Temporal Gyrus posterior division	2842	0	0	63.4	36.4
Frontal Pole	12429	0	0	62.1	37.5
Cingulate Gyrus posterior division	4666	0	0	63.4	37.6
Precentral Gyrus	9886	0	0	64.5	38.5

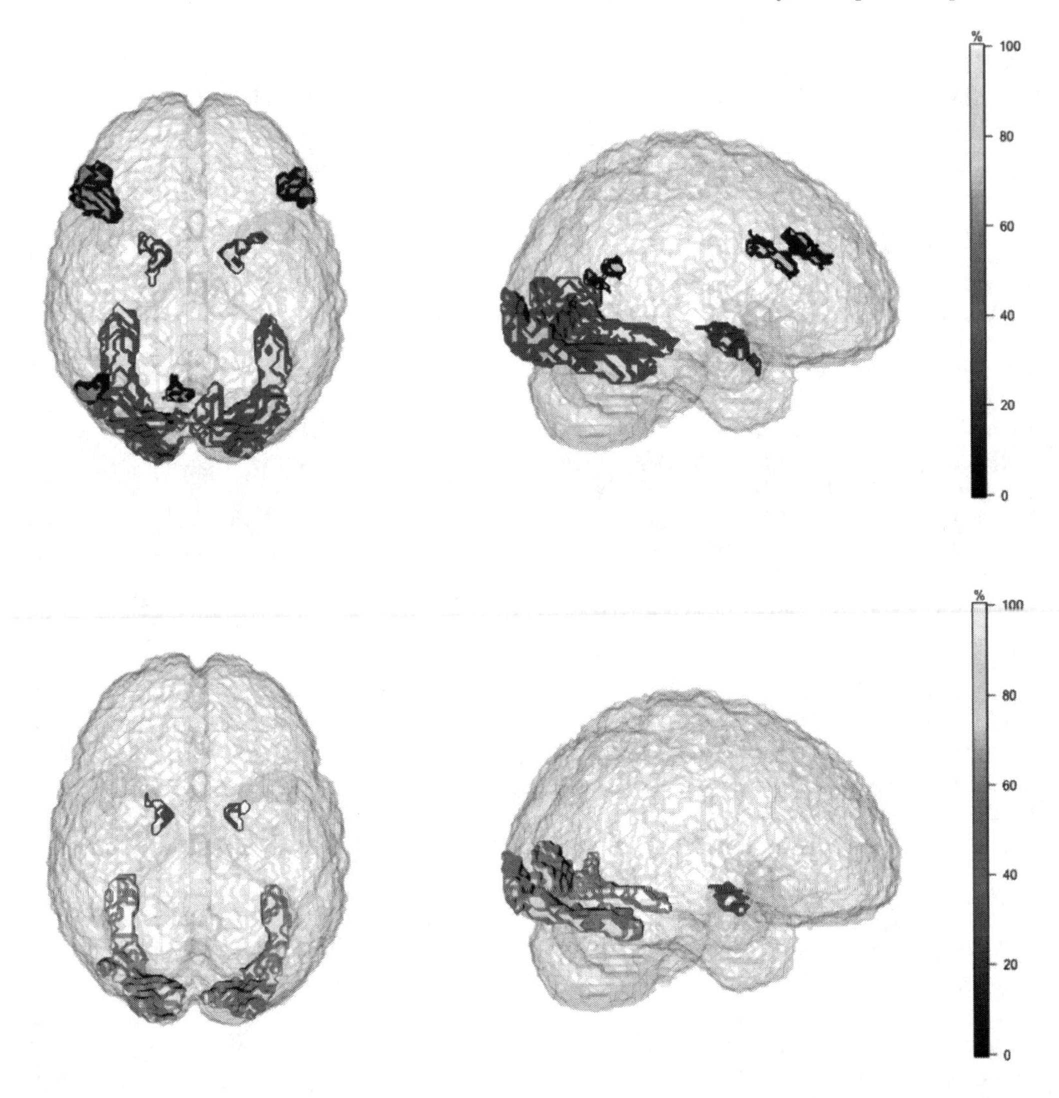

FIGURE 14.9
HCP data: 3D visualisation (left/right: view from above/side) of clusters obtained with a large threshold $\hat{\theta}_i > 0.8$ (top) and the concentration set threshold $\hat{\theta}_i > 1.008$ and belonging to selected cortical areas (bottom).

14.6 Conclusion

In large-scale hypothesis testing, (i) statistical significance should match practical significance and (ii) selection is unavoidable, and must be taken into account. On the one hand, although underused, the theoretical results enabling confident conclusions about effect sizes have been known for a very long time (Wellek, 2017). On the other hand, significant progress has been made recently on the development of statistical procedures for assessing the strength of evidence after selection (Taylor and Tibshirani, 2015). In this Chapter, existing solutions to both (i) and (ii) have been combined in a unified framework.

Our approach uses contextual information at two levels: it defines the minimum effect size of real interest, and it drives the user's exploration of the data. This allows users to explore where the truly interesting effects in the data can be found. The simultaneous approach allows a truly interactive approach to selective inference, that does not set any limits on the way the researcher chooses to perform the selection. The selection process does not have to be declared beforehand; it does not have to be well-described; it may be circular; it does not even have to be repeatable. The selection may be data-driven or knowledge-driven, or any mix of the two. Regardless of the selection process used, the researcher obtains a valid confidence bound for the number of proportion of truly relevant units in the final selected set. We have illustrated the way a user may approach the selection on the basis of a nueroimaging data set and on a toy example inspired by neuroimaging data. However, as recently argued, the same type of exploratory inference may also be valuable in genomics (Ebrahimpoor et al., 2020).

A recurring theme in this Chapter has been that we may gain power in test procedures if we are content with less detailed answers. If we are only interested in knowing whether an effect is at least meaningfully large (i.e., at least Δ in absolute value), we have more power if we take on that question directly, rather than by way of the regular confidence interval, because that interval gives more information than we need. Our approach adheres to principle for solving problems using a restricted amount of information formulated by Vladimir Vapnik (2013): "when solving a problem, try to avoid solving a more general problem as an intermediate step." If we are interested in counting truly relevant effects are present in a region, we should not try to pinpoint those effects. The latter question gives more information than we need, and would sacrifice power to produce irrelevant knowledge.

Acknowledgments

The authors are grateful to Thomas Nichols for providing help with Human Connectome Project Data and to anonymous reviewers for useful suggestions. Data were provided [in part] by the Human Connectome Project, WU-Minn Consortium (Principal Investigators: David Van Essen and Kamil Ugurbil; 1U54MH091657) funded by the 16 NIH Institutes and Centers that support the NIH Blueprint for Neuroscience Research; and by the McDonnell Center for Systems Neuroscience at Washington University.

Bibliography

Barch, D. M., G. C. Burgess, M. P. Harms, S. E. Petersen, B. L. Schlaggar, M. Corbetta, M. F. Glasser, S. Curtiss, S. Dixit, C. Feldt, et al. (2013). Function in the human connectome: task-fmri and individual differences in behavior. *Neuroimage 80*, 169–189.

Benjamini, Y. (2010). Simultaneous and selective inference: Current successes and future challenges. *Biometrical Journal 52*, 708–721.

Benjamini, Y. and Y. Hochberg (1995). Controlling the false discovery rate: a practical and powerful approach to multiple testing. *Journal of the Royal Statistical Society. Series B (Methodological) 57*(1), 289–300.

Benjamini, Y., Y. Hochberg, and P. B. Stark (1998). Confidence intervals with more power to determine the sign: two ends constrain the means. *Journal of the American Statistical Association 93*, 309–317.

Benjamini, Y. and D. Yekutieli (2005). False discovery rate: Adjusted multiple confidence intervals for selected parameters. *Journal of the American Statistical Association 100*, 71–81.

Berk, R., L. Brown, A. Buja, K. Zhang, and L. Zhao (2013). Valid post-selection inference. *Annals of Statistics 41*(2), 802–837.

Blanchard, G., P. Neuvial, and E. Roquain (2020, 06). Post hoc confidence bounds on false positives using reference families. *Ann. Statist. 48*(3), 1281–1303.

Bowring, A., F. Telschow, A. Schwartzman, and T. E. Nichols (2019). Spatial confidence sets for raw effect size images. *NeuroImage 203*, 116187.

Cohen, J. (1988). *Statistical Power Analysis for the Behavioral Sciences.* Abingdon, Oxfordshire: Routledge.

Cox, D. R. (1965). A remark on multiple comparison methods. *Technometrics 7*, 223–224.

de Groot, A. D. (2014). The meaning of "significance" for different types of research. *Acta Psychologica 148*, 188–194.

Ding, Y., Y. G. Li, Y. Liu, S. J. Ruberg, and J. C. Hsu (2018). Confident inference for snp effects on treatment efficacy. *Ann. Appl. Stat. 12*(3), 1727–1748.

Ebrahimpoor, M., P. Spitali, K. Hettne, R. Tsonaka, and J. Goeman (2020). Simultaneous enrichment analysis of all possible gene-sets: unifying self-contained and competitive methods. *Briefings in bioinformatics 21*(4), 1302–1312.

Finner, H. and K. Strassburger (2002). The partitioning principle: a powerful tool in multiple decision theory. *Annals of Statistics*, 1194–1213.

Gelman, A. (2016). The problems with p-values are not just with p-values. *The American Statistician, supplemental material to the ASA statement on p-values and statistical significance. 10.*

Gelman, A. and E. Loken (2013). The garden of forking paths: Why multiple comparisons can be a problem, even when there is no fishing expedition or p-hacking and the research hypothesis was posited ahead of time. *Technical report.*

Genovese, C. R. and L. Wasserman (2006). Exceedance control of the false discovery proportion. *Journal of the American Statistical Association 101*(476), 1408–1417.

Glasser, M. F., S. N. Sotiropoulos, J. A. Wilson, T. S. Coalson, B. Fischl, J. L. Andersson, J. Xu, S. Jbabdi, M. Webster, J. R. Polimeni, et al. (2013). The minimal preprocessing pipelines for the human connectome project. *Neuroimage 80*, 105–124.

Goeman, J., J. Hemerik, and A. Solari (2020). Only closed testing procedures are admissible for controlling false discovery proportions. *Annals of Statistics (to appear).*

Goeman, J., R. Meijer, and T. Krebs (2017). *hommel: Methods for Closed Testing with Simes Inequality, in Particular Hommel's Method.* R package version 1.1.

Goeman, J., A. Solari, and T. Stijnen (2010). Three-sided hypothesis testing: Simultaneous testing of superiority, equivalence and inferiority. *Statistics in Medicine 29*, 2117–2125.

Goeman, J. J., R. J. Meijer, T. J. Krebs, and A. Solari (2019). Simultaneous control of all false discovery proportions in large-scale multiple hypothesis testing. *Biometrika 106*(4), 841–856.

Goeman, J. J. and A. Solari (2011). Multiple Testing for Exploratory Research. *Statistical Science 26*(4), 584–597.

Hariri, A. R., S. M. Brown, D. E. Williamson, J. D. Flory, H. de Wit, and S. B. Manuck (2006). Preference for immediate over delayed rewards is associated with magnitude of ventral striatal activity. *Journal of Neuroscience 26*(51), 13213–13217.

Hayter, A. J. and J. C. Hsu (1994). On the relationship between stepwise decision procedures and confidence sets. *Journal of the American Statistical Association 89*(425), 128–136.

Hommel, G. (1988). A stagewise rejective multiple test procedure based on a modified Bonferroni test. *Biometrika 75*(2), 383–386.

Hsu, J. C. and R. L. Berger (1999). Stepwise confidence intervals without multiplicity adjustment for dose-response and toxicity studies. *Journal of the American Statistical Association 94*(446), 468–482.

Katsevich, E. and A. Ramdas (2020). Simultaneous high-probability bounds on the false discovery proportion in structured, regression, and online settings. *Annals of Statistics (to appear)*.

Kelley, K. (2007). Confidence intervals for standardized effect sizes: Theory, application, and implementation. *Journal of Statistical Software 20*, 1–24.

Marcus, R., P. Eric, and K. R. Gabriel (1976). Closed testing procedures with special reference to ordered analysis of variance. *Biometrika 63*(3), 655–660.

Rosenblatt, J. D., L. Finos, W. D. Weeda, A. Solari, and J. J. Goeman (2018). All-resolutions inference for brain imaging. *Neuroimage 181*, 1–11.

Simes, R. J. (1986). An improved Bonferroni procedure for multiple tests of significance. *Biometrika 73*(3), 751–754.

Simmons, J., L. Nelson, and U. Simonsohn (2011). False-positive psychology: Undisclosed flexibility in data collection and analysis allow presenting anything as significant. *Psychological Science 22*, 1359–1366.

Sorić, B. (1989). Statistical ?discoveries? and effect-size estimation. *Journal of the American Statistical Association 84*(406), 608–610.

Stefansson, G., W. Kim, and J. Hsu (1988). On confidence sets in multiple comparisons. In S. Gupta and J. Berger (Eds.), *Statistical Decision Theory and Related Topics IV*, Chapter 2, pp. 89–104. Springer-Verlag, New York, NY.

Su, W. J. (2019). The fdr-linking theorem. *arXiv:1812.08965*.

Swanson, D. (2020). Localizing differences in smooths with simultaneous confidence bounds on the true discovery proportion. *arXiv preprint arXiv:2007.15445*.

Taylor, J. and R. J. Tibshirani (2015). Statistical learning and selective inference. *PNAS 1122*, 7629–7634.

Tukey, J. W. (1960). Conclusions vs decisions. *Technometrics 2*(4), 423–433.

Tukey, J. W. (1991). The philosophy of multiple comparisons. *Statist. Sci. 6*(1), 100–116.

Vapnik, V. (2013). *The nature of statistical learning theory.* Berlin: Springer Science & Business Media.

Wellek, S. (2010). *Testing Statistical Hypotheses of Equivalence and Noninferiority.* Boca Raton, FL: Chapman and Hall/CRC.

Wellek, S. (2017). A critical evaluation of the current ‘p-value controversy’. *Biometrical Journal 59*, 854–872.

15

Testing SNPs in Targeted Drug Development

Ying Ding and Yue Wei and Xinjun Wang

University of Pittsburgh

Jason C. Hsu

The Ohio State University

CONTENTS

15.1 Introduction

Testing single nucleotide polymorphisms (SNPs) in drug development aims to find the right subgroup of patients that has differential treatment effect between a new treatment and a control treatment. This problem is at the heart of modern drug development and personalized medicine. Such a testing procedure is very distinct from the usual association literature related to SNPs, in which researchers seek to discover which SNPs

DOI: 10.1201/9780429030888-17

are associated with disease incidence or progression, e.g., through genome-wide association studies (GWAS). While the latter is important for our understanding of biology, the former is much more important, indeed essential, for understanding how to treat patients with the disease.

Finding SNPs that are predictive through comparing drug efficacy between treatment and control in a randomized controlled trial (RCT) is critical to decide which subgroup (with enhanced treatment effect) to target in drug development. Such development usually proceeds in the following three steps. (1) From epidemiological genetic studies, data bases, and knowledge about drug pathways, a list of candidate SNPs is compiled. (2) Subjects who consent to giving DNA samples in a clinical trial are then tested for these SNPs, to see if any subgroup defined by their alleles has a clinically meaningful enhanced efficacy than the complimentary subgroup. (3) The findings then guide the decision whether to pursue an enriched confirmatory trial or an all-comer confirmatory trial in drug development. Nowadays as the genotyping technologies rapidly enhance, in step (1), some modern RCTs conduct genotyping on the whole genome instead of on a list of pre-selected candidate genes.

Despite of the fundamental difference between the standard association test where no treatment or clinical outcome is involved and testing SNP's predictiveness of treatment efficacy from an RCT, the current practice tends to apply similar statistical approaches from GWAS to RCT's SNP testing. For example, the MAX3 test, originally proposed by Hothorn and Hothorn (2009) and So and Sham (2011) to test for each SNP whether it has a dominant, recessive, or an additive effect, where the maximum test statistic under three different genetic models (dominant, recessive, and additive) is used to denote the significance of a single SNP, has been adapted to test SNP's effect on treatment efficacy. Sometimes, a 2 degrees of freedom F-test from a co-dominant model (a commonly used genetic model with homozygous and heterozygous groups compared to the other homozygous groups separately) or a linear trend test is also used to test the significance of a single SNP.

15.2 Issues in Testing SNPs on Treatment Efficacy

In this section, we present several key statistical issues that are fundamental but sometimes overlooked in current practices for SNPs testing in RCTs.

15.2.1 The p-value versus confidence interval

In clinical trials, it is now understood that establishing mere statistical significance is insufficient. Since the publication of Ruberg (1995), it has gradually become accepted that what matters is a *clinically meaningful effect.* This concept is explicit in non-inferiority and superiority trials. P-values (computed from tests of no effect) cannot provide information on effect size, while confidence intervals can.

Consider two hypothetical studies with the same control and the same design including sample size. Take Type II diabetes for instance. For each patient, the most common clinical response of interest is his/her change in Hemoglobin A1c (HbA1c) from baseline, and the treatment efficacy is measured as the mean difference in clinical response between the new treatment (denoted as Rx) and the control (denoted as C). Suppose the first study gives a confidence interval for reduction in HbA1c of $(0.4, 0.6)$, while the second study gives a confidence interval of $(0.8, 1.2)$. Then, for testing the null hypothesis of no difference between Rx and C, the two studies report the same p-value. One way to see this is the first confidence interval is 0.5 ± 0.1, while the second confidence interval is 1.0 ± 0.2, doubling the point estimate and its standard error. However, clinically, treatment in the second study is far superior to the treatment in the first study as indicated by both the point estimate and its confidence interval. The importance of using confidence intervals for statistical inference is noted in ICH E9 (1997) as well as the FDA guidance on non-inferiority trials, where the former states a preference for confidence intervals and the latter states that non-inferiority trials should be analyzed using confidence intervals.

It may seem surprising that confidence intervals are not routinely produced in biomarker studies. While reasons may be multitude, we offer two plausible explanations in the next two sections.

15.2.2 Control of familywise error rate (FWER)

First, to obtain valid simultaneous confidence intervals, the corresponding multiple test must control FWER strongly. In the context of testing for genetic effect of a SNP, strong FWER control means the probability of incorrectly inferring dominant, recessive, or additive effect of an allele is guaranteed not to exceed level α, an obviously desirable property, especially in the context of drug development. To obtain a strong control of FWER, null hypotheses for multiple tests must be stated as **complements** (opposites) of the alternative hypotheses of genetic effects.

Denote the variants of a SNP by AA, Aa, and aa. Let μ_{AA}, μ_{Aa}, and μ_{aa} be a (pre-determined) suitable efficacy measure for the respective populations with AA, Aa, and aa genotypes. For example, the mean difference of the clinical response between Rx and C if the clinical response is a continuous outcome, can be a suitable efficacy measure. To infer dominant, recessive, or additive effect, one should test H_{01}: a is **not** dominant vs. H_{a1}: a is dominant; H_{02}: a is **not** recessive vs. H_{a2}: a is recessive; etc. In this formulation, inferring a recessive $\mu_{aa} > \mu_{\{Aa,AA\}}$ when the truth is A dominant counts as a Type I error for testing H_{02}, because A dominant is in the null space of H_{02}: a is not recessive $\mu_{aa} \leq \mu_{\{Aa,AA\}}$.

Such a null hypothesis formulation agrees with standard practice (accepted by the industry and the FDA) in clinical studies. In dose-response studies with low, medium and high dose groups, for example, the null hypothesis of interest is not a single overall hypothesis $\mu_L = \mu_M = \mu_H$, but instead a family of hypotheses H_{01}: low dose lacks efficacy $\mu_L < \delta$, H_{02}: medium dose lacks efficacy $\mu_M < \delta$, and H_{03}: high dose lacks efficacy $\mu_H < \delta$, where δ is a difference of interest.

A $100(1-\alpha)\%$ confidence interval readily reflects a level-α test: reject the null hypothesis if and only if the confidence interval does not intersect the null hypothesis. Theorem 4 of Berger et al. (1996) gives a formal proof. Using $100(1-\alpha)\%$ simultaneous confidence intervals to test multiple null hypotheses automatically controls FWER strongly, because covering all the true parameter values with a probability of 1-α implies the probability of rejecting at least one true null hypothesis is no more than α.

On the other hand, weak FWER control only controls the probability of an incorrect dominant, recessive, or additive inference when $\mu_L = \mu_M = \mu_H$. For weak FWER control, if the truth is A is dominant, then incorrectly inferring a recessive does not count as a Type I error. Hothorn and Hothorn (2009) and So and Sham (2011) and others, state the null hypotheses as a "complete null" format H_{00}: $\mu_{AA} = \mu_{Aa} = \mu_{aa}$, against the alternatives of dominant, recessive, or additive effects, so the multiple tests in those papers are meant to control FWER weakly, and cannot be expected to have simultaneous confidence intervals. Weak FWER control may have issues more serious than lack of corresponding confidence intervals, as we illustrate in the next section.

Multiple tests that compute p-values based on permutation are popular. Many such methods claim FWER control without distributional assumption, claiming in addition a power advantage by being able to capture the dependence structure without modeling.

However, Huang et al. (2006) have shown permutation tests cannot control FWER, even weakly, without distributional assumption. The assumption that is most readily checked is the marginal-determine-the-joint (MDJ) assumption in Xu and Hsu (2007), which states that equality of the marginal distributions must guarantee equality in the joint distribution (including equality of covariances). The MDJ condition does *not* hold for linear models without the normality assumption (Calian et al., 2008). When it does not hold, a permutation test can report wildly misleading p-values. Kaizar et al. (2011) shows an example, for the binary response case, where a permutation test can report a p-value of .14 when the true p-value is .99, as the MDJ condition does not hold.

15.2.3 Predictors versus response variables

A second possible reason that confidence intervals are not routinely produced in biomarker studies is that many statistical methods developed for biomarker studies take values of multiple biomarkers (such as gene expression levels or number of reads from RNA-Seq) as *responses*, comparing their values between disease and normal groups, or between responders and non-responders.

This practice may have a historical root. Van't Veer et al. (2002) conducted a pioneering study using gene expression levels for prognosis of breast cancer progression. MammaPrint, a custom microarray built on its findings, became the first FDA-approved microarray diagnostic device. It uses the expression level of 70 genes as prognostic of the likelihood that a patient will relapse. Gene expression levels are the *predictors*, while relapse or no relapse is the clinical *response*. To select good predictors, a standard statistical technique is to regress the response on the predictors. For example, the logistic regression can be used for binary response. However, to select genes for the prognostic classifier, Van't Veer et al. (2002) tested for significant "correlations" between gene expression levels and relapse status (see Ein-Dor et al. (2005) for a detailed description). Correlation analysis treats the roles of phenotype (a vector of zeros and ones, indicating relapse or no relapse) and genotype (a vector of expression levels of that gene) as interchangeable. Their test for significant "correlation" turns out to be equivalent to a two-sample permutation t-test comparing the expression levels of patients who have relapsed and those who have not. As the two-sample t-test is usually thought of as comparing continuous *responses* (gene expression levels in this case) between two discrete groups, this influential paper may contribute to the reason why biomarker values are often thought of as responses.

However, there is no common scale for differential expression levels, since for some genes, a small difference in expression level may produce a big biological change, while for other genes a big change in expression level may produce little biological change. Therefore, the practice of treating biomarker values as responses may have contributed to a lack of interest in confidence intervals.

The emergence of targeted treatment development has made the roles of predictors and responses unambiguous. Such a development process involves a new treatment, a control, with one or more biomarkers that may predict differential treatment efficacy. The roles are clear: clinical measurement such as change in HbA1c is the *response*, while biomarker values such as gene expression or SNP categories are *predictors*. Clinical response is measured on a medically meaningful scale. Confidence intervals for clinical response are the most useful summary of treatment effect.

15.2.4 Additive effect for clinical response

In the traditional association test of SNPs for quantitative traits, the linear trend test is frequently used. For such a trend test to hold, the difference between Aa and AA, and between aa and Aa, must be exactly the same. This is what "additive effect" usually means for the model that tests the genotypic difference of quantitative traits. However, given that the treatment efficacy is already a "relative" effect, such as the difference between Rx and C, and within each treatment, the clinical effect is often measured by a difference between the endpoint and baseline, it seems extremely unlikely that these differences will be exactly the same between Aa and AA, and between aa and Aa. In drug development, instead of exact additivity, the order of the genetic groups in terms of treatment efficacy is more relevant. Therefore, it is more reasonable to consider the additive effect for clinical response to be an increasing ordering among three genetic subgroups: $\mu_{aa} > \mu_{Aa} > \mu_{AA}$, i.e., a co-dominant effect. This suggestion has also been proposed in Ding et al. (2018).

Figure 15.1 gives a demonstration of possible treatment efficacy scenarios across three genetic groups. In fact, the first four scenarios, denoted by minor allele a having (1) complete null effect, (2) exact additive effect, (3) dominant effect, and (4) recessive effect, are very rare in the clinical response setting, while the last four scenarios (5-8) are a lot more likely to happen in practice.

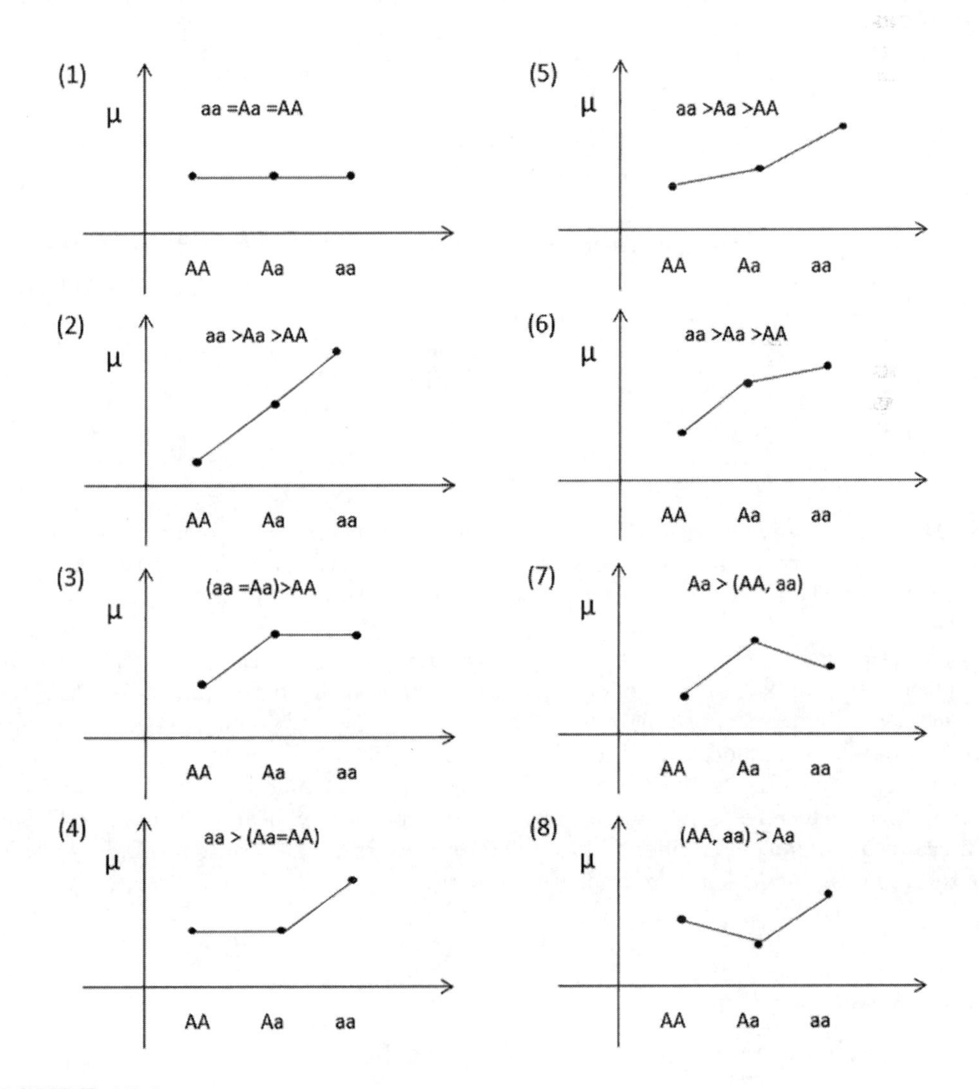

FIGURE 15.1
Possible treatment efficacy scenarios across three genetic groups AA, Aa and aa.

15.2.5 Least squares means versus marginal means

The above issues are fundamental in SNP testing for modern drug development and are regardless of what statistical model to fit the data or what estimation procedure to use for the inference. Additional issues remain in the detailed data analysis, for example, how to appropriately estimate the possible dominant, recessive or additive effect in subgroups and their combinations after a model being fitted.

In the setting of continuous clinical response, such as the change in HbA1c in Type II diabetes, or the change in Positive and Negative Syndrome Scale (PANSS) in schizophrenia, a linear model with independent normal errors is usually fitted. Denote by $Y_{i,h,r}$ the clinical response for the rth individual from the ith treatment group with genotype h, where $i = Rx$ or C, $h = AA, Aa, aa$, $r = 1, ..., n_{ih}$. Let μ_h denote the efficacy in group h (e.g., difference in mean response between Rx and C in subgroup h, i.e., $\mu_h = \mu_{Rx,h} - \mu_{C,h}$). To estimate dominant and recessive effects, it is important to use least squares (LS) means as opposed to marginal (MG) means when computing the efficacy for a combined subgroup. The LSmeans approach first estimates efficacy within each subgroup which is

$$\hat{\mu}_h^{LS} = (n_{Rx,h}^{-1} \sum_{r=1}^{n_{Rx,h}} Y_{Rx,h,r}) - (n_{C,h}^{-1} \sum_{r=1}^{n_{C,h}} Y_{C,h,r}), \; h = AA, Aa, aa,$$

and then combines the subgroups of interest weighted by their prevalence. For example, the LSmeans estimate for the combined subgroup $(Aa,\ aa)$ is

$$\frac{\pi_{Aa}}{\pi_{Aa} + \pi_{aa}} \hat{\mu}_{Aa}^{LS} + \frac{\pi_{aa}}{\pi_{Aa} + \pi_{aa}} \hat{\mu}_{aa}^{LS},$$

where π_h denote the population prevalence of genotype group h. While the MGmeans approach first combines data among subgroups of interest, and then estimate efficacy in the combined groups. For example, the marginal mean estimate for the combined subgroup $\{Aa,\ aa\}$ is $\hat{\mu}_{\{Aa,aa\}}^{MG} = \hat{\mu}_{Rx,\{Aa,aa\}}^{MG} - \hat{\mu}_{C,\{Aa,aa\}}^{MG}$, where

$$\hat{\mu}_{i,\{Aa,aa\}}^{MG} = (n_{i,Aa} + n_{i,aa})^{-1} (\sum_{r=1}^{n_{i,Aa}} Y_{i,Aa,r} + \sum_{r=1}^{n_{i,aa}} Y_{i,aa,r}), \; i = Rx \text{ or } C.$$

The MGmeans approach is incorrect as it does not account for imbalance between Rx and C across different subgroups and Simpson's Paradox can happen with this approach. Therefore, one should use LSmean estimates for estimating the dominant, recessive, and additive effects.

Some existing methods and packages use LSmeans but assume equal genotype prevalence. If a package does not ask for user's input of genotype prevalence, the user should check whether the package uses MGmeans, or LSmeans assuming equal prevalence. In either case, the method or package should be modified appropriately.

The above illustration of LSmeans is for continuous outcome fitted by a linear model with i.i.d. normal errors. In practice, we found that the LSmeans technique has been indiscriminately applied on different types of outcomes, which can cause misleading results. We refer to Chapter 13 for more in-depth discussion.

15.3 Methods that Test the Complete Null

Since the publication of articles such as Marcus et al. (1976), by the 1990s, tests of multiple hypotheses have stopped being formulated as tests of the complete null against multiple specific alternatives. As we illustrate in Section 15.2, a test of the complete null against a specific alternative can reject for the wrong reason (e.g., if neither the complete null hypothesis nor the specific alternative is true), and thus controlling the Type I error rate under the complete null does not necessarily control the FWER.

A possible reason why some authors have formulated the null hypotheses as $H_{00} : \mu_{AA} = \mu_{Aa} = \mu_{aa}$ is that it happens to be the intersection of the null hypotheses H_{01}: allele a is not dominant, H_{02}: allele a is not recessive, etc., and testing the intersection hypothesis is the first step in standard multiple test construction. This is similar to pairwise comparisons, where $H_{00} : \mu_{AA} = \mu_{Aa} = \mu_{aa}$ is the intersection of $H_{001} : \mu_{AA} = \mu_{Aa}$, $H_{002} : \mu_{AA} = \mu_{aa}$, and $H_{003} : \mu_{Aa} = \mu_{aa}$. But, as with pairwise comparisons, after H_{00} is rejected, careful further analyses are needed to control FWER strongly to infer specific effects. See Hsu (1996) section 5.1.6 for an example of a mistake in some standard statistical software.

Here we illustrate potential statistical issues with three popular complete null testing methods when used in the targeted treatment development process.

15.3.1 The F-test

First, consider the 2 degrees of freedom F-test studied in Lettre et al. (2007). The F-test tests the complete null against the existence of a non-zero contrast. Its rejection implies the existence of a contrast which can be statistically inferred to be non-zero, $(\frac{1}{2}\mu_{Aa} + \frac{1}{3}\mu_{AA}) - \frac{5}{6}\mu_{aa} > 0$, for example. Such a contrast may not be biologically interpretable. If one pivots the F-test to Scheffé's confidence set, then one may find none of the biologically meaningful effects is significant. If, whenever the F-test rejects, one forces an inference such as 'A' is dominant even though Scheffé's method does not find it significant, then any meaningful error rate control is lost. In drug development, correct inference matters, with an incorrect inference possibly worse than no inference at all. For example, when the truth is 'a' dominant, then inferring 'a' recessive leads to targeting only the aa subgroup, missing out on targeting the Aa subgroup. Worse, if the truth is 'A' dominant, then inferring 'a' recessive leads to targeting the aa subgroup instead of the complementary $\{AA, Aa\}$ subgroup. In the simulation study of Lettre et al. (2007), if the F-test rejects due to either of the incorrect inferences above, it is counted positively toward "power", which is very misleading.

15.3.2 The linear trend test

As we illustrated in Section 15.2, the linear trend statistic tests the complete null against an "additivity" alternative. In the SNP setting with a continuous clinical response, it is defined as $\mu_{Aa} - \mu_{AA} = \delta > 0$ and $\mu_{aa} - \mu_{AA} = 2\delta$, coding the genotype as a continuous variable with values $X = 0, 1, 2$ for AA, Aa and aa, respectively. Even in the setting of a single-arm study (i.e., just the treatment, without a control), the trend test can mislead. In the example depicted in Figure 15.2, the a allele has a dominant beneficial effect on change from baseline of the Positive Syndrome Score (PSS), which is the positive scale part of the PANSS, so the effect is non-linear. Yet the linear trend test is highly significant (p-value < 0.001 in Table 15.1). The linear trend test merely tests for non-zero slope, forcing a straight line to be fitted through the data, even if the response profile is clearly non-linear (as the lack-of-fit test shows in Table 15.1). The linear trend test can reject the complete null hypothesis of a flat response for a myriad of non-flat alternatives.

Some practice uses an alternative to the linear trend test, which is to test the contrast $\mu_{aa} - \mu_{AA}$ with H_0: $\mu_{aa} - \mu_{AA} \leq 0$ and H_a: $\mu_{aa} - \mu_{AA} > 0$ (e.g., in page 144 of Dean and Voss (1998)). We use this test to illustrate the fact that the linear trend test is not always more powerful than the contrast test.

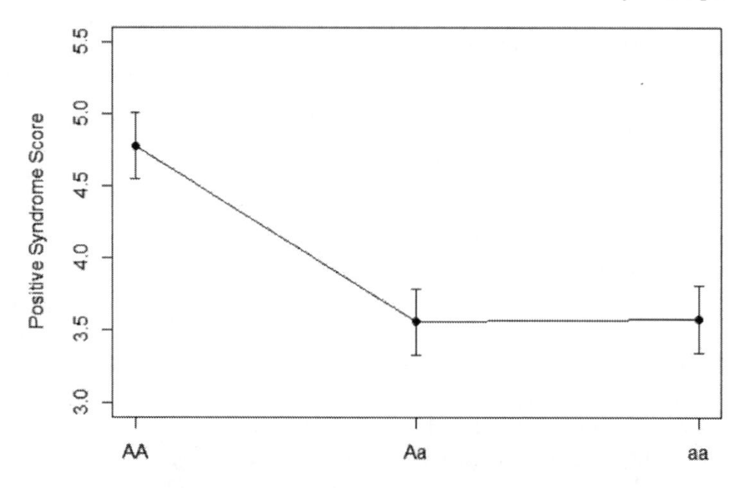

FIGURE 15.2
The mean ± SE for three genotype groups (AA, Aa, aa) of Positive Syndrome Score in an allele a dominant effect scenario.

TABLE 15.1
Model fit and linear trend test for the Positive Syndrome Score data.

Source	d.f.	Sum of Squares	Mean Squares	F-statistic	p-value
slope	1	36.720	36.720	13.27	< 0.001
Residual	151	417.755	2.767		
Lack of fit	**1**	**13.070**	**13.070**	**4.84**	**0.29**
Pure error	**150**	**404.685**	**2.698**		
Total	152	454.475			

When AA, Aa and aa are coded as -1, 0, and 1, in the situation of equal sample size of aa group and AA group, these two tests have equivalent parameter estimates, with the second test yielding a parameter estimate exactly twice of the slope estimate from the linear regression. This is because when aa and AA have the same sample size, the slope estimator from linear regression is given by $(\overline{Y}_{aa} - \overline{Y}_{AA})/2$ which is free of $\overline{Y}_{Aa}$, where $\overline{Y}_h$, $h = AA, Aa, aa$, is the average response within each genotype group. Only the intercept estimator is affected by $\overline{Y}_{Aa}$ (given by $\overline{Y}$, the average of all the responses). However, these two tests are not equivalent. Besides the fact that the second test can produce a confidence interval while the first one cannot, they do not produce exactly the same test statistics. For the data depicted in Table 15.1, we can see that F-statistic from the linear trend test is smaller than that from the contrast $\mu_{aa} - \mu_{AA}$ ($13.27 = 36.720/2.767$ vs. $13.61 = 36.720/2.698$). The reason is that the residual error from the linear regression contains two parts: lack of fit and pure error; and the linear trend test uses the entire residual error when computing the test statistic while the contrast test for $\mu_{aa} - \mu_{AA}$ uses only the "pure error" portion. In general, unless the sum of squares of lack of fit is zero, these two tests are not equivalent and the test for contrast $\mu_{aa} - \mu_{AA}$ is more "powerful" as compared to the linear trend test, when sample sizes of the aa and AA groups are the same or not very different.

The explanation above also shows that the linear trend test cannot differentiate between trend (i.e., additive with $\mu_{aa} - \mu_{Aa} = \mu_{Aa} - \mu_{AA}$), co-dominance (i.e., complete ordering with $\mu_{aa} > \mu_{Aa} > \mu_{AA}$ or $\mu_{aa} < \mu_{Aa} < \mu_{AA}$), and super-dominance (i.e., heterozygous group is better or worse than both homozygous groups). When sample sizes of the aa and AA groups are the same, changing the value of $\overline{Y}_{Aa}$ toward either $+\infty$ or $-\infty$ will not affect the slope estimate, but will increase the lack of fit error, and hence increase the overall residual error, leading the linear trend test to fail to reject. This indicates, in general, the linear trend test cannot differentiate between extreme super-dominance from a flat response, pivoting to confidence intervals of $(-\infty, \infty)$ for $\mu_{Aa} - \mu_{AA}$ and $\mu_{aa} - \mu_{Aa}$, which are non-informative. When sample sizes of the aa and AA groups are rather different, the linear trend test may reject more often than the test for $\mu_{aa} - \mu_{AA}$ contrast.

15.3.3 The MAX3 test

In traditional association studies, researchers are interested in testing whether there exist associations between quantitative traits and SNPs, assuming some commonly used genetic models: dominant, recessive or additive model. However, the true model is rarely known a priori, and model mis-specification can lead to a loss of power. For example, in a GWAS which screens millions of SNPs, different effects may be observed at varying loci. Without knowing the true model for each locus, as stated by Zheng et al. (2006), the commonly used additive model can be inefficient in detecting dominant or recessive effect. One intuitive approach is to consider the maximum of the test statistics based on dominant, recessive and additive models (MAX3), proposed by Freidlin et al. (2002) and González et al. (2008). Since the test statistics from three genetic models are not independent, Lettre et al. (2007) proposed a permutation-based procedure to get the p-value from the MAX3 model. Later, some analytical approximation approaches have been proposed to avoid permutation and to allow for covariates (Li et al., 2008) and different types of quantitative traits (Zhang and Li, 2015).

None of these approaches can be directly applied to SNP testing in RCTs where treatment arms are involved. Following the logic of MAX3, people may consider to formulate the test by using the statistic from testing the "treatment-by-SNP" interaction from each of the three genetic models and use the maximum of the three interaction test statistics to represent the SNP's effect. However, this MAX3 type of approach for testing SNP's predictiveness of treatment efficacy is problematic, as each genetic model still tests a complete null versus a specific alternative, and it is highly likely none of the specific alternatives or the complete null is true in the RCT setting for studying treatment efficacy.

15.4 Confident Effect through 4 Contrasts (CE4)

To appropriately take care of the issues raised in Section 15.2, we propose a novel formulation for the SNP testing problem in RCTs by using contrasts. Specifically, the method will provide simultaneous confidence intervals for dominant, recessive, and additive effects, on the clinical response, characterized by four contrasts.

15.4.1 Construct contrasts

To follow the genetic model convention, we code AA, Aa, aa as 0, 1, 2. Suppose having the 'a' allele is beneficial and we want to evaluate the effect size with respect to each effect: dominant, recessive or additive. The contrasts of interest are formulated as:

$$\begin{aligned} \theta_{(1,2):0} &= \mu_{\{Aa,aa\}} - \mu_{AA} \\ \theta_{2:(0,1)} &= \mu_{aa} - \mu_{\{AA,Aa\}} \end{aligned} \qquad (15.1)$$

$$\theta_{1:0} = \mu_{Aa} - \mu_{AA}$$
$$\theta_{2:1} = \mu_{aa} - \mu_{Aa},$$

where we use two contrasts ($\theta_{1:0}$ and $\theta_{2:1}$) to assess the additive effect and use $\mu_{\{Aa,aa\}}$, $\mu_{\{AA,Aa\}}$ to denote the efficacy in the combined $\{Aa, aa\}$ and $\{AA, Aa\}$ group, respectively.

Note that this formulation is general, independent of the outcome type, the definition of the efficacy measure μ, and the specific model that will be used to fit the data. However, the choice of the efficacy measure when patient population is a mixture of subgroups with heterogeneous treatment efficacy can be non-trivial. Some commonly used efficacy measures such as the Odds Ratio for the binary outcome, or the Hazard Ratio for the survival outcome are not suitable. This topic is carefully discussed in Chapter 13. Here, we assume that a suitable efficacy measure has already been chosen, and the efficacy on the combined groups has been also correctly formulated. For example, in the situation where the clinical response is continuous, such as the change of HbA1c from baseline in Type II diabetes, the difference in mean change between treatment and control, i.e., $\mu_g = \mu_g^{Rx} - \mu_g^C$, g denotes the genetic subgroup, is a suitable efficacy measure. In this case, the efficacy for the combined group is the weighted average of the efficacy from each subgroup, for example, $\mu_{\{Aa,aa\}} = (\pi_{Aa}\mu_{Aa} + \pi_{aa}\mu_{aa})/(\pi_{Aa} + \pi_{aa})$, where π_g denote the population prevalence of each genotype group. Again, continuous outcome is more straight-forward. For binary or survival outcome, we refer to Chapter 13 and Ding et al. (2016) and Lin et al. (2019) for detailed discussions.

Similarly, we can write out the parameters of interest if having the 'A' allele is beneficial:

$$\begin{aligned} \theta_{(0,1):2} &= \mu_{\{AA,Aa\}} - \mu_{aa}, \\ \theta_{0:(1,2)} &= \mu_{AA} - \mu_{\{Aa,aa\}}, \\ \theta_{1:2} &= \mu_{Aa} - \mu_{aa}, \\ \theta_{0:1} &= \mu_{AA} - \mu_{Aa}. \end{aligned} \tag{15.2}$$

It is easy to notice that the set of parameters in (15.2) is just the negative of the set of parameters in (15.1) (e.g., $\theta_{(0,1):2} = -\theta_{2:(0,1)}$). Therefore, the two-sided confidence intervals for the first set of parameters are sufficient to tell which allele is beneficial and the possible effect size with regard to each effect. For example, if the lower bounds of the four simultaneous confidence intervals of the parameters in (15.1) are all greater than zero, then it indicates that the 'a' allele is beneficial and the effects could be dominant, recessive and/or additive. Note that in the situation where the clinical response is continuous, the dominant, recessive and additive effects are not mutually exclusive.

To test the additive effect, instead of using two contrasts, one may consider using only one contrast to test for additivity. One draw-back of using three contrasts to assess clinical efficacy is that the complete ordering of the three genotype groups AA, Aa, and aa may not be fully determined. For example, the following data (Figure 15.3) may suggest $\mu_{aa} > \mu_{Aa} > \mu_{AA}$ since all the lower bounds of three simultaneous confidence intervals are greater than zero. However, the two vectors $(-0.06, 0.91, 1.15)$ and $(-0.06, 1.15, 0.91)$ (corresponding to the estimated subgroup efficacy for AA, Aa, and aa respectively) are both within the confidence region for $(\mu_{aa}, \mu_{Aa}, \mu_{AA})$. So co-dominance $\mu_{aa} > \mu_{Aa} > \mu_{AA}$ cannot be inferred in this case.

One may also consider using five contrasts $(\theta_{(1,2):0}, \theta_{2:(0,1)}, \theta_{1:0}, \theta_{2:1}, \theta_{2:0})$ to assess clinical response. The simultaneous confidence intervals can be computed similarly, while the confidence region will be closer to the confidence region from Scheffé's method corresponding to the F-test when the number of contrasts increases. We recommend using four contrasts to assess the clinical efficacy since it is the most parsimonious formulation of the problem that aligns with the biological hypotheses of interest.

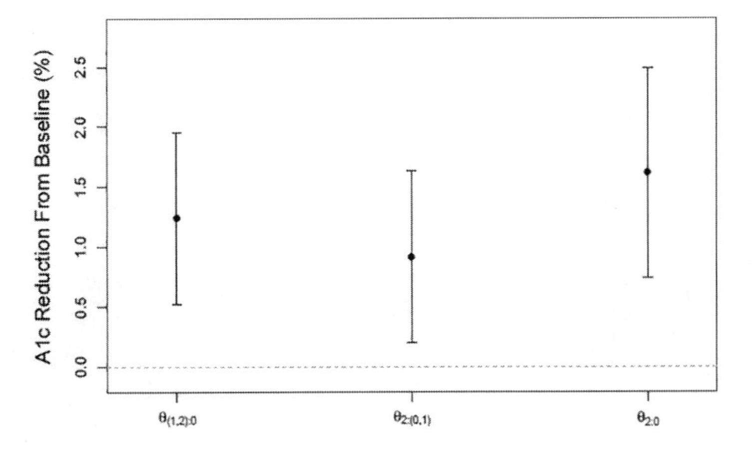

FIGURE 15.3
95% simultaneous confidence intervals for the three contrasts in HbA1c reduction from baseline.

15.4.2 Simultaneous confidence intervals

It has been proved that using simultaneous confidence intervals to test against the null hypotheses in (15.1) controls FWER strongly (see Theorem 4 of Hsu and Berger (1999)). Moreover, besides providing information on the size of the clinical effects, an essential advantage of the confidence interval formulation is that, since they are constructed from pivotal statistics, false coverage of test quantities can be controlled, regardless of whether their true values are zero or not. This advantageous feature has been also stated in Ding et al. (2018).

We propose to use four "two-sided" confidence intervals for the set of contrasts in (15.1) instead of using eight "one-sided" confidence intervals for both sets of contrasts in (15.1) and (15.2), as they are sufficient to indicate which allele is beneficial and the possible effect size of each effect. We name it as the "CE4" (confident effect with 4 contrasts) method.

Geometrically, each of the four equations $\theta_c = 0$ from (15.1) is a plane dividing the 3D efficacy space $(\mu_{AA}, \mu_{Aa}, \mu_{aa})$ into two halves, and each effect is the half space on the positive side of its corresponding plane. The left panel of Figure 15.4 illustrates the half spaces for *a* dominant and *a* recessive. To the left of the vertical plane is *a* recessive since $\mu_{aa} > \mu_{\{AA,Aa\}}$ holds on the left side of the vertical plane. Similarly, below the horizontal plane is *a* dominant since $\mu_{AA} < \mu_{\{Aa,aa\}}$ holds on the bottom side of the horizontal plane. Note that *a* dominant and *a* recessive are not mutually exclusive. The right panel of Figure 15.4 shows all four planes on the 3D space.

The specific form of simultaneous confidence intervals depends on the outcome type, the efficacy measure, and the statistical model used to fit the data. For example, in the continuous outcome case with difference of means (between Rx and C) being the efficacy measure, under the linear regression model with i.i.d. normal errors, the four simultaneous confidence intervals have the following form:

$$\theta_c \in [\hat{\theta}_c^{LS} - qs\sqrt{\nu_{cc}},\ \hat{\theta}_c^{LS} + qs\sqrt{\nu_{cc}}],\ c = \{(1,2){:}0,\ 2{:}(0,1),\ 1{:}0,\ 2{:}1\},$$

where $s^2\nu_{cc}$ is the estimated variance of the least squares estimator $\hat{\theta}_c^{LS}$, and the quantile q is a number such that the joint probability

$$Pr\{|\hat{\theta}_c^{LS} - \theta_c|/(s\sqrt{\nu_{cc}}) < q,\ c = \{(1,2){:}0,\ 2{:}(0,1),\ 1{:}0,\ 2{:}1\} = 1 - \alpha.$$

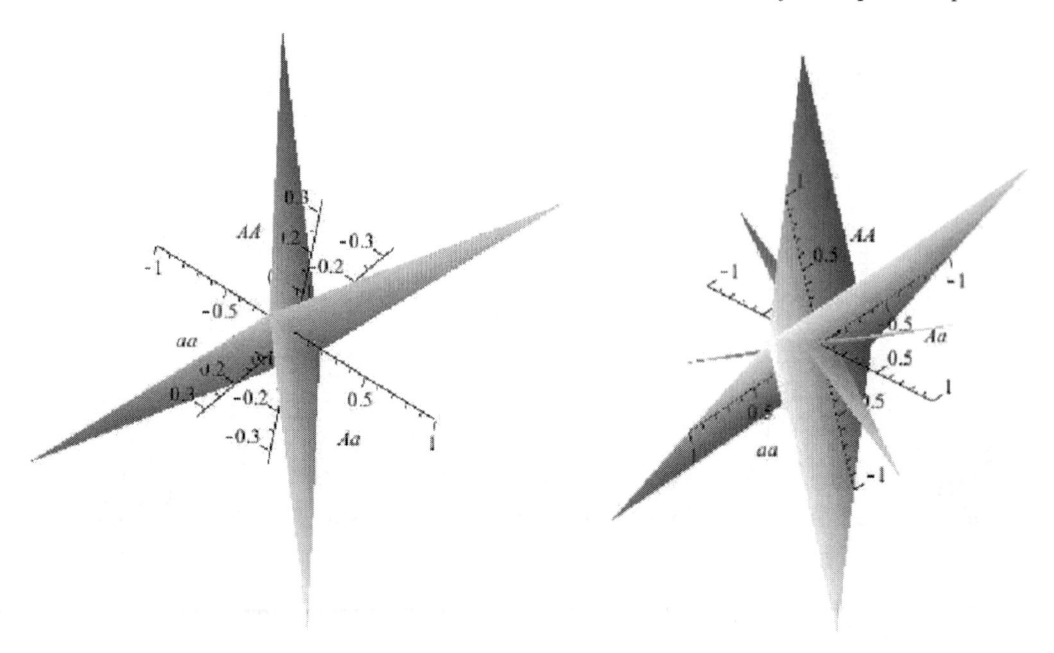

FIGURE 15.4
Left: two planes in the 3D efficacy space to denote the a recessive (left to the vertical plane) and a dominant (below of the horizontal plan) half planes; Right: the four planes that correspond to four CE4 contrasts in (15.1) setting to 0. The aa axis is on the $\mu_{AA} = \mu_{Aa}$ plane, and the AA axis is on the $\mu_{Aa} = \mu_{aa}$ plane.

In this case, the joint statistics $\{\hat{\theta}_c^{LS}\}_c$ have a multivariate t-distribution. Even though the distribution is singular, it is free of unknown parameters, so q can be computed by simulation. For example, the pseudo-Monte-Carlo algorithm of Genz and Bretz (1999), which is applicable to arbitrary correlation structure and is based on the multivariate t-distribution (the *qmvt* function in R), can be used.

The left panel of Figure 15.5 gives the geometric representation of CE4's four two-sided simultaneous confidence sets. In fact, it is a eight-sided cone made by two sets of four planes that correspond to the four lower and four upper bounds of the confidence intervals. As a comparison, the right panel gives Scheffé's confidence set, which is an infinite cylinder. In principle, if the number of contrasts increases, the cone will become closer to the cylinder.

For other types of outcome, the exact distribution for the joint statistics may be intractable. In that case, asymptotic distributions will need to be used. For example, in Wei et al. (2020), they consider a survival type of outcome and demonstrate that the ratio of quantile survival time (between Rx and C) is a suitable treatment efficacy measure. Under the Weibull model they fit, they use (asymptotic) multivariate Normal distribution to derive the simultaneous confidence intervals for their joint statistics, which are based on the differences in logarithm of the ratio of quantile survivals between subgroups and their combinations.

With simultaneous confidence intervals, we provide a meaningful quantity: the probability of making "correct and useful inference" (Appendix C of Hsu (1996)); for example, the probability of correctly inferring 'A' is dominant when the truth is 'A' is dominant. By specifying the null hypotheses as the complements of the alternative and provide the confidence intervals of all the tested effects, such a practice not only guarantees a strong control of FWER (i.e., "correct" inference), but also aligns directly with

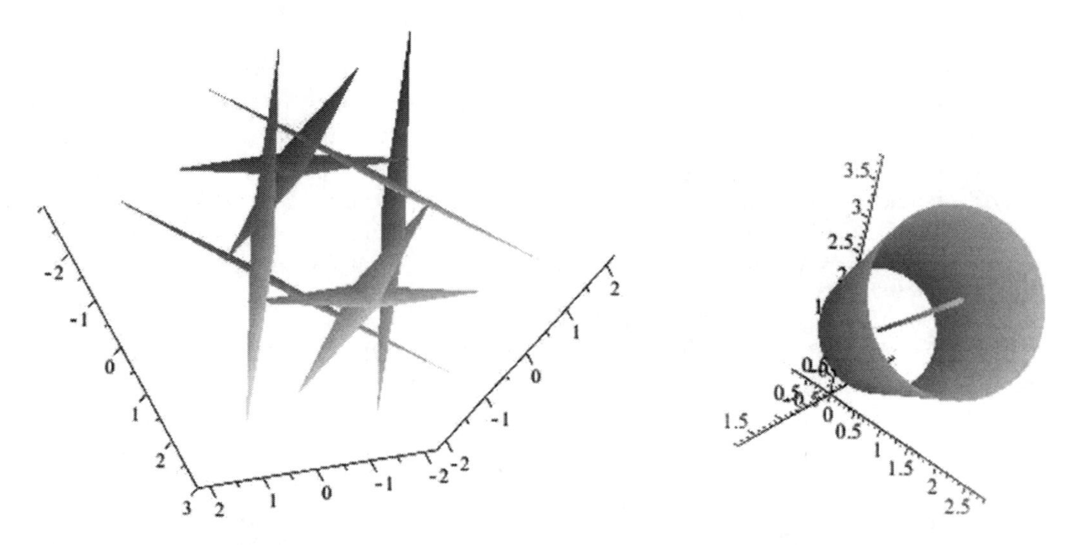

FIGURE 15.5
Left: geometric representation for CE4's simultaneous confidence sets in a 3D space; Right: geometric representation for Scheffé's confidence set.

decision making process (i.e., "useful" inference). In the situation of developing personalized medicine, this "correct and useful inference" will assure a "correct" patient subgroup(s) to target.

15.5 Multiplicity Adjustment across SNPs

In the discovery phase of a targeted treatment development, usually a large number of biomarkers will be tested. Therefore, in addition to adjusting for multiplicity of the contrasts within each SNP, multiplicity across the SNPs needs to be adjusted for as well.

15.5.1 Two types of error rates

There are two types of inferences, within a SNP and across SNPs. The decision rule is to select a SNP if it has at least one confidence interval not covering zero, after the confidence level been adjusted for across-SNP multiplicity. To differentiate between the two types of inferences, we refer to the within SNP inference on (15.1) and (15.2) as a *group* of inferences, and the across SNP inference as a *panel* of inferences. How the *panel* error rate is controlled will relate to how to adjust the confidence level of each *group* inference for multiplicity of the SNPs.

For *group* inference within a SNP, consequence of an incorrect inference is dire for a selected SNP as it may lead to targeting wrong patient population, so a strong FWER control seems appropriate. For inference across multiple SNPs, controlling a less stringent error rate such as *per family* error rate is acceptable since it allows flexibility in the exploration of multiple candidate SNPs and it is very possible that different SNPs may lead to similar patient population for targeting.

Suppose a study consists of testing K SNPs. For inference within the kth SNP, denote by V_k the number of confidence intervals that fail to cover their true values. Let $I_{\{V_k>0\}}$ be the indicator function that at least one of the confidence intervals for the kth SNP fails to cover its true contrast value. Then α_k, the FWER (i.e., *group-wise* error rate) for the kth SNP, is $\alpha_k = P\{V_k > 0\} = E[I_{\{V_k>0\}}]$.

For inference across a panel of K SNPs, let $V_\star$ denote the number of SNPs that have at least one of its confidence intervals failing to cover its true contrast value. Then $E[V_\star]$, the *per panel* error rate, is the expected number of SNPs with some incorrect confidence intervals,

$$E[V_\star] = E\left[\sum_{k=1}^{K} I_{\{V_k>0\}}\right] = \sum_{k=1}^{K} P\{V_k > 0\} = \sum_{k=1}^{K} \alpha_k. \tag{15.3}$$

15.5.2 Additive multiplicity adjustment

Suppose the desired per panel error rate is m. The m value is a pre-specified positive integer, reflecting the number of mistakes that one would tolerate. By (15.3), it is the sum of the group-wise error rates of all SNPs. We suggest a simple adjustment to control the per panel error rate $E[V_\star]$, the *additive* adjustment, setting α_k for each SNP to be $\frac{m}{K}$. This makes the group-wise error rate α_k the same for all SNPs, $k = 1, ..., K$. In practice, SNPs are sometimes pre-separated to different tiers to reflect their likelihood to be the potential candidate biomarkers (based on historic data this far). In that case, different α_k may be assigned to different SNPs, with the top tier SNPs having larger α_k than the second or third tier SNPs.

It is important to emphasize that this is *not* the Bonferroni probabilistic inequality adjustment $\alpha_k = \frac{\alpha}{K}$ for controlling FWER for the panel, as m is a positive integer while α is always a small value such as 0.05. However, this additive adjustment relates to Bonferroni adjustment as follows. To control the per panel FWER at α, the Bonferroni probabilistic inequality adjustment for controlling FWER would set the non-coverage rate for each SNP at $\alpha_k = \frac{\alpha}{K}$, implying $E[V_\star] = K \times \frac{\alpha}{K} = \alpha$. Thus, setting $\alpha_k = \frac{m}{K}$ to allow $E[V_\star] = m$ is equivalent to "relaxing" the Bonferroni multiplicity adjustment by a factor of $\frac{m}{\alpha}$. Take $\alpha = .05$ and $m = 10$ for example. The Bonferroni multiplicity adjustment $\alpha_k = \frac{\alpha}{K}$ allows only $K \times \frac{0.05}{K} = 0.05$ false discoveries on average. While allowing $m = 10$ false discovery on average is to relax the Bonferroni adjustment by a factor of $\frac{m}{\alpha} = \frac{10}{0.05} = 200$.

15.5.3 The p-value under CE4 method

People might be interested whether the proposed CE4 method can produce a p-value for each SNP. The answer is yes, as we illustrate here. Indexing the kth SNP by the superscript $^{(k)}$, let $G^{(k)}$ denote the cumulative distribution function (CDF) of the pivotal quantity

$$T_\star^{(k)} = \max_c \frac{|\hat{\theta}_c^{(k)} - \theta_c^{(k)}|}{s^{(k)}\sqrt{v_{cc}^{(k)}}}, \quad c \in \{(1,2){:}0,\ 2{:}(0,1),\ 1{:}0,\ 2{:}1\},$$

then every $U_\star^{(k)} = G^{(k)}(T_\star^{(k)})$ has a Uniform(0,1) distribution.

Note that in the actual testing procedure, we do not need to compute $U_\star^{(k)}$, which is not computable due to unknown $\theta_c^{(k)}$ (even under the null). Since $V_k > 0$ if and only if $U_\star^{(k)} > 1 - \frac{m}{K}$, so $P\{V_k > 0\} = \frac{m}{K}$, regardless of the dependence among $U_\star^{(k)}$. It is also important to emphasize that since (15.3) is an equality, so the additive multiplicity adjustment is **exact**, not conservative, in controlling the per panel error rate at m.

Define

$$U^{(k)} = G^{(k)}\left(\max_c \frac{|\hat{\theta}_c^{(k)}|}{s^{(k)}\sqrt{v_{cc}^{(k)}}}, c = \{(1,2){:}0,\ 2{:}(0,1),\ 1{:}0,\ 2{:}1\}\right),$$

then $1 - U^{(k)}$ can be thought of as the p-value for the kth SNP. This p-value corresponds to the smallest quantile q, or equivalently the largest α, that makes all the four confidence intervals (within each SNP) still cover zero. Under the normal outcome case, if the simultaneous confidence intervals are computed using the *glht* function in the R package `multcomp`, then this p-value is the smallest (within SNP) "adjusted" p-value from the four contrasts. It is equivalent to the probability calculated by *pmvt* function in R with the lower and upper integration bounds set by $-\max_c \frac{|\hat{\theta}_c^{(k)}|}{s^{(k)}\sqrt{v_{cc}^{(k)}}}$ and $\max_c \frac{|\hat{\theta}_c^{(k)}|}{s^{(k)}\sqrt{v_{cc}^{(k)}}}$, respectively.

15.5.4 Why not controlling false discovery rate (FDR)

People may wonder why we propose to control for *per panel* error rate instead of the commonly used false discovery rate (FDR) in genetic studies. In GWAS, it is plausible *biologically* that the vast majority of the SNPs are not associated with the specific disease. However, when treatments are involved, the biological processes become more complex, and in the next section, we will demonstrate that the complete null (i.e., zero null) hypothesis of no association will be statistically false for all SNPs if it is false for one SNP. This phenomenon was first observed in Ding et al. (2018), where the treatment efficacy was simulated based on a single causal SNP with no random error being added. The SNPs that can cause differential treatment efficacy in different genotype groups are called "causal" SNPs. It was found that all other SNPs also appear "associated" with the clinical outcome when analyzed in a SNP-by-SNP fashion. The reason is that most SNPs are not "orthogonal" to each other, and thus any SNP will appear somewhat associated with the clinical outcome as long as the distribution for proportions of being $\{AA, Aa, aa\}$ in this SNP and in the causal SNP are not independent, which is almost all the cases. With almost all no-association null hypotheses being false if one is false, FDR control which controls the expected *proportion* of false discovery has little meaning. Controlling the *per panel* error rate is more meaningful and interpretable.

15.6 Illustration and Application of CE4

In this section, we use "realistic" simulations to demonstrate how to use our CE4 method to identify patient subgroups in targeted treatment development.

The SNP data we used here are from real subjects, who are from the Age-Related Eye Disease Study (AREDS) data, a randomized clinical trial for an eye disease, the age-related macular degeneration (AMD) (Group et al., 1999). Among those participants who had DNA collected and genotyped, we randomly selected 1000 Caucasian participants and randomly "assigned" them in a 1:1 ratio to the new treatment Rx and a standard care C. We selected two chromosomes, chromosome 19 and 11 for our analysis and we explain the reason for such a choice below.

We simulated outcomes to mimic an Alzheimer's Disease (AD) trial. AD is a devastating illness which, unless a treatment is discovered, is expected to affect 17 million Americans by 2050. Clinical response to AD treatments is usually measured as the reduction in Alzheimer's

Disease Assessment Scale-cognitive (ADAS-cog) from baseline, which is typically modeled linearly with i.i.d. normal random errors. It is well-known that the ApoE4 protein, encoded by the *APOE* gene on chromosome 19, is the best known genetic risk factor for late-onset of AD. Therefore, we selected a variant $rs429358$ from the *APOE* gene region on Chromosome 19 as the causal SNP, and assumed the minor allele of this variant has a dominant or recessive beneficial effect on Rx. We kept the three genotype groups (defined by the causal SNP) balanced between Rx and C. Thus, the SNP data we have are real (from real subjects) and the outcomes are simulated from realistic artificial SNP effects. Finally, the purpose of including chromosome 11 is to have SNPs that are expected to be in less linkage disequilibrium (LD) with the causal SNP on chromosome 19. With this setting, we can explore how our CE4 method works on the causal SNP and SNPs that are in different LD with the causal SNP.

15.6.1 All no-association nulls are statistically false, if one is false

Table 15.2 presents two hypothetical effects assigned to the causal SNP $rs429358$, dominant beneficial effect or recessive beneficial effect for allele a, indicated by more negative values for both Aa and aa groups or the aa group alone in the Rx arm (more negative means better efficacy). A SNP has absolutely no effect if $\mu_{AA} = \mu_{Aa} = \mu_{aa}$, where in this case, efficacy $\mu_g = \mu_g^{Rx} - \mu_g^{C}$ $(g = AA,\ Aa,\ aa)$ is the difference of mean change between Rx and C. So the SNP has an effect if at least one of $\theta_{(1,2):0}, \theta_{2:(0,1)}, \theta_{1:0}, \theta_{2:1}$ in (15.1) is non-zero, or equivalently, $\max_c |\theta_c| > 0, c = \{(1,2){:}0,\ 2{:}(0,1),\ 1{:}0,\ 2{:}1\}$. Therefore, the commonly referred to as the *complete null* hypothesis $H_{00} : \mu_{AA} = \mu_{Aa} = \mu_{aa}$ can be equivalently stated as the *zero-null* hypothesis $H_{00} : \max_c |\theta_c| = 0$.

Similar as in Ding et al. (2018), in this set of simulation we did not add any random error to the outcome. Therefore, the subjects who carry the same genotype for the causal SNP $rs429358$ and receive the same treatment have identical outcome Y values. Then we calculated $\max_c |\theta_c|$ for all the SNPs on chromosome 19 and 11. We filtered out SNPs with number of subjects in any of the genotype-by-treatment combination cell ≤ 2 to ensure the model to run. In total, we have $118,604$ SNPs from chromosome 19 and $257,792$ SNPs from chromosome 11 being analyzed.

For most SNPs, the populations are *unbalanced* in design. Therefore, we applied the LSmeans technique to calculate what the parameters in model would be in a *balanced* population. Specifically, for each SNP, we calculated $\mu_{AA}, \mu_{Aa}, \mu_{aa}$ based on the LSmeans estimates for parameters from the following linear model:

$$\begin{aligned} y_i = \beta_0 &+ \beta_1 I(Trt_i = 1) + \beta_2 I(M_i = 1) + \beta_3 I(M_i = 2) \\ &+ \beta_4 I(Trt_i = 1) I(M_i = 1) + \beta_5 I(Trt_i = 1) I(M_i = 2) + \epsilon_i, \end{aligned}$$

TABLE 15.2
True response Y for subgroups defined by $rs429358$ under two hypothetical scenarios.

Effect	Treatment Group	Genotype			Max effect
		AA	Aa	aa	
Dominant	Rx	−0.5	−2.5	−2.5	2
	C	0	0	0	
Recessive	Rx	−0.5	−0.5	−2.5	2
	C	0	0	0	

where $Trt_i = 1$ (Rx) or 0 (C), $M_i = 0$ (AA), 1 (Aa) or 2 (aa). Then we calculated $\pi_{AA}, \pi_{Aa}, \pi_{aa}$ as follows. Denote the counts for AA, Aa, and aa (for Rx and C combined) as n_{AA}, n_{Aa}, n_{aa}. The allele frequency π_A for A is calculated as

$$(2 \times n_{AA} + n_{Aa})/[2 \times (n_{AA} + n_{Aa} + n_{aa})],$$

and then

$$\pi_{AA} = \pi_A \times \pi_A,\ \pi_{Aa} = 2 \times \pi_A \times (1 - \pi_A),\ \pi_{aa} = 1 - \pi_{AA} - \pi_{Aa}.$$

We then computed $\theta_c, c = \{(1,2){:}0,\ 2{:}(0,1),\ 1{:}0,\ 2{:}1\}$, for θ_c defined in (15.1) and recorded $\max_c |\theta_c|$ for each SNP. These θ_c are considered as the true effect of the contrasts in a balanced population.

Figure 15.6 summarizes the distributions of $\max_c |\theta_c|$. As can be seen, besides the causal SNP, every non-causal SNP picked up some non-zero effect. The causal SNP picked the maximum effect ($\max_c |\theta_c| = 2$), which is supposed to be. This is readily explained from a geometrical point of view: considering SNPs as categorical predictors, for a non-causal SNP to be independent of $rs429358$ and *not* pick up any of its effect, its percentages of individuals in the AA, Aa, and aa categories must remain exactly the same for each of the AA, Aa, and aa categories of $rs429358$, which is very unlikely or impossible for a given snapshot of population. Comparing the distributions from two different chromosomes, the non-causal SNPs on chromosome 19 picked up more of the $rs429358$ effect than the SNPs on

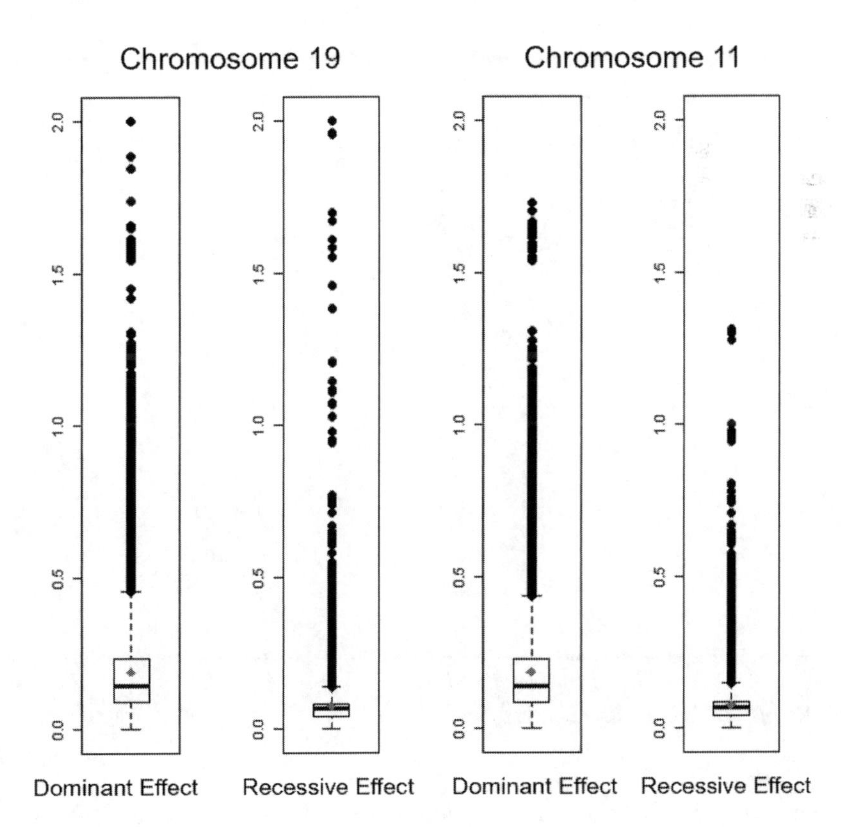

FIGURE 15.6
Distribution of apparent effects ($\max_c |\theta_c|$) of all SNPs from chromosome 19 and 11 under dominant or recessive effect, respectively. The dot represents the mean of the distribution and the line represents the median.

chromosome 11, as one would expect. Also, comparing the distributions from two different effect scenarios, on average, the non-causal SNPs picked up more of the $rs429358$ effect under dominant scenario than recessive scenario.

Therefore, it appears that testing for *association* by testing against the zero-null hypothesis $H_{00} : \max_c |\theta_c| = 0$ is not an appropriate formulation. While such null hypotheses might be biologically plausible, statistically they are false, rendering control of Type I error rate difficult to interpret. Thus, in our CE4 formulation, the null hypotheses are all stated as the complement of alternative hypotheses (inequalities instead of equations).

15.6.2 One run of realistic simulation

In this set of realistic simulation, random errors ϵ_i's, generated from a Normal distribution with mean 0 and variance 2^2, were added to each outcome from the dominant effect scenario in Table 15.2. Then we applied CE4 to analyze all the 118,604 SNPs from chromosome 19. We set $m = 10$, allowing on average 10 out of about 120,000 SNPs with at least one confidence interval failing to cover its true value, which is equivalent to setting the α_K level at 8.43×10^{-5} $(= \frac{m}{K} = \frac{10}{118{,}604})$.

A total of 48 SNPs were identified by CE4 as at least one of their CE4 confidence intervals not covering 0. Among those 48 SNPs, 37 of them are from the *APOC1-APOE-PVRL2-TOMM40* region, including the causal SNP. The other 11 SNPs belong to four different gene regions, which are distance away from the causal gene region. The left panel of Figure 15.7

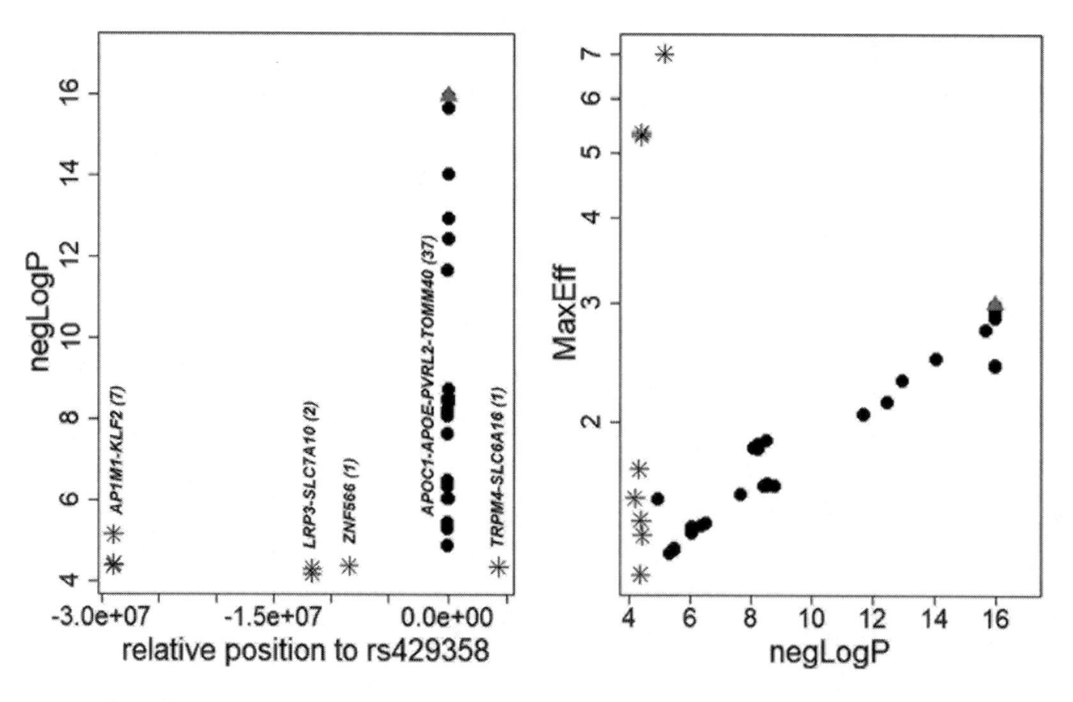

FIGURE 15.7
Top 48 identified SNPs from one chromosome-wide realistic simulation. Left: the negative log10 of the p-value versus relative position to the causal SNP; right: the maximum CE4 effect versus the negative log10 of the p-value. The triangle indicates the causal SNP and the dots indicate other SNPs that are from the same gene region as the causal SNP. The stars indicate the all other SNPs that are not from the causal SNP's gene region.

plotted the positions of these SNPs relative to the causal SNP, with y-axis (negLogP) showing the significance level of each SNP (only within-SNP multiplicity adjusted, not across-SNP), where p-value was calculated as explained in Chapter 15.5.3. The right panel of Figure 15.7 plotted MaxEff vs negLogP, where MaxEff (maximal effect) is defined as $\max_c |\hat{\theta}_c|$, the maximum absolute value among the estimated CE4 contrasts that do not cover zero. The causal SNP has the smallest p-value with MaxEff of 2.99. Note that some top SNPs have large MaxEff values. For example, SNP $rs17709615$ from the *AP1M1-KLF2* region has the largest MaxEff of 6.99, while its p-value is relatively large ($= 7.04 \times 10^{-6}$). We caution against the situation when a large effect size is seen, since such a large effect for treatment efficacy is clinically unlikely. For this specific SNP, it is not surprising to see the corresponding confidence interval for $\theta_{2:1}$ is quite wide.

To further investigate the relationship between the top SNPs and the causal SNP, we proposed a novel SNP *cross-talk* plot. It is based on a ternary diagram using barycentric coordinates to display the proportion of three variables that sum to one. Specifically, we projected the percentages of the AA, Aa, and aa categories of the causal SNP in each of these categories of a given top SNP onto the triangular diagram, and connect the points with lines. If the SNP is highly correlated with the causal SNP in terms of the distribution of AA, Aa, and aa, the percentages will be close to (1,0,0), (0,1,0) and (0,0,1), and thus the connected line segments will be long and lie closely to the two edges of the triangle. Otherwise, the three dots will be close to each other to give a short angle. For example, on the left panel of Figure 15.8, the black solid or dashed line segment, which completely or almost completely coincide with the $AA \rightarrow Aa \rightarrow aa$ edge, indicates the perfect or strong correlation with the causal SNP. While the small segment or the dot indicates nearly or complete independence with the causal SNP. On the right panel, we present such cross-talk plots for all the 48 top SNPS. Note that all 37 SNPs from *APOC1-APOE-PVRL2-TOMM40* region are highly correlated with the causal SNP, indicated by the long line segments, which explains why they have been identified by CE4. For the 11 SNPs from other regions, their line segments are all short, indicating they might have been identified due to randomness.

Then we further investigated the CE4 results by focusing on three top SNPs, the causal SNP $rs429538$, another SNP $rs73052335$ from the same gene region *APOE* as the causal SNP, and a SNP $rs10413861$ from a different gene region *KLF2*. Figure 15.9 demonstrates the observed mean clinical response in each treatment arm, the treatment effect profiles,

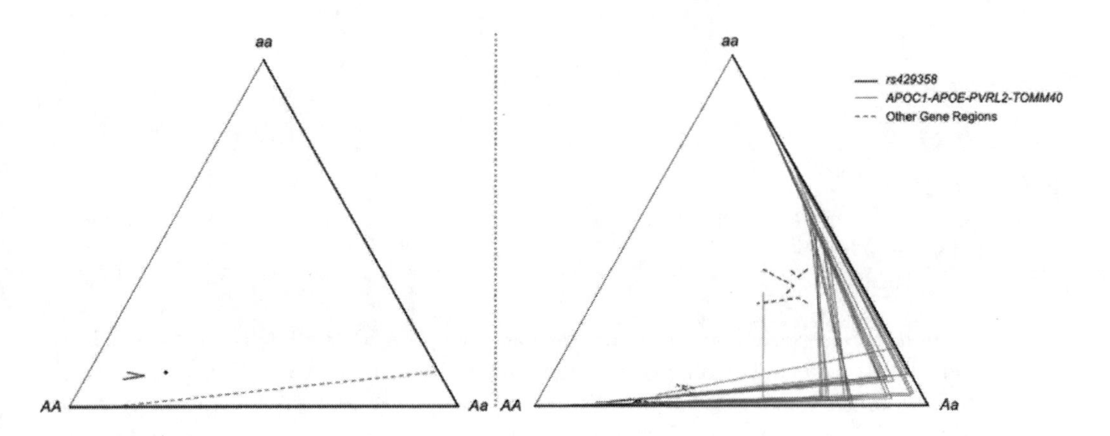

FIGURE 15.8
The SNP cross-talk plot. Left: Four hypothetical cases to demonstrate from perfect correlation to complete independence with the causal SNP; right: the cross-talk plot for top 48 SNPs identified by CE4.

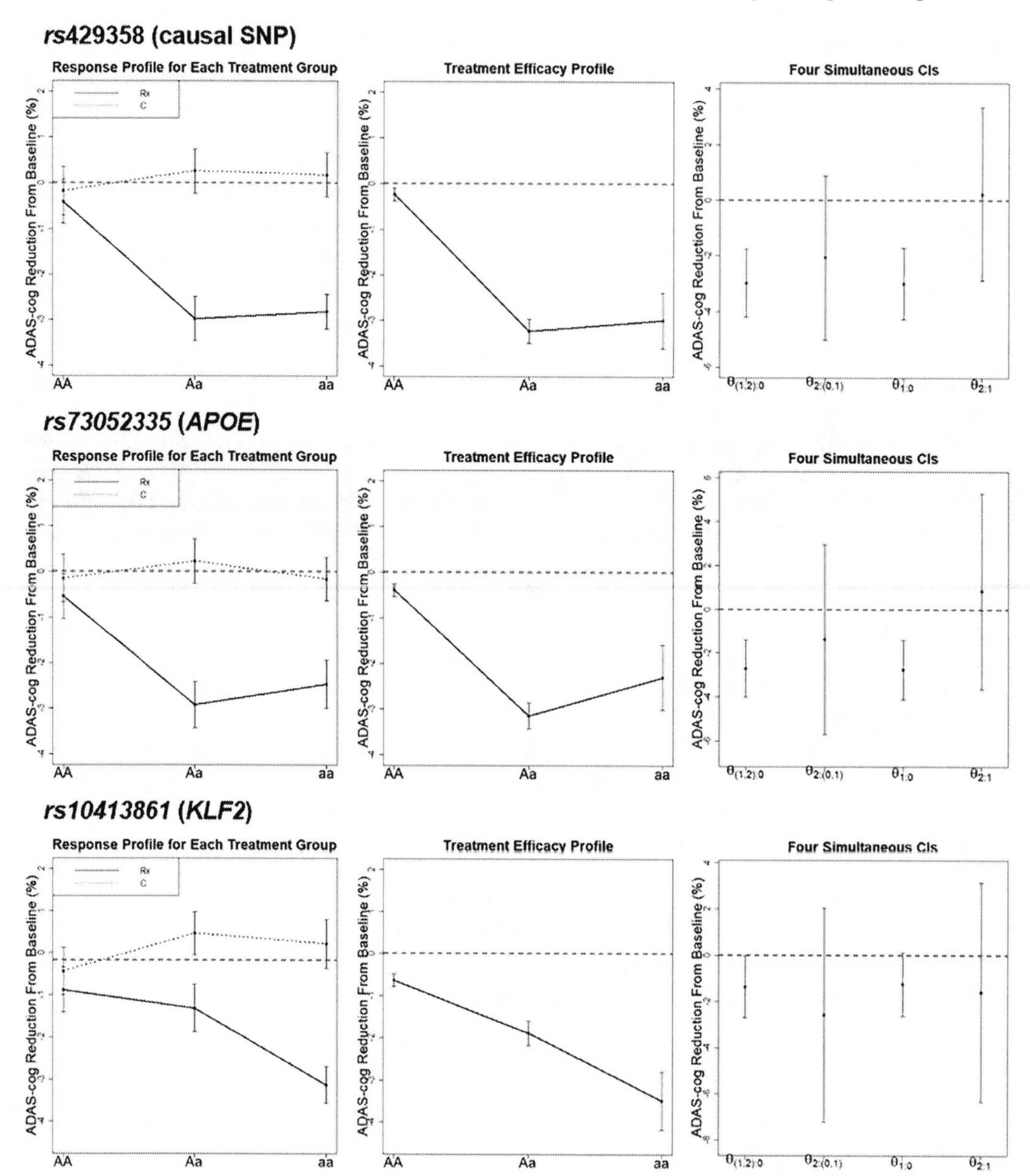

FIGURE 15.9
Mean clinical responses, treatment efficacy profiles and the CE4 simultaneous confidence intervals for three SNPs.

and the simultaneous confidence intervals for each of these SNPs. For the causal SNP, it can be seen that the efficacy profile may suggest a dominant beneficial effect of a. The CE4 simultaneous confidence intervals confirm that the targeted group is $\{Aa, aa\}$ combined (i.e., allele a has a beneficial dominant effect) since the confidence intervals of $\theta_{(1,2):0}$ and $\theta_{1:0}$ are below the zero line. Very similar observations are found for SNP $rs73052335$, which is due to its strong correlation with the causal SNP. The targeted patient population suggested by this SNP largely overlaps with the true population (that is determined by the causal SNP), with a "sensitivity" of 82.4% (i.e., 82.4% of the true targeted population can be identified by this SNP) and a "specificity" of 98.7% (i.e., 98.7% of the true non-targeted population can be identified by this SNP). For the third SNP, the treatment efficacy profile seems to suggest an additive effect of allele a, while only one confidence interval $\theta_{(1,2):0}$ is below the zero line ($\theta_{1:0}$ is marginally covering 0), suggesting to target $\{Aa, aa\}$ combined. However, in terms of accuracy for the identified targeted population, this SNP only has a sensitivity of 26.6% and a specificity of 77.4%, which is not surprising given its weak correlation with the causal SNP.

15.6.3 100 repeated runs of realistic simulation

Finally, to understand the robustness of the CE4 method, we repeated this chromosome-wide realistic simulation for 100 times.

In these 100 repeated runs, the SNP data are all the same but the outcomes are different due to randomness from the model. By setting $m = 10$, on average there are 37.5 SNPs identified per run with a total of 1235 unique SNPs being picked at least once. The causal SNP was picked in all 100 runs and 82 out of 100 times the causal SNP was ranked number one, indicating that our CE4 is robust in identifying the true causal SNP. Figure 15.10 summarizes all the identified SNPs from all 100 runs in terms of their relative position to the causal SNP and their frequencies of being picked up. We found that 97% of the 1235 SNPs were only picked less than 5 times, which are highly likely due to randomness. While for SNPs close to the causal SNP and located in the same gene region, the probability of being selected is much higher, among which 10 SNPs were identified for more than 80% of the times. From this repeated chromosome-wide simulations, we confirmed that there are possibilities that some SNPs are picked by random error but the true causal SNP and its surrounding SNPs can be identified with very high probabilities by CE4. Moreover, due to

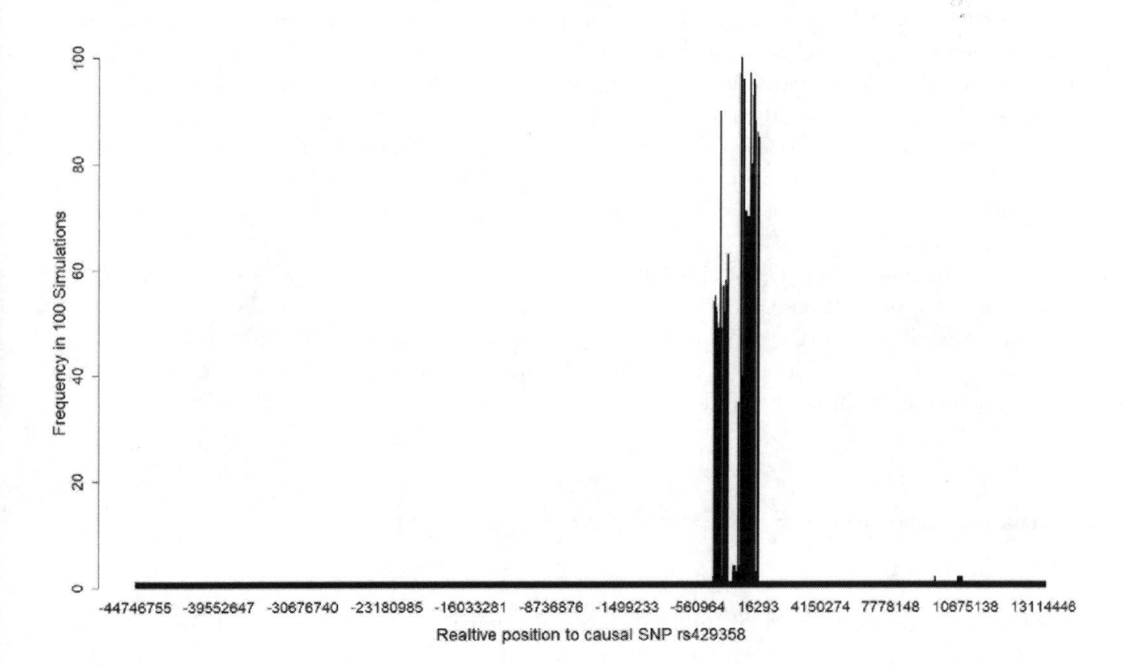

FIGURE 15.10
The frequency and relative position (to the causal SNP) of all identified SNPs in 100 simulations.

the existence of LD among SNPs, it is very unlikely that an isolated SNP will be the true causal SNP.

Based on the observations from our repeated realistic simulations, we recommend the following guidance towards the selection of "candidate" SNPs from those identified by CE4: (1) There are multiple SNPs (e.g., at least more than 3) being picked from the same gene region; (2) The MaxEff should not be unrealistically large; and (3) The targeted group should be a reasonable proportion (not too small or large, e.g., 5% – 95%) of the total population.

15.7 Summary and Discussion

New drug development involves assessing the efficacy of a new treatment Rx relative to a control treatment C, based on a clinical meaningful outcome. This makes testing SNPs for use as potential biomarkers in modern drug development more complex than the traditional association detection for a quantitative trait. Therefore, traditional genetic models or tests for association detection have issues and cannot be simply applied to such a SNPs testing problem in the drug development situation.

Our proposed new formulation of SNPs testing with the CE4 method, derived from the fundamental multiple testing principle, assesses all possible effects of a SNP on the efficacy of a new drug. The pivotal statistics, on which the simultaneous confidence intervals are based, guarantee the false coverage control to be exact and unaffected by the true test quantity values.

Our methodology adjusts for multiplicity taking dependence into account, both within each SNP and across the SNPs. It rigorously combines two error rate controls, strong FWER control within each SNP, and per family (per panel) error rate control across the SNPs. Such error control is appropriate in a drug development process, as it allows flexibility in the exploration of multiple candidate markers, while being confident in the patient subgroup to target from any selected marker(s). We believe such a novel and rigorous multiplicity adjustment contributes to the reduction in the so-called "reproducibility crisis" in which many discoveries in markers or effective subgroups turn out to be false positive findings. We refer to Chapter 14 on a related topic regarding the multiple testing approaches for "selective inference".

The independent linear model we use to illustrate this CE4 framework can be viewed as a starting model. In the situation where observations are correlated or the outcome is not normally distributed, other models should be considered. However, the key elements of our proposed method for SNPs testing are not specific to a particular model and the method can be extended to fit other models.

Bibliography

Berger, R. L., J. C. Hsu, et al. (1996). Bioequivalence trials, intersection-union tests and equivalence confidence sets. *Statistical Science 11*(4), 283–319.

Calian, V., D. Li, and J. C. Hsu (2008). Partitioning to uncover conditions for permutation tests to control multiple testing error rates. *Biometrical Journal 50*, 756–766.

Dean, A. M. and D. Voss (1998). *Design and Analysis of Experiments.* New York, Springer.

Ding, Y., Y. G. Li, Y. Liu, S. J. Ruberg, J. C. Hsu, et al. (2018). Confident inference for snp effects on treatment efficacy. *The Annals of Applied Statistics 12*(3), 1727–1748.

Ding, Y., H.-M. Lin, and J. C. Hsu (2016). Subgroup mixable inference on treatment efficacy in mixture populations, with an application to time-to-event outcomes. *Statistics in Medicine 35*, 1580–1594.

Ein-Dor, L., I. Kela, G. Getz, D. Givol, and E. Domany (2005). Outcome signature genes in breast cancer: is there a unique set? *Bioinformatics 21*, 171–178.

Freidlin, B., G. Zheng, Z. Li, and J. L. Gastwirth (2002). Trend tests for case-control studies of genetic markers: power, sample size and robustness. *Human heredity 53*(3), 146–152.

Genz, A. and F. Bretz (1999). Numerical computation of multivariate t-probabilities with application to power calculation of multiple contrasts. *Journal of Statistical Computation and Simulation 63*, 361–378.

González, J. R., J. L. Carrasco, F. Dudbridge, L. Armengol, X. Estivill, and V. Moreno (2008). Maximizing association statistics over genetic models. *Genetic Epidemiology: The Official Publication of the International Genetic Epidemiology Society 32*(3), 246–254.

Group, A.-R. E. D. S. R. et al. (1999). The age-related eye disease study (areds): design implications areds report no. 1. *Controlled clinical trials 20*(6), 573.

Hothorn, L. and T. Hothorn (2009). Order-restricted scores test for the evaluation of population-based case-control studies when the genetic model is unknown. *Biometrical Journal 51*, 659–669.

Hsu, J. (1996). *Multiple comparisons: theory and methods.* London, CRC Press.

Hsu, J. C. and R. L. Berger (1999). Stepwise confidence intervals without multiplicity adjustment for dose response and toxicity studies. *Journal of the American Statistical Association 94*, 468–482.

Huang, Y., H. Xu, V. Calian, and J. C. Hsu (2006). To permute or not to permute. *Bioinformatics 22*, 2244–2248.

ICH E9 (1997). *Statistical Principles for Clinical Trials* (Draft ICH (International Conference on Harmonisation) Guideline ed.). London: CPMP (Committee for Propritary Medical Products), EMEA (The European Agency for the Evaluation of Medical Products).

Kaizar, E. E., Y. Li, and J. C. Hsu (2011). Permutation multiple tests of binary features do not uniformly control error rates. *Journal of the American Statistical Association 106*, 1067–1074.

Lettre, G., C. Lange, and J. N. Hirschhorn (2007). Genetic model testing and statistical power in population-based association studies of quantitative traits. *Genetic Epidemiology: The Official Publication of the International Genetic Epidemiology Society 31*(4), 358–362.

Li, Q., G. Zheng, Z. Li, and K. Yu (2008). Efficient approximation of p-value of the maximum of correlated tests, with applications to genome-wide association studies. *Annals of human genetics 72*(3), 397–406.

Lin, H.-M., H. Xu, Y. Ding, and J. C. Hsu (2019). Correct and logical inference on efficacy in subgroups and their mixture for binary outcomes. *Biometrical Journal 61*(1), 8–26.

Marcus, R., P. Eric, and K. R. Gabriel (1976). On closed testing procedures with special reference to ordered analysis of variance. *Biometrika 63*(3), 655–660.

Ruberg, S. J. (1995). Dose response studies. II. Analysis and interpretation. *Journal of Biopharmaceutical Statistics 5*, 15–42.

So, H. and P. Sham (2011). Robust association tests under different genetic models, allowing for binary or quantitative traits and covariates. *Behav Genet 41*, 768–75.

Van't Veer, L. J., H. Dai, M. J. Van De Vijver, Y. D. He, A. A. Hart, M. Mao, H. L. Peterse, K. Van Der Kooy, M. J. Marton, A. T. Witteveen, et al. (2002). Gene expression profiling predicts clinical outcome of breast cancer. *nature 415*(6871), 530–536.

Wei, Y., J. C. Hsu, W. Chen, E. Y. Chew, and Y. Ding (2020). A simultaneous inference procedure to identify subgroups from RCTs with survival outcomes: Application to analysis of AMD progression studies. *arXiv preprint arXiv:2003.10528*.

Xu, H. and J. C. Hsu (2007). Using the partitioning principle to control the generalized family error rate. *Biometrical Journal 49*, 52–67.

Zhang, W. and Q. Li (2015). Nonparametric risk and nonparametric odds in quantitative genetic association studies. *Scientific reports 5*(1), 1–12.

Zheng, G., B. Freidlin, J. L. Gastwirth, et al. (2006). Comparison of robust tests for genetic association using case-control studies. In *Optimality*, pp. 253–265. Institute of Mathematical Statistics.

Index

Note: **Bold** page numbers refer to tables and *italic* page numbers refer to figures.